THE
JUVENILE
HORMONES

THE JUVENILE HORMONES

Edited by
Lawrence I. Gilbert

Northwestern University
Evanston, Illinois

PLENUM PRESS · NEW YORK AND LONDON

Library of Congress Catalog Card Number 76-21097
ISBN 0-306-30959-9

Proceedings of an International Symposium on the Chemistry, Metabolism,
and Modes of Action of the Juvenile Hormones of Insects held at
Lake Geneva, Wisconsin, November, 1975

© 1976 Plenum Press, New York
A Division of Plenum Publishing Corporation
227 West 17th Street, New York, N.Y. 10011

Printed in the United States of America

Preface

The juvenile hormones of insects are unique molecules in terms of their chemical nature (methyl esters of sesquiterpene epoxides) and action (both as modulators of morphogenesis during the larval life of insects and as a gonadotropic agent in many female adults). Although a symposium dedicated to the chemistry, metabolism and effects of juvenile hormone at a number of levels would be merited on the basis of its interest to the chemist, physiologist, endocrinologist, developmental biologist and entomologist, the juvenile hormones are special in the sense that juvenile hormone mimics (juvenoids, insect growth regulators, analogs) are currently being utilized to control various insect pests. Indeed, a number of commercial firms are currently developing new compounds with juvenile hormone activity that might possess a narrow spectrum of activity and which would be relatively biodegradable. Thus, a symposium on the juvenile hormones is also merited on a practical basis since juvenoids are already becoming constituents of our environment and it is apparent that in order to design effective mimics of the natural juvenile hormones, one should understand the means by which juvenile hormone elicits its effects.

As will become evident to the reader, the great majority of data presented at this symposium have not been published previously and the symposium itself was organized along natural divisions dealing with the chemistry, metabolism and multi-level modes of action of the juvenile hormones. Special lectures were presented by Professors C.M. Williams, B.W. O'Malley and W.S. Bowers, the latter for the first time presenting his studies elucidating the structure and effects of the first natural products with anti-juvenile hormone activity. Three papers are included in the volume that were not presented at the symposium because the author was invited but unable to attend (Sehnal; Trautmann) or because lack of program time prevented the presentation of his work (Yagi). At the beginning of each section the session chairman has summarized the papers presented in his session and in some cases has critically reviewed the research area. It is hoped that this will be of some use to the reader.

This symposium was made possible by a major contribution from
The Roche Institute of Molecular Biology (Hoffman-LaRoche, Inc.)
and lesser but critical contributions from Ciba-Geigy, Zoecon Corp.,
FMC Corp., Merck Sharp & Dohme Research Labs., Plenum Press and
Waters Associates. I wish to thank my students (J. Wielgus,
S. Smith, R. Vince. P. Johnston, K. Katula) for their expert
handling of both transportation and slide projection. A special
note of gratitude is due my secretary, Pauline Bentley, who contrib-
uted significantly to the organization of this important scientific
meeting and who did the complete typing of these proceedings.
Finally, in a professional sense I dedicate this symposium to
Sir Vincent Wigglesworth who has made enormous contributions to
this research area and who was unable to join us at Lake Geneva in
November, 1975. In a personal sense I dedicate this volume to my
wife, Doris P. Gilbert, who has made enormous contributions to me.

Lawrence I. Gilbert
Evanston, Illinois
February, 1976

Contents

CONTENTS

V.   EFFECTS OF JUVENILE HORMONE AT THE MOLECULAR LEVEL (PROTEIN SYNTHESIS)

CONCLUDING LECTURE

JUVENILE HORMONE.....IN RETROSPECT AND IN PROSPECT

C.M. Williams

The Biological Laboratories, Harvard University
Cambridge, Massachusetts 02138

RETROSPECTIONS ON THE CONTROL OF THE CORPORA ALLATA

Figure 1 reproduces a summary of endocrine dynamics which,
I venture to think, identifies most of the phenomena that will be
dealt with in this Symposium. For present purposes, let the
endocrine gland be the corpora allata (CA) and the endocrine
agent be juvenile hormone (JH). Attention is directed to the
"controlling mechanisms" modulating in this case the synthesis
and secretion of JH by the CA. Since JH has never been shown to
accumulate in the glands themselves, the indications are that CA
activity is adjusted by the modulation of synthesis rather than
secretion (Tobe and Pratt, 1974).

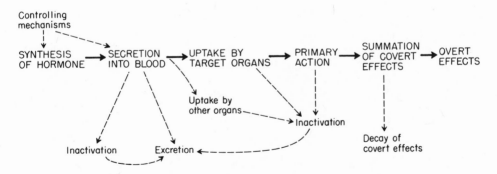

Fig. 1. A generalized model of the dynamics of hormone action.
(From Ohtaki et al., 1968).

In the substantial literature on the subject there is a
wealth of evidence that the CA show sustained but not necessarily
constant activity during most of larval life and that their
excision from immature larvae leads to precocious metamorphosis
(Bounhiol, 1938; Fukuda, 1944). Except in species that diapause
as mature larvae, the CA are turned off late in larval life. In
most insects under normal conditions the timing of this event is
species-specific. Still later, after adult development is complete
or nearly complete, the CA are usually turned on again.

There is general agreement that the brain is the highest
center in controlling the CA and that this control is exercised
by the way of its nervous connections to the corpora cardiaca-
corpora allata complexes (CC+CA) (Scharrer, 1957). Since the
nerves in question have been shown to contain both ordinary and
neurosecretory axons (Stumm-Zollinger, 1957), the brain's control
of the CA could, in principle, consist of any combination of
excitatory nerve impulses, inhibitory nerve impulses, excitatory
neurosecretions, and inhibitory neurosecretions.

A further complication is that some of the axons from the
brain terminate in the CC rather than the CA (Nijhout, 1975c).
And since the CC contain intrinsic neurosecretory cells, there
remains the possibility of a relay mechanism in the CC. This
prospect has been difficult or impossible to test since the CC in
most species cannot be excised without detaching the CA from
their nervous connections with the brain.

In addition to brain-centered controls, it seems altogether
likely that the CA can be influenced by blood-borne factors
arising elsewhere in the body. These could include, not only
substrate precursors of JH, but also other specific hormones and
neurosecretions.

In view of all these complications it is not surprising to
encounter in the literature a rather bewildering array of experimental findings (for reviews see Engelmann, 1970; Wigglesworth,
1970; Wyatt, 1972; Doane, 1973; de Wilde and de Loof, 1973;
Gilbert and King, 1973; Sláma et al., 1974; Willis, 1974; Steel,
1975). Attention has centered for the most part on the control of
the CA in adult insects, mainly in regard to the regulation of
reproductive cycles, yolk deposition, and adult diapause. Though
it is difficult to generalize, most investigators have found evidence for brain-centered nervous and/or neuroendocrine control of
the CA.

The same is true in the less numerous studies carried out on
larval insects. Thus, in a recent study of wax moth larvae,
Granger and Sehnal (1974) conclude (page 415) that "the CA are
activated by a neurohumoral factor from the brain during larval

development and that they are inactivated by nervous inhibition just before metamorphosis." For insects in general Sehnal and Granger (1975) suggest (page 113) that "the CA are controlled by both activating and inhibitory stimuli which reach the glands via the hemolymph or via the nerves."

Of one thing we can be certain: the CA do not act autonomously to control their own secretion. To use Wigglesworth's (1948) apt phrase (page 12) "it is not the CA, itself, which counts the instars." Sehnal (1971) has summarized his views as follows (page 33): "Apparently an insect does not 'count moults' before pupation. Rather, it 'measures' its growth rate and when this growth rate is maximal under the conditions of nutrition that exist, it turns off its CA and pupates. How the insect knows when its growth rate is maximal remains to be discovered."

INHIBITION OF PUPATION BY JH

In studies carried out in connection with his Ph.D. dissertation at Harvard, Fred Nijhout undertook a detailed examination of the control of pupation in larvae of the tobacco hornworm, Manduca sexta. This study made use of thousands of carefully timed larvae reared in the "Harvard Manduca factory" which is capable of turning out several hundred pupae per day. I venture to think that this Symposium will document the debt of thanks we owe to the unsung heroes who developed a practical method for the large-scale rearing of Manduca on an artificial diet (Yamamoto, 1969; Bell, private communications; Bell and Joachim, 1975).

Figure 2, taken from the paper by Nijhout and Williams (1974b), presents the results of bioassays of JH in the hemolymph of diet-reared hornworms throughout the final (5th) larval instar. It will be observed that the high titer distinctive of the first half of the instar declines during the second half. By the time the larva comes to weigh 7.5-8.0 g, the titer has become too low to bioassay.

Larval hornworms can be continuously exposed to high levels of JH by rearing them on an artificial medium containing the hormone or analogs thereof. Experiments of this sort have been carried out by Safranek (1974). The first four of the five larval instars were little affected by chronic exposure to JH, the only visible change being a blanching of the usual bluish-green coloration of the integument and the prolongation of the first four instars by a total of about one day. By contrast, larvae during the final (5th) stage were much affected. The duration of the instar was remarkably prolonged, the larvae in most cases developing into giants weighing up to 18 g, all of which finally died without forming normal pupae. Clearly, the final larval instar, unlike the earlier four, is impressively sensitive to exogenous JH.

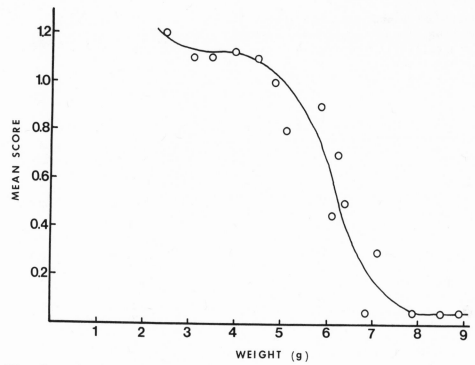

Fig. 2. The titer of juvenile hormone in the hemolymph of <u>Manduca</u> larvae at various weights in the fifth instar. The hemolymph of three larvae in each weight class was pooled and assayed on five test larvae according to the "black larval assay" described by Truman <u>et al</u>. (1973). (From Nijhout and Williams, 1974b).

How can we account for this differential sensitivity of the final larval instar? We have substantial though circumstantial evidence that after the molt to the final instar the brain loses its ability to secrete prothoracicotropic hormone (PTTH). Moreover, the brain cannot reacquire this ability in the presence of JH. Chronic exposure to JH during the final instar therefore prevents PTTH from turning on the prothoracic glands.

This fact has been exploited by insects that diapause as mature larvae. Thus, as discussed elsewhere (Nijhout and Williams, 1974b; see also Yagi and Fukaya, 1974; Chippendale and Yin, 1975), the indications are that the CA remain continuously active in diapausing larvae and are shut off only after exposure to the environmental conditions that potentiate escape from diapause. Evidently, in non-diapausing larvae the shutting-off of the CA during the second half of the final instar does two things: it

allows the brain to secrete PTTH and thereby stimulate the prothor-
cic glands and, secondly, it permits the resulting ecdysone to
activate the pupal gene-set.

THE FIRST CRITICAL SIZE

Under normal culture conditions Manduca invariably goes
through five larval instars prior to pupation. The fifth instar
differs from the other four in that, as we have seen in Fig. 2,
the CA get shut off as a necessary prelude to pupation. Nijhout
(1975a) has inquired as to how the larva identifies the final
larval instar - i.e., the instar during which the CA will be inactivated. In the experiments in question he exploited the fact that
the head capsule is discarded at each larval molt and replaced by
a new and larger one which does not undergo further growth during
the intermolt. The width of the head capsule thus becomes a
useful parameter of the size that the larva attains at the outset
of successive instars.

By temporarily starving third or fourth instar larvae,
Nijhout obtained a continuous series of undersized fifth stage
larvae whose head capsules were also proportionately undersize.
When the subsequent fate of these larvae was examined, he found
the remarkable relationships summarized in Fig. 3. It will be
seen that all larvae with head capsules larger than 5.4 mm formed
normal pupae at the next molt. By contrast, those with head
capsules measuring less than 5.0 mm always molted into supernumerary larval instars which could pupate only after undergoing
further growth and molting. As indicated in Fig. 3, the midpoint in the transition was a head capsule size of 5.1 mm.

On the basis of these findings, Nijhout (1975a) concludes
(page 221) that "a larva does not 'count instars' but rather, it
continues to grow and molt until it reaches a certain threshold
size." In the case of Manduca this "first critical size" corresponds to a head capsule width of at least 5.1 mm. And since the
commitment to turn off the CA is apparently made during or immediately after the molt, Nijhout's data show that for the larva as
a whole the first critical size is about 0.6 g.

THE SECOND CRITICAL SIZE

When a molting larva exceeds the first critical size, the
decision is made to inactivate the CA during the ensuing instar;
the latter thereby becomes the final one. But in order for this
decision to be implemented, the larva must reach a "second critical
size." And it is this event - the attainment of the second critical size - that sets in motion a concatenation of further events

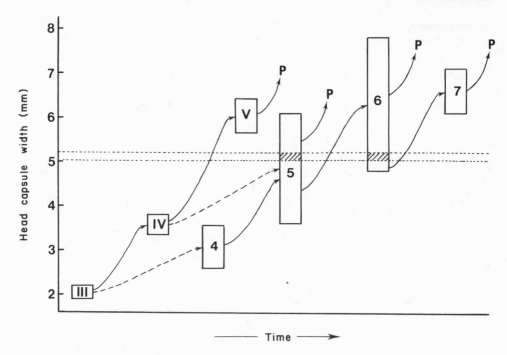

Fig. 3. The vertical dimensions of the boxes indicate the range
of larval head capsule widths found in each group. Roman numerals
indicate larvae growing under standard laboratory conditions.
Arabic numerals indicate larvae derived from third and fourth
instar larvae that had been temporarily starved (dashed lines).
Instead of pupating, larvae with head capsules smaller than the
threshold size (5.1 mm) molt to a sixth or seventh larval instar.
Larvae pupate (P) after their head capsules have attained or
exceeded a width of 5.1 mm. (From Nijhout, 1975a).

that leads to the inactivation of the CA.

The existence of the second critical size was first recognized
and subjected to detailed study by Nijhout and Williams (1974a,b).
Unlike the absolute dimensions of the first critical size, what
the second critical size will be is determined at the outset of
the final larval instar and is directly proportional to the size
of the larva at that time. Thus the second critical size for a
normal, well-fed larva whose head capsule is 6 mm in width will be
larger than that for a malnourished individual whose head width
barely exceeds the threshold value of 5.1 mm. Consequently, the
second critical size is subject to individual variation to a far
greater extent than is the first critical size. Under the standardized laboratory and rearing conditions employed by Nijhout

and Williams (1974a,b), the second critical size was approximately
5 g.

BRAIN-CENTERED MECHANISM FOR INACTIVATION OF THE CA

When a larva attains the weight of about 5 g, a diffusable
agent that we call the "inhibitor" is released from some unknown
source in the abdomen. The inhibitor acts on the brain. Then,
via its nerves to the CA, the brain turns off further secretion of
JH. We have evidence that the brain does so by curtailing the
secretion of a CA-stimulating factor ("allatotropin," Scharrer,
1958) and simultaneously secreting a CA-inhibiting factor; for the
latter we suggest the name "allatohibin." We suspect that both
are neurosecretions which, in Manduca, are axonally transported
to the CA.

The need to postulate two rather than just one controlling
factor is made necessary by the behavior of "loose" Manduca CA
when separated from the brain and bioassayed by implantation.
Under that circumstance, active CA lost most but not all of their
activity presumably because they are deprived of allatotropin.
In similar tests of inactive CA, the glands reacquire a certain
low level of activity because they are no longer acted upon by
allatohibin. At the present time nothing is known about the
molecular nature of allatotropin or allatohibin.

A larva continues to feed and increase in mass for about
2 1/2 days after attaining the critical weight of 5 g; thus it
comes to weigh 9-10 g before the first prodromal signs of pupation become evident. We originally thought that, once the critical weight of 5 g was attained, the brain switched off the CA in
an abrupt manner and that, as shown in Fig. 2, about 1 1/2 days
was required for the decay of the JH that had already been secreted.
Nijhout's (1975b) further investigation has made this hypothesis
untenable.

The experiments in question were carried out on approximately
100 young fifth instar larvae weighing 3.0-3.5 g; the abdomen of
each individual was isolated by ligation between the second and
third abdominal segments and the anterior end excised. Bioassays
of the hemolymph at the time of ligation revealed a substantial
JH titer equivalent to 1 ng of C_{18}-JH per ml. Similar measurements
performed on groups of abdomens at successive periods after ligation revealed a surprising finding - namely, that the half-life of
endogenous, circulating JH is only 25 min and that within 2 hr
it is cleared from the blood.

There is no reason to believe, and much reason to disbelieve,
that the half-life of circulating JH is any longer after fifth

instar larvae attain the second critical weight of 5 g. Yet, as
shown in Fig. 2, unligated larvae require about 1 1/2 days for
endogenous, circulating JH to fall to undetectable levels. Clearly,
the CA are continuing to secrete JH during this period, albeit at a
steadily declining rate. Evidently, about 1 1/2 days are required
for the complete inactivation of the CA by the brain.

PERSISTENCE OF JH EFFECTS

Up to this point we have been considering the chain of events
that culminates in the disappearance of circulating JH. As shown
in Fig. 1, it would not be surprising to find JH to be more stable
after its uptake or binding to target cells. Moreover, even after
its inactivation, the biochemical and biophysical effects of prior
exposure to JH (the so-called "covert effects" in Fig. 1) might
require substantial time to decay.

Just such persistent effects of prior exposure to JH have
been documented by Fain and Riddiford (1976a,b) for fourth and
fifth instar Manduca both in vivo and in vitro. The phenomenon
has been further examined by Nijhout (1975b) on fifth instar
hornworms weighing 3 and 4 g. Here again, the abdomens were
isolated by ligation at time zero. The abdomens were stored
for 1 to 4 days at 25°C and then perfused with β-ecdysone to provoke molting. The results were clear-cut. All abdomens aged for
only 1 day showed larval molting. In order for a pupal-type molt
to take place, the abdomens had to be aged for at least 2 to 3
days prior to the infusion of β-ecdysone. Thus the persistent
effects of JH were evident long after the 2 hr required for the
hormone to be cleared from the blood.

In interpreting these findings, Nijhout considered the two
possibilities mentioned above - namely, that JH may have a prolonged half-life when bound to target cells or, alternatively,
that the covert effects of prior exposure may persist long after
molecular JH has been eliminated from the entire insect. A combination of the two phenomena could also be entertained.

Nijhout (1975b) attempted to distinguish between these
possibilities by preparing extracts of the blood-free carcasses
of young fifth instar larvae. Control extracts showed traces of
JH activity in the bioassay. However, the large amount of lipid
co-extracted with the JH severely interfered with the quantitative
aspects of the bioassay even after partial purification of the
extract by thin-layer chromatography. Similar extracts prepared
from carcasses of abdomens aged for 4 hr after ligation showed no
trace of JH. Though the aged, blood-free abdomens clearly contained
less JH than did the controls, it was impossible to conclude that
they contained none.

Nijhout (1975b) went on to perform a further series of
illuminating experiments in which abdomens were isolated from
larvae weighing 3.0-3.5 g and immediately parabiosed to larger
intact larvae weighing 6.8-8.0 g. Within two days, both the
large larva and the abdominal partner initiated a pupal-type molt.
In effect, the presence of the large pupating larva hastened the
decay of the persistent effects of JH in the partner abdomen with
which it shared the same blood. In interpreting this finding,
Nijhout (1975b) concluded that, in addition to supplying ecdysone
to turn on the molt, the large pupating larva supplied an additional diffusable factor which erased the persistent effects of
JH in the pupal partner. He conjectured that this additional
factor might be the JH specific esterase which Sanburg et al.
(1975a,b) have shown, puts in its appearance at precisely this
time in the larger partner. Since the JH specific esterase
attacks the hormone itself, all this suggests that the persistent
effects of prior exposure to JH are due, at least in part, to a
corresponding persistence within the target cells of tightly
bound hormone. Under this point of view, JH specific esterase
is a fail-safe device to erase receptor-bound JH from the tissues:
much earlier, as we have seen, the non-specific esterases have
required less than 2 hr to dispose of hemolymph JH once the CA
are shut off.

THE TWO SURGES OF ECDYSONE

Among the target organs cleared of the persistent effects of
JH is the brain itself. The brain is thereby set free to secrete
PTTH when it sees its very next photoperiodic "gate." The prothoracic glands respond by secreting a surge of ecdysone, as demonstrated by Truman and Riddiford (1974) and subsequently confirmed
in the radioimmune assays reported by Bollenbacher et al. (1975).

The first surge of ecdysone has remarkable consequences. As
diagrammed in Fig. 4, it causes an abrupt change in behavior. The
larva ceases to feed, purges its gut, and begins the "wandering
stage." It will dig into the ground if allowed to do so. Simultaneously, the wandering larva shows an impressive change in its
sensitivity to JH, as Prof. Riddiford will detail in her communication tomorrow morning (see Riddiford, this volume). Suffice it to
say that once the integumentary cells have seen the first surge of
ecdysone, they are committed to pupation. But in order for them
to go ahead and actually implement the pupal program, they must
be acted upon some 2 days later by a second surge of ecdysone, the
latter being provoked by a corresponding gated release of PTTH.

It is important to emphasize that pupation cannot take place
until the integumentary cells see the second surge of ecdysone.
Until that arrives, the wandering larva remains in a kind of

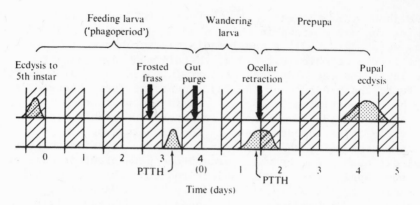

Fig. 4. The timing of morphological events (top panel) and the
two periods of PTTH and ecdysone release (bottom panel) during
the final (5th) instar and prepupal stage of <u>Manduca</u> at 25°C
under a 12L:12D photoperiod. The clear and cross-hatched periods
denote photophase and scotophase, respectively. (Taken from
Nijhout and Williams, 1974a, as modified from Truman and Riddiford,
1974.)

developmental limbo: it cannot secrete a new larval cuticle,
it cannot secrete a pupal cuticle, it cannot resume feeding - it
is indeed a "wandering larva." In more ways than one it is the
most fascinating stage in the life history.

THE DETERMINATION OF CRITICAL SIZE

It is probable that most insects have mechanisms for continuously monitoring body size. We have seen that in the case of
<u>Manduca</u> larvae there are two critical sizes that must be attained
before pupation can take place. The first critical size identifies the instar in which the CA will be inactivated; thus indirectly, it identifies what will prove to be the final larval instar.
It is an assessment that a certain absolute size has been attained
during or immediately after the molt. Two to three days later the
rapidly growing larva attains the second critical size that triggers the above-mentioned chain of events which lead to the inactivation of the CA.

Nijhout (1975a) has inquired as to how an insect might be
able to assess its own size. He comes to the conclusion that the
simplest, if not the only, way in which an individual can monitor
its own size is by some sort of allometry - by detecting, for
example, a critical rate of firing of receptors that are progressively stretched as the insect grows to a critical size. But this

is only one of scores of models that could be entertained. Now
that the phenomenon has been identified, it merits and doubtlessly
will receive much additional attention.

THE INHIBITOR

There remains to be considered the diffusable inhibitor which
begins to be released when the larva attains the second critical
size of about 5 g. Sivasubramanian and Williams (1976) used an
in vitro system for detecting its presence in the hemolymph. Each
culture consisted of 20 active brain+CC+CA complexes obtained from
early fifth instar hornworms. The presence of the inhibitor was
recognized by its ability to turn off the active complexes. Control cultures were prepared in Grace's medium; experimental cultures, in 1 part Grace's medium and 1 part hemolymph or subfrac-
tion thereof. After 48 hr the complexes were rinsed and recultured
in fresh Grace's medium. After 24 hr this fresh medium was extracted for JH, the extract partially purified, and subjected to
bioassay.

In this manner we were able to determine that an agent capable
of turning off the active complexes begins to appear in the blood
after about 48 hr of the fifth instar. It peaks on day 4 - i.e.,
in hemolymph derived from larvae weighing 7-8 g; it then declines
on day 5 and becomes undetectable on day 6.

Further studies (Sivasubramanian and Williams, 1976) suggest
that the inhibitor is a heat-insensitive, trypsin-sensitive substance whose molecular weight is similar to that of bacitracin
(ca 1400 daltons). It may be noted that these are properties
identical to those reported for an "antigonadotropin" extracted
from the ovaries of Rhodnius (Liu and Davey, 1974). If, as seems
likely, the inhibitor proves to be a polypeptide, it could contain
about a dozen ordinary-size amino acids. Once the inhibitor is
obtained in pure form, its sequencing and synthesis should be
straightforward. But obtaining it in pure form is likely to be
a formidable undertaking.

PROSPECTS

Everything we learn about the insect endocrine system has
the potentiality of being, not only of theoretical interest, but
also of practical concern. Thus the isolation and characterization of the JH of the Cecropia silkmoth culminated in the theory
and practice of the "third generation pesticides" in which synthetic JH analogs are used to block metamorphosis. Agents of this
sort are most promising in the case of species whose deleterious
effects are limited to the adult stage. This happens to be so in

the case of most, but not all, species that transmit human and animal diseases. For example, a mosquito or tsetse fly poses no problem until it comes to be an adult.

The situation is quite otherwise in the case of the major insect consumers of man's food and fiber where the larval stages do most of the damage. Clearly, the need is for new strategies in the control of larval insects.

There are ample theoretical grounds for believing that anti-hormones will provide at least part of the answer. Thus, an anti-ecdysone would block larval molting as well as metamorphosis and thereby enforce a developmental standstill reminiscent of diapause. An anti-JH would oppose the production or action of endogenous JH and provoke a precocious, lethal metamorphosis of immature larvae; moreover, in species requiring JH for egg maturation, an anti-JH might sterilize females that had already attained the adult condition.

The search for anti-insect-hormones has up to now been pre-occupied with the screening of competitive inhibitors bearing a chemical resemblance to ecdysone or JH (Svoboda et al., 1972; Sláma et al., 1974; Robbins et al., 1975; Thompson et al., 1975). As we learn more about the mechanisms by which the brain, prothoracic glands, and CA are turned off and on, a theoretical basis is thereby established for the discovery of agents that upset the normal controls. Consequently, instead of being dismayed by the complexities of the mechanisms controlling, for example, the pupation of the tobacco hornworm, we should rejoice. By that very complexity the insect pests and vectors of disease have opened themselves to attack by additional tools and tactics.

ACKNOWLEDGEMENTS

I wish to thank Drs. Frederick Nijhout, David Denlinger, Edwin Marks, and Mr. Louis Safranek for critical reading of the manuscript. This work was supported by the Rockefeller Foundation and by grants from the National Science Foundation and the National Institutes of Health.

REFERENCES

Bell, R.A., and Joachim, F.G., 1975, Ann. Ent. Soc. Am., in press.
Bollenbacher, W.E., Vedeckis, W.V., Gilbert, L.I., and O'Connor, J.D., 1975, Develop. Biol. 44:46.
Bounhiol, J.J., 1938, Bull. Biol. Fr. Belg., Suppl. 24:1-199.
Chippendale, G.M., and Yin, C.-M., 1975, Biol. Bull. 149:151.
Doane, W.W., 1972, in "Insects: Developing Systems" (S.J. Counce and C.H. Waddington, eds.) pp. 291, Academic Press, New York.

Engelmann, F., 1970, "The Physiology of Insect Reproduction,"
   Pergamon Press, Oxford.
Fain, M.J., and Riddiford, L.M., 1976a, Gen. Comp. Endocrinol.,
   in press.
Fain, M.J., and Riddiford, L.M., 1976b, Develop. Biol., in press.
Fukuda, S., 1944, J. Fac. Sci., Imp. Univ. Tokyo Ser. IV 6:477.
Gilbert, L.I., and King, D.S., 1973, in "The Physiology of Insecta"
   (M. Rockstein, ed.) pp. 249, Academic Press, New York.
Granger, N.A., and Sehnal, F., 1974, Nature 251:415.
Liu, T.P., and Davey, K.G., 1974, Gen. Comp. Endocrinol. 24:405.
Nijhout, H.F., 1975a, Biol. Bull. 149:214.
Nijhout, H.F., 1975b, Biol. Bull. 49:568.
Nijhout, H.F., 1975c, Int'l. J. Insect Morph. Embryol., in press.
Nijhout, H.F., and Williams, C.M., 1974a, J. Exp. Biol. 61:481.
Nijhout, H.F., and Williams, C.M., 1974b, J. Exp. Biol. 61:493.
Ohtaki, T., Milkman, R.D., and Williams, C.M., 1968, Biol. Bull.
   135:322.
Robbins, W.E., Thompson, M.J., Svoboda, J.A., Shortino, T.J.,
   Cohen, C.F., Dutky, S.R., and Duncan, O.J., III, 1975, Lipids
   10:353.
Safranek, L.L., 1974, "Hormonal Control of Pigmentation in the
   Tobacco Hornworm, Manduca sexta." Honor's Thesis, Harvard
   University.
Sanburg, L.L., Kramer, K.J., Kézdy, F.J., Law, J.H., and Oberlander,
   H., 1975a, Nature 253:266.
Sanburg, L.L., Kramer, K.J., Kézdy, F.J., and Law, J.H., 1975b,
   J. Insect Physiol. 21:873.
Scharrer, B., 1957, in "Proc. 2nd Int. Symp. Neurosecretion,"
   Lund: 79, Springer-Verlag, Berlin.
Sehnal, F., 1971, Endocrin. Exp. 5:29.
Sehnal, F., and Granger, N.A., 1975, Biol. Bull. 148:106.
Sivasubramanian, P., and Williams, C.M., 1976, in preparation.
Sláma, K., Romañuk, M., and Šorm, F., 1974, "Insect Hormones and
   Bioanalogues," Springer-Verlag, New York.
Steel, C.G.H., 1975, Nature 253:267.
Stumm-Zollinger, E., 1957, J. Exp. Zool. 134:315.
Svoboda, J.A., Thompson, M.J., and Robbins, W.E., 1972, Lipids
   7:553.
Thompson, M.J., Serban, N.N., Robbins, W.E., Svoboda, J.A.,
   Shortino, T.J., Dutky, S.R., and Cohen, C.F., 1975, Lipids
   10:615.
Tobe, S.S., and Pratt, G.E., 1974, Nature 252:474.
Truman, J.W., and Riddiford, L.M., 1974, J. Exp. Biol. 60:371.
Truman, J.W., Riddiford, L.M., and Safranek, L., 1973, J. Insect
   Physiol. 19:195.
Wigglesworth, V.B., 1948, J. Exp. Biol. 25:1.
Wigglesworth, V.B., 1970, "Insect Hormones," Oliver & Boyd,
   Edinburgh and London.
de Wilde, J., 1973, in "The Physiology of Insecta" (M. Rockstein,
   ed.) pp. 97, Academic Press, New York.

Willis, J.H., 1974, Ann. Rev. Entomol. 19:97.
Wyatt, G.R., 1972, in "Biochemical Action of Hormones" (G. Litwack,
    ed.), Vol. 2, pp. 385-490, Academic Press, New York.
Yagi, S., and Fukaya, M., 1974, Appl. Entomol. Zool. 9:247.
Yamamoto, R.T., 1969, J. Econ. Ent. 62:1427.

SUMMARY OF SESSION I:

CHEMISTRY OF THE JUVENILE HORMONES AND JUVENILE HORMONE ANALOGS

J. Siddall

Zoecon Corporation, 975 California Avenue
Palo Alto, California 94304

The following four papers, though grouped under the heading
of "Chemistry of the Juvenile Hormones and Juvenile Hormone
Analogs", present topics ranging from the nonspecificity and differential activity of the three molecular forms of natural JH to
various methods for determination of their in vivo titers and to
observations on their biosynthesis and storage. Questions of the
active conformation of JH and its relevance for the design of
practical, useful analogs for pest control are also addressed.
The reader may well conclude that more questions are raised than
are answered.

The paper of Dahm and co-authors concerns the biosynthetic
origins, titers and functions of the natural hormones in an
attempt to explore the question, "Which is the morphogenetic
hormone of larval insects?". They also present completely new
information showing storage of JH by accessory male sex glands and
suggest a vestigial role of JH in adult male silk moths for reproductive functions.

These authors quote selected works from the classical literature bearing on the stage and order unspecificity of the corpus
allatum secretion and conclude that since published results do not
allow an unambiguous identification of the morphogenetic hormone,
one must reconsider the evidence necessary to support a claim that
a given chemical is a juvenile hormone. In light of their three
criteria, first of production of the chemical by corpora allata,
second of titer fluctuation in synchrony with the physiological
processes controlled, and third of the demonstration of activity
by replacement chemotherapy in allatectomized or glandless insects,
we may usefully review the status of any or all of the three

15

natural juvenile hormone molecules. For all three molecules, the
first criterion, production by corpora allata, is well satisfied.
However, the second condition of titer fluctuation in synchrony
with morphogenesis has not been demonstrated adequately for either
JH I or JH II and not at all for JH III. Detailed work on replacement chemotherapy in Manduca sexta third instar allatectomized
larvae, which is the essence of the third criterion, was presented
at an informal session of the conference by G.B. Staal but is not
included in this book. In these allatectomized larvae, JH III
was 1,000 times less potent than JH I or JH II for maintenance of
larval-larval molting (inhibition of metamorphosis to precocious
pupae) but JH III was nevertheless able to regulate larval morphogenesis. Continuous quantitative administration through the diet
proved to be an experimentally reliable and reproducible method of
hormone therapy during the later stages of larval development.

The status of JH III as a morphogenetic hormone is also
questioned by Dahm and co-authors since such unphysiologically
large quantities of it are necessary to mimic the effects of JH I
in several assays including the above mentioned work of G.B. Staal.
Noteworthy is the observed pre-disposition of in vitro cultured
adult glands of several species including the holometabolous
Tenebrio molitor L., and Leptinotarsa decemlineata (Say) to biosynthesize only JH III even when supplied with homomevalonate, a
proposed precursor of JH I and II. These glands were strongly
stimulated by mevalonate added to the culture medium and it will
be interesting to know what JHs are produced by larval glands of
these species in vitro or in vivo.

The advent of three separate analytical chemical methods for
titer determination through electron capturing chemical derivatives
of the hormone molecules (Schooley et al., this volume; Dahm et al.,
this section; and L.W. van Broekhoven et al., Z. Naturforsch. 30:
726, 1975) is a welcome phenomenon. For it seems that extensive
titer determinations will be needed to decipher whether fluctuations in the level of one single hormone, or changes in the ratio
of two hormones are responsible for control of morphogenesis.
Along these lines, work described in this book by Dahm and colleagues
can be compared with results of Schooley and colleagues for M. sexta
larval and adult titers. Although levels measured by the former
group are generally lower, there is agreement that JH I and II are
at comparable and high levels in early fourth instar larvae while
JH II and JH III are at lower but equal titers on day 1 of the
fifth (last) instar with JH I still predominant. Again in fresh
adult females JH II and JH III are present in equal titers but now
JH I has fallen to a minor level. Through this limited picture
in M. sexta, JH II appears to be curiously present throughout, but
JH I and inversely JH III appear to rise or fall in keeping with
their suspected roles as the morphogenetic and gonadotrophic hormones respectively. It is likely that such a simple idea of

separate roles for the different molecular forms of JH will prove
incorrect, and in this context the observed selective synthesis
of JH III by cultured glands of several species (see above) and
of JH II by glands of Galleria mellonella suggests these species
as prime candidates for titer studies in vivo, and for cross
transplantation of glands followed by JH titer study. Bearing
on the question of what controls the qualitative nature of hormones
produced by corpora allata (i.e. JH I or II or III or mixtures),
almost no information appears in this section on "Chemistry" and
while it is of fundamental importance, such information may be
extremely difficult to obtain in future work. Theories are
offered by Schooley and co-authors (this volume).

The development of competitive binding protein assays (CBPA)
for JHs by Goodman et al. (this section) and by Kramer et al.
(this volume) complements the chemical titer methods above, but
appears to trade rapidity and relative simplicity for sensitivity
and for the important ability to measure separately the levels of
the three different hormones. Though not noted by the authors,
their CBPA provides a fine method for scouting out those optimum
stages for chemical titer work before embarcation on the much more
tedious chemical methods.

The two remaining papers of this chapter focus on more
practically oriented questions of design of JH analogs (insect
growth regulators with JH activity) for pest control. An attempt
to derive information on the conformation of the active form of
the morphogenetic hormone is made by Henrick et al., using
relative activities of a number of rigid and semi-rigid analogs of
JH tested in assays where metabolism and transport are likely not
to modify intrinsic activity. The latter assumption may, however,
be invalid for the topical Tenebrio and Galleria pupal assays, but
the salient requirements for virtual planarity of atoms 1 through
6 of the JH chain for maximum potency are derived from comparisons
of pairs of chemically similar analogs with differences of three
orders of magnitude in potency, in a series of 2,4-dienoic acid
ethyl esters. These authors conclude that the receptor binding
site for morphogenetic hormone action should accommodate this planar
array of C-1 through C-6 which accounts for one half of the JH
molecule.

In a paper on phenyl ethers as insect growth regulators,
Zurflueh describes improved laboratory bioassays and field tests
against mosquitoes for both R-20458 and for a recent analog
Ro10-3108 which also gave some control of Adoxophyes orana, mealy
bugs and scale insects in the field. Interesting activity against
lepidopterous insects is associated with the presence of an isobutyl side chain at the epoxide terminal of a series of phenyl
ethers of monoterpenoid chains (Table 8). From these and other

benzofuran analogs of phenyl ethers emerges a picture of highest
lepidopteran activity only when an epoxide and central double
bond are present together, just as in the natural juvenile hormones.
Though the promising analog Ro10-3108 is discussed in light of
possible fields of application, the prognosis for its commercial-
ization is left unclear by Zurflueh.

ON THE IDENTITY OF THE JUVENILE HORMONE IN INSECTS

K.H. Dahm, G. Bhaskaran, M.G. Peter, P.D. Shirk,
K.R. Seshan, and H. Röller

Texas A&M University, Institute of Developmental
Biology, College Station, Texas 77843

INTRODUCTION

The secretion of the corpora allata, the juvenile hormone
(JH), has a morphogenetic function during the larval life of an
insect and may act as a gonadotropin in the adult. On the basis
of transplantation experiments, it had been concluded that the
hormone of the corpora allata is stage and order unspecific
(Wigglesworth, 1970; Novák, 1966). Consequently, the name "juvenile hormone" was retained irrespective of its endocrinological
function during a specific postembryonic stage. Piepho (1950)
even had suggested that the juvenile hormones of all insects be
chemically related, if not identical.

During the past eight years three natural compounds, claimed
to be juvenile hormones, have been isolated from insect sources.
In order of their historical appearance, we will call them JH I
(1, Fig. 1) (Röller et al., 1967; Dahm et al., 1967), JH II (2)
(Meyer et al., 1968), and JH III (3) (Judy et al., 1973a).
Numerous experiments have demonstrated that these compounds can
indeed be produced by corpora allata, and there is extensive
literature about their biological activity in various life stages
and different orders of insects. In view of this, it is surprising
that at the present time we cannot say with certainty which of the
three is the morphogenetic hormone of any insect species. With a
single exception (Lanzrein et al., 1975), all identifications are
based on work with adult insects or their corpora allata in vitro.
Consequently, they yield at best information about the gonadotropic
hormone but not necessarily about the morphogenetic hormone. Our
present dilemma in identifying the hormone of a specific species
and postembryonic stage has its roots in the finding that the three

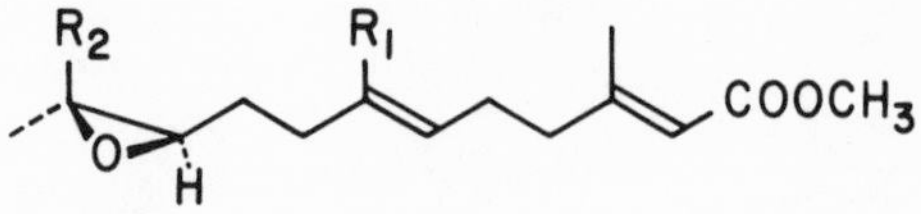

$$I \quad R_1 = R_2 = -C_2H_5 \qquad JH-I$$
$$2 \quad R_1 = -CH_3, \quad R_2 = -C_2H_5 \quad JH-II$$
$$3 \quad R_1 = R_2 = -CH_3 \qquad JH-III$$

Fig. 1. Structures of the known juvenile hormones. In all experiments with synthetic hormones, the racemates were used.

hormones, as predicted, are closely related chemically, but that contrary to earlier assumptions they can be readily distinguished by their physiological activities.

In an extension of an experiment reported earlier (Röller and Dahm, 1968), we injected various amounts of JH I, JH II and JH III, diluted in olive oil, into full-grown eye-class 1 larvae of Galleria mellonella (scoring according to Piepho, 1938). A dose of 1 µg JH I caused more than 50% of the injected larvae to molt into larval-like intermediates. Ten µg JH II were less effective; and after application of 100 µg JH III most animals finally molted into normal pupae. Obviously JH III, even in a most unphysiological dose of 100 µg, is not able to mimic the effect of three brain-corpora cardiaca - corpora allata complexes transplanted from third or fourth instar larvae into full-grown larvae (wandering last larval instar) as reported by Piepho (1942). For the evaluation of JH activity, the Tenebrio assay (Karlson and Nachtigall, see Röller and Dahm, 1968; Bjerke and Röller, 1974) and the Galleria wax test (de Wilde et al., 1968) have been used quite frequently. Both tests do not involve transportation of the applied material by the circulatory system; it is deposited close to the target tissue which it reaches by diffusion. Complications owing to JH-degrading enzymes in the hemolymph should, therefore, be absent from these test systems. Nevertheless, in the Galleria wax test the specific activity of JH III is lower than that of JH I or JH II by a factor of 100. In the Tenebrio assay the difference is even larger, the specific activities of JH I and JH III are separated by four orders of magnitude (Table 1). The latter result is particularly puzzling since JH III is the only hormone that could be isolated from extracts of reproductive adults of ·this species (Trautmann et al., 1974a) or from in vitro cultures of their corpora allata (Judy et al., 1975). A plausible, but not necessarily correct explanation of the apparent discrepancy between occurrence and morphogenetic activity of JH III would be the assumption that at least some insect species use different JH homologs as morphogenetic and gonadotropic hormones.

TABLE 1. SPECIFIC ACTIVITIES OF RACEMIC JUVENILE HORMONES IN THE GALLERIA WAX TEST AND IN THE TENEBRIO ASSAY.

	GU/µg	TU/µg
JH I	200×10^3	800
JH II	200×10^3	30
JH III	2×10^3	0.05

GU = Galleria Unit; amount required to elicit a positive response in 50% of the animals scored. TU = Tenebrio Unit; amount required to elicit a positive response in 40% of the treated pupae.

With one exception (Lanzrein et al., 1975), JH I and JH II have been identified only in lepidopterous insects (Röller et al., 1967; Meyer et al., 1968; Dahm and Röller, 1970; Röller and Dahm, 1974) or in cultures of their corpora allata (Roller and Dahm, 1970; Judy et al., 1973a; Jennings et al., 1975a). JH III alone was found in orthopteran, coleopteran and a hymenopteran species (Trautmann et al., 1974a,b) and also in cultures of orthopteran and coleopteran corpora allata (Judy et al., 1973b, 1975; Müller et al., 1974). Disregarding all other evidence, these findings might suggest that the original JH of insects is JH III, and that JH I and JH II are special evolutionary achievements in Lepidoptera and possibly some other insect orders.

It should be possible to test the two hypotheses outlined above by transplantation experiments. Corpora allata producing exclusively JH III should not be able to replace the original gland in an insect where only JH I or JH II show appreciable morphogenetic activity. The reverse experiment would yield no information since, as far as known, the biological activities of JH I or JH II are always equal or greater than that of JH III. Examination of the literature from this point of view reveals that within the Lepidoptera, the stage unspecificity of the corpora allata secretion is well documented (Piepho, 1942; Williams, 1959; Fukuda, 1962, 1963).

In a series of classical experiments demonstrating order unspecificity, Piepho (1950) implanted ten corpora allata with attached corpora cardiaca of Tenebrio larvae or four corpora allata of adult Carausius morosus Br. (Orthoptera, Phasmida) into full grown Galleria larvae and observed the same effects as after

transplantation of corpora allata from larvae of <u>Achroia grisella</u>
(Fabr.), <u>Anagasta kuehniella</u> (Zeller), and <u>Bombyx mori</u> (L.).
However, the juvenilizing action of these glands was weak and not
comparable to that of implanted corpora allata from <u>Galleria</u> larvae
(Piepho, 1942). Inhibition of metamorphosis was restricted to the
regenerating epidermis at the implantation site which is particularly sensitive to JH. The observed effects might have been easily
induced by JH III. A striking experiment, on the other hand, has
been reported by Yashika (1960). He implanted ten brain - corpora
allata complexes of adult <u>Ctenolepisma villosa</u> (Apterygota) into
diapausing pupae of <u>H. cecropia</u> from which the brain with its
corpora cardiaca and corpora allata had been extirpated within
five hours after pupation. Seven out of ten pupae experienced a
second pupal molt while the other three developed into, what
Yashika called, "imperfect adults". All 14 controls, implanted
with brain only, developed into normal adults. In view of the
relative specific activities of the three JH's in other lepidopterous insects, it is very doubtful whether JH III producing corpora
allata would cause the effects observed by Yashika.

   Since interpretation of published results does not lead to
an unambiguous identification of the morphogenetic corpus allatum
hormone, it is necessary to reconsider the nature of the evidence
essential for identification of a chemical entity with the physiologically defined hormone. It appears that three criteria must be
met to identify the JH chemically without any ambiguity. First,
the compound in question must be produced by the corpora allata.
Second, its titer in the circulatory system must rise and fall
in synchrony with the processes controlled by the hormone. Third,
when under appropriate experimental conditions the natural hormone
has been deleted from the animal, the artificially supplied chemical
must be able to substitute fully for the authentic hormone.

   When it could be assumed that the juvenile hormone in all
insects is identical, results of experiments with various species
could be generalized for interpretation. Since the concept of the
single hormone is challenged, it is necessary that all the criteria
are met experimentally in one defined stage of a specific insect
species. Considerations and experiments concerning the origin,
titer, and function of JH are presented in the following.

                         RESULTS AND DISCUSSION

        Juvenile Hormones Produced by Corpora Allata <u>In Vitro</u>

   The identification of methyl 10,11-epoxy-farnesoate as a JH
(JH III) is based mainly on the finding that it is produced by
adult corpora allata of several species under <u>in vitro</u> conditions.

The isolation of small amounts of this compound from total extracts
of plant feeding insects would not have, without prejudice, led to
the conclusion that it functions as a hormone in the respective
species. An important piece of evidence missing in these investigations is the demonstration that JH III, while present in some
developmental stages, is absent in others where its absence can be
postulated on the basis of general physiological considerations.
With regard to the in vitro systems, it is quietly assumed that
corpora allata under such artificial conditions produce the same
endocrine products as in vivo. We cannot yet explain why corpora
allata of Manduca sexta in vitro incorporate mevalonate into JH II
and JH III (Schooley et al., 1973) while H. cecropia seem to be
unable to use this compound as a precursor for JH I and JH II
(Metzler et al., 1971). During our work on the biosynthesis of
JH in Cecropia, we made another observation which might be relevant
to the question of possible differences in the biosynthetic activity of corpora allata in vivo or in vitro. Male moths usually are
vigorous to an age of about five days; thereafter they deteriorate
rapidly. They normally contain besides JH I 10 to 30% JH II.
However, very old males which had been injected with [S-methyl-^{3}H]-
methionine and incubated for a few days contained more labeled
JH II than JH I (Fig. 2). It is reasonable to assume that the
aging moths fail to synthesize enough of the special precursor for
the ethyl side chains of the hormones, in consequence of which the
corpora allata produce relatively more JH II. With this consideration in mind we re-examined corpora allata of female M. sexta
under in vitro conditions (Judy et al., 1973a).* JH III was
the major product, but the time course of JH-production proved
to be interesting (Table 2). During the first days after explantation, the glands release more JH II than JH III into the
medium, but JH II synthesis ceases soon while the rate of JH III
synthesis initially increases. The result seemed to indicate
that a specific precursor for JH II and possibly JH I is in
short supply in the in vitro system. Since the studies of the
Zoecon group (Schooley et al., 1973; Jennings et al., 1975b) as
well as our own investigations (Gowal et al., 1975; Peter and
Dahm, 1975) had indicated that in vitro as well as in vivo the

---

* In this and all other in vitro experiments we used as the basic
culture medium Grace's insect tissue culture medium without methionine (GIBCO, Grand Island Biol. Co., NY), supplemented with 1%
bovine albumin (Fraction V from bovine plasma, Metrix, Div. of
Armour Pharmaceutical Co., Chicago, IL) (Judy et al., 1973a). In
most experiments (S-methyl-^{14}C)-methionine (spec. activities in
the range between 47 and 58 Ci/mol) was added in order to label the
hormones. After incubation, the media were extracted several times
with ethyl acetate, at which point depending on the experiment
unlabeled JH's were added as carriers. The hormones were purified
by TLC on methanol-washed Silica Gel HF-254 (E. Merck, Darmstadt,
Germany) with benzene/5% ethyl acetate and resolved by HPLC-II
(see Table 10).

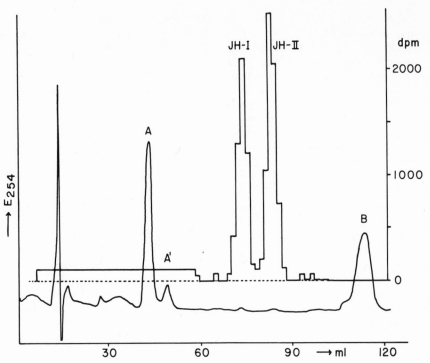

Fig. 2.   Juvenile hormones from a 9 day-old male Cecropia moth.   The
moth had been injected on day 6 with 50 µCi [S-methyl-^{3}H]-methio-
nine.   After the usual purification procedure (see Table 10), the
hormones were separated on an 0.23 x 300 cm Porasil T (Waters Associates, Inc., Milford, MA) column with hexane/2.8% ethyl acetate/
0.04% 2-propanol.   The peaks A, A', and B on the UV-trace are
those of reference compounds: A and B are 0.4 µg each of nonanal-
and acetone 2,4-dinitrophenylhydrazone, A' is an impurity in A.
The preparation contained 1.0 µg (5200 dpm) JH I, 1.2 µg (6400 dpm)
JH II, and no JH III (GLC-analysis and liquid scintillation counting).

ethyl side chains of JH I and JH II are introduced through incorporation of 3'-homomevalonate or its biochemical equivalent, it
seemed logical to supply the corpora allata cultures with this
intermediate.   Addition of 1 mg homomevalonolactone (synthesized
in our laboratories) per ml medium stimulated JH II biosynthesis
considerably, and JH I was produced in amounts comparable to those
of JH III.   Even when equal amounts of homomevalonolactone and
of mevalonolactone were added, JH II synthesis was stimulated
much more than that of JH III; also, JH II was the major product
as long as the glands remained active (Table 3).   A similar effect
was observed when corpora allata of female adult Heliothis virescens
(Fabr.) were kept under the same conditions.   These glands have

TABLE 2. PRODUCTION OF JUVENILE HORMONES BY CORPORA ALLATA OF
FEMALE ADULT MANDUCA SEXTA IN VITRO.

Day in culture	Experiment I [dpm/day x gland pair]		Experiment II [dpm/day x gland pair]	
	JH II	JH III	JH II	JH III
1st – 2nd	770	630	1,300	970
3rd – 4th	300	590	480	1,280
5th – 6th	70	1,470		
7th – 12th	∿2	∿2		
13th – 16th	nil	nil		

The culture medium contained 2 µCi [S-methyl-^{14}C]-methionine
per ml.

TABLE 3. PRODUCTION OF JUVENILE HORMONES BY CORPORA ALLATA OF
FEMALE ADULT MANDUCA SEXTA IN VITRO.

Day in culture	[dpm/day x gland pair]		
	JH I	JH II	JH III
1st – 4th	220	7,300	2,550
5th – 8th	nil	690	34
9th – 12th	nil	20	6

The culture medium contained 2 µCi [S-methyl-^{14}C]-methionine,
1 mg mevalonate, and 1 mg homomevalonate per ml.

been reported to produce in vitro JH I and JH II (Jennings et al., 1975a); in our system they also release JH III (Table 4). Addition of the mevalono- plus homomevalonolactone to the medium had little effect on the production of JH III, but increased the synthesis rate of JH II four-fold and of JH I seven-fold.

TABLE 4. PRODUCTION OF JUVENILE HORMONES BY CORPORA ALLATA OF FEMALE ADULT HELIOTHIS VIRESCENS IN VITRO.

Experiment I			Experiment II: mv + hmv			
Day in	[dpm/day x gland pair]		Day in	[dpm/day x gland pair]		
culture	JH I	JH II	JH III	culture	JH I	JH II	JH III
1st - 4th	76	530	300	1st - 3rd	550	2,180	450
5th - 8th	66	480	190	4th - 6th	510	2,080	300

All culture media contained 2 µCi [S-methyl-^{14}C]-methionine per ml. mv + hmv: the medium contained 1 mg mevalonate and 1 mg homomevalonate per ml.

Koyama et al. (1973) have shown that pig liver farnesyl pyrophosphate synthetase can utilize the normal isoprenyl pyrophosphates and their C_6-homologs rather indiscriminately to produce farnesyl pyrophosphate homologs. In contrast, the corpora allata system seems to be very specific. Under the experimental conditions, JH II would probably not have been distinguished from its 10'-desmethyl-7'-methyl isomer, but all JH variants with an ethyl instead of a methyl group at C-3 would have been easily detected. Cum grano salis we come to the conclusion that even in the presence of homomevalonolactone the corpora allata cannot be forced to the biosynthesis of a host of sesquiterpene-like compounds, but that they produce only those structures for which they are naturally programmed.

This appears to be confirmed by results of in vitro experiments in which the corpora allata produce exclusively JH III. Corpora allata of adult Periplaneta americana released 10 ng JH III per day and gland pair during the first few days after explantation. The specific incorporation ratio of label from [S-methyl-^{14}C]-methionine and 2-[^{14}C]-mevalonolactone is between 10 and 100%, of 1,2-[^{14}C] acetate about 10%, and of 1-[^{14}C]-propionate about 1% (Table 5).

TABLE 5. PRODUCTION OF JH III BY CORPORA ALLATA OF MALE
PERIPLANETA AMERICANA IN VITRO.

Precursor		JH III
[2 µCi/ml medium]	[µCi/µmol]	[dpm/day x gland pair]
[S-methyl-^{14}C]-methionine	58	5,000
2-[^{14}C]-mevalonolactone	11	1,900
1,2-[^{14}C]-acetate	54	3,300
1-[^{14}C]-propionate	53	260

The culture medium was supplemented with 50 µg (unlabeled)
methionine per ml.

Propionate should not be incorporated directly (Judy et al., 1973a;
Peter and Dahm, 1975) but only after degradation to smaller fragments. Corpora allata of males and females release similar amounts
of JH III except when females bear ootheca ready for deposition
(Table 6). Addition of mevalonolactone increases the JH III yield
about ten-fold while homomevalonolactone has no effect. With the
exception of JH III, no trace of a compound with JH-structure was
ever detected; the detection limit in some experiments was less
than 0.1% JH III.

The same selectivity for incorporation of mevalonate and
rejection of homomevalonate was found in adult corpora allata
cultures of Blaberus discoidalis (Serville) (Table 7), P. fuliginosa
(Serville), T. molitor L., and diapausing and non-diapausing beetles
of Leptinotarsa decemlineata (Say). After incubation, the cultures
contained only JH III. When the time course of JH production was
determined, the corpora allata of all investigated species decreased
JH synthesis drastically within the first two weeks in spite of the
fact that they can be kept in culture for months without visible
deterioration. So far we have found no way to reactivate inactive
glands in vitro.

Successful culture of corpora allata from immature insects has
not yet been reported by other laboratories. The glands are less
active than those of adults but they also can be stimulated by
addition of mevalonolactone/homomevalonolactone to the medium
(Fig. 3). So far, JH I and JH II have been produced only by

TABLE 6. PRODUCTION OF JH III BY CORPORA ALLATA OF ADULT
PERIPLANETA AMERICANA IN VITRO.

Day in culture	JH III [dpm/day x gland pair]				
	♂	♂, mv	♀	♀, mv	♀, ♂
1st - 4th	2,600	23,100	5,600	40,600	24,000
5th - 8th	1,900	2,600	2,000	5,800	15,200
9th - 12th	1,020	170	830	360	9,700
13th - 16th	620				
17th - 20th	160				

All culture media contained 2 µCi [S-methyl-^{14}C]-methionine
per ml. mv: 1 mg mevalonate/ml medium. o: female, bearing
ootheca ready for deposition.

TABLE 7. PRODUCTION OF JH III BY CORPORA ALLATA OF ADULT
BLABERUS DISCOIDALIS IN VITRO.

Corpora allata of	JH III [dpm/day x gland pair]	
		mv + hmv
Female adult	2,500	4,700
Male adult	3,900	5,600

All culture media contained 2 µCi [S-methyl-^{14}C]-methionine
per ml. mv + hmv: The medium contained 1 mg each of mevalonate
and homomevalonate per ml.

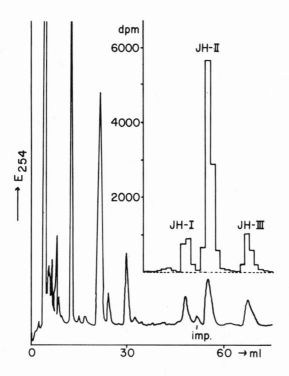

Fig. 3. High pressure liquid chromatography of [^{14}C]-labeled JH
from corpora allata cultures of M. sexta larvae. Eight brain –
corpora cardiaca – corpora allata complexes of freshly molted 5th
instar larvae had been cultured for three days in the presence of
2 μCi [S–methyl-^{14}C]–methionine, 1 mg mevalonate, and 1 mg homomevalonate per ml incubation medium. Unlabeled juvenile hormones
(containing an impurity) had been added to the preparation as
carriers: they appear as peaks in the UV-absorption curve.

corpora allata of lepidopterous larvae (Table 8). While our results
corroborate the view that JH I and JH II are restricted to a few
species or orders of insects, they still cannot be taken as firm
evidence. Free mevalonate and homomevalonate may after all not be
intermediates in the biosynthesis of the natural hormones. Also,
if homomevalonate is an unnatural product, e.g. in P. americana, why
should the JH-synthesizing enzymes discriminate against it so
strongly? In order to determine the specificity of the concerned
enzymes more rigorously, we have begun to investigate JH biosynthesis in corpora allata homogenates. Under conditions described
by Reibstein and Law (1973), homogenates and supernatants of M.
sexta corpora allata produced JH II and JH III (Fig. 4), while in
similar preparations from P. americana only JH III was detected.

TABLE 8. PRODUCTION OF JUVENILE HORMONES BY CORPORA ALLATA OF
INSECT LARVAE IN VITRO.

| | | [dpm/day x gland pair] | | |
Species	Instar	JH I	JH II	JH III
Galleria mellonella	7th	nil,<5	250	nil, <5
Hyalophora cecropia	4th	280	980	nil,<30
Manduca sexta	5th	68	400	72
Manduca sexta[a]	5th	nil,<1	7	4
Periplaneta americana, male	[b]	nil,<5	nil,<5	630
Periplaneta americana, male[a]	[b]	nil	nil	210
Periplaneta americana, female	[b]	nil,<5	nil,<5	830
Periplaneta americana, female[a]	[b]	nil	nil	310
Blaberus discoidalis[c]	[b]	nil,<5	nil,<5	700[d]
Blaberus discoidalis[a,c]	[b]	nil	nil	1000

The incubation medium contained per ml 2 µCi [S-methyl-^{14}C]-
methionine, 1 mg mevalonate, and 1 mg homomevalonate.

[a] Experiment without mevalonate and homomevalonate.

[b] Advanced stage but younger than last larval instar

[c] Sex was not determined.

[d] Yield lower than anticipated; the culture was contaminated with
bacteria.

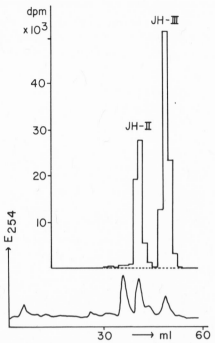

Fig. 4.   Juvenile hormones produced by a homogenate of 10 corpora cardiaca - corpora allata pairs from adult female M. sexta. The incubation medium (Reibstein and Law, 1973) contained 0.5 μCi S-adenosyl-[S-methyl-³H]-methionine (spec. act. 7.5 Ci/mmol).

Identification and Quantitative Determination
of Juvenile Hormones by a Chemical Method

    Since our in vitro experiments had revealed that JH III is not necessarily the major hormone of M. sexta, it was of interest to determine the JH composition in hemolymph of various postembryonic stages of this species. We decided to develop a procedure by which all three hormones are converted to halogen-containing derivatives which would allow final identification by gas chromatography with electron capture detection. Utilizing 10,11-bis-trifluoroacetoxy derivatives, Judy et al. (1973a) had already identified JH III in the hemolymph of fourth instar larvae, but attempts to verify the presence of JH I and JH II gave inconclusive results. In the meantime the same research group has improved the method by exploiting a differ- ent derivative and is now also in a position to analyze blood samples for all three hormones (Schooley et al., this volume). Selection of the larval stages used in our investigations was based on a JH titer curve which Judy (personal communication) obtained by means of the Galleria wax test. According to this

study the JH activity is high in early fourth instar larvae, moderately high in early fifth instar larvae, and very low in late fifth
instar larvae. In addition to the penultimate and the last larval
stages we included male and female adults in our experiments
(Peter et al., 1976).

    Hemolymph was collected in water of 0° and immediately extracted
with ethyl acetate. At this point [methoxy-^{3}H]-ttt-JH-0 (Fig. 5, 4:
$R_1=R_2=R_4=-C_2H_5$, $R_3=-CH_3$, [^{3}H]-labeled in the methyl ester group)
was added to each sample as an internal standard. The extracts of
hemolymph from adult females contained large amounts of oily material; most of it could be removed by precipitation in methanol at
-20°C. The hormones (4) in the supernatant of this preparation,
and in the other extracts directly, were purified by thin layer
chromatography (TLC) and converted with perchloric acid in methanol
to the 10-hydroxy-11-methoxy derivatives 5. After another TLC-
purification, the appropriate fractions were treated with 2,4-
dichlorobenzoyl chloride in pyridine. The derivatives 6 were purified by two subsequent TLC-separations and resolved by high pressure
liquid chromatography (HPLC) (Fig. 6). The retention volumes of
the hormone derivatives were calculated from that of the radiolabeled ttt-JH-0-derivative. The eluates of the HPLC separations
were collected in such a way that each hormone derivative was
spread over several fractions. The individual fractions were
evaporated to dryness, the residues redissolved in 25 µl 2,2,4-
trimethylpentane/20% ethyl acetate and analyzed by gas liquid
chromatography (GLC) with an electron capture detector (Fig. 7).

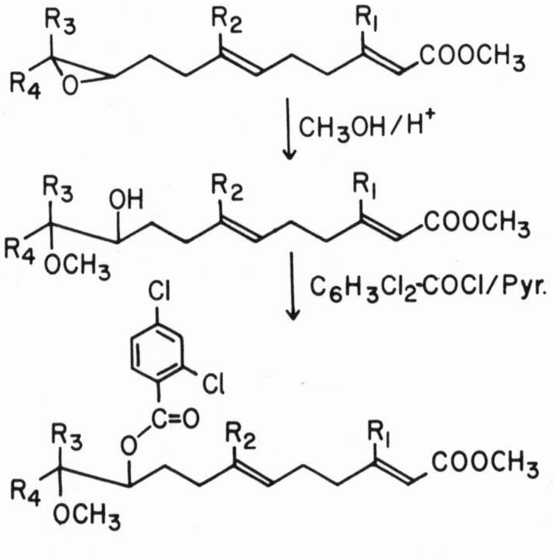

Fig. 5. Conversion of juvenile hormones to derivatives suitable
for gas liquid chromatography with an electron capture detector.

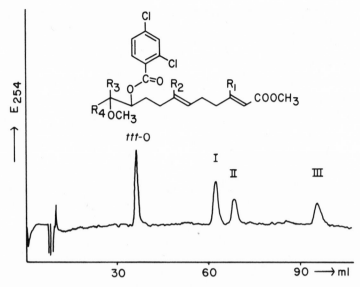

Fig. 6. Separation of the juvenile hormone derivatives 6 (0.5–
1 μg each) by high pressure liquid chromatography. [0.4 x 60 cm
μPorasil (Waters Associates Inc., Milford, MA), hexane/3.5% ethyl
acetate/0.02% 2-propanol.] ttt-0, II and III are the derivatives
of the respective JH. During analysis of natural hormones, the
elution volumes were calculated from that of the [^{3}H]-labeled
reference compound (ttt-0).

This procedure assured that material which on GLC had the retention
time of one of the hormone derivatives 6 had also on the HPLC
column the appropriate retention volume. Each chromatographic
identification incorporates consequently the resolving power of
both systems independently. To further assure accurate identifications, each fraction was analyzed on three of the following
GLC columns: 3% SE-30, 3% OV-1, 3% OV-17, 3% XE-60 or 10% UC-W98.
Authentic samples, prepared by the same method and identified by
mass spectrometry, served as reference compounds. Under the conditions used for the analysis, a peak of 5 mm height was produced by
5 pg of the JH III derivative. The recovery of small quantities
throughout the procedures was checked in pilot experiments where
0.25 – 1.0 ng of the hormones had been added to blood extracts of
late 5th instar larvae which are devoid of natural hormone. In
order to detect possible contamination of samples with synthetic
hormones, each set of experiments included controls with and without JH.

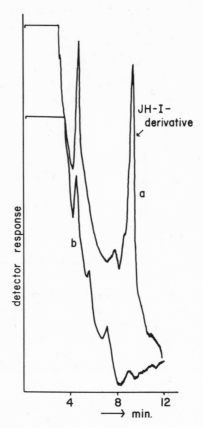

Fig. 7.   JH I in hemolymph of M. sexta larvae.   The hormones have
been extracted from hemolymph, converted to the 10-(2,4-dichloro-
benzoyloxy)-11-methoxy-derivatives, and separated by high pressure
liquid chromatography.   The fractions corresponding to the peak
center of the JH I derivative (calculated from the radiolabeled
reference compound, see Fig. 6) were injected in a gas chromatograph
equipped with an electron capture detector [0.3 x 180 cm glass
column, 3% OV-1 on 60/80 mesh Gas Chrom Q (Applied Science Lab.,
Inc., State College, PA), column temp. 275°C].   a) is a preparation
from early fourth instar larvae; b) the corresponding preparation
from late fifth instar larvae.

     The highest concentration of JH was found in hemolymph of
early fourth instar larvae; it was considerably lower in early fifth
instar larvae (Table 9).   No JH could be detected in wandering fifth
instar larvae;   the detection limit in this experiment was < 0.003
ng/ml hemolymph.   This result corresponds well with the earlier mentioned titer determinations by the Galleria wax test (K. Judy,
personal communication) and a Manduca assay (L. Riddiford, personal

TABLE 9. JUVENILE HORMONES IN HEMOLYMPH OF <u>MANDUCA</u> <u>SEXTA</u>

No. animals	Stage	Hemolymph collected [ml]	JH	Concen- tration [ng/ml]
82	early 4th larval instar (1-2 day-old)	3.5	I	0.62
			II	1.1
			III	0.07
28	early 5th larval instar (1-2 day-old)	5.1	I	0.24
			II	0.11
			III	0.12
17	5th larval instar full-grown (4-5 day-old)	10.8	I	<u>nil</u>
			II	<u>nil</u>
			III	<u>nil</u>
116	female adults (1-2 day-old)	4.9	I	0.06
			II	0.14
			III	0.11
60	male adults (1-2 day-old)	1.5	I	0.10
			II	0.04
			III	<0.02

<u>nil</u>: not detected, $\leq$0.003 ng/ml.

communication). It should be taken into account that hemolymph of late fifth instar larvae has a high JH esterase activity which makes the presence of any JH at this stage and age unlikely (Weirich et al., 1973). In all experiments, the three JH's were found together in varying concentrations. At present no interpre- tation is possible with regard to the JH composition in the diff- erent stages. Before any conclusions can be reached, one has to know whether the ratios JH I:JH II:JH III are closely controlled or whether in allatectomized larvae physiological effects of JH mixtures depend on the ratios of their components. It has been determined, however, that the morphogenetic activity of JH III

in Manduca pupae is 300 times lower than that of JH I and JH II
(Riddiford and Ajami, 1973). In females of Manduca, JH controls
egg maturation (Sroka et al., 1975) and in males, presumably, the
function of accessory sex glands, but the relative gonadotropic
activities of JH I, JH II and JH III have not been determined.

In hemolymph of the roach Nauphoeta cinerea, depending on
stage and age, varying concentrations of JH I, JH II and JH III
were also detected (Lanzrein et al., 1975). From the relative
concentrations in nymphs and in adults it is suggested that JH I
and JH II may be responsible for the morphogenetic and JH III for
the gonadotropic action. Hopefully, these identifications can be
confirmed, since in total extracts of several adult roaches
including Nauphoeta, only JH III was found (Trautmann et al.,
1974a). JH III was also the only hormone detected in larval and
adult corpora allata cultures of a number of roach species. We
will now apply the new method for detection of juvenile hormone
in selected species. Of particular interest to us are L. decemlineata as a representative of Coleoptera and some insects of the
orders Apterygota, Isoptera and Diptera. In early fourth instar
larvae of H. cecropia we have found 0.15 ng JH II per ml hemolymph
but were unable to detect JH I or JH III.

Accumulation and Storage of Juvenile Hormone in the Male Accessory Sex Glands of Hyalophora cecropia

The adult males of H. cecropia and of some related saturniid
moths are distinguished by their ability to produce and store
microgram quantities of JH (Röller and Dahm, 1968; Meyer et al.,
1968; Röller et al., 1969; Dahm and Röller, 1970; Röller and Dahm,
1974). Consequently, Cecropia was the insect of choice for
studying the biosynthesis of the hormones in vivo (Metzler et al.,
1971, 1972; Gowal et al., 1975; Peter and Dahm, 1975). It has
been shown that JH I is indeed synthesized by the corpora allata
(Röller and Dahm, 1970) and that no hormone is sequestered in the
abdomens of adults when the animals are allatectomized as pupae
(Williams, 1963). To date, it has not been possible to detect a
biological function of the large quantities of JH in adult saturniid moths. Allatectomized male and female Cecropia, crossmated,
produced viable offspring (Williams, 1959).

The isolation of JH from insects other than H. cecropia and
some related species is at the very least a tedious process. Since
corpora allata in vitro do not necessarily produce the same compounds as in vivo, we had planned to use allatectomized male
Cecropia moths as hosts for corpora allata of different insects
and to rely on their JH collecting capability for accumulating the
hormone. The procedure for isolating the JH from saturniid moths
has been perfected during the past ten years and it is now possible

to recover amounts as low as 0.1 μg from single animals (Table 10).
When the hormones have been labeled in vivo by injection of radiolabeled methionine (Metzler et al., 1971), the fractions of the
second HPLC separation contain little radioactive material besides
the hormones. The hormones recovered from this separation are
gas chromatographically pure (Fig. 8).

TABLE 10.   ISOLATION OF JUVENILE HORMONES FROM ABDOMENS OF MALE
HYALOPHORA CECROPIA.

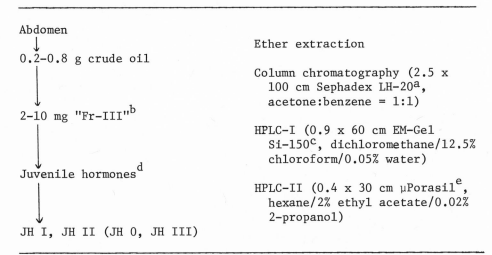

Abdomen
↓
0.2-0.8 g crude oil   Ether extraction

| Column chromatography (2.5 x
↓   100 cm Sephadex LH-20[a],
2-10 mg "Fr-III"[b]   acetone:benzene = 1:1)

| HPLC-I (0.9 x 60 cm EM-Gel
↓   Si-150[c], dichloromethane/12.5%
Juvenile hormones[d]   chloroform/0.05% water)

| HPLC-II (0.4 x 30 cm μPorasil[e],
↓   hexane/2% ethyl acetate/0.02%
JH I, JH II (JH 0, JH III)   2-propanol)

[a] Pharmacia Fine Chemicals AB, Uppsala, Sweden
[b] Nonanal- and acetone 2,4-dinitrophenyl-hydrazone serve as
internal standards. The JH's emerge with the nonanal dinitrophenyl-
hydrazone (total volume about 24 ml). "Fr-III" contains as a
precaution against unexpected variations also the preceding ∿3 and
following ∿6 ml.

[c] E. Merck, Darmstadt, Germany.

[d] The JH's are found in the fraction with 45-80% the retention volume
of the internal reference compound methyl p-hydroxybenzoate.

[e] Waters Associates, Inc., Milford, MA.

When 1-2 day-old male Cecropia moths are injected with 50 μCi
[S-methyl-^{3}H]-methionine and sacrificed 48 hr later, usually JH
with [^{3}H]-activity between 10,000 and 100,000 dpm can be isolated.
After depriving males of their corpora allata by decapitation
before they were three hr old, we implanted corpora allata of other
males or females and injected the radiolabeled methionine as a

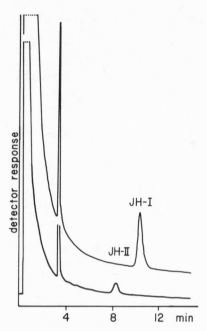

Fig. 8. Gas chromatograms of JH I and JH II isolated from a single
male Cecropia moth by the procedure described in Table 10. The
peak at 3 min represents 0.12 μg methyl palmitate added as an internal standard. [0.4 x 180 cm glass column, 3% QF-1 on 100/120
mesh Gas Chrom Q (Applied Science Lab., Inc., State College, PA),
column temp. 160°C, flame ionization detector].

tracer. The implanted corpora allata appeared to function properly;
large amounts of labeled JH could be isolated (Table 11). We have
never been able to isolate either labeled or unlabeled JH from
females; bioassay data indicate that they may contain only nanogram amounts of the hormone. The surprising result in this experiment was that allatectomized moths synthesized labeled JH. Since
a wealth of endocrinological research had indicated that JH is
produced in the corpora allata and not in other tissues, it seemed
most likely that the labeled JH is not produced de novo but that
the methyl ester group of the hormone, already present in the
animal, is replaced by the methyl group of the labeled methionine.
Degradation of JH in a number of insects involves loss of the
methyl ester group as one of the first steps (Slade and Zibitt,
1972), while male Cecropia methylate the acid to the hormone in high
yield (Metzler et al., 1972). Therefore it was likely that the
methyl exchange was in some way connected with the ability of the
moths to accumulate the hormone.

TABLE 11.   [^{3}H]-LABELED JH I FROM DECAPITATED MALE CECROPIA MOTHS.

Preparation	Implant	JH I [dpm]
Intact male	none	55,000
Intact female	none	nil, 20
Decapitated male	none	7,000
Decapitated male	1 pair male cc-ca	71,000
Decapitated male	4 pair male cc-ca	128,000
Decapitated male	1 pair female cc-ca	21,000

Corpora cardiaca - corpora allata complexes of adult Cecropia moths were implanted in 5-16 hr old males which had been decapitated within 3 hr after adult eclosion. Two days later the moths were injected with 50 µCi [S-methyl-^{3}H]-methionine. They were sacrificed after a 48 hr incubation period.

When males were allatectomized as pupae during the early stages of adult development and were processed as adults in the usual fashion after being injected with the labeled methionine, no trace of labeled or unlabeled hormone could be detected. In some preparations of this nature, adult male or female corpora allata were implanted shortly after adult eclosion. These animals contained afterwards JH I as well as JH II (Table 12). It cannot yet be concluded whether female corpora allata in their natural environment also produce more JH II than JH I or whether the preponderance of JH II is a result of the experimental conditions. Some other surgical procedures in male or female pupae were without effect on the JH production.

In order to study the methyl ester exchange we injected unlabeled or [^{14}C]-labeled JH I and JH III together with [^{3}H]-labeled methionine in decapitated males (Table 13). When 5 µg each of unlabeled JH I and JH III were injected, the yield of [^{3}H]-JH I was as high as in intact males while the yield of [^{3}H]-JH II was similar to that in controls. The yield of [^{3}H]-JH III was relatively low. The same discrimination in favor of JH I was apparent when [methoxy-^{14}C] labeled JH I and JH III were injected in intact or decapitated males. About 20% of the [^{14}C]-JH I, but only 1% of [^{14}C]-JH III were recovered unchanged. The [^{3}H]-incorporation ratio in these experiments was in the usual range.

TABLE 12. JH I AND JH II FROM ADULT CECROPIA AFTER ALLATECTOMY OR OTHER SURGICAL PROCEDURES DURING THE PUPAL STAGE.

Preparation	JH I [dpm]	JH II [dpm]
male, allatectomized	nil	nil
male, N.C.C.[a] I and II cut	106,000	26,000
male, 1 pair ovaries implanted	71,000	10,000
female, N.C.C.[a] I and II cut	nil	nil
female, 4 pair male gonads implanted	nil	nil
male, allatectomized, [2 pair male adult cc-ca implanted][b]	10,800	9,900
male, allatectomized, [2 pair female adult cc-ca implanted][b]	320	1,090

The surgical procedures except implantation of cc-ca complexes were performed on pupae prior to initiation of adult development. The implanted ovaries and male gonads were taken from pupae of the same developmental stage. Fifty μCi [S-methyl-^{3}H]-methionine were injected 48 hr after adult eclosion. The moths were sacrificed after an incubation period of 48 hr. nil: not detectable, less than 20 dpm.

[a] nervi corporis cardiaci.

[b] The corpora cardiaca - corpora allata complexes were implanted during the first day after adult eclosion.

Obviously, some tissue or tissues in the moth are able to degrade and to resynthesize the juvenile hormone after it is secreted by the corpora allata. It had generally been assumed that JH is sequestered in the fat body (Williams, 1963), but when rechecking the literature on this point we found that convincing evidence had never been presented.

When we dissected male moths and tried to isolate JH from the various tissues, we quickly discovered that not the fat body but tissues of the reproductive tract are responsible for the accumulation of the hormone (Shirk et al., 1976). In order to facilitate

TABLE 13. REPLACEMENT OF THE METHYL ESTER GROUP OF JH BY THE S-METHYL GROUP OF METHIONINE IN INTACT AND DECAPITATED MALES OF HYALOPHORA CECROPIA.

Preparation (adult males)	Material injected besides 50 μCi [^{3}H]-methionine	JH I [dpm]		JH II [dpm]	JH III [dpm]	
		[^{3}H]	[^{14}C]	[^{3}H]	[^{3}H]	[^{14}C]
Intact	none	82,000	--	11,000	nil	--
Decapitated	none	2,800	--	1,900	--	--
Decapitated	5 μg JH I + 5 μg JH III	80,000	--	1,400	6,300	--
Decapitated	10^4 dpm [^{14}CH$_3$O]–JH I and [^{14}CH$_3$O]–JH III	11,500	2,500	8,900	1,300	131
Intact	same as above	52,000	1,400	8,600	167	14

The moths were decapitated within 3 hr after adult eclosion, injected at an age of 2 days, and sacrificed at an age of 4 days. nil: not detectable, less than 40 dpm.

the isolation of JH, it was labeled routinely by injection of
[S-methyl-^{3}H]-methionine into the animals 48 hr prior to dissection.
In a number of experiments we analyzed for JH not only radiochemi-
cally and gas chromatographically but also with the Galleria wax
test (Table 14). JH could be identified by radiolabel and by gas
chromatography only in tissues of the reproductive tract. According to the bioassay its concentration in the remainder of the body
is at least three orders of magnitude lower. In another experiment
(Table 15), the male accessory sex glands were dissected out and
extracted separately from the remainder of the abdomens. JH was
found exclusively in the extracts of the male accessory glands.
The hormones appear to be associated with the luminal contents
rather than the gland tissue (Table 16).

With regard to the biochemistry of the JH accumulation, we
have already obtained a number of miscellaneous results which eventually will give a complete picture of the process. Some saturniids
other than H. cecropia also accumulate JH in the male accessory
sex glands. No trace of JH was detected by bioassay in the accessory sex glands of male M. sexta. The methyl exchange is not
required for the accumulation; in vivo and in vitro the accessory
glands are able to assimilate JH without chemical change. At the
same time they have in vivo as well as in vitro the ability to
assimilate the corresponding acid, methylate and store it, and also
to exchange the methoxy group in JH already stored.

                                CONCLUSION

Our discovery that the accessory sex glands of male Cecropia
have the unique ability to store large amounts of JH merits further
intensive study. The endocrine situation in male Cecropia is similar to that in females in that the reproductive functions are independent of JH-control. The corpora allata of males are appreciably
larger than those of females and have a much higher biosynthetic
activity, as shown by our transplantation experiments. However,
the marked difference between the sexes in JH content must to a
large part be attributed to the exceptional nature of the male
accessory sex glands. Although JH in the Cecropia moth has apparently no function, its accumulation may be a physiological relic,
albeit exaggerated, of a process which in some other insect species
is important for the maturation or activity of the accessory sex
glands.

We have shown that larval and adult corpora allata of Manduca
sexta under certain in vitro conditions produce JH I, JH II and
JH III. Hemolymph of larvae and adults contains the same hormones
in quantities corresponding to the JH-activity profile established
by bioassays. The occurrence of all three hormones in various
relative concentrations might suggest specific roles for the

TABLE 14. JUVENILE HORMONE IN DIFFERENT TISSUES OF ADULT MALE HYALOPHORA CECROPIA.

Moth #	Preparation	Biol. activity [GU]^a	JH I [dpm]	JH I [μg]	JH II [dpm]	JH II [μg]
1	Reproductive tract	20 x 10^6	32,000	1.4	12,000	0.4
	Remainder of carcass	<1 x 10^3	nil	--	nil	--
2	Accessory sex glands plus seminal vesicles	5 x 10^6	6,200	2.1	590	0.3
	Remainder of carcass	1 x 10^3	nil	--	nil	--

Two 48 hr old male moths were each injected with 2 μCi [S-methyl- H]-methionine. They were dissected and extracted after a 48 hr incubation period. nil: not detectable, less than 35 dpm.

a Galleria Unit (see Table 1).

TABLE 15.   JUVENILE HORMONE IN THE ACCESSORY SEX GLANDS OF MALE
HYALOPHORA CECROPIA.

Moth #	Preparation	JH I [dpm]	JH I [μg]	JH II [dpm]
1	Accessory sex glands	3,100	95%[a]	860
	Remainder of abdomen	nil	58%[a]	nil
2	Accessory sex glands	4,500	2.8	1,800
	Remainder of abdomen	nil	nil	nil
3	Accessory sex glands	42,900	1.8	nil
	Remainder of abdomen	nil	nil	nil
4	Accessory sex glands	4,000	3.9	350
	Remainder of abdomen	nil	nil	nil
5	Accessory sex glands	44,000	2.2	5,300
	Remainder of abdomen	nil	nil	nil

Five 24 hr old moths were injected each with 25 μCi [S-methyl-^{3}H]-methionine. They were dissected and extracted after a 48 hr
incubation period.

[a] % recovery: based on the recovery of 10 μg unlabeled JH I added
to the crude extracts of moth 1. nil: not detectable, less than 200
dpm and/or 0.2 μg.

individual hormones in morphogenesis and reproduction. However,
the low morphogenetic activity of JH III coupled with the fact that
it is only as effective as JH I or JH II in reproductive assays
warrants cautious interpretation of the data. Whether or not the
three hormones have qualitatively different morphogenetic functions
may best be investigated in insects with polymorphic larvae and
adults.

TABLE 16. BIOSYNTHETICALLY LABELED JH I AND JH II FROM LUMINAL
CONTENT AND GLAND TISSUE OF ACCESSORY SEX GLANDS FROM MALE
HYALOPHORA CECROPIA.

Moth	Preparation	JH I	JH II
#		[dpm]	[dpm]
1	Luminal content	10,700	3,600
	Gland tissue	1,800	400
2	Luminal content	24,600	3,800
	Gland tissue	1,100	100

Male moths were prepared as described in Table 15. The
accessory sex glands were placed in insect Ringer solution and cut
in pieces. The luminal contents were separated from gland tissue
as well as possible by pressing the pieces with glass rods.

Corpora allata in vitro are surprisingly specific with regard
to JH-synthesis. Homomevalonate and/or mevalonate in the medium
are used as precursors and even stimulate JH-biosynthesis. However, production of JH I and of JH II was observed only in cultures
of lepidopterous corpora allata, while those of other species produced exclusively JH III. JH-homologs or isomers other than JH I,
JH II and JH III were never detected. We may conclude that the
in vitro production of specific juvenile hormones indicates intrinsic biosynthetic capabilities of corpora allata from different
species and more likely than not reflects the situation in vivo.
Whether or not this conclusion is valid will be shown by chemical
identification of juvenile hormones in hemolymph of such species
whose corpora allata in vitro produce only one hormone.

ACKNOWLEDGEMENTS

We are indebted to the National Science Foundation (Grants
GB-7941 and BMS7201892) and the Robert A. Welch Foundation
(Grant A598) for supporting our work. Gratefully we acknowledge
the competent and dedicated assistance of our technical staff:
Marilyn Burditt, Yang-Chih Hrung, Jane Moore, Peggy Olson,
Vicki Riddle, Judy Whitley and Nila Williams. We also would like
to thank our many colleagues and friends for fruitful discussions
and pre-publication information about their work.

REFERENCES

Bjerke, J.S., and Röller, H., 1974, in "Invertebrate Endocrinology
    and Hormonal Heterophylly" (W.J. Burdette, ed.), pp. 130-139,
    Springer-Verlag, New York.
Dahm, K.H., and Röller, H., 1970, Life Sci. (Part II) 9:1397.
Dahm, K.H., Trost, B.M., and Röller, H., 1967, J. Am. Chem. Soc.
    89:5292.
de Wilde, J., Staal, G.B., de Kort, C.A.D., de Loof, A., and Baard,
    G., 1968, Koninkl. Nederl. Akad. Wetensch., Series C, 71:321.
Fukuda, S., 1962, Annot. Zool. Jpn. 35:199.
Fukuda, S., 1963, Annot. Zool. Jpn. 36:14.
Gowal-Rodé, H., Abbott, S., Meyer, D., Röller, H., and Dahm, K.H.,
    1975, Z. Naturforsch. 30c:392.
Jennings, R.C., Judy, K.J., Schooley, D.A., Hall, M.S., and
    Siddall, J.B., 1975a, Life Sci. 16:1033.
Jennings, R.C., Judy, K.J., and Schooley, D.A., 1975b, J. Chem.
    Soc., Chem. Comm. 1975:21.
Judy, K.J., Schooley, D.A., Dunham, L.L., Hall, M.S., Bergot, B.J.,
    and Siddall, J.B., 1973a, Proc. Nat. Acad. Sci. USA 70:1509.
Judy, K.J., Schooley, D.A., Hall, M.S., Bergot, B.J., and Siddall,
    J.B., 1973b, Life Sci. 13:1511.
Judy, K.J., Schooley, D.A., Troetschler, R.G., Jennings, R.C.,
    Bergot, B.J., and Hall, M.S., 1975, Life Sci. 16:1059.
Koyama, T., Ogura, K., and Seto, S., 1973, Chem. Soc. Japan, Chem.
    Letters 1973:401.
Lanzrein, B., Hashimoto, M., Parmakovich, V., Nakanishi, K.,
    Wilhelm, R., and Lüscher, M., 1975, Life Sci. 16:1271.
Metzler, M., Dahm, K.H., Meyer, D., and Röller, H., 1971, Z.
    Naturforsch. 26b:1270.
Metzler, M., Meyer, D., Dahm, K.H., Röller, H., and Siddall, J.B.,
    1972, Z. Naturforsch. 27b:321.
Meyer, A.S., Schneiderman, H.A., Hanzmann, E., and Ko, J.H., 1968,
    Proc. Nat. Acad. Sci. USA 60:853.
Müller, P.J., Masner, P., Trautmann, K.H., Suchý, M., and Wipf,
    H.-K., 1974, Life Sci. 15:915.
Novák, V.J.A., 1966, "Insect Hormones", pp. 80-119, Methuen and
    Co., Ltd., London.
Peter, M.G., and Dahm, K.H., 1975, Helv. Chim. Acta 58:1037.
Peter, M.G., Dahm, K.H., and Röller, H., 1976, Z. Naturforsch.,
    in press.
Piepho, H., 1938, Biol. Zbl. 58:356.
Piepho, H., 1942, Arch. Entwmech. Org. 141:500.
Piepho, H., 1950, Biol. Zbl. 69:1.
Reibstein, D., and Law, J.H., 1973, Biochem. Biophys. Res. Commun.
    55:266.
Riddiford, L.M., and Ajami, A.M., 1973, J. Insect Physiol. 19:749.
Röller, H., and Dahm, K.H., 1968, Rec. Prog. Hormone Res. 24:651.
Röller, H., and Dahm, K.H., 1970, Naturw. 57:454.

Röller, H., and Dahm, K.H., 1974, in "Invertebrate Endocrinology and Hormonal Heterophylly" (W.J. Burdette, ed.), pp. 235-238, Springer-Verlag, New York.

Röller, H., Dahm, K.H., Sweeley, C.C., and Trost, B.M., 1967, Angew. Chemie. 79:190. 1967, Angew. Chemie, Internat. Edit. 6:179.

Röller, H., Bjerke, J.S., Holthaus, L.M., Norgard, D.W., and McShan, W.H., 1969, J. Insect Physiol. 15:379.

Schooley, D.A., Judy, K.J., Bergot, B.J., Hall, M.S., and Siddall, J.B., 1973, Proc. Nat. Acad. Sci. USA 70:2921.

Shirk, P.D., Dahm, K.H., and Röller, H., 1976, Z. Naturforsch, in press.

Slade, M., and Zibitt, C.H., 1972, in "Insect Juvenile Hormones", (J.J. Menn and M. Beroza, ed.), pp. 155-176, Academic Press, New York.

Sroka, P., Barth, R.H., Gilbert, L.I., and Staal, G.B., 1975, J. Insect Physiol. 21:463.

Trautmann, K.H., Masner, P., Schuler, A., Suchý, M., and Wipf, H.-K., 1974a, Z. Naturforsch. 29c:757.

Trautmann, K.H., Schuler, A., Suchý, M., and Wipf, H.-K., 1974b, Z. Naturforsch. 29c:161.

Weirich, G., Wren, J., and Siddall, J.B., 1973, Insect Biochem. 3:397.

Wigglesworth, V.B., 1970, "Insect Hormones", pp. 42-84, W.H. Freeman and Company, San Francisco.

Williams, C.M., 1959, Biol. Bull. 116:323.

Williams, C.M., 1963, Biol. Bull. 124:355.

Yashika, K., 1960, Memoirs of the Col. Sci., U. of Kyota, Series B. 27:83.

STRUCTURE ACTIVITY RELATIONSHIPS IN SOME JUVENILE HORMONE ANALOGS[*]

C.A. Henrick, G.B. Staal and J.B. Siddall

Zoecon Research Laboratories, 975 California Avenue
Palo Alto, California 94304

INTRODUCTION

Although the chemical structures of the three known natural juvenile hormones do provide considerable helpful information to a chemist for the synthesis of active analogs, the almost infinite number of conformations that they can adopt is a severely limiting factor in this. When a juvenile hormone elicits a particular biological effect, one anticipates by analogy with more advanced work in related fields, that the hormone adopts a particular conformation during the process of binding to a specific receptor protein. The question of what is the particular shape of a juvenile hormone in its active form is perhaps impossible to answer but is nonetheless important to an understanding of hormone action. Although we do not have the luxury of the relatively rigid tetracyclic molecules of the mammalian sex hormones, some insights may be gained by examination of the biological activity of closely related hormone analogs, provided that suitable bioassay systems are used. We wish to present some of our results from a detailed study of structure-activity relationships within a group of potent juvenile hormone analogs, in order to gain some insights into the likely conformations of the active forms of the juvenile hormones. In so doing we recognize the limitations to the amount of conformational information derivable from whole animal bioassay.

---

[*] Contribution No. 45 from Zoecon Research Laboratories.

MATERIALS AND METHODS

Bioassays were performed on synchronized fresh pupae of the greater wax moth (Galleria mellonella) and the yellow mealworm (Tenebrio molitor) as previously described (Henrick et al., 1973). The activities are expressed as ID_{50} values (dose required to produce 50% inhibition of metamorphosis).

All the new compounds mentioned in Tables 1-6 gave satisfactory elemental analyses (within 0.4% of the theoretical values) and were characterized also by their NMR, ir, and mass spectra. All the analogs described in the tables are racemic compounds; the prefix (±) is omitted.

Table 1. The syntheses of compounds 1-10 have been previously described (Henrick et al., 1973; 1976).

Table 2. The preparations of analogs 11-15 and 17-20 have been previously described by Henrick et al. (1973; 1975a,b,c; 1976). The ester 16 was prepared by condensing 10-methoxy-6,10-dimethyl-2-undecanone with diethyl 3-ethoxycarbonyl-2-propenylphosphonate (NaH, dimethylformamide, room temperature; see Henrick et al., 1973).

Table 4. The preparation and physical properties of ethyl (E)-3,5-ethano-7,11-dimethyl-2,4-dodecadienoate (25), [uv max (hexane) 284 nm (ε24,800)]; 26, [uv max (hexane) 282 nm (ε25,200)]; 27, and of 28 will be described elsewhere (Henrick et al., 1975d).

Table 5. Reaction of 4,8-dimethylnonylmagnesium bromide with 3-ethoxy-2-cyclohexen-1-one gave the enone 32. Conversion of 4,8-dimethyl-1-nonyne to the lithium reagent (with methyllithium) and reaction with 3-ethoxy-2-cyclohexen-1-one gave, after acidic hydrolysis, 3-(4,8-dimethyl-1-nonynyl)-2-cyclohexen-1-one. Hydrogenation of this enyne over Lindlar catalyst, followed by isomerization of the (Z)-double bond with p-toluenesulfonic acid in benzene under reflux, gave the (E)-dienone 31.

Table 6. (E)-3-(4,8-Dimethyl-1-nonenyl)-2-buten-4-olide (33). Bromination of 3-methyl-2-butenoic acid with 2 equiv of N-bromosuccinimide in CCl_4 [with 2,2'-azobis(isobutyronitrile) added] and treatment of the reaction mixture with anhydrous Na_2CO_3 gave, in 10% yield, 3-(bromomethyl)-2-buten-4-olide, bp 99°C (0.1 mm). Heating the bromide with triethylphosphite (120°C for 2 hr) gave the corresponding phosphonate, bp 157-8°C (0.5 mm) which was condensed with 3,7-dimethyl-1-octanal (using hexamethyl-disilazyllithium in tetrahydrofuran) to give the lactone 33; bp (bath, short path) 100°C (0.1 mm); uv max (hexane) 252 nm (ε22,900).

Wittig reaction of triphenylphosphonium 4-methylpentylide with

1,3-diacetylbenzene (tetrahydrofuran; -70°C, 1 hr then 25°C for
3 days) gave, in low yield, 1-acetyl-3-(1,5-dimethyl-1-hexenyl)-
benzene. Hydrogenation of the double bond (10% Pd/C in pentane)
followed by condensation with triethylphosphonacetate (NaH in
dimethylformamide, 24°C, 24 hr then 70°C, 7 hr), and purification
by preparative TLC, gave the E analog 34.

     Treatment of 3-methylthiophene with n-butyllithium in tetra-
hydrofuran-ether at 0°C (2 hr) gave a mixture of the 5- and 2-
lithio reagents. Reaction with 6-methoxy-2,6-dimethyl-1-bromo-
heptane gave 5-(6-methoxy-2,6-dimethylheptyl)-3-methylthiophene
(ca. 80%) along with the corresponding 2-(6-methoxy-2,6-dimethyl-
heptyl)-3-methylthiophene (ca. 20%). Metalation of this mixture
with n-butyllithium, followed by reaction of the lithio reagents
with CO_2, esterification of the acids with dimethylformamide
diethyl acetal, and purification by preparative TLC gave the ethyl
thiophenecarboxylate 35.

     Wittig reaction of 3-bromobenzaldehyde with triphenylphos-
phonium 7-methoxy-3,7-dimethyloctylide and isomerization of the
double bond in the product with benzenethiol (see Henrick et al.,
1975a) gave (E)-1-bromo-3-(8-methoxy-4,8-dimethyl-1-nonenyl)-
benzene. Carbonation of the corresponding Grignard reagent with
CO_2 in tetrahydrofuran, followed by esterification with diazoethane,
gave the benzoate 36.

     Similarly, Wittig reaction of triphenylphosphonium 7-methoxy-
3,7-dimethyloctylide with 3-methoxybenzaldehyde, followed by isomerization of the product with benzenethiol at 80°C (2.5 hr) in
the presence of 2,2'-azobis(isobutyronitrile) (see Henrick et al.,
1975a) gave the substituted benzene analog 37.

                          RESULTS AND DISCUSSION

     Some of the implications of the act of measuring biological
activity require brief discussion before consideration of structure-
activity relationships (see Staal, 1972). For a particular hormone
or a particular chemical analog, a discussion of biological potency
should ideally include several factors which may influence biological activity such as the rates of penetration into and metabolic
removal from various tissues, the distribution of the chemical
between free solution and protein bound forms in the hemolymph, and
the exchange of these forms to provide intracellular distributions
of free and bound forms of the chemical. However, within a series
of closely related chemical analogs, for example those of Table 1,
all of which contain a polar ester group attached to a hydrocarbon
chain, some of the above factors may be assumed to have relatively
minor importance. Differences in rates of penetration into the
cuticle and from the cuticle into epidermal cells are likely to

TABLE 1. ID_{50} VALUES ON SENSITIVE SYNCHRONIZED INSTARS.[a]

No.		Galleria mellonella µg/pupa	Tenebrio molitor µg/pupa
1		10	54
2		0.76	0.19
3		4.4	34
4[b]		0.040	0.25
5		5.0	4.4
6		0.015	0.10
7		0.050	0.10
8		>100	>100
9		6.7	4.2
10		70	84

[a]The syntheses of these compounds are described by Henrick et al., 1976. [b]Hydroprene (trademark Altozar IGR; ZR 512; Henrick et al., 1973).

depend mainly on the balance of hydrophobic to hydrophilic characteristics which are fairly constant in a given series. In this work, the majority of chemicals under consideration are carboxylic esters (the four ketones in Table 5 are treated separately), and we may assume that all these ethyl esters can be degraded to inactive carboxylic acids by hemolymph esterases, though not at the same rate for each different analog. However, if esterases were to be a major determinant of biological potency in this series, one might expect the ketones of Table 5, of which compound 30 is isosteric with an ethyl ester, to show greatly increased potency. The fact that this is not the case in Galleria mellonella (compare the ketones 29 and 30 with the ester 4) could argue against a major role of esterases in determining potency. Moreover one would expect juvenile hormone esterases to degrade most rapidly those analogs which bear closest resemblance to the natural juvenile hormones. This effect could tend to lower the persistence and hence the potency of analogs whose shape most closely resembles that of the active hormone, and to raise the persistence and potency of analogs which poorly imitate the hormone's conformations.

In the opposite sense an analog whose shape closely resembles that of the hormone could preserve endogenous natural hormones by competitive inhibition of degradative enzymes. That such protective effects can be ruled out in the bioassays used here is based on the assumption that endogenous juvenile hormones are completely absent from pupal tissues at the critical period of determination for expression of adult characters which follows the time of application in these bioassays rather closely.

In view of these factors it is likely that the observation of vast differences in potency between closely related structures has more fundamental implications, and suggests that a more potent analog has greater affinity for the presumed hormone receptors. However, it is quite conceivable that two chemicals with equal binding affinity for a given hormone receptor will not be equally effective in eliciting the biological response, when present at the same concentration of bound forms. Such questions cannot be answered by relative potency measurements and must await experimental measurement in subcellular systems.

Another limitation to our considerations stems from the assumption that the natural hormones will take up a particular conformation which allows tight binding to their receptors, and that highly active hormone analogs will tend to adopt the same conformation. Such assumptions may be invalid since the addition of a new structural feature to a hormone molecule could well change the conformation of the bound form to take advantage of new binding interactions not present in the original hormone. One example of this has been observed in the ~1.9 fold enhanced binding of JH 0 (Table 3, 24) relative to JH I (23) with a purified carrier binding protein from larval hemolymph of Manduca sexta (see Goodman et al., Kramer et al., this volume). In the same system the hormone JH III (21) bound only one-third as well as JH I, and if such effects play any role

TABLE 2. ID_{50} VALUES ON SENSITIVE SYNCHRONIZED INSTARS.

No.		Galleria mellonella µg/pupa	Tenebrio molitor µg/pupa
11[a]		>100	>100
12[b]		3.3	2.6
13[c]		73	>100
14[d]		84	13
15[b]		0.17	8.9
16		>100	>100
17[b,e]		5.7	0.0040
18[a]		80	0.056
19[a]		>100	0.35
20[a]		>10	>100

[a]Henrick et al., 1976. [b]Henrick et al., 1973. [c]Henrick et al.,
1975b. [d]Henrick et al., 1975c. [e]Methoprene (trademark Altosid
IGR; ZR 515; Henrick and Siddall, 1975).

TABLE 3. ID_{50} VALUES ON SENSITIVE SYNCHRONIZED INSTARS.

No.			Galleria mellonella μg/pupa	Tenebrio molitor μg/pupa
21[a]		JH III	12	4.5
22[a]		JH II	0.13	4.3
23[b]		JH I	0.060	0.70
24[c]		JH O	1.0	0.15

[a]Anderson et al., 1972.   [b]Henrick et al., 1972.   [c]Anderson et al., 1975.

in pupal bioassay, they are completely obscured in the results for Galleria mellonella pupae in Table 3.   These same workers found that one compound in the 2,4-dienoate series, methoprene (Table 2, 17) did not compete with JH I for sites on the carrier binding protein, and that hydroprene (Table 1, 4) did not compete with JH I in a similar preparation of carrier protein. If these results can be extrapolated to G. mellonella, it is likely that hemolymph carrier proteins may play no significant role in determining the observed potencies in this series of hormone analogs.

Structure Activity Relationships

In the majority of chemical analogs within the series under discussion (Tables 1-6), the carbon skeleton bears either a close resemblance or is identical to that of the natural hormones. All of the tabulated chemicals containing asymmetric carbon atoms are

TABLE 4. ID_{50} VALUES ON SENSITIVE SYNCHRONIZED INSTARS.

No.		Galleria mellonella µg/pupa	Tenebrio molitor µg/pupa
4		0.040	0.25
5		5.0	4.4
25[a]		0.0029	0.0085
26[a]		>10	10
27[a]		>100	>100
28[a]		45	∿100

[a]Henrick et al., 1975d.

racemic mixtures. The presence of methyl branches along the chain is clearly important for high biological activity in the majority of juvenile hormone analogs. Compounds lacking all the methyl branches have very low activity (see Wakabayashi et al., 1969) and even compounds such as 1 which possess only the C-3 methyl branch still have low activity (see Bell et al., 1973). The addition of another methyl group at C-11 in passing from compound 1 to compound 2 markedly improves activity. The further addition of a central methyl branch at C-7 (compare 2 with 6) leads to 50-fold enhancement

TABLE 5.   ID$_{50}$ VALUES ON SENSITIVE SYNCHRONIZED INSTARS.

No.		Galleria mellonella µg/pupa	Tenebrio molitor µg/pupa
29[a]		0.070	0.0032
30[a]		0.90	0.0053
31[b]		>100	42
32[b]		>100	>100

[a]Henrick et al., 1976. [b]Henrick et al., 1975e.

in Galleria mellonella potency. Within the experimental error of these bioassays, the differences of two to three fold produced by changing the terminal methyl group to a terminal ethyl group may not be significant for compound 4 versus compound 6, but such differences are striking in the natural hormone series in going from JH III to JH II (Table 3).

Especially important in the 2,4-dienoate series is the presence or absence of a methyl branch at C-3, associated with a ca. 100-fold change in potency (compound 3 vs. 4) in both bioassays. This change may be typical only of 2,4-dienoate compounds (Henrick et al., 1976) since it has been found that the C-3 methyl group is not very important for analogs of methyl 10,11-epoxyfarnesoate tested against the silkworm, Bombyx mori (Ohtaka et al., 1972; Kiguchi et al., 1974) and against several other insect species, including Tenebrio molitor, Tribolium castaneum, and Galleria mellonella

TABLE 6. ID_{50} VALUES ON SENSITIVE SYNCHRONIZED INSTARS.

No.	Galleria mellonella μg/pupa	Tenebrio molitor μg/pupa
4	0.040	0.25
33[a]	>100	∿10
34[a]	>100	0.020
35[a]	>100	>100
36[a]	>100	>100
37[a]	>100	4.5

[a]Henrick et al., 1975e.

(Mori, 1971; 1972; Mitsui et al., 1973).   Movement of C-3 methyl branch to C-2 essentially removes the morphogenetic activity (compound 4 vs. compound 8) although an additional methyl at C-2 (compound 18 vs. compound 17) merely reduces potency.   In contrast an additional methyl at C-4 (compound 19 vs. 17) greatly reduces the observed activity, as does the joining of these two branches to form a 2-carbon bridge linking C-2 to C-4 as in compound 28,

Table 4. The compound 9 containing a cyclopropane ring at C-2,3
has moderate activity and can be compared electronically with 3 and
4, and also sterically with the 2,3-dihydro analog 11, Table 2,
which is inactive.

Despite the interesting changes in potency associated with the
presence or absence of such alkyl branches, particularly for the
C-3 ethyl branched compound 5, we can deduce almost nothing about
the active conformation of these analogs, with the exception of
the fairly rigid compound 28.

In many series of juvenile hormone analogs from different
laboratories (see Henrick et al., 1976 and references therein),
the presence of a carbonyl group in the form of an ester, ketone,
or amide has been found to be important, in general, for high
potency. Because rotation can occur freely around the bond between
the carbonyl carbon C-1 and its neighbor C-2, one cannot predict
the conformation of the carbonyl group in the active form of the
hormone. However, by fixing the carbonyl group in a ring system
such as in compounds 31 and 32, the biological activity is essentially removed and this change is especially striking in comparison
to the isomeric ketone 29 (see 31). Dreiding models of the cyclic
ketones 31 and 32 show that the carbonyl group could lie between
approximately 45°C above and below the plane of the adjacent
double bond. Although other factors such as excessive bulk of the
ring could be responsible for the lack of activity in the cyclic
ketones 31 and 32, it is probable that their carbonyl groups are
held in conformations which are very unfavorable for biological
activity. Similar considerations attend the comparison of the
lactone compound 33 with the acyclic analog 4 which is at least
40 times more potent in the Tenebrio molitor test, and over
2,500 times in the Galleria mellonella bioassay than is 33.

The analogs 34, 35, and 36 containing relatively rigid ring
systems provide results which are difficult to interpret since
additional bulk has been introduced in a possibly detrimental way.
However, in cases where the use of a rigid ring does not involve
additional bulk, for example in the cyclic isomer of an acyclic
analog, one may derive considerable information about conformation.
This appears to be the case in what is perhaps the most striking
comparison in the whole series under discussion, namely the cyclic
compounds of Table 4 (Henrick et al., 1975d). When the "parent"
compound 4 is homologated to the 3-ethyl branched analog 5, potency
is reduced by 10 to 100-fold. But when the 3-ethyl branch is
joined to carbon 5 to form the 5-membered ring of compound 25,
a remarkable increase of 1,700-fold occurs in the observed potency
(between compound 5 and 25) in the Galleria mellonella test. The
introduction of one additional CH_2 group to give the compound 26
almost removes the morphogenetic activity. Despite this extra
methylene group, the carbon atoms 1 through 6 of the dienoate

system can be almost co-planar in both compounds. Clearly, the
acyclic analog 4 could also take up this planar conformation around
carbons 1 through 6 of the 2,4-dienoate group. One may conclude
that the cyclopentene compound 25 probably approaches the ideal
combination of a planar conformation with appropriate but not
excessive bulk in the region normally occupied by the 3-methyl
group of the natural hormone.

The importance of the distance between C-3 and C-7 methyl
branches, illustrated in the Tables by comparison of the compound
20 with 17 in the Tenebrio molitor test, and by comparison of the
compound 10 with 4 in both bioassays, is noteworthy when combined
with the observed activity of the cyclic compound 25. Other
examples of the importance of this isoprenoid nature of the carbon
chain have been discussed by Henrick et al. (1976), and it would
be interesting to know whether the position of the central methyl
branch is still important in compound 25. Providing that the
earlier stated assumptions are valid, one can conclude that the
natural hormones, and the analogs 4 and 25, may adopt similar
active conformations in which the carbon atoms 1 to 6 are almost
co-planar.

REFERENCES

Anderson, R.J., Henrick, C.A., Siddall, J.B., and Zurflueh, R.,
    1972, J. Amer. Chem. Soc. 94:5379.
Anderson, R.J., Corbin, V.L., Cotterell, G., Cox, G.R., Henrick,
    C.A., Schaub, F., and Siddall, J.B., 1975, J. Amer. Chem. Soc.
    97:1197.
Bell, E.W., Gast, L.E., Cowan, J.C., Friedrich, J.P., and Bowers,
    W.S., 1973, J. Agr. Food Chem. 21:925.
Henrick, C.A., Schaub, F., and Siddall, J.B., 1972, J. Amer. Chem.
    Soc. 94:5374.
Henrick, C.A., Staal, G.B., and Siddall, J.B., 1973, J. Agr.
    Food Chem. 21:354.
Henrick, C.A., and Siddall, J.B., 1975, U.S. Patent 3,904,662
    (Sept 9); 3,912,815 (Oct 14).
Henrick, C.A., Willy, W.E., Baum, J.W., Baer, T.A., Garcia, B.A.,
    Mastre, T.A., and Chang, S.M., 1975a, J. Org. Chem. 40:1.
Henrick, C.A., Willy, W.E., McKean, D.R., Baggiolini, E., and
    Siddall, J.B., 1975b, J. Org. Chem. 40:8.
Henrick, C.A., Willy, W.E., Garcia, B.A., and Staal, G.B., 1975c,
    J. Agr. Food Chem. 23:396.
Henrick, C.A., Labovitz, J.N., and Corbin, V.L., 1975d, Zoecon
    Corporation, Palo Alto, paper in preparation.
Henrick, C.A., Baggiolini, E., Willy, W.E., Garcia, B.A., and
    McKean, D.R., 1975e, Zoecon Corporation, Palo Alto,
    unpublished information.

Henrick, C.A., Willy, W.E., and Staal, G.B., 1976, J. Agr. Food Chem. 24:in press.

Kiguchi, K., Ohtaki, T., Akai, H., and Mori, K., 1974, Appl. Entomol. Zool. 9:29.

Mitsui, T., Nobusawa, C., Fukami, J., Mori, K., and Fukunaga, K., 1973, Appl. Entomol. Zool. 8:27.

Mori, K., 1971, Bull. Soc. Entomol. Suisse 44:17.

Mori, K., 1972, Agr. Biol. Chem. 36:442.

Ohtaki, T., Kiguchi, K., Akai, H., and Mori, K., 1972, Appl. Entomol. Zool. 7:161.

Staal, G.B., 1972, in "Insect Juvenile Hormones: Chemistry and Action" (J.J. Menn and M. Beroza, ed.), pp 69-94, Academic Press, New York.

Wakabayashi, H., 1969, J. Med. Chem. 12:191.

PHENYLETHERS AS INSECT GROWTH REGULATORS: LABORATORY AND

FIELD EXPERIMENTS

R.C. Zurflueh

SOCAR Ltd., Chemical Research Laboratory
CH-8600 Duebendorf, Switzerland

INTRODUCTION

The usefulness of most reported insect growth regulators
(IGR's) having juvenile hormone (JH) activity is limited by their
instability under environmental conditions. Reports on the acute
toxicity of several candidate compounds to mammalian and wildlife
species showed that they are only slightly toxic. It would therefore be valuable to have more stable compounds with otherwise
similar chemical and "insecticidal" properties. This report shows
our progress towards the development of such compounds.

R-20458 [6,7-epoxy-1-(p-ethylphenoxy)-3,7-dimethyl-trans-2-
octene], which was reported by Pallos et al. (1971) was shown by
Hoppe et al. (1974) in field trials to be very active against
Culex pipiens. It is as potent as methoprene, the active ingredient of Altosid SR 10 which successfully controls floodwater
mosquitoes, and is at present the only IGR registered for commercial use. R-20458 was slightly inferior to methoprene in standard
laboratory bioassays on Culex sp. The good performance of R-20458
in the field was therefore somewhat surprising, and it could be
explained only after the development of new laboratory tests that
conformed more with the practice of mosquito control. These experiments are being reported here.

In a recent communication from our laboratory by Hangartner
et al. (1976) it was shown that Ro 10-3108 [6,7-epoxy-1-(p-ethylphenoxy)-3-ethyl-7-methylnonane] is a new IGR with promising pros-
pects in the area of plant protection. The compound proved to be
sufficiently stable under environmental conditions and gave good
control of natural populations of the summer fruit tortrix moth,

mealy bugs and scale insects in the field. The good field performance, the selectivity and favorable toxicology are considered here.
Phenylethers having an isobutylgeranyl sidechain were reported by
Jacobson et al. (1972) and shown to be of moderate interest. In
the present communication these compounds are shown to be highly
active against lepidopterous insects.

MATERIALS AND METHODS

Chemicals

The phenylethers were prepared by standard methods which involve alkylation of the various phenols with the appropriate allylic bromide followed by epoxidation, and by hydrogenation where
applicable. The synthesis of each compound has been described
in detail elsewhere.

Compound 1; Ro 8-9801, R-20458, ENT 70,221, 6,7-epoxy-1-
(p-ethylphenoxy)-3,7-dimethyl-trans-2-octene (Pallos et al., 1971).
The "flowable" formulation (see Table 2) was obtained from
Stauffer Chemical Company.

Compound 2; Ro 20-3600, ENT 70,357 (Bowers, 1969).

Compound 3 to 7, 17 and 18 were described by Chodnekar et al.,
(1973).

Compound 8; Ro 6-9550 (Pfiffner, 1971).

Compound 9; Ro 10-3109, ENT 34,979 (Wright et al., 1974).

Compound 10; Ro 10-3108, 6,7-epoxy-1-(p-ethylphenoxy)-3-
ethyl-7-methylnonane (Hangartner et al., 1976).

Compound 11; Ro 10-2202; the related all-trans isomer is
identical with ZR-512, hydroprene (Henrick et al., 1973).

Compound 12; Ro 10-6425; the related all-trans isomer is
identical with ZR-515, methoprene (Henrick et al., 1973).

Compound 13 and 14 were described by Hangartner and Zurflueh
(1975).

Compound 15 and 16 were described by Chodnekar et al. (1975).

Biological Trials

Laboratory Bioassays. All bioassays were performed on sensitive stages (synchronized) of insect species. The activities were expressed as -log ED$_{95}$ values which is the effective dose required to elicit abnormal development in at least 95% of the treated insects. The ED$_{95}$ values are rough estimates only because they are usually determined from just two replicates; however, they provide a measure of relative activities.

Effect on metamorphosis of the codling moth (Carpocapsa pomonella): The bottoms of petri dishes were treated with an acetone solution of the test compound to give a dosage of 10^{-5} - 10^{-10} g of active ingredient/cm^2. After 1 hr one last instar larva was introduced into each dish (10 replicates), fed with artificial diet and incubated at 26°C and 60% R.H. After eclosion of the adults, the activity was calculated after Abbott (1925) as percent reduction of the total number of adults showing no visible morphogenetic defects (wing deformation <50%).

Effect on metamorphosis of the red flour beetle (Tribolium castaneum): Wheat flour (10 g) was moistened with acetone solutions of the test compound to give a dosage of 10^{-5} - 10^{-8} g of active ingredient/g flour. The treated flour was infested with 20 mature beetles and placed into small plastic cups. After 5 days the P-generation was removed and the cups were incubated at 29°C and 75-80% R.H. After eclosion of the F$_1$-generation, the activity was calculated after Abbott (1925) as percent reduction of the F$_1$-generation as compared with the controls.

Procedures for bioassay of the last larval instars of Adoxophyes orana have been described previously (Hangartner et al., 1976).

Procedures for bioassays on fresh Tenebrio molitor pupae and last larval instar Aedes aegypti have also been previously described (Dorn et al., 1976).

Evaluation of the persistence of different formulations of R-20458 tested against fourth instar larvae of Culex pipiens: The assays were carried out in glass-beakers, each containing 1.5 L of tap-water and approximately 100 g of soil. Aqueous solutions or granules of formulated test compounds were added to give a concentration of 0.1 ppm of active ingredient. The beakers were covered with a UV-transparent plastic foil and incubated either in a laboratory or in a UV-chamber. After 1 week each plot was infested with 20 late fourth instar Culex pipiens larvae and kept under optimal conditions (28°C, 14 hr photoperiod fluorescent light). The assessment was carried out as soon as adult emergence of more than 90% was observed in untreated plots.

Evaluation of the persistence of different IGR's tested under various conditions against fourth instar larvae of Culex pipiens: a) Without illumination - Glass-beakers containing 100 ml tap water were treated with acetone solutions of test compounds and covered with a UV-transparent plastic foil. They were incubated for 6 days in the ecophyts at 28°C and 85% R.H., without any illumination. 20 late fourth instar larvae were then introduced into each beaker and were kept under optimal conditions (28°C, 14 hr photoperiod) until development was completed. The effect was expressed as the percentage of unemerged adults as compared with untreated plots. b) With UV-light - The same procedure as described in (a) was used except that incubation took place in a UV-chamber. c) In the presence of soil - The same procedure as described in (a) was used except that two tablespoons of soil were added to the water before treatment and incubation.

Field Experiments

Effect on metamorphosis of the summer fruit tortrix moth (Adoxophyes orana): Four 6-year old apple trees per variant were sprayed with a 0.1% active ingredient emulsion to the run off point. Three days after treatment, 2 branches of each tree were infested with 5 last instar larvae trapped in a gauze bag.

After eclosion of the adults, the activity was calculated after Abbott (1925) as percent reduction of normal imagos (i.e. adults with no visible morphogenetic defects). Location: Valais, Switzerland.

Effect on a natural, extremely dense, non-synchronized larval population of Culex and Culiseta mosquitoes: The trials were carried out in ditches (length: 100-150 m, width: 0.8-3 m, depth: 0.5 - 1 m) with heavily polluted water (sanitation system of the surrounding villages). The liquid formulations were applied with a motorsprayer (approximately 3000 l/ha). The granules were sprayed by hand over the water surface. Evaluations were made by collecting pupae at regular intervals (1-3 days) and placing them into plastic cups fitted with a gauze bottom. The cups were fixed on a float in such a way that only the lower part was floating in the water. Another cup served as cover. The hatched normal mosquitoes as well as the living and dead larvae were counted when the adults in the untreated plots nearby had emerged. The activity was indicated by stipulating the number of days during which the reduction of adult emergence was 95% or better. Trial 1 was carried out in August (high temperature), whereas trials 2 and 3 were performed later in the season. Location: near Barcelona, Spain.

RESULTS AND DISCUSSION

Table 1 shows the results of a persistency test of IGR's tested against fourth instar Culex pipiens larvae. The compounds had been exposed to such environmental factors as hydrolysis, photodecomposition and microbial attack. All compounds tested resulted in full control after an incubation time of 6 days, but only as long as the treated water was kept in the dark. Under UV-light or on contact with soil, all the activity was lost in the case of Ro 20-3600. R-20458 tested as an emulsifiable concentrate proved to be sensitive to UV-light and soil, and Altosid SR 10 shows a weakness against UV which had been noticed before (Schooley et al., 1975; Quistad et al., 1975).

TABLE 1.   PERSISTENCE OF IGR'S TESTED AGAINST FOURTH INSTAR LARVAE OF CULEX P.P.

	% Reduction of adult emergence		
	without illumination	UV irradiation	soil
R 20458 **1** Ro 8-9801 (EC)	100	89	37
2 Ro 20-3600 (EC)	100	0	0
Altosid™ SR 10	100	42	100

Incubation time: 6 days at 28°C;
Dosage 0.1 ppm.

The results illustrated in Table 2 demonstrate the importance of the correct formulation. The granular as well as the flowable (microcap-type) formulation protect the active ingredient from degradative factors, and in the presence of UV-light their residual activity was found to be superior to Altosid SR 10. The 95% control obtained with the flowable formulation of R-20458 after 3 weeks incubation time under UV-light in the presence of soil and at a relatively high temperature is especially noteworthy.

The good stability of R-20458 if properly formulated was confirmed in large-scale field trials in Spain where a natural, extremely dense, mixed population of Culex and Culiseta sp. was treated in irrigation and sanitation ditches (Table 3).

TABLE 2.   PERSISTENCE OF DIFFERENT FORMULATIONS OF R-20458 TESTED
AGAINST FOURTH INSTAR LARVAE OF <u>CULEX</u> <u>P.P</u>.

	% Reduction of adult emergence			
	daylight water + soil, 23° incubation time		UV water + soil, 29° incubation time	
	1 week	3 weeks	1 week	3 weeks
R-20458 EC	79	–	32	0
R-20458 Granules	97	78	100	68
R-20458 Flowable	100	100	100	95
Altosid™ SR 10	100	100	78	21

Dosage 0.1 ppm a.i.

TABLE 3.   FIELD EXPERIMENT WITH R-20458 AGAINST <u>CULEX</u> <u>PIPIENS</u> AND
<u>CULISETA</u> <u>SP</u>. IN HEAVILY POLLUTED WATER.

		Dosage a.i.		number of days during which the reduction of adult emergence was between 95% and 100%		
		ppm	kg/ha	trial 1	trial 2	trial 3
R 20458	EC	0.25	2.5	3	3	
R 20458	Granules	0.25	1.2	14	36	
		0.1	0.9		35	
		0.05	0.5			>19
Altosid™ SR 10		0.017	0.12	4	10	
		0.05	0.4			>19
Abate®	EC	0.25	2.5	20		

    Altosid SR 10 although not marketed for control of standing
water mosquitoes was used as a standard of the IGR-type, and Abate
was included as a conventional larvicide standard. The EC-formula-
tion, although applied at a relatively high dosage, gave control
for three days only, whereas the granular formulation had a residual activity of two to five weeks even at the lower dosage.
Altosid SR 10 if applied at the same concentration as R-20458 also
resulted in good residual activity. Abate gave full control for
three weeks but eliminated practically all aquatic organisms in
the trial plots whereas the IGR's showed little effect on nontarget

organisms. The above demonstrates the potential of R-20458 which
can be considered to be a very effective mosquito larvicide having
the advantage over conventional larvicides in that it is practi- *
cally harmless to many other aquatic animals as well as the user.
The results were obtained in heavily polluted water and are therefore especially interesting for future use in urban sanitation
programs.

The good performance of phenylethers having a para-ethyl
group and the high activity of the methylenedioxy compounds
(Bowers, 1968; 1969), at least in the laboratory , stimulated us
to search for structurally related molecules. Excellent activity
was found with the coumaranes or 2,3-dihydrobenzofuranes (Table 4).
In our standardized assays on Tenebrio molitor and Aedes aegypti,
compounds 3 - 7 are much more active than the synthetic mixture
of isomers of JH I (compound 8). By replacing the epoxide with
an ethoxy group (compound 5) we observe a slight increase in activity against T. molitor but a considerable loss of activity
against mosquitoes. This finding parallels the one with the
corresponding p-ethyl phenyl ether, the so-called JH 25 (Sarmiento
et al., 1973), which we found to be quite specific for Tenebrionidae, and which was reported to have a narrow spectrum of activity
(see Hammock et al., 1975). The effect of C-2 methyl substitution
on activity is shown in compounds 6 and 7. Adding one methyl
only results in a more effective compound (6) which in the assay
on Aedes aegypti is 10-fold more active than compound 7. The
potential of this class of compounds has not been fully evaluated
yet but some representatives were found to be rather UV labile.

Since the target insects are susceptible to IGR's only during
a short period in their life cycle, good residual activity is more
crucial for these agents than for conventional rapidly acting
insecticides. Therefore it cannot be emphasized enough how important stability is for the overall performance of an IGR. Table 5
shows that hydrogenation of the double bond in R-20458 and its
diethyl analog causes a decrease in activity with some of our test
insects. Nevertheless we felt that this loss of activity might
be compensated in the field by the increased stability. Thus, a
series of physico-chemical methods was developed which should
give us a quick first answer regarding the possible performance
under field conditions of compounds with interesting biological
activity. The methods included a test for UV-stability, for hydrolytic stability and more practically oriented tests like stability
in aqueous emulsion outdoors and persistency on bean leaves. The
results obtained with six compounds are reported by Hangartner
et al. (1976); they showed that compound 10 (Ro-3108) exhibited
the best overall stability.

_____

* Acute toxicity data of R-20458 were presented by Pallos and
Menn (1972).

TABLE 4. MORPHOGENETIC ACTIVITY OF 2,3-DIHYDROBENZOFURAN-DERIVATIVES.

	$-\log ED_{95}{}^a$	
	Tenebrio molitor 10^{-x} g/pupa	Aedes aegypti 10^{-x} g/ml
3	8.5	7.5
4	7.5	7.5
5	9	4.5
6	8.5	8.5
7	7	7.5
8	5.5	5.5

a) Effective dose resulting in abnormal development of 95–100% of treated insects.

TABLE 5. EFFECT OF HYDROGENATION ON MORPHOGENETIC ACTIVITY OF p-ETHYLPHENYL ETHERS.

		$-\log ED_{95}{}^a$		
		Tenebrio molitor 10^{-x} g/pupa	Tribolium castaneum 10^{-x} g/g flour	Aedes aegypti 10^{-x} g/ml
9 Ro 10-3109		8.5→7.5	5.5→5	8→7.5
10 Ro 10-3108		8.5→8.5	6.5→6.5	7→6.5

a) Effective dose resulting in abnormal development or failure to reproduce of 95–100% of treated insects.

Table 6, column a, presents the laboratory data we obtained with IGR's in tests on the summer fruit tortrix moth, a representative member of the economically important family <u>Tortricidae</u>. Ro 10-3108 shows the highest activity in this test. Even at a concentration of 10 ng/cm^2, 100% reduction of normal imaginal development was observed. Among the next best compounds in this assay is Ro 10-2202 a mixture of isomers of ZR-512 or hydroprene, one of two standards used in our tests. Hydroprene was reported by Henrick <u>et al</u>. (1973) to have good activity on Lepidoptera and methoprene successfully controlled the hemlock looper in field trials as reported by Staal (1975).

TABLE 6.   EFFECT OF VARIOUS IGR'S ON LAST INSTAR LARVAE OF <u>ADOXOPHYES</u> <u>ORANA</u> IN THE LABORATORY AND IN THE FIELD.

Code	Compound Structure	a Laboratory trial −log ED$_{95}$[a] 10^{-x} g/cm^2	b Field trial % reduction of normal adults
1 R-20458 Ro 8-9801		x = 6.5	35
2 Ro 20-3600	c,t c,t	7.5	78
10 Ro 10-3108	c,t	8.5	100
11 Ro 10-2202	t c,t	7.5	.
12 Ro 10-6425	MeO t c,t	6.5	.
8 Ro 6-9550	COOMe c,t c,t c,t	5.5	.

a) Effective dose resulting in abnormal development of 95–100% of treated insects.

Ro 20-3600, the compound found by Bowers (1969), shows good activity too and R-20458 proves to be rather specific for Diptera by being at least 100 times less potent than Ro 10-3108 against

this lepidopterous pest. All compounds tested exhibited much
higher activity than the synthetic JH I.

The potency of Ro 10-3108 was proven in a field experiment
with a carefully synchronized population of summer fruit tortrix
moth. Table 6, column b, shows that Ro 10-3108 had enough residual
activity to affect all the trapped larvae. The performance of the
other compounds tested (1 and 2) reflects their activity obtained
in the laboratory. If tested as a 0.1% active ingredient emulsion
on natural (unsynchronized) populations of the summer fruit tortrix
moth (Adoxophyes orana), the grape berry moth (Lobesia botrana)
and the larch tortrix moth (Zeiraphera diniana) a close relative
of the spruce budworm, Ro 10-3108 was fully effective as shown by
Hangartner et al. (1976), a result which in our opinion demonstrates
the good stability of this compound on foliage.

Scale insects and mealybugs can be considered ideal targets
for control by IGR's because they build up a large damaging population only over a long period, their development is seasonally
synchronized and reinfestation from untreated surroundings is slow.
Furthermore the active ingredient applied to the bark is not
exposed to the degradation of leaf enzymes and the target does
not move away. The finding by the same authors that one application of Ro 10-3108 is sufficient to control field populations of
San Jose scales (Quadraspidiotus perniciosus) and citrus snow
scales (Unapsis citri) for over 6 months opens a promising new
field for the practical realization of insect control by IGR's.
Other fields of possible application of Ro 10-3108 include stored
product pests (Hoppe and Suchy, 1975) and aphids (Hangartner
et al., 1976; Meier et al., 1975).

Although Ro 10-3108 has a relatively broad spectrum of activity it is only moderately active against Diptera as shown in
tests against flies (Musca domestica and Lucilia sericata) and
mosquitoes. This selectivity makes the compound safe for most
beneficial insects which happen to belong to dipteran or hymenopteran groups. In an experiment with San Jose scale and its
parasite Prospaltella perniciosi, Frischknecht and Mueller (1976)
found that 45 days after spraying with a 0.1% emulsion of Ro 10-3108
the parasite population was still fully active, whereas the host
population was controlled. Similar results, using the same host-
parasite situation, have been reported by Scheurer and Ruzette
(1974) using two different IGR's.

Table 7 gives some toxicological data on 10-3108. The LD_{50}
values for mice and rats are exceedingly high. Skin irritation,
eye irritation and acute inhalation toxicity is favorable and the
compound proved to be essentially non-toxic to guppies (Lebistes
reticulatus) and rainbow trout (Salmo gairdneri). The degradation

TABLE 7. TOXICOLOGICAL DATA ON Ro 10-3108.

Acute Toxicity	Rats	$LD_{50} > 32000$ mg/kg p.o.
	Rats	$LD_{50} > 16000$ mg/kg i.p.
Five Applications (daily interval)	Mice	$LD_{50} > 5640 \pm 450$ mg/kg p.o.
	Rats	$LD_{50} > 8000$ mg/kg p.o.
Skin Irritation	Guinea pig	No effect at 1%. Slight irritation at 3% (repeated appl.).
Eye Irritation	Rabbit	No effect up to 10%
Acute inhalation Toxicity	Rats	> 1.3 mg/l – 4 hr exposure. Slight transient eye and respiratory irritation during exposure.
Fish Toxicity	Guppy	LC_{50} (96 hr) > 5000 ppm
	Rainbow trout	LC_{50} (96 hr) > 5000 ppm

of Ro 10-3108 in polluted water was studied by Dorn et al. (1976).
Their findings indicate that this IGR is highly biodegradable and
that it will hardly leave persistent residues in the environment.

Because of the generally accepted belief that epoxides are
detrimental to stability (Henrick et al., 1973; Staal, 1975;
Hammock et al., 1975), many laboratories looked for efficient
procedures to replace this function. In most cases this was done
by substituting a tertiary ether function (Henrick et al., 1963;
Sorm, 1971; 1974). Whereas in the case of the dienoates replacement of the epoxide by the methoxy group generally increased activity (Henrick et al., 1973), the situation is different for phenylethers. 7-Alkoxy geranyl ethers are in some cases more potent
against specific insects but have a much narrower spectrum of
activity (see Hammock et al., 1974). We therefore concentrated
on compounds having the epoxide function which might be important
for an optimal fit at the hormone receptor site or sites, or with
a carrier protein. However, the epoxide had to be rendered less
susceptible to environmental degradation and probably metabolic
deactivation in the insect, by changing the steric requirements
for an attack at the oxygen function. Thus, a number of compounds
were synthesized having bulky substituents at C-7. Table 8 shows
the effect on the activity if the methyl or ethyl group is replaced
by an isobutyl group in three series of phenylethers. Looking at
the Tenebrio-, Tribolium-, and Aedes-assays, a considerable loss
of activity is noted by making this modification. The decrease in
activity is especially marked for the saturated compounds 14, 16
and 18. However in the case of the lepidopterous insects
Adoxophyes and Carpocapsa the situation is reversed and we observe

a strong increase in activity by moving from the methyl or ethyl
compounds 2, 1 and 3 to the corresponding isobutyl compounds 13
to 18. Just as important is the fact that the saturated compounds
which can be expected to be more stable, are only slightly less
active. The isobutyl group therefore constitutes a kind of
lepidopterous active principle and one might speculate that this
has something to do with the level and specificity of epoxide
hydratase in different orders of insects.

TABLE 8. EFFECT OF THE ISOBUTYL GROUP ON MORPHOGENETIC ACTIVITY
OF PHENYLETHERS.

Structure	$- \log ED_{95}$ [a]				
	Tenebrio molitor 10^{-X} g/pupa	Tribolium castaneum 10^{-X} g/g flour	Aedes aegypti 10^{-X} g/ml	Adoxophyes orana 10^{-X} g/cm²	Carpocapsa pomonella 10^{-X} g/cm²
2	7.5	6	8.5	7.5	5.5
13	7	4.5	6.5	9	8.5
14	5.5	4	6	8.5	7.5
1	8.5	5.5	8	6.5	5.5
15	7.5	5	5.5	8.5	7.5
16	7	4	5.5	8	7.5
3	8.5	4.5	7.5	6.5	8.5
17	6.5	3.5	5.5	8	8.5
18	5.5	$\leqq 3$	5.5	7	6.5

[a] Effective dose resulting in abnormal development or failure to reproduce
of 95–100% of treated insects.

ACKNOWLEDGEMENTS

I am indebted to T. Hoppe and W. Jucker (Dr. R. Maag Ltd.) for skillfully performing the mosquito experiments in the laboratory and in the field respectively, to W. Hangartner (Dr. R. Maag Ltd.) for the experiments with the summer fruit tortrix moth, to S. Dorn (Dr. R. Maag Ltd.) for biological testing, to A. Pfiffner and M. Suchý for the synthesis of several compounds, to W. Roth for formulation work, to H.P. Bächtold (F. Hoffmann-La Roche) and the Huntingdon Research Center (England) for toxicological data and to H.K. Wipf and M. Gruber for comments on the manuscript.

REFERENCES

Abbott, W.S., 1925, J. Econ. Entomol. 18:265.
Bowers, W.S., 1968, Science 161:895.
Bowers, W.S., 1969, Science 164:323.
Chodnekar, M., Loeliger, P., Pfiffner, A., Schwieter, U., Suchý, M., and Zurflueh, R. (to F. Hoffmann-La Roche Co.), 1973, Belg. Pat. 798,506.
Chodnekar, M., Loeliger, P., Pfiffner, A., Schwieter, U., Suchý, M., and Zurflueh, R. (to F. Hoffmann-La Roche Co.), 1975, Belg. Pat. 818,829.
Dorn, S., Oesterhelt, G., Suchy, M., Trautmann, K.H., and Wipf, H.K., 1976, J. Agric. Food Chem., in press.
Feutrill, G.I., and Mirrington, R.N., Tetrahedron Letters, 1970, 2527.
Frischknecht, M., and Mueller, P.J., 1976, Mitt. Schweiz. Ent. Ges., submitted for publication.
Gill, S.S., Hammock, B.D., and Casida, J.E., 1974, J. Agric. Food Chem. 22:386.
Hammock, B.D., Gill, S.S., and Casida, J.E., 1974, J. Agric. Food Chem. 22:379.
Hammock, B.D., Gill, S.S., Hammock, L., and Casida, J.E., 1975, Pestic. Biochem. Physiol. 5:12.
Hangartner, W., and Zurflueh, R. (to F. Hoffmann-La Roche Co.), 1975, Belg. Pat. 830,263.
Hangartner, W., Suchý, M., Wipf, H.K., and Zurflueh, R., 1976, J. Agric. Food Chem. 24:169.
Henrick, C.A., Staal, G.B., and Siddall, J.B., 1973, J. Agric. Food Chem. 21:354.
Hindley, N.C., and Andrews, D.A. (to F. Hoffmann-La Roche Co.), 1974, Belg. Pat. 806,775.
Hoffmann, L.J., Ross, J.H., and Menn, J.J., 1973, J. Agric. Food Chem. 21:156.
Hoppe, T., and Suchý, M., 1975, EPPO Bull. 5:193.
Hoppe, T., Isler, H., and Vogel, W., 1974, Mosq. News 34:293.

Jacobson, M., Beroza, M., Bull, D.L., Bullock, H.R., Chamberlain, W.F., McGovern, T.P., Redfern, R.E., Sarmiento, R., Schwarz, M., Sonnet, P.E., Wakabayashi, N., Waters, R.M., and Wright, J.E., 1972, in "Insect Juvenile Hormones, Chemistry and Action" (J.J. Menn and M. Beroza, eds.), pp. 249, Academic Press, New York.

Pallos, F.M., and Menn, J.J., 1972, in "Insect Juvenile Hormones, Chemistry and Action" (J.J. Menn and M. Beroza, eds.), pp. 303, Academic Press, New York.

Pallos, F.M., Menn, J.J., Letchworth, P.W., and Miaullis, J.B., 1971, Nature (London) 232:486.

Pauling, H. (to F. Hoffmann-La Roche Co.), 1972, Belg. Pat. 783,055.

Pfiffner, A., 1971, in "Aspects of Terpenoid Chemistry and Biochemistry" (T.W. Goodwin, ed.), pp. 95, Academic Press Inc., London.

Quistad, G.B., Staiger, L.E., and Schooley, D.A., 1975, J. Agric. Food Chem. 23:299.

Sarmiento, R., McGovern, T.P., Beroza, M., Mills, G.D., and Redfern, R.E., 1973, Science 179:1342.

Scheurer, R., and Ruzette, M.A., 1974, Z. Angew. Entomol. 77:218.

Schooley, D.A., Bergot, B.J., Dunham, L.L., and Siddall, J.B., 1975, J. Agric. Food Chem. 23:293.

Šorm, F., 1971, Mitt. Schweiz. Ent. Ges. 44:7.

Šorm, F., 1974, in "Insect Hormones and Bioanalogues" (K. Sláma, M. Romañuk, and F. Šorm, eds.), pp. 170, Springer, New York.

Staal, G.B., 1975, Ann. Rev. Entomol. 20:417.

Vogel, W., Masner, P., and Frischknecht, M., Mitt. Schweiz. Ent. Ges., submitted for publication.

Wright, J.E., McGovern, T.P., Sarmiento, R., and Beroza, M., 1974, J. Insect Physiol. 20:423.

A COMPETITIVE BINDING PROTEIN ASSAY FOR JUVENILE HORMONE

W. Goodman, W.E. Bollenbacher, H.L. Zvenko, and
L.I. Gilbert

Department of Biological Sciences, Northwestern
University, Evanston, Illinois 60201

INTRODUCTION

Despite the discovery of juvenile hormone (JH) four decades ago (Wigglesworth, 1934) and the structural identification of the molecule nearly a decade ago (Röller et al., 1967), the lack of a simple, rapid, and sensitive quantification procedure for JH has hindered progress in the field of insect endocrinology. The minute quantities of hormone, the paucity of insect tissue and the high potential for degradation contribute to the technical difficulties encountered when titering JH. At present, two methods are being used to determine JH titers, analytical chromatography (van Broek-hoven et al., 1975; Dahm et al., this volume; Schooley et al., this volume) and bioassay (Staal, 1972; Bjerke and Röller, 1974).

Competitive protein binding assays, including radioimmuno-assays and those utilizing naturally occurring binding proteins, present a unique approach to the problem of JH quantification. The advantages of this method are simplicity, high specificity and theoretical ultrasensitivity. Moreover, this type of assay can be performed in less than 24 hr and requires little complex equipment. Although an immunochemical quantification assay for the three naturally occurring homologs of JH has been developed (Lauer et al., 1974), the limited sensitivity and availability of the JH antiserum led us to investigate the potential assay capabilities of the naturally occurring JH binding protein found in the hemolymph of Manduca sexta (Kramer et al., 1974; Goodman and Gilbert, 1974). The hemolymph of this widely used laboratory insect can be utilized to perform rapid, sensitive and specific competitive protein binding assays for the determination of JH.

MATERIALS AND METHODS

Hormones and Analogs. [^{3}H]-JH I (11.3 Ci/mmol, radiopurity 94%) was purchased from New England Nuclear Corp. Homologs and analogs used in the specificity studies were JH 0, methyl-trans, trans,cis-11-methyl-3,7-diethyl-10,11-epoxy-2,6-tridecadienoate; JH I, 95% methyl-trans,trans,cis-3,11-dimethyl-7-ethyl-10,11-epoxy-2,6-tridecadienoate; JH II, 95% methyl-trans,trans,cis-3,7,11-trimethyl-10,11-epoxy-2,6-tridecadienoate; JH III, 80% methyl trans,trans-3,7-11-trimethyl-10,11-epoxy-2,6-dodecadienoate; ZR-512, ethyl-3,7,11-trimethyl-2,4-dodecadienoate; ZR-515, iso-propyl-11-methoxy-3,7,11-trimethyl-2,4-dodecadienoate, from Zoecon Corp.; JH I, mixed isomers, 17% trans,trans,cis (Kramer et al., 1974), from Ayerst Laboratories; and R-20458, 1-(4'-ethylphenoxy)-3,7-dimethyl-6,7-epoxy-trans-octene, from Stauffer Chemical Co. JH acid, 10,11-epoxy-3,11-dimethyl-7-ethyl-2,6-tridecadienoic acid, was synthesized from the JH I mixed isomers (Hammock et al., 1975). Fatty acid standards were obtained from Applied Sciences Laboratories. Methyl epoxy stearic acid was purchased from Analabs, while methyl epoxy palmitic acid was prepared by J.A. Marshall (Northwestern University). Precocene II (6,7-dimethoxy-2,2-dimethyl chromene) was a gift from W.S. Bowers (Cornell University).

Purity of [^{3}H]-JH I, unlabeled homologs and analogs ZR-515 and R-20458 was determined by thin layer chromatography (Silica Gel 60, F-254 EM) and high-speed liquid chromatography (LC). LC purification was carried out using a Waters Associates M6000 solvent delivery system equipped with a U6K injector and a Schoeffel 770 variable wave length spectrophotometer. Chromatographic separations were performed on a μPorasil column (1/4" x 1', Waters Associates) using a solvent system of hexane:diethyl ether, 96:4 (v/v), at a flow rate of 1.5 to 2.0 ml/min (600 psi) and were monitored at 220 nm. ZR-512, fatty acids and fatty acid derivatives were analyzed for purity by thin layer chromatography. ZR-512 was found to be approximately 90% pure. Quantification of the homologs and analogs was carried out using a Gilford 2400 spectrophotometer. JH homologs were monitored at 220 nm in methanol (molar absorptivity ε_{220} = 13,830; Trautmann et al., 1974; K.H. Dahm, personal communication; J.A. Marshall, personal communication), while ZR-512 and ZR-515 were monitored at 260 nm (Henrick et al., 1975; K. Judy, personal communication).

Buffers, Solvents and Glassware. The assay buffer used to resolubilize delipidated protein was 0.025 M phosphate and 0.05 M KCl, pH 7.5 (buffer I). Buffer I containing 10% ethanol (buffer II) was used to resolubilize [^{3}H]-JH I, JH standards, and unknowns. The stock solution of [^{3}H]-JH I for the competitive protein binding assay (JH-CPBA) was prepared by adding [^{3}H]-JH I to a vial, evaporating the solvent, and adding enough buffer II to produce a final specific activity of 18,000 cpm per 0.05 ml. This solution was sonicated for 30 sec at 30 watts to insure complete solubilization.

Organic solvents for extractions and LC separations were glass distilled (Burdick and Jackson). Ethanol was purchased from U.S. Industrial Chemicals Co. Glassware was precoated with a 10% polyethylene glycol solution (MW 20,000; Calbiochem), rinsed and dried (45°C). Hamilton syringes were used throughout for the transfer of JH standards and unknowns.

Charcoal. The charcoal-dextran suspension was prepared by the addition of charcoal to a 0.66% dextran solution at a final concentration of 0.3%. The mixture was then stirred for 30 min. In all cases the charcoal was prepared immediately preceding use in the assay.

Animals. Manduca sexta larvae were reared at 26°C under a 16 light:8 dark photoperiod and were fed an artificial diet developed by Yamamoto (1969) and modified by Bell (personal communication). At the onset of the second instar, larvae were transferred from mass rearing chambers to individual 30 ml plastic cups containing artificial diet.

Preparation of Hemolymph Binding Protein. Late fourth instar larvae exhibiting head capsule slippage were removed from the diet, washed and then bled by making a small incision in the proleg. The hemolymph was collected on ice, centrifuged at 1000 x g for 10 min to remove hemocytes and recentrifuged at 12,000 x g to remove debris (Fig. 1). After centrifugation, the hemolymph was treated with GSH (15-20 µmol/ml) and diisopropylphosphorofluoridate (DFP) (Hammock et al., 1975). Hemolymph was collected and prepared just prior to the assay.

When further purification of the binding protein was desired, 100 ml of hemolymph was prepared as described above. The hemolymph was then concentrated, incubated with 0.1 µCi of [^{3}H]-JH I and passed through a Sephadex G-150 column (150 x 2.5 cm) using a buffer of 0.05 M phosphate and 0.05 M KCl, pH 6.7 (Fig. 2). The effluent was monitored by liquid scintillation radioassay and the active fractions pooled. The pooled fractions were dialyzed against a buffer containing 0.01 M sodium acetate and 0.05 M KCl, pH 5, and passed through a carboxymethyl cellulose column (CM-52, 5 x 2.5 cm) equilibrated with the dialysis buffer. The binding protein fractions (first protein peak) from this column were collected, pooled and then dialyzed against a buffer system containing 0.05 M Tris, 0.001 M EDTA, 0.01 M 2-mercaptoethanol, 0.05 M KCl and 5% glycerol at pH 7.5. This fraction was concentrated using an Amicon ultrafiltration system (UM 10 filter). The concentrate was treated with DFP as previously described and stored in 0.2 ml aliquots at -76°C until needed. No detectable loss of binding activity was noted over a two-month period.

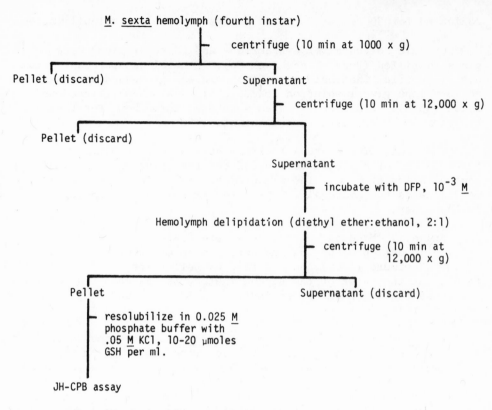

Fig. 1. Preparation of hemolymph proteins for juvenile hormone competitive protein binding assay.

Delipidation of the hemolymph of partially purified binding protein was performed in ether:ethanol, 2:1 (v/v), at 4°C (Fig. 1). For hemolymph, 2.0 ml of prepared sample was added to 50 ml of the ether:ethanol solvent, mixed and then incubated for 30 min. The mixture was centrifuged at 12,000 x g for 10 min and the supernatant discarded. The pellet was washed with delipidation solvent and then dried under a gentle stream of purified air. The pellet was resolubilized in 0.5 ml of buffer I containing 10-20 μmol of GSH per ml. This stock was routinely diluted to a final concentration of 3.0% by volume and immediately used for the JH-CPBA.

Partially purified binding protein was delipidated as described above with the incubation period modified to 10 min. The pellet was resolubilized in buffer I without GSH. To determine the appropriate concentrations of binding protein to be used in the assay, partially purified binding protein was serially diluted from 1:1

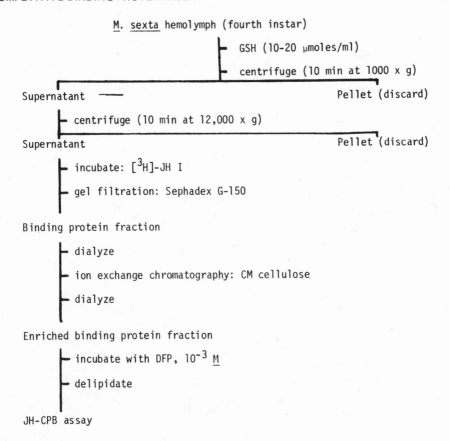

M. sexta hemolymph (fourth instar)
- GSH (10-20 µmoles/ml)
- centrifuge (10 min at 1000 x g)

Supernatant ——   Pellet (discard)
- centrifuge (10 min at 12,000 x g)

Supernatant   Pellet (discard)
- incubate: [³H]-JH I
- gel filtration: Sephadex G-150

Binding protein fraction
- dialyze
- ion exchange chromatography: CM cellulose
- dialyze

Enriched binding protein fraction
- incubate with DFP, 10^{-3} M
- delipidate

JH-CPB assay

Fig. 2. Preparation of binding protein for juvenile hormone competitive protein binding assay.

to 1:500 and assayed for binding activity against the [³H]-JH I stock solution (Fig. 3). That dilution of the protein which bound 35-40% of the [³H]-JH I (approximately 6500 cpm) was then used for the assay. Since the binding activity of partially purified binding protein was not reproducible, it was essential that the dilution necessary for optimal binding activity be redetermined for each preparation of partially purified protein.

Assay. Either JH standards at concentrations from 0 to 128 ng or unknowns were added to the assay tubes (6 x 50 mm; 3 tubes per concentration), evaporated under nitrogen and resolubilized in 0.05 ml of buffer II. Next, 0.05 ml of the [³H]-JH I stock solution was added. Diluted, delipidated hemolymph or partially

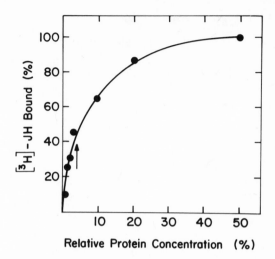

Fig. 3. Binding activity of serially diluted stock JH binding
protein. Arrow represents concentration of protein which will
bind 35-40% of the total radioactivity in the assay mixture.

purified binding protein (0.1 ml) was added to the assay tubes and
gently vortexed. The tubes were incubated overnight at 4°C. The
assay was then terminated by adding 0.1 ml of the charcoal-dextran
suspension to each tube and vortexing immediately. Charcoal was
added to the assay tubes during a 1.0 min period; routinely, 10
tubes could be terminated during the time allotted. Following the
addition of charcoal, tubes were incubated for 5.0 min and then
centrifuged at 2800 x g for 10 min to pellet the charcoal. After
centrifugation, 0.1 ml of the supernatant was removed from the
assay tube and monitored by liquid scintillation radioassay.
Samples were counted in 3.0 ml of scintillation fluid composed of
xylene:triton X-100:ethanol:ethylene glycol, 600:275:106:37 (v/v),
containing 3 g/l of pre-blend dry fluor 2a20 (98% PPO and 2% POPOP)
(Fricke, 1975). Liquid scintillation counting was conducted with
a Packard TriCarb scintillation spectrophotometer, Model 3380,
having an efficiency of 33%.

     Extraction of JH. Hemolymph from fourth instar larvae was
extracted for endogenous or exogenously added JH (JH I or [^{3}H]-
JH I) by adding 1 volume of methanol to 2 volumes of hemolymph to
precipitate proteins (Fig. 4). The mixture was vortexed and 2
volumes of hexane were then added, resulting in a final ratio of
hemolymph:methanol:hexane, 2:1:2 (v/v). The hemolymph-solvent
mixture was again vortexed and centrifuged at 12,000 x g for 10 min
at 4°C to generate a partition. The hexane epiphase, containing
the JH, was removed and saved. A second hexane extraction followed.
This epiphase was pooled with the first, and taken to dryness under

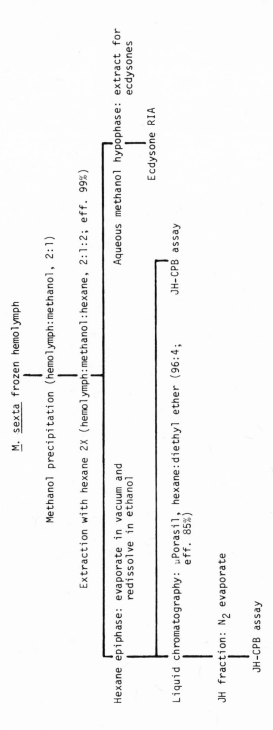

Fig. 4. Flow diagram of the isolation and assay of juvenile hormones and ecdysones from M. sexta hemolymph.

vacuum at room temperature. The aqueous methanol hypophase from
both extractions was saved for ecdysone assay. The epiphase residue was resuspended in ethanol and aliquots were assayed for JH
(JH-CPBA or liquid scintillation radioassay). Samples were purified by LC as previously outlined. Retention volumes for the
respective JH homologs were collected and assayed by JH-CPBA,
liquid scintillation radioassay, or ultraviolet absorbance (LC).

RESULTS

Hemolymph Versus Partially Purified Binding Protein

Gel permeation chromatography as well as Scatchard plot
analysis (Scatchard, 1949) of the JH-CPBA indicate that at least
two proteins in the hemolymph of M. sexta are responsible for JH
binding (Fig. 5A). Partial purification of the high affinity
binding protein eliminates the low affinity interaction (Fig. 5B).
Although the partially purified product represents four or five
proteins as determined by gel electrophoresis (Goodman and Gilbert,
unpublished observations), Scatchard plot analysis reveals the
existence of a single class of JH binding proteins. As the
Scatchard plot also demonstrates, hemolymph and partially purified
binding protein have different slopes for the high affinity sites,
indicating that the affinities of the two fractions for JH are
different. Elimination of the hemolymph low affinity interaction
by gel permeation and ion exchange chromatography (Fig. 2) results
in a decreased apparent K_D for the partially purified binding
protein. The decrease in apparent K_D effects an increase in the
sensitivity of the assay.

Optimization of Assay Conditions

To optimize the assay conditions for maximum sensitivity,
partially purified binding protein is diluted to a concentration
that will bind 35-40% of the total [^{3}H]-JH I in the assay mixture
(Fig. 3). When the proper dilution has been established for the
protein, a standard curve is plotted by comparing the radioactivity
remaining in the supernatant to the log of the amount of JH present
(Fig. 6). Improper dilution of the protein will yield either a
very shallow curve or a normal curve lacking significant differences in the higher concentration ranges.

Although the equilibrium between JH and the binding protein
is established rapidly, longer incubation periods are required to
decrease variability. At 4°C, approximately 90% of the maximum
binding is attained within 5-6 hr. Incubation periods of 16 hr
or longer yield maximum binding (Fig. 7). Assays incubated at

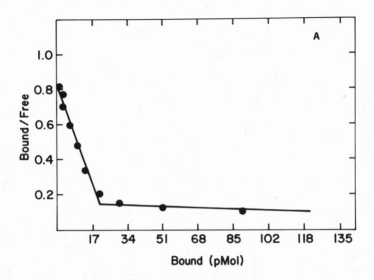

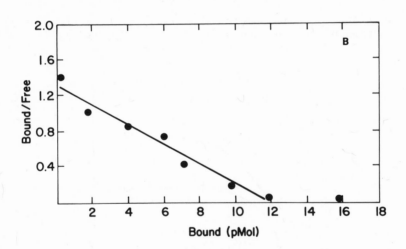

Fig. 5. Scatchard plot analysis of JH–CPBA using fourth instar hemolymph (A) and partially purified binding protein (B). The abscissa is expressed in pmol of JH bound per 0.1 ml of assay mixture. The apparent K_D for the high affinity interaction in (A) is 2.85×10^{-7} M and (B) 9.35×10^{-8} M. Lines were fitted by the method of least squares analysis.

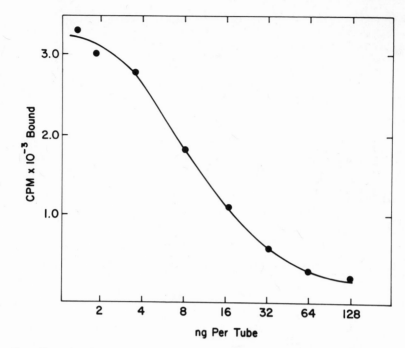

Fig. 6. Standard curve showing the displacement of [³H]-JH I as a function of the log concentration of unlabeled JH I.

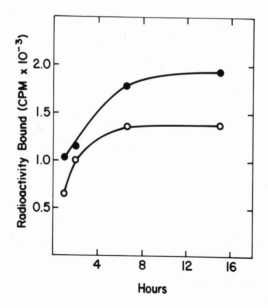

Fig. 7. Effect of temperature and length of incubation on binding of JH to binding protein. (●——●) 4°C and (o——o) room temperature.

room temperature attain equilibrium more rapidly, but the elevated
temperatures favor greater dissociation of the JH–binding protein
complex. Thus, incubations for 16 hr at 4°C result in greatly
enhanced sensitivity and reproducibility.

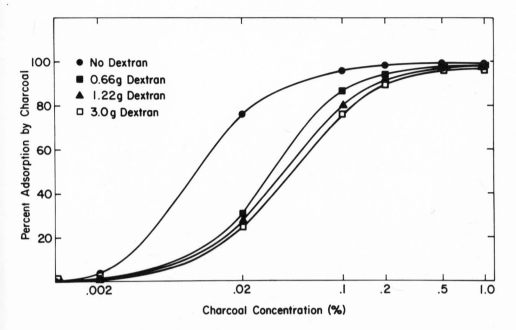

Fig. 8. Charcoal adsorption of [³H]–JH I as a function of charcoal
concentration. Bovine serum albumin is used as a nonspecific
binding protein.

     Separation of bound and unbound hormone is dependent upon
the adsorptive capacity of the charcoal. Figure 8 shows the relative charcoal adsorption as a function of the charcoal–dextran
concentrations when a nonspecific binding protein, bovine serum
albumin, is present at a concentration equivalent to that of the
binding protein in the assay. As the concentration of charcoal is
increased, more nonspecifically–bound as well as unbound JH is
adsorbed. At concentrations of 0.2% to 0.3%, virtually all of the
nonspecifically–bound and unbound JH is adsorbed. Dextran in this
assay serves two purposes. First, when small amounts of binding
protein are added to the assay, there is a tendency for the charcoal to adsorb the protein. When charcoal is mixed into a dextran
solution, there is a decrease in adsorptive sites on the charcoal,
thus decreasing the possibility that protein, rather than unbound
hormone, is adsorbed. Secondly, we have found that dextran–coated

charcoal pellets more efficiently than does uncoated charcoal, resulting in greater sample reproducibility.

The use of dextran-coated charcoal is acceptable only if relatively complete adsorption of the unbound hormone takes place, with minimal adsorption of the bound hormone. Since a 0.3% charcoal suspension in a 0.66% dextran solution was found to optimally adsorb the nonspecifically-bound hormone, the stripping effect of this concentration of charcoal on binding protein was determined. During a 5 min incubation period, less than 5% of the bound counts was removed, using the protocol previously described.

Specificity

When homologs of JH were tested for their displacement activity against labeled JH I (t,t,c), a pattern of displacement based on polarity was evident with either hemolymph or partially purified binding protein (Fig. 9). Analogs which exhibit marked morphogenetic activity in Manduca (Truman et al., 1973) show no detectable displacement activity. Moreover, the primary hemolymph metabolite of JH, JH acid (Slade and Zibbit, 1972), was ineffective in displacing JH I. Precocene II, a molecule with reported anti-JH activity (Bowers, this volume), displayed no displacement activity, suggesting that it does not block hemolymph transport of JH. Since a hexane extract of hemolymph contains a complex mixture of lipids, these lipids may interfere with binding of JH in the JH-CPBA. Therefore, several fatty acids and fatty acid derivatives having structural similarities to JH were tested for their ability to compete with [^{3}H]-JH I in the assay. In all cases, they proved ineffective (Table 1). Evidence that hexane-extractable lipids other than those tested above do not interfere with the assay is presented below.

Extraction of JH from Biological Material

Extraction efficiency and molecular stability were monitored by JH-CPBA of exogenously added unlabeled JH I and radiotracer analysis of exogenously added [^{3}H]-JH I. The efficiency of extraction was 99% and the qualitative homogeneity of the JH molecule was unaltered (Fig. 9).

In developing a rapid method for JH quantification, the ability to quantitatively determine JH levels directly from the hexane extracts of hemolymph is critical. JH-binding protein interactions could be seriously hindered by the large amounts of lipid present in the hexane extract. To demonstrate that JH was indeed being accurately measured, hexane extracts of hemolymph were partially purified by LC and the recoveries determined. Using [^{3}H]-JH I and

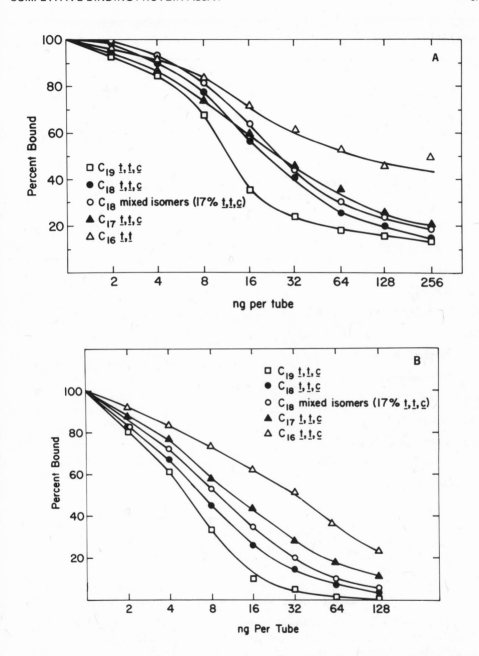

Fig. 9. Competitive displacement of [^{3}H]-JH I from binding protein by JH homologs. Hemolymph (A) and partially purified binding protein (B). Each point represents the average of 3 assays (3 replicates per concentration).

TABLE 1.   RATIO OF ASSOCIATION CONSTANTS[a] FOR HEMOLYMPH AND PARTIALLY PURIFIED BINDING PROTEIN.

Compound[b]	Hemolymph	Partially Purified Binding Protein
1. JH 0	2.82	3.36
2. JH I	1.00	1.00
3. JH I (mixed isomers)	.58	.56
4. JH II	.49	.35
5. JH III	.08	.08
6. JH Acid (C_{18})	--	--
7. ZR 512	--	--
8. ZR 515	--	--
9. Stauffer Compound R20458	--	--
10. Methyl Epoxy Palmitate	--	--
11. Methyl Epoxy Stearate	--	--
12. Precocene II	--	--
13. Palmitic Acid	--	--
14. Oleic Acid	--	--
15. Linolenic Acid	--	--

[a] Korenman (1970)

$$RAC = \frac{R\ (RA)}{R + 1-(RA)}$$

where: $R = \dfrac{\text{Free }[^3H]-JH}{\text{Bound }[^3H]-JH}$

and

$RA = \dfrac{\text{Molar Conc. JH I}}{\text{Molar Conc. Test Compound}}$ required to decrease bound $[^3H]$-JH I by 50%

[b] Compounds 1-6 at the highest concentrations are below their respective critical micellar concentrations (Goodman and Gilbert, unpublished observations).

Compounds 7-15 exhibit no displacement activity at the highest concentration used.

unlabeled JH I as markers, efficiency of recovery was determined to be 85% by both recovery of [^{3}H]-JH I and JH-CPBA of unlabeled JH I (Fig. 9). Since both determinations yield the same efficiency of recovery, hexane extracts of hemolymph can be assayed directly without further purification.

To verify the accuracy of the assay, the recovery of exogenously added JH I from control hemolymph (fourth instar, head capsule slippage) and goat serum was monitored by JH-CPBA. Table 2 demonstrates that recoveries of exogenous JH I were well within the range normally encountered in this type of assay (Murphy, 1970; Strott, 1975).

TABLE 2.   RECOVERY OF EXOGENOUSLY ADDED JH I FROM HEMOLYMPH BY JH-CPBA.

Amount Added (ng)	Amount Recovered (ng)	% Recovery
5.6	6.4	114
7.5	7.6	101
8.0	7.6	95
8.0	6.5	81
10.0	9.0	90
20.0	18.9	95
30.0	30.4	101
37.0	42.0	114
41.6	38.7	93
50.0	42.0	84

Mean   96.8%
Standard deviation   10.3
Standard error   3.3

Comparison of JH-CPBA to Other JH
Quantification Techniques

Manduca hemolymph (21 ml) was collected from fourth instar
0-3 hr larvae and extracted for JH. The hexane extract was divided
into two fractions, one partially purified by LC and the other
assayed by JH-CPBA (Fig. 9). LC revealed two peaks, having retention volumes the same as JH I and JH II, which were of approximately
equal area (determined by the height x width at half-height method).
Quantification of these peaks against external standards confirmed
that each represented approximately 2.8 ng of hormone per ml of
hemolymph. As summarized in Table 3, the total displacement activity based on JH I equivalents represented 4.2 ng/ml. Subtracting
the 2.8 ng/ml of JH I derived by LC from the total JH-CPBA activity,
1.4 ng/ml of JH II was obtained. At the concentration assayed,
JH II displaces approximately half the [^{3}H]-JH I displaced by
unlabeled JH I, indicating that 1.4 ng/ml represented half of the
actual amount of JH II present. The corrected value of 2.8 ng/ml
correlates with the LC quantification, which revealed equal amounts
of JH I and JH II in the sample. When the amounts of both JH I
and JH II are summed, 5.6 ng/ml of JH is found in the hemolymph
of the 0-3 hr fourth instar animal. This figure is in close agreement with physicochemical determinations presented by Schooley
et al. (this volume), and with bioassay results reported by Fain
and Riddiford (1975) (Table 3).

TABLE 3.   QUANTIFICATION OF JH IN FOURTH INSTAR M. SEXTA HEMOLYMPH
BY THREE DIFFERENT METHODS.

	Bioassay[a] (0-5 hr)	GLC/ECD[b] (24 hr)	JH-CPBA (0-3 hr)
JH I equivalents	11.0		4.2[c]
JH I		3.4	2.8
JH II		2.8	2.8[d]
JH III		0.7	
Total (JH I + JH II + JH III)		6.9	5.6[d]

Figures are expressed in ng per ml of hemolymph. [a] Black larval
assay (Fain and Riddiford, 1975). [b] Gas liquid chromatography/
electron capture detection (Schooley et al., this volume). [c] Un-
corrected value. [d] Corrected JH-CPBA for JH II based on LC determination.

The quantification of JH I, JH II, and JH III by the JH-CPBA is extremely complex and was conducted in this case only to demonstrate the validity of this assay in comparison to other techniques. In practice, the assay data should be expressed in JH I equivalents.

DISCUSSION

The discovery of a naturally occurring high affinity JH binding protein in the hemolymph of M. sexta has led to the development of a competitive protein binding assay for JH. Competition between the labeled and unlabeled forms of the hormone for a limited number of high affinity binding sites provides the basis for this assay. When increasing amounts of unlabeled hormone are added to a fixed quantity of binding sites and labeled hormone, a decrease in the amount of bound radioactive hormone is observed. By comparing the characteristic displacement of a known amount of hormone standard to the displacement of an unknown, the quantity of hormone in the unknown can be determined, provided that a number of conditions are met.

Either hemolymph or partially purified binding protein can, with little manipulation, lend themselves to a competitive protein binding assay for JH. Less than 2 ml of hemolymph, collected from 10 larvae, provides enough binding activity to perform an average assay. Moreover, hemolymph taken from a large population of animals over a six-month period binds JH in a consistent and predictable manner. On the other hand, we have found it more expeditious and convenient to partially purify larger amounts of hemolymph for binding protein. These can be stored in small aliquots for future use to avoid continual maintenance of small hemolymph supplies. Partial purification also eliminates the low affinity interaction of the lipoprotein which can influence the high affinity binding to JH. The purification of this single class of binding sites results in a decreased apparent K_D and the slightly enhanced sensitivity of the assay (Baulieu, 1970).

It is becoming increasingly evident that despite its solubility in an aqueous medium (Kramer et al., 1974) JH in physiological concentrations shows a marked adsorption to surfaces. To overcome this problem, several strategies were employed which would either increase solubility or reduce surface binding sites. A buffer with decreased polarity (10% ethanol) was developed which facilitated JH solubility, yet was compatible with binding protein when diluted to the final incubation volume. Glassware, when untreated, offers a particularly active surface for JH adsorption; however, treatment with 10% polyethylene glycol minimizes the problem. Since many plastic polymers exhibit the same surface-active binding problems, the use of such materials is restricted to the transfer of bound JH. This precaution is emphasized by Westphal et al. (1975) in their study on steroid hormone-plastic polymer interactions.

Maintaining the integrity and homogeneity of the JH molecule
in the assay presents potential problems. The hemolymph of late
fourth instar M. sexta contains virtually no JH-specific esterase
activity, although nonspecific esterases are present (Sanburg
et al., 1975; Nowock and Gilbert, 1976). Partially purified protein, as well as the hemolymph, showed no JH-specific esterase
activity. However, both protein preparations exhibited nonspecific
esterase activity. DFP, an esterase inhibitor, was used to block
the nonspecific hydrolysis of the JH molecule and thus preserve
the integrity of the molecule in the assay. JH standards are
routinely monitored by thin layer chromatography or LC to insure
homogeneity. The importance of this procedure cannot be over--
emphasized, since competition between labeled and unlabeled hormone
is concentration-dependent and slight deviations in quantity will
result in suboptimal competition.

Failure to remove endogenous JH from the binding protein
source will also result in suboptimal competition. Delipidation
of the partially purified protein as well as the hemolymph is
performed to remove bound and unbound JH.

JH is assumed to interact with the specific binding protein
at a strictly defined hydrophobic site on the protein. The binding
site has recently been studied, utilizing the displacement activities of various homologs and analogs of JH to characterize the
active site (Gilbert et al., 1976; Kramer et al., this volume).
The basic requirements for binding to the specific site include a
methyl ester at one end of the molecule and an epoxide function
at the opposite end. Aliphatic side chains and the trans,trans
configuration about the double bonds also appear to be necessary.
Analogs which display morphogenetic activity in Manduca do not
possess the necessary determinants for binding and therefore exhibit no displacement activity. The limited number of molecules
which fulfill the determinant criteria for binding make the assay
quite specific for the three naturally occurring hormones.

An important aspect of the binding protein is its characteristic affinity towards the various JH homologs. The least polar
homologs exhibit the greatest activity, and increasing the polarity
by shortening the side chains leads to a decreasing displacement
activity. JH 0, a synthetic homolog yet to be identified in insects,
exhibits the greatest displacement activity, while JH III, the most
polar of the homologs, displays the least activity. The concept of
polarity binding interactions has been considered elsewhere
(Goodman and Gilbert, A.C.S. Symposium, Aug. 1975; Gilbert et al.,
1976; Goodman and Gilbert, in preparation).

The differential displacement of the three naturally occurring hormones has considerable significance for the JH–CPBA. It is now accepted that the circulating titer of JH comprises all three hormones in varying ratios depending, in part, upon the developmental stage (see Gilbert et al., 1976). The marked preferential binding of JH I over JH II and JH III indicates that the values obtained from the assay are lower than the sum of the homologs when JH II and JH III are present in the extract. This necessitates the expression of data in JH I equivalents.

To date, three basic methods of JH quantification have been developed: analytical chromatographic methods, based on physico-chemical properties of the molecule; bioassay, based on the biological response of an animal to the hormone; and a competitive protein binding assay, based on the interaction between the specific binding site and the ligand. Each of these methods has inherent limitations.

Analytical chromatography (gas liquid chromatography/electron capture detection) offers a unique and ultrasensitive procedure for the detection and quantitative discrimination between the three naturally occurring homologs. At present, as little as 3 to 5 pg of JH can be detected with the same degree of sensitivity for all three homologs. Despite the obvious advantages of the method, it remains technically too difficult for routine analysis of JH titers. Extensive prepurification and derivatization processes limit the number of samples that can be assayed. Fluctuations in the lipid composition of extracts taken at different developmental stages often renders predetermined purification methods at one stage unsuitable for another (Schooley et al., this volume).

Bioassays for JH have been developed that cover a wide range of insects with an equally wide range of evaluation methods (see Staal, 1972; Sláma et al., 1974). In contrast to the analytical chromatographic method, most bioassays are simple, relatively inexpensive, and do not require extensive prepurification and derivatization. Under optimal conditions, the most sensitive bioassay, the Galleria wax test, can detect as little as 5 pg of JH I (de Wilde et al., 1968). Bioassays, however, require long incubation periods and a large population of test animals that are developmentally responsive to JH. The scoring of bioassays is subjective and often leads to significant differences in interlaboratory results (Bagley and Bauerfeind, 1972). Bioassays measure JH titers in terms of JH activity on a particular target tissue, usually expressed as equivalents of the JH standard. Since test animals exhibit marked differences in response to each of the naturally occurring hormones, the JH equivalents may not reflect the absolute quantity of JH present in an unknown sample. Further problems associated with bioassay technology are discussed by Staal (1972).

The JH-CPBA does not provide the sensitivity that is afforded by the preceding assays. Under optimal conditions this assay can detect quantities of JH as low as 1 ng; however, even after LC purification hexane extracts of hemolymph or goat serum have a 2-3 ng background. When hexane extracts are titered, the lowest quantity of JH I that can be detected is 3 ng (5×10^{-8} M). Despite the lack of ultrasensitivity, the JH-CPBA has several important advantages. A large assay, including replicates, can be conducted in a relatively short period of time. Interference from JH analogs and structurally related compounds has not been encountered. The accuracy as well as precision of the JH-CPBA is far superior to the bioassay (no statistical data presented for the gas liquid chromatography/electron capture detection method). The cost per tube is minimal, and once enough hemolymph has been collected and stored, maintenance of a <u>Manduca</u> colony is no longer necessary.

The JH-CPBA represents a simple, rapid and sensitive method for the quantification of JH. With the development of competitive protein binding assays, as well as the more sophisticated physico-chemical techniques, research into all areas of JH research will be greatly facilitated.

ACKNOWLEDGEMENTS

We thank Drs. J.B. Siddall and G.B. Staal of Zoecon Corp., Dr. J. Menn of Stauffer Chemical Co., and Dr. A.J. Manson of Ayerst Laboratories for supplying the homologs and analogs. We are especially indebted to Dr. W. Vedeckis (Baylor College of Medicine) for advice and technical assistance. The assistance of Mr. L. Feiss is gratefully acknowledged.

This work was supported by grants AM-02818 from the National Institutes of Health and GB-27574 from the National Science Foundation. W.E. Bollenbacher was supported by a post-doctoral fellowship from the National Institutes of Health.

REFERENCES

Bagley, R.W., and Bauerfeind, J.C., 1972, <u>in</u> "Insect Juvenile Hormones: Chemistry and Action" (J. Menn and M. Beroza, ed.), pp. 113-151, Academic Press, New York.
Baulieu, E.E., Raynaud, J.P., and Milgrom, E., 1970, <u>Acta Endocrinologica</u> 147(suppl.):104.
Bjerke, J.S., and Röller, H., 1974, <u>in</u> "Invertebrate Endocrinology and Hormone Heterophylly" (W. Burdette, ed.) pp. 130-139, Springer-Verlag, New York.

van Broekhoven, L.W., van der Kerk-van Hoff, A.C., and Salemink, C.A., 1975, Z. Naturforsch. 30:726.

Fain, M.J., and Riddiford, L.M., 1975, Biol. Bull. 149:506.

Fricke, U., 1975, Anal. Biochem. 63:555.

Gilbert, L.I., Goodman, W.G., and Nowock, J., 1976, in "Colloque International du C.N.R.S.: Biosynthesis, Metabolism and Cellular Actions of Invertebrate Hormones," in press.

Goodman, W.G., and Gilbert, L.I., 1974, Amer. Zool. 14:1289.

Hammock, B., Nowock, J., Goodman, W., Stamoudis, V., and Gilbert, L.I., 1975, Molec. Cell. Endocrinol. 3:167.

Henrick, C.A., Willy, W.E., Garcia, B.A., and Staal, G.B., 1975, Agri. Food Chem. 23:396.

Korenman, S.G., 1970, Endocrinology 87:1119.

Kramer, K.J., Sanburg, L.L., Kézdy, F.J., and Law, J.H., 1974, Proc. Nat. Acad. Sci. USA 71(2):493.

Lauer, R.C., Solomon, P.H., Nakanishi, K., and Erlanger, B.F., 1974, Experientia 30:558.

Murphy, B.E., 1971, in "Principles of Competitive Protein-Binding Assays" (W.D. Odell and W.H. Daughaday, ed.), pp. 108-133, J.B. Lippincott Co., Philadelphia.

Nowock, J., and Gilbert, L.I., 1976, in "Applications of Invertebrate Tissue Culture in Biology, Medicine and Agriculture" (E. Kurstak and K. Maramorosch, ed.), Academic Press, New York.

Röller, H., Dahm, K.H., Sweeley, C.C., and Trost, B.M., 1967, Angew. Chem. 79:190.

Sanburg, L.L., Kramer, K.J., Kézdy, F.J., Law, J.H., and Oberlander, H., 1975, Nature (London) 253:266.

Scatchard, G., 1949, Ann. N.Y. Acad. Sci. 51:660.

Slade, M., and Zibitt, C.H., 1972, in "Insect Hormones: Chemistry and Action" (J. Menn and M. Beroza, ed.), pp. 155-176, Academic Press, New York.

Sláma, K., Romañuk, M., and Šorm, F., 1974, "Insect Hormones and Bioanalogues," Springer-Verlag, New York.

Staal, G.B., 1972, in "Insect Juvenile Hormones: Chemistry and Action" (J. Menn and M. Beroza, ed.), pp. 69-94, Academic Press, New York.

Strott, C.A., 1975, in "Methods in Enzymology, vol. XXXVI" (B.W. O'Malley and J.G. Hardman, ed.), pp. 34-48, Academic Press, New York.

Trautmann, K.H., Schuler, A., Suchý, M., and Wipf, H.-K., 1974, Z. Naturforsch. 29c:161.

Truman, J.W., Riddiford, L.M., and Safranek, L., 1973, J. Insect Physiol. 19:195.

Westphal, U., Burton, R.M., and Harding, G.B., 1975, in "Methods in Enzymology, vol. XXXVI" (B.W. O'Malley and J.G. Hardman, ed.), pp. 91-104, Academic Press, New York.

Wigglesworth, V.B., 1934, Quart. J. Microscop. Sci. 79:191.

Yamamoto, R.T., 1969, J. Econ. Ent. 62:1427.

SUMMARY OF SESSION II:

BIOSYNTHESIS AND METABOLISM OF JUVENILE HORMONE

K. Judy

Zoecon Corporation, 975 California Avenue, Palo Alto,
California 94304

The papers contained in this section focus on two major
questions concerning the juvenile hormones (JH). How does the
titer of JH vary, both qualitatively and quantitatively during the
life stages of representative insect species? By what pathway or
pathways and from what precursors are the JHs biosynthesized? The
article by Trautmann et al. was not presented at the Conference but
is included here by special arrangement.

Since the structure of JH was first described, the question
of qualitative and quantitative JH titer determination has been
a crucial one. Nanograms of labile lipid hormone concealed in
milligrams of miscellaneous body lipids make conventional extraction and separation procedures virtually worthless. The first
successful JH identifications were possible only because, for
unknown reasons, certain male Saturniid moths accumulate up to a
microgram of JH in their accessory reproductive glands (see Dahm
et al., this volume). Subsequent JH identification became possible
when it was determined that corpora allata can produce their product in vitro where it accumulates relatively free of contaminating
substances. However, amounts of JH produced in vitro may not
reliably correlate with in vivo titers so this method has little
or no quantitative value. On the other hand bioassays for JH yield
quantitative data but are unable to distinguish qualitatively
between JHs. A need clearly exists for a relatively rapid procedure which can determine both the qualitative nature of JHs present
in a hemolymph sample and the amounts of each hormone present.
Ideally, this technique should be able to monitor changes within
a single insect over periods of time and should be sensitive enough
to detect even subphysiological levels of JH. Although still short
of this ideal, some recently developed methodologies such as those

96

described below by Trautmann et al. and Schooley et al. (see also Dahm et al., this volume; Goodman et al., this volume) offer considerable improvement over techniques of just one or two years ago.

The biosynthesis of JH has attracted interest because of the unusual nature of these hormones. The three known JHs are sesqui-terpene-type compounds containing an epoxide function. JH I and II are particularly interesting because they contain ethyl branches rather than the methyl side chains of conventional terpenes. Sev-eral hypotheses have been advanced for the biosynthesis of this type of structure, but only recently has any positive evidence been obtained. It now appears that JH is assembled as a sesqui-terpene from isoprene subunits, but some insects also produce homoisoprenoid intermediates which account for the ethyl branches of JH I and II. These intermediates may arise via homomevalonate, a compound not previously found in nature. Formation of the epoxide and methyl ester groups are probably the final steps in JH biosynthesis.

The paper by Trautmann et al. provides a condensation of several previously published investigations by this group. Their use of an isotope dilution technique applied to whole-body extrac-tions produced the first JH identifications from Coleoptera and Hymenoptera as well as titer data on eight species including 3 species of beetles and 3 species of cockroaches. Their detection limit of 0.1 ng hormone per gram fresh body weight is adequate for "gonadotropic" JH determinations from adult insects, but it is apparently not sufficiently sensitive to detect the "morphogenetic" JH of larval or nymphal stages. Also, this method requires rela-tively large amounts of insect material (0.4 - 1.2 kg were used in the determinations reported here) making it rather difficult to test carefully staged or timed insects. However, despite these limitations, Trautmann and colleagues succeeded in demonstrating that JH III is the most widespread form of the hormone for regulat-ing reproductive processes.

As if to demonstrate the rate at which technological improve-ments appear in modern research, the paper by Schooley et al. presents a new, previously unpublished and very sensitive method-ology for JH titer determination. Their procedure involves extrac-tion of lipids from samples of 10 g or less of hemolymph or whole bodies followed by a two-step derivatization which introduces a strongly electron capturing pentafluorophenoxyacetate (PFPA) group. Final separation and quantitation is accomplished by gas chroma-tography with electron capture detection (GC/ECD).

Schooley et al. report the first identification of JH from
Diptera as well as some comparative studies on the absolute amounts
of the three known JHs in certain life stages of a single species.
Like Trautmann and his colleagues, these investigators note that
JH III is the predominant form in adults while larval stages
(except for the Diptera) produce relatively more JH I and/or II.

The second portion of the Schooley et al. contribution deals
with the question of JH biosynthesis, specifically the early and
intermediate steps. In previous publications this group has hypothesized that homomevalonate, formed from acetate and propionate,
is a key intermediate in JH I and II biosynthesis. In the present
correspondence the authors indicate that exogenously added homomevalonate may have difficulty entering corpora allata in vitro.
To circumvent such difficulties they now use cell-free homogenates
of corpora allata in vitro rather than organ cultures as used
previously, and some preliminary findings are discussed.

The paper by Reibstein et al. summarizes the research of Law's
group into the terminal steps of JH biosynthesis. Making extensive use of corpus allatum homogenates, these workers have collected
data to support the concept that JH III is formed from farnesyl
pyrophosphate via 10,11-epoxy farnesoic acid. While they cannot
exclude the possibility, these authors find no evidence for the
intermediacy of methyl farnesoate as claimed by Pratt and coworkers.
The Law group has clearly established the role of S-adenosyl methionine in providing the methyl carbon for esterification of epoxy
farnesoic acid to JH III. Furthermore, they show that the epoxidation of farnesoic acid requires NADPH and molecular oxygen. The
same cell-free in vitro system which converts farnesyl pyrophosphate to JH III is capable of converting the appropriate homologs
of farnesyl pyrophosphate to JH I and II.

The article by Pratt et al. describes a device for the continuous monitoring of labeled JH production by corpora allata in
vitro in medium containing labeled precursors. Using their perfusion chamber the authors present preliminary data from studies
with Periplaneta americana that demonstrate a regular decline in
gland activity with time in vitro. Initial rates of activity in
vitro depend upon the intrinsic level of activity in the glands at
the time of extirpation. Glands whose activity level had declined
with time in vitro could be stimulated to produce increased amounts
of JH if farnesoic acid was provided in the medium.

In their discussion the authors raise several cogent points
concerning the value of their perfusion culture system to studies
of the control of JH production. They state that the lack of any
provision for a gas phase near the glands as well as the inability
of any culture medium to duplicate the changing conditions in
normal hemolymph may impose serious limitations on this method.

Furthermore, normal regulation of corpus allatum activity very likely involves transmission of nerve impulses and/or neurosecretory substances along axons leading to and from the glands, but these connections are severed in the in vitro system. Many other factors could conceivably affect the rate of hormone production in vivo or in vitro but the authors do recognize that "experimental solutions to these problems are unlikely to be simple."

The final article in this section, that by Tobe and Pratt, concerns the significance of a ratio between "stimulated" and "unstimulated" in vitro corpus allatum activity, and an attempt to localize the site of JH production by autoradiography. With regard to the first topic the authors impart considerable significance to what they quaintly call "FEAR," the "Fractional Endocrine Activity Ratio." Simply stated, FEAR is a ratio between the amount of JH produced in a given period of time by Schistocerca gregaria corpora allata under certain in vitro conditions, and the amount of JH produced by the same or similar glands in medium containing farnesoic acid. Although this technique for artificially stimulating JH III production is interesting, its value for studies on the natural rate of gland activity (and hence the concept of the FEAR ratio) should be seriously questioned. In the second part of their paper Tobe and Pratt present an autoradiographic study suggesting that JH is made in all of the cells of the corpus allatum, and that it is probably not manufactured in their nuclei. However, because of the obvious limitations inherent in the method, the conclusions presented here can be regarded as no more than preliminary.

It was the general feeling of those attending the Conference that the areas of JH identification and biosynthesis should progress very rapidly in the next few years. Publications describing qualitative and quantitative titers of JH in representative insects should be forthcoming and with them the evidence to support or refute hypotheses concerning differential roles for the three JHs.

The ability to accurately measure JH titers in hemolymph should advance our understanding of corpus allatum regulation, although this aspect of insect endocrine physiology may prove to be extremely complex. At the very least, useful information will become available as to what constitutes a "physiological" dose of JH for experimental treatments, and how endogenous JH levels may be affected by administration of various JH analogs, inhibitors or other potential control agents.

Concerning JH biosynthesis, again the immediate future looks bright. With more laboratories using in vitro and cell-free systems the identity of the intermediates in JH biosynthesis should soon be defined. Studies on the biochemical properties of

each enzyme in the path could continue almost indefinitely, but
the location of a critical step susceptible to inhibition by
externally applied chemicals could prove most rewarding.

DETERMINATION OF THE PHYSIOLOGICAL LEVELS OF JUVENILE HORMONES IN

SEVERAL INSECTS AND BIOSYNTHESIS OF THE CARBON SKELETONS OF THE

JUVENILE HORMONES

D.A. Schooley, K.J. Judy, B.J. Bergot, M.S. Hall,
and R.C. Jennings

Zoecon Corporation, 975 California Avenue, Palo Alto,
California 94304

INTRODUCTION

The purpose of this paper is to summarize certain studies undertaken in our laboratories which attempt to elucidate the relative physiological importance of the three known JHs as indirectly evidenced by their titers in several life stages of diverse insects, and the biogenetic origin of the carbon skeletons of the structurally unusual homosesquiterpenoids JH I and JH II. The following narrative account of our efforts to answer these questions is a blend of published information, unpublished details, and studies still in progress.

RESULTS AND DISCUSSION

Analysis of the Juvenile Hormones at Physiological Levels by Gas-Liquid Chromatography with Electron Capture Detection

Since our discovery of the third naturally occurring juvenile hormone, the sesquiterpenoid JH III (Judy et al., 1973), reports have appeared on the isolation of the JH of a number of species of insects of diverse orders. These studies, conducted by several research groups using both in vitro and in vivo methodologies, have amassed a considerable body of evidence suggesting that (1) JH I, II and III appear to be the only structural forms of JH in insects, and (2) excluding the Lepidoptera, JH III appears to be the predominant JH. The latter conclusion must be qualified carefully, since only 10+ species of non-lepidopterous insects have

been examined in detail, since most investigations have concentrated on adult female animals, and especially since Lanzrein
et al. (1975) have reported the co-occurrence of JH I, II, and
III in a cockroach.

    The above considerations serve to emphasize the pressing need
for a sensitive, specific assay method for simultaneous qualitative and quantitative determination of the three known juvenile
hormones at physiological levels in insects. We have been seeking
to develop such a method for over two years in order to answer
questions such as: what are the species variations in abundance of
the three known JHs and what are the JH titer changes during the
life cycle of selected insects, not only in terms of total biological activity, but also in terms of relative distribution of the
three hormonally active species?

    In developing such an assay based on chromatographic methods,
we are able to provide a degree of qualitative discrimination
between the three JHs that is absent in classical JH-bioassay
titer methods. However, a strictly physico-chemical assay method
will be "blind" to detection of any new juvenile hormone structures unless such new hormones are certain simple homologs or
isomers of JH I-III. However, if supplemented by bioassay data,
such a chromatographic analysis could conceivably identify insects
which produce still unknown forms of JH, that is, by demonstrating
a lack of the known JHs or by marked differences between measured
titers of the three known JHs and the total JH activity in the
insect as determined by bioassay techniques.

    There is little published information concerning prior
methodology for such an assay. Bieber et al. (1972) have described
an isotope dilution assay utilizing gas-liquid chromatography-mass
spectrometry (GLC-MS) to determine the content of JH I (only) in
H. cecropia. Use of the mass spectrometer as a highly specific
detection device was provided by focusing only on ions highly characteristic of JH, the so-called multiple ion detection (MID) mode
of operation. Trautmann et al. (1974a;b; and this volume) have
described a method for determination of JH I, II, and III in
various insects using isotope dilution analysis for quantitation.
Their method relies on large scale extraction of kilogram quantities of insect bodies, followed by isolation procedures similar to
those of Dahm and Röller (1970), with final qualitative and quantitative analysis achieved by capillary GLC/MS.

    A GLC-electron capture detection (ECD) method has been reported
that demonstrated the feasibility of introducing electrophoric
(electron capturing) moieties into the JH molecule (which of itself
has little inherent EC properties), allowing highly sensitive
analytical detection. This procedure, developed by L.L. Dunham of
Zoecon Corporation and described in preliminary form by Judy et al.

(1973), employed acidic hydration of the epoxy ring in JH to form the diol derivative, followed by acylation of the diol to the strongly electrophoric bis-trifluoroacetate (TFA) derivative. With this method, the probable presence of JH III in <u>Manduca</u> <u>sexta</u> hemolymph was demonstrated. However, the bis-TFA derivative is hydrolytically unstable, and the conditions used for chemical hydration (1:1 or 5:3 tetrahydrofuran-water containing 0.02 M $HClO_4$ or H_2SO_4) have been found by us to give inexplicably erratic yields of diol in the microgram-submicrogram scale, accompanied by non-polar products suggestive of the intramolecular acid-catalyzed cyclization of synthetic epoxy methyl farnesoate (JH III) (van Tamelen, 1968). Other investigators (Hammock, 1975, and personal communication) have confirmed the erratic results obtained with this reaction on a micro-scale, in direct contrast to the smooth conversion noted on the milligram scale.

We now report our progress in developing an improved GLC/ECD assay for the three known JHs. The method, outlined in Fig. 1, has achieved our aim of detecting picogram quantities of JH, thereby requiring only a few grams of insect bodies or a few milliliters of hemolymph. As with any gas chromatographic analysis employing acylation of a hydroxyl group by an electrophoric functionality, specificity of analysis remains the chief problem, but sensitivity of the method appears totally adequate for the biological investigations currently contemplated. Due to the dramatic fluctuation in relative titers of different lipid classes even within the life span of a single insect species, development of a <u>generally</u> <u>applicable</u> method of assay for these lipoidal hormones is extremely challenging and problematical. We are presently analyzing a number of different life stages of selected insects from several orders to determine the breadth of applicability of our method, and full experimental details will be published when it is apparent that a reliable, general method is at hand.

The details of our most promising scheme are, however, outlined as follows: Measured quantities of insect blood or bodies (1-10 ml hemolymph, 5-20 g bodies) are extracted with methanol (or acetonitrile) to which is added an internal standard, the n-propyl ester of JH I containing tritium radiolabel. The standard [prepared by transesterification of carrier-diluted [7-<u>ethyl</u>-^{3}H]-JH I (New England Nuclear) according to the procedure of Mori <u>et al</u>. (1973)] allows determination of recoveries at each step, and simultaneously provides a detectable peak for quantitation purposes in the final GLC/ECD analysis. The filtered extract is partitioned between pentane and water, a procedure inspired by Miller <u>et al</u>. (1975) who utilized it as the first step in a broadly applicable residue analysis scheme for the commercial JH analog methoprene. The concentrated pentane extract contains tri-, di-, and monoglycerides, sterols and their esters, free fatty acids, and (presumably) cuticular hydrocarbons in addition to JH. The biomass of crude lipid

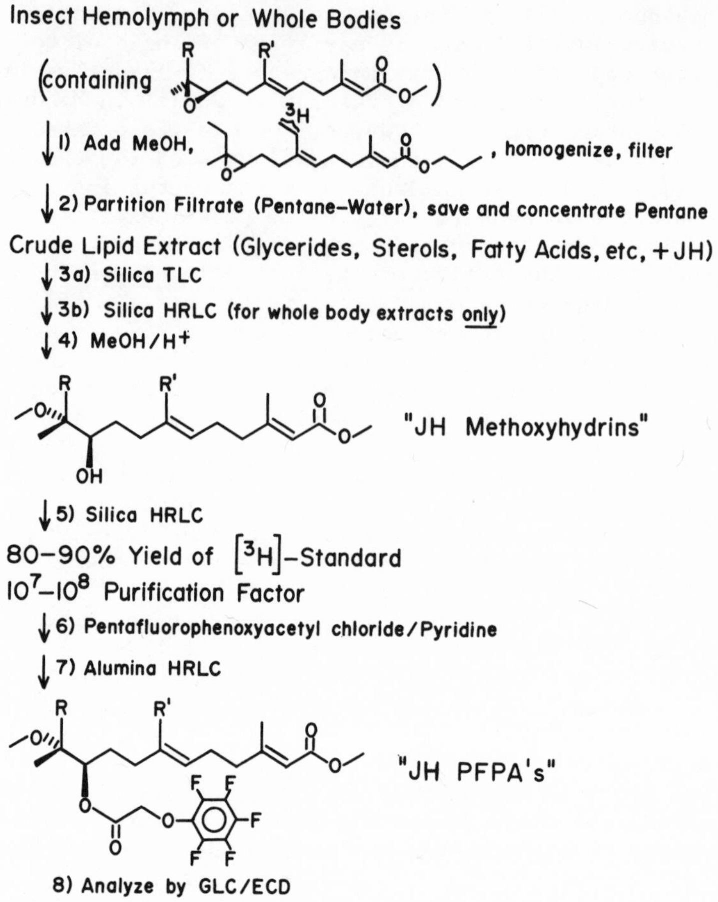

Insect Hemolymph or Whole Bodies

Fig. 1. Flow diagram of steps involved in GLC/ECD analysis for the three known juvenile hormones.

usually comprises only 1-2% of the wet weight of whole insect bodies, an efficiency far exceeding previous methods involving ether extraction and low temperature precipitation.

The extract is next subjected to a separation by lipid classes on silica gel TLC, and after elution of the JH zone, hemolymph samples are ready for derivatization, but whole body extracts usually require additional purification. The latter samples are additionally subjected to high-resolution liquid chromatography (HRLC) using a microparticulate silica column. We prefer HRLC separations for this application due to higher resolution and recoveries than TLC; however, one preliminary TLC step is deemed necessary to remove polar substances that would accumulate in the

HRLC column and gradually degrade performance. In this and subsequent HRLC procedures, the JHs (or derivatives) are isolated as a group, rather than as individual components. We feel that this approach is distinctly safer than a method requiring "blind" separation of the three JHs according to their anticipated retention volumes.

The purified JH zones are subjected to methanolysis (see Boar and Damps, 1973) and the resultant 11-methoxy-10-hydroxy derivatives ("methoxyhydrins") subjected to a second HRLC purification. The polarity change afforded by this reaction allows the vast majority of remaining lipid to be easily separated from the methoxyhydrins.

Acylation of the purified methoxyhydrins with pentafluorophenoxyacetyl chloride in pyridine yields the corresponding esters. As in any ECD/derivatization procedure, purification of the pentafluorophenoxyacetate derivatives (PFPAs) is mandatory since even nanogram amounts of undesired substances bearing primary or secondary hydroxyl groups will become "tagged" with the electrophoric reagent. Alumina HRLC is effective for purifying these derivatives.

Qualitative and quantitative analysis is achieved using a "hybrid" bio-medical gas chromatograph (Hewlett-Packard Model 402) equipped with glass columns and Nickel-63 electron capture detectors (Tracor, 15 mCi; Tracor electrometer required). The JH-PFPA derivatives are well resolved on standard 3% loaded stationary phase columns. Of many candidate electrophoric derivatives, the PFPAs appeared to be among the very most sensitive, and with standard solutions, one (1) picogram can be detected with signal to noise ratio of about 6 or 8:1. It is advisable to chromatograph processed biological samples on two or more stationary phases of markedly different selectivity in order to ensure a valid result.

Table 1 summarizes some of the preliminary assay results obtained to date. It should be noted that problems do remain with the method; for example, with hemolymph of newly emerged adult female Manduca sexta an unknown contaminant obscures the apparent presence of at least some JH I on all GLC columns used. However, we are able to confirm the presence of JH II and III in M. sexta adult female hemolymph in roughly equal concentration (2 ng/ml), an in vivo result confirming the in vitro synthesis of JH II and III by M. sexta corpora allata (CA). Manduca pupae are seen to be essentially devoid of JH, while larvae contain appreciable titers; in particular, IV instar day 1 larvae contain high titers of JH I, II, and III. The predominance of JH I and II over JH III in these larvae is especially interesting compared to the situation in adults, and could be viewed as support of the hypothesis of Lanzrein et al. (1975), who suggest that JH I and II play predominantly morphogenetic

TABLE 1. JH TITERS OBSERVED IN SEVERAL INSECTS BY GLC/ECD.

Species	Stage[a]	Method[b]	JH I	JH II	JH III
Manduca sexta	IV, day 0-1	WB	0.57 ng/gm	0.98	0.3
	IV, day 1	H	3.4 ng/ml	2.8	0.7
	V, day 1	H	0.6 ng/ml	0.4	0.4
	V, day 7 (prepupae)	H	0.3 ng/ml	0.3	0.3
	pupae, day 8	H	<0.004[c] ng/ml	0.04	0.1
	adult ♀, 0-24 hr	H	$\leq$0.2[d] ng/ml	2	2
Samia cynthia	IV	H	0.08	0.82	0.16
	V	H	<0.01	0.12	0.03
Diatraea grandiosella	diap. larvae	H	1.0	3.1	1.2
	VI, non-diap. larvae	H	$\leq$0.07	0.15	0.5
	pupae	WB	$\leq$0.05 ng/gm	0.05	0.13
Sarcophaga bullata	adult ♀	WB	0.21 ng/gm	0.33	0.88
	II	WB	0.17 ng/gm	0.07	0.67
Musca domestica	adult ♀	WB	0.15 ng/gm	0.56	0.66

[a] Roman numerals indicate larval instar; diap. = diapausing.

[b] WB = whole body extraction (titer in nanograms/gram), H = hemolymph extraction (titer in nanograms/milliliter).

[c] <0.004 ng/ml = <1.4 x 10^{-11} M (<14 pM).

[d] Interference observed due to unknown peak, even on reanalysis on various GLC stationary phases.

roles in the pre-adult insect, while JH III has primarily a gonadotropic function in adults.

We now report the first results available on dipteran JH.
Low titers of JH I, II, and III were found in Musca domestica
(adult females) and Sarcophaga bullata (adult females and larvae).
The Sarcophaga data, however, do not appear to support the Lanzrein
hypothesis; in fact, JH III predominates in both larvae and adult
Sarcophaga.

Analysis of Samia cynthia larvae is quite interesting, as the
hemolymph contains mostly JH II, with JH I and III occurring at
only 10% and 20% respectively of the titer of JH II. Using specific
biological activities of 5 pg/Galleria unit (GU) for JH I and II,
and 70 pg/GU for JH III, we calculated a titer of (only) 4500 GU/ml
in IV instar larvae, ten times lower than the bioassay titer determinations of de Wilde et al. (1971).

Clearly, more data are necessary from many more insect species
before a definitive statement can be made regarding the relative
physiological importance of the three hormones.

Biosynthesis of the Carbon Skeletons of the
Homoisoprenoid Juvenile Hormones

Studies on the biosynthetic origin of the unusual carbon
skeletons of the homosesquiterpenoids JH I and II were undertaken
by us several years ago out of basic curiosity concerning the
origin of the "extra" carbon atoms. As the only known homoterpen-
oids in nature, the existence of these compounds represents a
considerable novelty in the annals of biosynthesis. However, it
is evident from the preceding section that such studies assume
added interest since at least certain insect species in the course
of development regulate not only the overall quantitative JH titer,
but also the qualitative balance between the three hormones. It
seems that the corpora allata possess an elaborate enzyme system
capable of controlling this relative balance of homoisoprenoid
and isoprenoid hormones secreted in the blood at any given developmental stage.

The in vitro system used to produce, isolate, and identify
JH II and JH III from M. sexta (Judy et al., 1973) afforded an
ideal opportunity to investigate simultaneously the biosynthesis
of the homosesquiterpenoid JH II along with its more "structurally
normal" congener JH III. We were able to demonstrate (Schooley
et al., 1973) the incorporation of propionate, mevalonate, and
acetate into the carbon skeleton of JH II, and mevalonate and
acetate but not propionate into JH III. Partial degradations were
performed (as shown in Figures 2, 3, and 4) to show that distribution

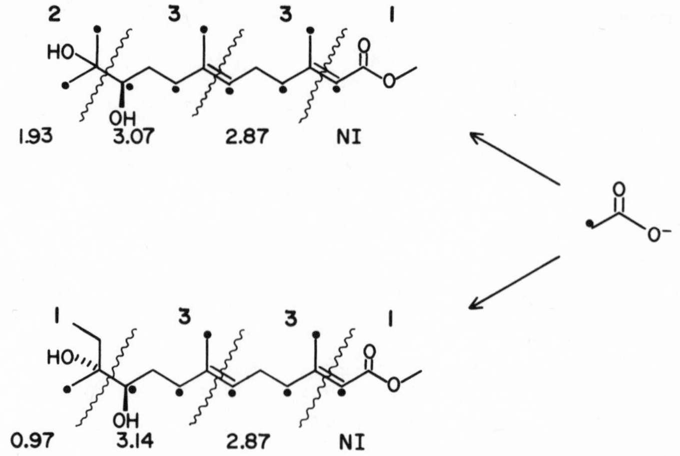

Fig. 2. Results of partial degradation of JH II and III biosynthesized from [2-^{14}C]acetate (summarized from Schooley et al.,
1973). Cleavage of each molecule gave three separately isolable
fragments, and determination of specific radioactivity of each
fragment allowed comparison of theoretical values (above structures)
for label distribution to experimentally determined distribution
(values below structures). One fragment of the molecule is non-
isolable (NI).

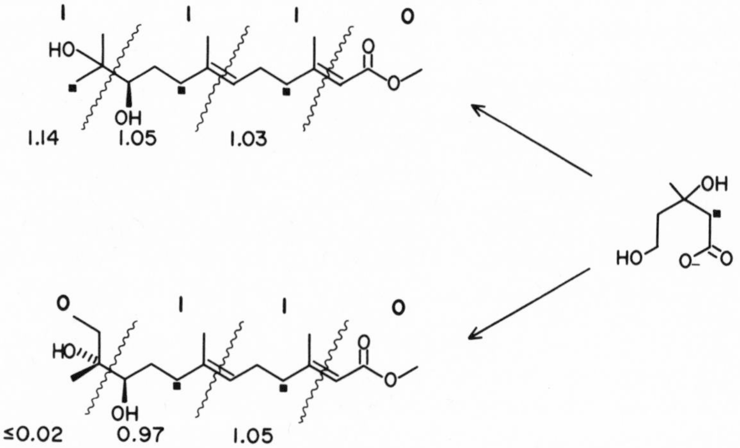

Fig. 3. Results of partial degradation of JH II and III biosynthesized from [2-^{14}C]mevalonate (other details in legend of
Fig. 2).

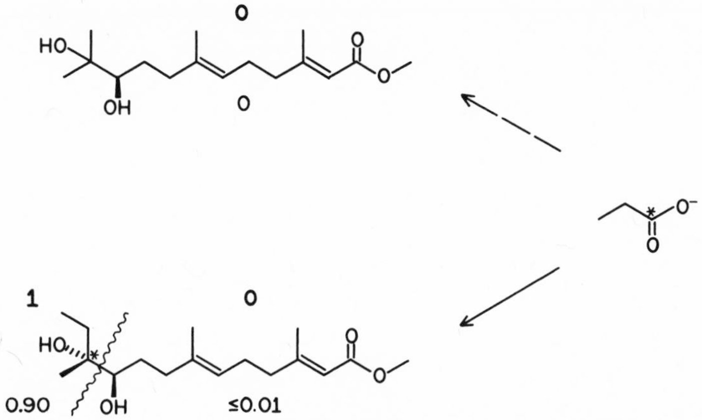

Fig. 4. Results of partial degradation of JH II biosynthesized from [1-^{14}C]propionate (other details in legend of Fig. 2).

of radiolabel was in agreement with a hypothesis in which the "extra" carbon atom of JH II had its origin in propionate via the probable intermediacy of homomevalonate. Thus a synthesis of all data from Figures 2-4 suggest that the skeleton of JH II is formed from one propionate and eight acetate units. Since only two units of mevalonate (= six acetates) are utilized to create this chain, it follows that the remaining two acetate units and propionate form the "missing" homoisoprenoid segment. On the other hand, the carbon skeleton of JH III would appear to be fabricated in usual isoprenoid fashion.

We subsequently extended these studies to investigate the origin of the bishomosesquiterpenoid carbon skeleton of JH I. Using the in vitro methodology, it was shown that corpora allata (CA) cultures of Heliothis virescens produce JH I and II in an approximate 1:3 ratio, and that [1-^{14}C] propionate incorporates into both hormones (Jennings et al., 1975b). Partial degradation of JH I and II biosynthesized from [1-^{14}C]propionate gave results (see Fig. 5) in accord with the supposition that the 7-ethyl group in JH I also had its origin in propionate (as well as the 11-ethyl groups in both hormones). This finding was independently verified by Peter and Dahm (1975) who performed more rigorous degradation of JH I biosynthesized in vivo by Hyalophora cecropia from [^{14}C]propionate. They demonstrated that specifically C-7 and C-11 of the JH I skeleton arise from C-1 of propionate.

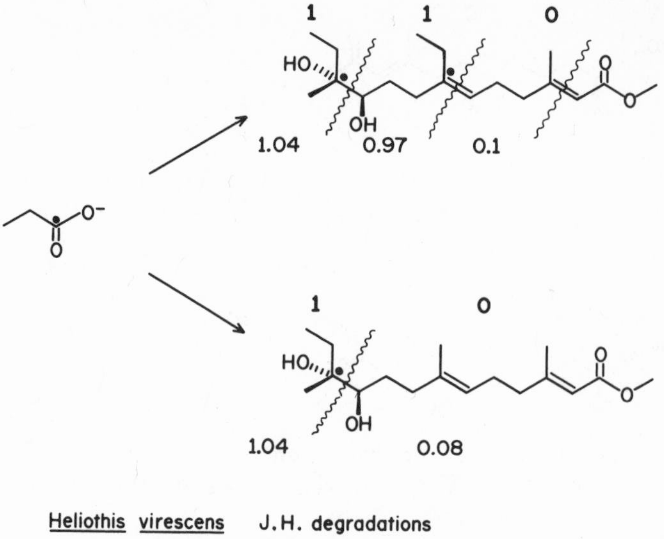

Heliothis virescens   J.H. degradations

Fig. 5. Results of partial degradation of JH I and II biosynthesized from [1-^{14}C]propionate by cultured corpora allata of
Heliothis virescens (summarized from Jennings et al., 1975). Comparison of theoretical vs experimental radiolabel distribution data
is denoted as in Fig. 2-4.

Our earlier hypothesis for JH II biosynthesis from propionate
is presented in expanded form in Fig. 6, and is based on simple
homologs of intermediates whose existence has been exhaustively
investigated in studies on isoprenoid biosynthesis. Combinations
of four subunits (two known and two hypothetical) should suffice
for enzymic generation of the carbon skeletons of all three hormones.
In addition to the ubiquitous mevalonate-derived isopentenyl pyrophosphate (IPP) and dimethylallyl pyrophosphate (or monoprenylpyro-
phosphate, [MpPP]), the scheme supposes the existence of the homologous compounds Z-3-methyl-2-pentenyl pyrophosphate (HMpPP, "homo-
monoprenyl pyrophosphate") and 3-ethyl-3-butenyl pyrophosphate
(HIPP, "homoisopentenyl pyrophosphate"). Varying amounts of these
precursors in the presence of the enzyme farnesyl pyrophosphate
synthetase (also termed prenyl transferase), leads first to diprenyl
pyrophosphate (DpPP) or its mono- or bis-homologs. Next, prenyl
transferase catalyzed condensation of DpPP, HDpPP, or BHDpPP with
IPP will yield triprenyl (farnesyl) pyrophosphate or its homologs,
compounds possessing the skeletal structure of JH III, II, and I.
Ample evidence suggests that this is not an unreasonable scheme,
since detailed studies on the substrate specificity of prenyl transferase from both plant and animal sources by several groups
(primarily that of Ogura at Tohoku University) have shown that

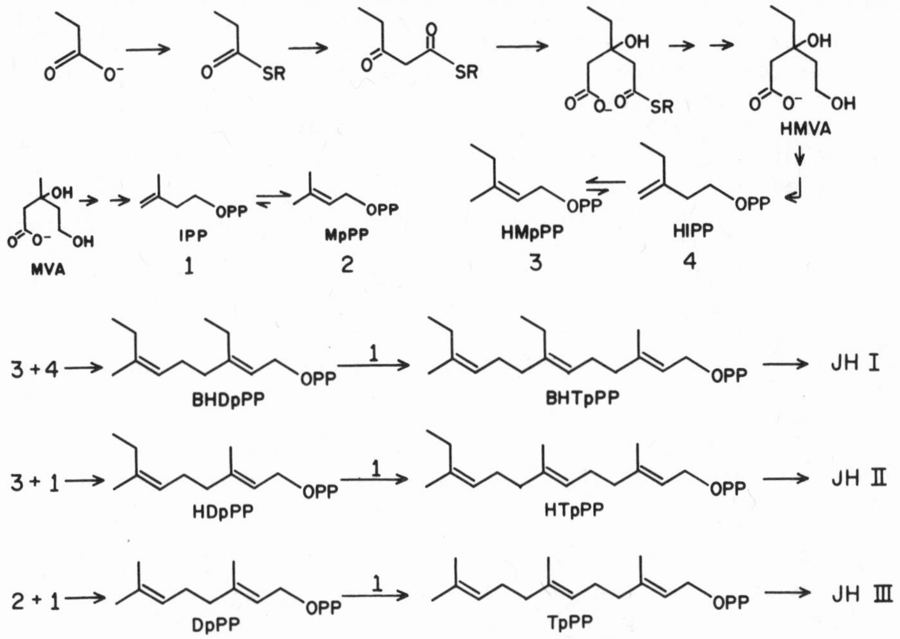

Fig. 6. Hypothetical scheme for carbon skeleton formation for the three known juvenile hormones. 1 and 2 are known isoprenoid intermediates. 3 and 4 are hypothetical homoisoprenoid intermediates. Propionate presumably is activated as the coenzyme A or other thioester.

numerous higher homologs of MpPP (in addition to HMpPP) are accepted by prenyl transferase. In addition, Ogura et al. (1972) have shown that pig liver prenyl transferase will convert MpPP and HIPP, (but no other higher homologs of IPP) to homologs of di- and tri-prenyl pyrophosphates (Fig. 6) bearing methyl at the left-hand terminus. The same group incubated pig liver prenyl transferase with HMpPP, HIPP, and IPP, and were able to isolate HTpPP (Fig. 6). Under somewhat forcing conditions even BHTpPP and a homolog bearing an ethyl group at C-3 were found (Koyama et al., 1973).

It is tempting to speculate that the structural identity of JH produced by the corpora allata is a function predominantly of the precursors (1-4 in Fig. 6) made available to prenyl transferase. Separate control of enzymes for the synthesis of IPP-MpPP versus HIPP-HMpPP would provide a simple means for modulation of qualitative JH levels during development. However, the substrate specificities and other properties of certain key enzymes would also play an important role in qualitative regulation. For example, no one has yet reported a juvenile hormone bearing an ethyl group at C-3, nor has an isomer of JH II bearing ethyl at C-7 and methyl at C-11 been detected. It is conceivable that traces of such compounds co-occurring with JH I and/or II could have so far escaped detection. However, it may also be that prenyl transfer-ase of insect CA will not readily condense HIPP with DpPP (or its homologs) to produce 3-ethyl branched JHs. Prenyl transferase is believed to have separate sites for formation of di- and tri-prenyl pyrophosphates (Ogura et al., 1969; 1974; Reed and Rilling, 1975), and the site for formation of triprenyl units in the CA enzyme could be highly specific in accepting only IPP as the chain extending unit. With regard to the apparent lack of a natural JH having ethyl at C-7 and methyl at C-11, it may be speculated that the appropriate prenyl transferase will accept HIPP as an extending unit only when HMpPP is the initiating unit. Alternatively, the isomerase interconverting HMpPP and HIPP may strongly favor forma-tion of the chain initiating HMpPP unit. Holloway and Popják (1968) have shown that pig liver isomerase favors the formation of MpPP over IPP by 9:1 at equilibrium.

However, such speculation is premature, especially since there is no firm proof for the natural occurrence of HMpPP or HIPP in insect sources. As we feel this is one of the most important questions in JH biosynthesis, our own research efforts are now focused on these suspected homoisoprenoid intermediates.

In a recent communication we described the specific incorpo-ration of [5-^{3}H]-RS-homomevalonate into JH II produced by M. sexta CA in vitro (Jennings et al., 1975a). Demonstration of such incor-poration does not, however, prove the natural intermediacy of homomevalonate. We now describe additional details of our studies on homomevalonate (HMVA).

Using in vitro corpora allata cultures of M. sexta, we sought to investigate the effects of added, non-radioactive HMVA on JH II and III production from [1-^{14}C]propionate and L-[methyl-^{3}H]methio-nine. The results are summarized in Table 2. Control cultures lacking HMVA were incubated, extracted, and analyzed according to our published procedures. Isolated hormones were radioassayed by double label counting. Tritium incorporation from methionine indicated a 2.1:1 ratio of JH II:III representing a slightly "stimulated" production of JH II in the presence of propionate. The S-methyl group of methionine has previously been shown to incorporate into the methyl of the ester moiety of JH II and III with essentially no dilution of specific radioactivity (Judy et al., 1973) and hence serves as a quantitative indication of the nano-moles of hormones produced. The dilution of specific radioactivity of propionate on incorporation into JH II can be readily calcu-lated based on the known specific radioactivities of the precursors.

TABLE 2. EFFECT OF NON-RADIOACTIVE HOMOMEVALONATE ON JH PRODUCED
BY CA CULTURES OF <u>MANDUCA SEXTA</u> FROM LABELED PRECURSORS.

Precursors	Products	
Control		
[1-^{14}C]Propionate (0.1 mM, 48 mCi/mmol)	[^{3}H]JH III	1
L-[<u>methyl</u>-^{3}H]Methionine, (0.34 mM, 145 mCi/mmol)	[^{3}H,^{14}C]JH II $$\frac{0.41 \text{ nmol } [^3\text{H}]}{0.34 \text{ nmol } [^{14}\text{C}]} = 1.2$$	2.1
Control + Homomevalonate (1 mM)	[^{3}H]JH III	1
	[^{3}H,^{14}C]JH II $$\frac{0.53 \text{ nmol } [^3\text{H}]}{0.23 \text{ nmol } [^{14}\text{C}]} = 2.3$$	5.5

Numbers to the right of product indicate ratios of occurrence based
on [^{3}H] content. Calculation below JH II gives relative dilution
of propionate incorporation relative to methionine.

The observed utilization of precursors for synthesis of the isolated
JH II implies a dilution value of only 1.2 for propionate. Similar
analysis of simultaneously incubated cultures, differing only by
the addition of 1.0 mM homomevalonate as the potassium salt gave
the following results: the ratio of JH II:III increased to 5.5:1,
while the dilution value for incorporation of [1-^{14}C]propionate
into JH II increased to 2.3. Homomevalonolactone re-isolated from
glands and culture medium was found to be devoid of any radioactivity. Although this experiment failed to trap any [^{14}C]HMVA, it
did show that exogenous HMVA could <u>partially</u> suppress propionate
incorporation into JH II. These two results suggest that if HMVA
is a natural intermediate, it cannot easily enter the natural pool.
Not only does HMVA (1 mM) stimulate JH II production with respect

to JH III, but a doubly labeled substance was recovered with HRLC
retention volume identical to JH I, present in about half the molar
amount of JH III. This radioactive band was not rigorously characterized as JH I. It is noteworthy that in the control double
label experiment, propionate incorporated very efficiently into
JH II (dilution value = 1.2).

We subsequently prepared [5-^{3}H]homomevalonolactone with
specific activity ~4500 mCi/mmol. Presenting this substrate, as
HMVA potassium salt, to M. sexta CAs under standard in vitro
conditions, together with L-[methyl-^{14}C]Met, gave doubly labeled
JH II. Incorporation of [5-^{3}H]HMVA is apparently specific
(Jennings et al., 1975a), providing further evidence that this
substrate is a natural intermediate. We now report additional
details of these incorporation studies. As shown in Table 3, the
ratio of specific radioactivities of precursors ([5-^{3}H]HMVA and
L-[methyl-^{14}C]Met) was 98:1, yet the [^{3}H]/[^{14}C] ratio in isolated
biosynthetic JH II was found to be 1.02.

This indicates a dilution value of 96 for labeled HMVA
incorporation as compared to 1.2 for propionate. Exogenous HMVA
may not be able to enter the "natural pool" within the gland due
to cell membrane permeability barriers. On the other hand, HMVA
could exist only as an enzyme-bound intermediate, or the natural
homoisoprenoid precursor may be some unknown metabolite of HMVA.
At this time we favor the hypothesis that HMVA has difficulty with
membrane permeability.

To circumvent some of the problems discussed above, we have
begun to investigate the utility of cell-free corpus allatum
homogenates as developed by Reibstein and Law (1973) and
Reibstein (1974) for studying the nature of the homoisoprenoid
intermediates produced by M. sexta. We find production of approximately equal amounts of JH II and III (only) by M. sexta CA
homogenates and/or 12,000 x g supernates supplemented with
[methyl-^{3}H]-S-adenosyl-L-methionine (SAM). This has been borne
out through several replications and is in contrast to the isolation of JH I, II and III reported by Reibstein (1974). We
attribute this discrepancy to the different analytical methodologies employed and believe that gas-liquid radiochromatography as
used by Reibstein is more likely to generate artifacts. We find
the rate of [^{3}H]JH production, in the absence of any added carbon
skeleton source such as farnesoic acid, to be in accord with that
reported by Reibstein, namely, about one picomole per gland pair per
day (gpd). However, this is far lower than the 28-33 pmol/gpd
found by Judy et al. (1973) for intact M. sexta glands. Addition
of dithiothreitol (1 mM) to CA homogenate incubations failed to
enhance the rate of production. Incubation of [7-ethyl-^{3}H]JH I
(10 pmol) with a homogenate of 10 CA revealed rapid degradation of
[^{3}H]JH I to predominantly one substance with TLC mobility similar

TABLE 3. PRODUCT/PRECURSOR RELATIONSHIP FOLLOWING ADDITION OF [5-^{3}H]HMVA TO CA CULTURES OF MANDUCA SEXTA.

Precursors	Products
[5-^{3}H]Homomevalonate, ~4500 mCi/mmol	[methyl-^{14}C]JH III (devoid of [^{3}H])
L-[methyl-^{14}C]Methionine, 48 mCi/mmol	[methyl-^{14}C, 9-^{3}H] JH II
Specific activity ratio of homomevalonate/methionine = 98	JH II product: [^{3}H/^{14}C]ratio = 1.02; i.e., homomevalonate dilution on incorporation = 98/1.02 = 96

Dilution of HMVA occurring on incorporation is determined by comparing product and precursor specific activity ratios, using L-[methyl-^{14}C]methionine incorporation as a mass marker.

to that of epoxyfarnesoic acid; 45% of the JH I being metabolized in 4 hr and 85% in 20 hr. Addition of the esterase inhibitors diisopropylparaoxon (0.01 mM, recommended and provided by B.D. Hammock) or diisopropylfluorophosphate (DFP, 0.2 mM) failed to enhance [^{3}H]JH production measurably. This suggests that the esterase liberated on corpus allatum homogenization is not a general esterase, but rather resembles the JH-specific esterase shown to be resistant to DFP inhibition by Sanburg et al. (1975). A similar result has been obtained with CA homogenates of the cockroach Blaberus giganteus (Hammock, 1975). Why homogenization of the gland responsible for synthesizing JH should release an enzyme specific for JH degradation is open to conjecture, but since the CAs are removed together with the corpora cardiaca and a portion of the aorta, there is a possibility that the CAs may not be the source of the esterase.

In a further attempt to stabilize the biosynthesized JH by trapping, we have added 0.5 µg each of JH I, II, and III to 100 µl of CA homogenate. Yield of [^{3}H]JH was increased, but HRLC analysis revealed the presence of radioactive JH I, II, and III in approximate 6:3:2 ratio. This would seem to be due to hydrolysis of exogenous JH by an esterase, and subsequent re-esterification by methyl transferase using [^{3}H]SAM. It could also imply that JH I is hydrolyzed by the esterase faster than JH II and III, or that the corresponding acid is methylated more readily.

The cell-free system thus has its drawbacks in JH biosynthetic studies, but if attention is focused on the earlier stages of biosynthesis, it may prove quite adequate. Further investigations along these lines are in progress.

SUMMARY

Preliminary details of an extremely sensitive assay for JH I, II, and III at physiological levels are described. The method employs efficient extraction and chromatographic purification steps. Two micro-derivatization reactions convert the epoxide ring of the juvenile hormones to their respective 11-methoxy-10-pentafluorophenoxyacetate esters, allowing detection of picogram quantities by gas-liquid chromatography with electron capture detection. Studies on the biosynthetic origin of the carbon skeletons of the juvenile hormones, especially JH II, are discussed, with particular emphasis on the involvement of propionate and homomevalonate as precursors of the ethyl side branch. New details are described indicating the possibility of low permeability of the corpus allatum to homomevalonate. Some details are given of attempts to improve rate of production of JH by corpus allatum homogenates.

ACKNOWLEDGEMENTS

We are grateful to L.L. Dunham, J.B. Siddall, and H. Röller for helpful discussion, and to L.E. Staiger for drafting the figures. Extracts of D. grandiosella were kindly provided by G.M. Chippendale, University of Missouri, as part of an ongoing collaborative study. We thank D. Moss, A. Breaux, and R. Troetschler for providing other insect samples. Partial support of this work by the National Science Foundation (Grant #BMS74-19048) is gratefully acknowledged.

REFERENCES

Bieber, M.A., Sweeley, C.C., Faulkner, D.J., and Peterson, M.R., 1972, Anal. Biochem. 47:264.

Boar, R.B., and Damps, K., 1973, Chem. Commun. 1973:115.

Dahm, K.H., and Röller, H., 1970, Life Sci. 9:1397.

Hammock, B.D., 1975, Life. Sci. 17:323.

Holloway, P.W., and Popják, G., 1968, Biochem. J. 106:835.

Jennings, R.C., Judy, K.J., and Schooley, D.A., 1975a, Chem. Commun. 1975:21.

Jennings, R.C., Judy, K.J., Schooley, D.A., Hall, M.S., and Siddall, J.B., 1975b, Life Sci. 16:1033.

Judy, K.J., Schooley, D.A., Dunham, L.L., Hall, M.S., Bergot, B.J., and Siddall, J.B., 1973, Proc. Nat. Acad. Sci. USA 70:1509.

Koyama, T., Ogura, K., and Seto, S., 1973, Chem. Lett. 1973:401.

Lanzrein, B., Hashimoto, M., Parmakovich, V., Nakanishi, K., Wilhelm, R., and Lüscher, M., 1975, Life Sci. 16:1271.

Miller, W.W., Wilkins, J.S., and Dunham, L.L., 1975, J. Assoc. Off. Anal. Chem. 58:10.

Mori, K., Tominaga, M., Takigawa, T., and Matsui, M., 1973, Synthesis 1973:790.

Ogura, K., Koyama, T., and Seto, S., 1969, Biochem. Biophys. Res. Commun. 35:875.

Ogura, K., Koyama, T., and Seto, S., 1972, Chem. Commun. 1972: 881.

Ogura, K., Saito, A., and Seto, S., 1974, J. Am. Chem. Soc. 96:4037.

Peter, M.G., and Dahm, K.H., 1975, Helv. Chim. Acta 58:1037.

Reed, B.C., and Rilling, H.C., 1975, Biochem. 14:50.

Reibstein, D., 1974, Ph.D. Dissertation, University of Chicago.

Reibstein, D., and Law, J.H., 1973, Biochem. Biophys. Res. Commun. 55:266.

Sanburg, L.L., Kramer, K.J., Kézdy, F.J., and Law, J.H., 1975, J. Insect Physiol. 21:873.

Schooley, D.A., Judy, K.J., Bergot, B.J., Hall, M.S., and Siddall, J.B., 1973, Proc. Nat. Acad. Sci. USA 70:2921.

van Tamelen, E.E., 1968, Acc. Chem. Res. 1:111.

Trautmann, K.H., Schuler, A., Suchý, M., and Wipf, H.-K., 1974a, Z. Naturforsch. C. 29:161.

Trautmann, K.H., Masner, P., Schuler, A., Suchý, M., and Wipf, H.-K., 1974b, Z. Naturforsch. C. 29:757.

de Wilde, J., de Kort, C.A.D., and de Loof, A., 1971, "Proceedings of the First Swiss Symposium on Juvenile Hormones," special issue of Mitt. Schweiz. Entomol. Ges. 44:79.

ISOLATION AND IDENTIFICATION OF JUVENILE HORMONES BY MEANS OF A RADIOACTIVE ISOTOPE DILUTION METHOD: EVIDENCE FOR JH III IN EIGHT SPECIES FROM FOUR ORDERS

K.H. Trautmann, M. Suchý, P. Masner,
H.-K. Wipf,* and A. Schuler

Dr. R. Maag AG, Dielsdorf, Switzerland and
Socar AG,* Dübendorf, Switzerland

INTRODUCTION

So far three naturally occurring juvenile hormones (JH) are known: JH I, JH II and JH III. From the extracts of three species of giant silkmoths, Hyalophora cecropia (Röller, 1967; Meyer, 1968), Hyalophora gloveri (Dahm, 1970) and Samia cynthia (Röller, 1972), methyl (2E, 6E)-(10R, 11S)-10,11-epoxy-7-ethyl-3,11-dimethyl-2,6-tridecadienoate (JH I) and the 7-methyl-analog, methyl (2E, 6E)-(10R)-10,11-epoxy-3,7,11-trimethyl-2,6-tridecadienoate (JH II) have been isolated and identified. Judy et al. (1973) extracted methyl (2E, 6E)-(10R)-10,11-epoxy-3,7,11-trimethyl-2,6-dodecadienoate (JH III) in addition to the known JH II from the culture medium of corpora allata of the sphingid moth, Manduca sexta. The JH III has also been found in an orthopteran, Schistocerca (Judy et al., 1973; Pratt and Tobe, 1974), and by Müller et al. (1974) in Periplaneta americana by using in vitro techniques.

These results raise the question as to whether these JH also occur in other insects, particularly from other orders. If not, other unknown hormones must be present. We have therefore developed an analytical method which makes use of radioactive dilution analysis and permits the simultaneous qualitative and quantitative determination of the three presently known hormones. In the following we describe this method and report the presence of JH III in eight insect species from four different orders.

MATERIALS AND METHODS

Animals. Colorado beetles, Leptinotarsa decemlineata, were
reared on potato plants under long day conditions (25°C, 60-70% rh,
18 hr photoperiod). The beetles were in the reproductive phase.

Mealworm beetles, Tenebrio molitor L., were kept in crowded
conditions (25°C, 60-70% rh) and fed on bran. For the extractions,
adults were taken at the age of 4 to 6 and 8 to 16 days, respectively.

Cockchafers, Melolontha melolontha, were caught near Güt-
tingen (Canton Thurgau, Switzerland) in May 1973; they were in the
reproductive phase.

Desert locusts, Schistocerca gregaria, were maintained in mass
colonies (30°C, 50-60% rh) on young wheat plants. For the analysis
only older adults were used.

Three roach species, Blatta orientalis, Leucophaea maderae
and Nauphoeta cinerea were mass reared (25°C, 60-70% rh) on dog
food pellets and water. Only adult roaches in the reproductive
phase were taken for our investigation.

Worker honey bees, Apis mellifera, of various ages were
received from the Swiss Federal Dairy Research Station, Bee Section, Liebefeld, Switzerland.

The silkmoths, Hyalophora cecropia, were obtained as diapausing pupae from California and kept at about 25°C until emergence.

Purification of Solvents and Reagents. All solvents used
were slowly distilled over a 40 cm column packed with glass rings.
Silica gel plates (0.5 mm) and alumina plates (0.25 mm) were
purified with methanol and reactivated either by air drying or
heating for 1 hr at 100°C. In order to avoid contamination with
plasticizers, waterfree sodium sulfate was washed with ether.
The florisil used as an adsorbent for column chromatography was
purified by ethyl acetate in a column and reactivated for 14 hr
at 120°C in vacuo.

Ether Extraction and Chromatographic Purification. Cockchafers
were chosen as the example for the description of extraction procedure and chromatographic purification, the methods being representative for all species investigated. Only in the case of the
giant silkmoth could the purification procedure be simplified
because of the extremely high hormone/lipid ratio of its ether
extract.

A total of 1,200 g of deep frozen Cockchafers were extracted
as follows: 150 ml of diethyl ether and 30 g of sodium sulfate
were added to 60 g of animals and the mixture homogenized for 5
to 10 min in a high speed Virtis homogenizer (Model 45) under ice.
The homogenates were pooled in a 6 L flask and shaken for 2 hr at
a frequency of 150 per minute and then vacuum filtered through a
big fritted glass funnel (2.5 L); the filtration residue was
thoroughly washed twice with 200 and 300 ml of ether, respectively.
After vacuum evaporation of the solvent, a green oil (35 g) was
obtained. After dissolution of the oil in 300 ml of ether and
addition of 1.606 x 10^6 dpm (corresponding to 45 ng) labeled JH I,
a low temperature precipitation at -78°C was performed (Dahm and
Röller, 1970) by adding 400 ml of precooled methanol and filtering
the resulting precipitate at the same temperature. Vacuum evaporation of the filtrate yielded a residue of 6.6 g.

The filtration residue was dissolved in 60 ml of n-hexane
and extracted nine times with 30 ml of hexane saturated acetonitrile. The combined acetonitrile extracts were evaporated in
vacuo and the residue subjected once more to the same partitioning
method by using 50 ml of hexane and eight portions of 30 ml each
of acetonitrile. Residue: 1.56 g.

The residue was put on a methanol as well as chloroform prewashed Sephadex LH-20 column (i.d.: 2.2 cm; length: 145 cm) and
eluted with the latter solvent. After elution of 160 ml, 4.3 ml
fractions were taken. Vacuum evaporation of the radioactive
fractions (fract. 5-15) yielded a residue of 0.13 g, which was
further purified by column chromatography on 35 g of florisil
(i.d.: 1.8 cm; length: 40 cm). After a prerun with 200 ml of 2%
ethyl acetate in n-hexane, the elution was continued with 320 ml
of 5% ethyl acetate in hexane and 10 ml fractions were taken.
Because of the partial separation of JH III from JH I and JH II,
both the radioactive fractions (fract. 5-15, Pool I) and the
following JH III containing fractions (fract. 16-27, Pool II) were
separately pooled. The residue of Pool I and Pool II was 6 and 5
mg, respectively.

For TLC, both residues were spotted separately on a silica
gel plate (Merck F_{254}, 0.5 mm) as 6 cm wide bands and chromatographed twice for the whole distance with n-hexane:chloroform:ethyl
acetate (7:7:1). Detection of the inseparable JH I and JH II was
accomplished by radio-scanning and of JH III by comparison with
the appropriate reference substance. The appropriate zones were
carefully scraped off and the JH III regions of Pool I and Pool II
combined. After elution with ethyl acetate and vacuum evaporation
of the solvent, the residues were re-chromatographed on an alumina
plate (Merck F_{254}, 0.25 mm) for the whole distance twice with
benzene:ethyl acetate (19:1). Detection of the JH regions was
carried out as described above and again the scrapings of the JH III

region of Pool I and Pool II were combined. Elution was performed with only small amounts of ethyl acetate (1-2 ml) in small elution tubes which contained a wadding plug thoroughly prewashed with chloroform as well as with ethyl acetate. The eluates were evaporated to small volumes in vacuo (water bath temperature ca. 20°C) and then transferred to small test tubes (i.d. 4.5 mm; length 60 mm) with the aid of glass capillaries. Solvent evaporation was accomplished at room temperature in an evacuated desiccator, and each residue was dissolved in exactly 50 µl of hexane. For the determination of radioactivity yield in the JH I and JH II containing Pool I hexane solution, an aliquot of 5 µl was taken and the radioactivity measured by standard liquid scintillation counting. Yield of radioactivity: 67.1%. Both GC-injection solutions were stored at -25°C.

GC-Analysis. Qualitative and quantitative determination of the three hormones in the hexane solutions was carried out by high resolution glass capillary GC on two different columns (Ucon and Emulphor). Sample injection on the Ucon-column was performed according to Grob (1972) and on the emulphor-column without split and solvent. Quantitative analyses were made only on the Ucon-column. For all analyses, a Fractovap G 1 (Carlo Erba) gas chromatograph was used which was equipped with an automatic digital integrator CRS 208 (Infotronics).

Separation Conditions. A 20 m glass capillary column (i.d. 0.32 mm) was coated with Ucon 50 HB 5100 (H. Jäggi, 9043 Trogen, Switzerland). Temperatures: injector 200°C; column: the split was opened 1 3/4 min after injection and after a total of 2 1/2 min the column was heated to 150°C at 90% heating power and then kept under isothermal conditions. Carrier gas: H_2 1.25 atm, 1.5 ml/min; detector gas: H_2 1.25 atm and air 1.5 atm.

A 20 m glass capillary column (i.d. 0.32 mm) was coated with Emulphor (H. Jäggi, 9043 Trogen, Switzerland). Temperatures: injector 190°C, column 170°C (isothermal, without split). Carrier gas: He 2.0 atm, 4 ml/min; detector gas: H_2 1.5 atm and air 2.0 atm.

GC-MS Analyses. GC-MS analyses were made by combination of the above mentioned Emulphor-column with a Varian MAT mass spectrometer.

RESULTS AND DISCUSSION

For the development of a radioactive dilution method permitting the determination of all three juvenile hormones, the tritium labeling of JH I with very high specific radioactivity was necessary. This was attained by exchanging the carboxyl proton of the JH I acid with tritium by means of carrier-free tritiated water and esterifying the tritiated acid with diazomethane (Fig. 1). The specific radioactivity obtained by this method (Trautmann et al., 1973) was 4.34 Ci/mmol.

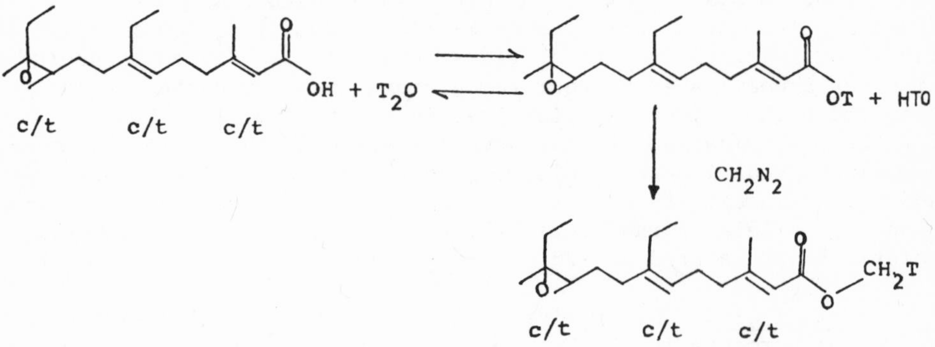

Fig. 1. Tritium labeling of methyl 10,11-epoxy-7-ethyl-3,11-dimethyl-2,6-tridecadienoate (JH I).

When compared to the endogenous hormone content, a very small amount of labeled JH I (49.3 ng corresponding to 1.6×10^6 dpm) was added to the ether extract of the insect species. This served as an internal standard as well as radioactive tracer during the purification procedure. The addition of one radioactive JH for the simultaneous determination of all three hormones was only possible because of their very similar behavior during the purification procedure (Fig. 2). Since part of the JH III was separated from JH I and JH II by florisil chromatography, it was necessary to collect not only the radioactive fractions (Pool I) but also the following ones (Pool II). The remaining JH III in Pool I was further separated by two thin layer chromatographic steps by which complete separation was achieved. JH I and JH II were not separable from each other under the conditions chosen.

GC-analyses of JH I/JH II in Pool I and JH III in Pool II were performed by two high resolution glass capillary columns (Ucon and Emulphor). The three hormones and their isomers were separated by both GC-systems (Table 1 and 2). The identity of a peak with one of the three hormones was confirmed in all cases by use of retention times as well as by co-injection of the corresponding reference substance and by GC-MS analysis. Calculation of the endogenous hormone content in the ether extract was carried out for all three hormones by the following equation:

$$He = (C/a) - Hex$$

He   endogenous hormone

Hex   exogenous hormone

a   radioactivity yield

C   hormone determined by GC

In all cases where the endogenous hormone content exceeds the exogenous one by the factor of 100, the subtraction of Hex can be neglected as the error becomes less than 1%. The calculation for JH II and JH III is also based on the radioactivity yield. This was established by spiking an ether extract of Tenebrio larvae with JH I and JH III and subjecting it to the same purification procedure. Both hormones were recovered with the same yield as the radioactive JH I.

The detection limit for all three hormones depends on the sensitivity of the GC-system, hormone yield and volume of GC-injection solution. Taking the highest sensitivity of 1.6 ng JH per cm peak height, an average yield of 65% and a minimum volume of 50 µl for the injection solution, only 150 ng of each hormone has to be present in an ether extract for its definitive detection. This amount is below the hormone content of one day old male adult of H. cecropia which contain, according to our investigations, 540 ng of JH I and 150 ng of JH II, respectively. The error of this radioactive dilution method is about 10% for each hormone; it is acceptable in view of the very low JH level in insect species.

Whenever possible, animals in the reproductive phase were chosen since it is known that the level of JH is highest during egg maturation (Johnson et al., 1971), and the presence of the hormone is also necessary both for the proper functioning of the genital accessory glands of the male (Wigglesworth, 1936) and for pheromone production in one or both sexes (Loher, 1960). The results obtained for eight insect species from four different orders are summarized in Table 3. Only JH I and JH II were found in H. cecropia, whereas in all other species only JH III could be detected. The amount of JH III is expressed in ng per g body weight.

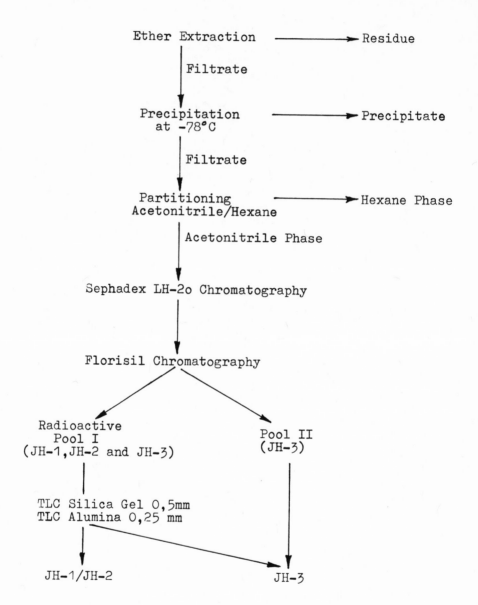

Fig. 2.   Isolation of JH I, JH II and JH III from ether extract of insect species.

TABLE 1.   GC – RETENTION TIMES OF JH I, JH II AND JH III
           (COLUMN 2, EMULPHOR)

JH	Retention Times (min)
JH I (2-trans; 6-trans; 10-cis)	11.6
JH II (2-trans; 6-trans; 10-cis)	9.5
JH III (2-trans; 6-trans)	6.9

TABLE 2.   GC – RETENTION TIMES OF CIS/TRANS-ISOMERS OF JH I
           (COLUMN 2, EMULPHOR)

JH	Retention (min)
2-cis; 6-cis; 10.11-cis	7.6
2-cis; 6-cis; 10.11-trans	7.9
2-cis; 6-trans; 10.11-cis	9.0
2-cis; 6-trans; 10.11-trans	9.2
2-trans; 6-cis; 10.11-cis	10.1
2-trans; 6-cis; 10.11-trans	10.5
2-trans; 6-trans; 10.11-cis (JH I)	11.6
2-trans; 6-trans; 10.11-trans	11.9

TABLE 3. CONTENT OF JH III AND THE BIOLOGICAL ACTIVITY OF THE GC-SOLUTION CORRESPONDING TO JH III IN INSECT SPECIES OF VARIOUS ORDERS.

Insect species	Total animal weight (g)	Average weight per animal (g)	JH III content* (ng/g body weight)	Biological Activity (GU/g body weight)
Tenebrio molitor L. young adults 4 to 6 days old.	579	0.14	--	
reproductive adults 8 to 16 days old.	1022	0.15	7.4	30
Leptinotarsa decemlineata reproductive adults	500	0.14	11.1	286
Melolontha melolontha reproductive adults	1200	1.04	4.1	
Blatta orientalis reproductive adults	842	0.39	3.1	333
Leucophaea maderae reproductive adults	820	2.09	3.5	161
Nauphoeta cinerea mainly reproductive adults	1042	0.50	6.1	83
Schistocerca gregaria older adults	409	2.76	0.5	10
Apis mellifera workers of various ages	740	0.12	2.8	62

* The dash suggests the possible occurrence of JH III, but it is below the detection limit of 0.1 ng per g fresh weight.

From the order Coleoptera, Melolontha melolontha (Scarabaeidae), Leptinotarsa decemlineata (Chrysomelidae) and Tenebrio molitor (Tenebrionidae) were investigated. In all cases where the animals were in the reproductive stage, only JH III could be detected in concentrations ranging from 4.1 to 11.1 ng per g weight. Surprisingly, no JH was found in young, 3 to 6 day old adults of Tenebrio as well as in larvae of the same species. These results are consistent with those of Judy et al. (1975) who used cultures of corpora allata. The fact that the hormone level in larvae is below the detection limit of our method (0.1 ng/g) parallels the situation observed in larvae of the grasshopper, Locusta migratoria (Johnson, 1973), and is further supported by in vivo experiments with the corpora allata of Periplaneta americana (Müller et al., 1974) where glands of reproducing females synthesized much more hormone than those of larvae. These negative results demonstrate the great importance of the insect's developmental stage for the isolation of JH.

From the order Blattidae reproducing adults of Blatta orientalis (Blattoidae), Leucophaea maderae (Blaberidae) and Nauphoeta cinerea (Blaberidae) were studied. Again only JH III could be detected in each of the three species in concentrations not markedly different from that found in beetles. These results are consistent with those found in in vitro experiments with corpora allata from reproductive Periplaneta americana (Blattidae) (Müller et al., 1974). However, Lanzrein et al. (1975) recently found small amounts of JH I and JH II in the hemolymph of reproductively active female adults of Nauphoeta cinerea with JH III being the major hormone. The amounts of JH I and JH II in the hemolymph of 5 day old females were found to be 0.04 and 0.01 ng per animal, respectively. These very low concentrations are below the detection limit of our method (0.1 ng/g) and represent only 1% of the total JH content.

From the order Orthoptera the desert locust, Schistocerca gregaria (Acrididae), was investigated and only JH III was found. Its occurrence in Schistocerca vaga has already been reported (Judy, 1973). The low JH III content of 0.5 ng per g might be explained either by the inherent moderate biosynthetic activity of the corpora allata, or by the possibility that some of the animals analyzed had already passed the reproductive stage.

As a representative of the order of Hymenoptera the worker honey bee, Apis mellifera (Apidae), was investigated. JH III was detected in a concentration of 2.8 ng per g.

With the exception of Schistocerca gregaria where the amount of JH III was very low, the structure of JH III for all other species investigated was further confirmed by GC-MS analysis. Due to its relatively low intensity, the molecular ion was only observed in the spectrum of the extract from Leptinotarsa decemlineata.

However, in the upper mass range the following characteristic
fragments generally occurred: m/e 235 ($M_{\cdot}^{+}$ - .OCH_3), 234 ($M_{\cdot}^{+}$ -
CH_3OH), 219 ($M_{\cdot}^{+}$ - CH_3OH, .CH_3) and 206 ($M_{\cdot}^{+}$ - CH_3OH, CO). In the
intermediate mass range m/e 163, 153, 149, 135, 121 and 114 are
the most intense ions. Of these m/e 114 ($\diagup$ $COOCH_3$)$_{\cdot}^{+}$ is a particularly characteristic rearrangement ion.

An aliquot of the two GC-injection solutions which corresponds
to the purified fractions of JH I/JH II and JH III, respectively,
was diluted with olive oil and tested biologically in the Galleria
mellonella wax test (De Wilde et al., 1968) for each of the
species investigated. The results were compared with the corresponding standards. Under our conditions 1.4 pg JH I or 62 pg per
pupa JH III correspond to one Galleria unit (GU). As the solution
corresponding to the JH I/JH II fraction contained labeled JH I
at a concentration of 0.6 pg/μl (0.3 pg per Galleria pupa), we
would expect about 20% activity in the wax test if no endogenous
JH I or JH II is present in the extract. For the three holometabolous species (Apis mellifera, Leptinotarsa decemlineata and
Tenebrio molitor), an activity of 30% was found which may indicate,
at the most, some traces of an active substance. For the four more
primitive species, 0-4% activity was found. In no case was the
presence of endogenous JH I or JH II clearly indicated by GC
(detection limit 0.1 ng/g). In the solutions corresponding to the
JH III fraction, the presence of JH III was confirmed approximately
at the level indicated by quantitative GC for all species investigated (Table 1); only in the case of Blatta orientalis is the
biological activity higher - by a factor of six - than expected
from GC-analysis.

CONCLUSIONS

Evidence of JH III in eight insect species from four different
orders and, in the case of Coleoptera and Blattoidea also in three
and two different families respectively, demonstrates the broad
distribution of JH III in adult insects. Its widespread and abundant occurrence might be explained by the fact that it is derived
directly from the terpenoid biosynthetic pathway, whereas JH I and
JH II contain ethyl branches on the carbon skeleton that require
unique precursors for their biosynthesis. Since Lanzrein et al.
(1975) could detect, in addition to the major JH (JH III), small
amounts of JH I and JH II in reproductive females of Naphoeta
cinerea, it is suggested that our negative results concerning the
absence of JH I and JH II respectively, might not be conclusive.
However, it can be concluded from the GC-detection limit that the
possible content of each hormone cannot exceed about 0.1 ng per
g of body weight.

There were no signs of the existence of unknown hormone(s) in
the insect species investigated in this study. However, the investigation of a highly biologically active ether extract from
Attacus atlas moths (Paguia et al., 1975) by means of our radioactive isotope dilution method revealed the presence of an unknown
active substance which possesses similar chromatographic behavior
on TLC to JH III. 93% of the total biological activity was found
to be associated with this substance. Based on these results and
taking into consideration that Attacus atlas is closely related to
the JH I and JH II containing H. cecropia, one might speculate that
a certain hormonal diversification has accompanied the evolution
of Saturniidae. On the other hand, the pronounced presence of the
same JH molecule in very primitive roaches and highly specialized
honey bees and butterflies indicates that the biochemistry of JH
may be rather similar throughout the class Insecta.

SUMMARY

By means of radioactive dilution analysis (detection limit
ca. 0.1 ng/g of body weight) allowing the qualitative and quantitative determination of all three presently known juvenile hormones
(JH I to III) the following eight insect species from four orders
were investigated mainly in the reproductive stage: Tenebrio
molitor, Leptinotarsa decemlineata, Melolontha melolontha
(Coleoptera); Schistocerca gregaria (Orthoptera); Blatta orientalis,
Leucophaea maderae, Nauphoeta cinerea (Blattoidea); Apis mellifera
(Hymenoptera).

In all these species only one of the three known juvenile
hormones was found, namely, methyl (2E, 6E)-10,11-epoxy-3,7,11-
trimethyl-2,6-dodecadienoate (JH III), in amounts of 0.5 to 11 ng
per g of body weight. The results of the chemical analyses were
confirmed biologically by the Galleria wax test.

The results demonstrate the widespread occurrence of JH III in
adult insects of different orders.

ACKNOWLEDGEMENTS

We thank Dr. Gerig, Swiss Federal Dairy Research Station,
Liebefeld, Switzerland, for providing the honey bees. The authors
are also grateful to Dr. Neuner and Mr. Zarske, Givaudan AG,
Dubendorf, Switzerland, for their valuable help performing the GC-
MS analyses. We are also grateful to Dr. Vonder Muhll, Dr. R. Maag
AG, Dielsdorf, for stimulating discussions.

REFERENCES

Dahm, K.H., and Röller, H., 1970, Life Sci. 9:1397.
De Wilde, J., Staal, G.B., De Kort, C.A.D., De Loof, A., and
    Baard, G., 1968, Proc. Kon. Nederl. Akad. Wetensch., Ser. C
    71:321.
Grob, K., 1972, Chromatographia 5:3.
Johnson, R.H., and Hill, L., 1971, 6th Conference of European
    Comparative Endocrinologists, Montpellier, p. 101.
Johnson, R.H., and Hill, L., 1973, J. Insect Physiol. 19:1921.
Judy, K.J., Schooley, D.A., Dunham, L.L., Hall, M.S., Bergot, B.J.,
    and Siddall, J.B., 1973a, Proc. Nat. Acad. Sci. USA 70:1509.
Judy, K.J., Schooley, D.A., Hall, M.S., Bergot, B.J., and Siddall,
    J.B., 1973b, Life Sci. 13:1511.
Judy, K.J., Schooley, D.A., Troetschler, R.G., Jennings, R.C.,
    Bergot, B.J., and Hall, M.S., 1975, Life Sci. 16:1059.
Lanzrein, B., Hashimoto, M., Parmakovich, V., Nakanishi, K.,
    Wilhelm, R., and Lüscher, M., 1975, Life Sci. 16:1271.
Loher, W., 1960, Proc. Roy. Soc. B 153:380.
Meyer, A.S., Schneiderman, H.A., Hanzmann, E., and Ko, H.I.,
    1968, Proc. Nat. Acad. Sci. USA 60:853.
Müller, P.J., Masner, P., Trautmann, K.H., Suchý, M., and Wipf,
    H.-K., 1974, Life Sci. 15:915.
Paguia, P., Masner, P., Trautmann, K.H., and Schuler, A., 1975,
    Experientia (in press).
Pratt, G.E., and Tobe, S.S., 1974, Life Sci. 14:575.
Röller, H., and Dahm, K.H., 1972, Proc. Conf. Workshop on Hormonal
    Heterophyly.
Röller, H., Dahm, K.H., Sweely, C.C., and Trost, B.M., 1967,
    Angew. Chem. Int. Ed. 6:179.
Trautmann, K.H., Schuler, A., Suchý, M., and Wipf, H.-K., 1974,
    Z. Naturforsch. C 29:161.
Wigglesworth, V.B., 1936, Quart. J. Microscop. Sci. 79:91.

ENZYMATIC SYNTHESIS OF JUVENILE HORMONE IN MANDUCA SEXTA

D. Reibstein, J.H. Law, S.B. Bowlus, and
J.A. Katzenellenbogen

Department of Biochemistry, University of Chicago,
Chicago, Illinois 60637 and Department of Chemistry,
University of Illinois, Urbana, Illinois 61801

INTRODUCTION

Following the successful elucidation of the structure of the
Cecropia hormone by Roller et al. (1967), experiments designed to
explore the mode of synthesis of this molecule in insects were
immediately initiated. Much progress in our understanding of
juvenile hormone biosynthesis has come from studies with excised
corpora allata maintained in tissue culture supplemented with
isotopically labeled precursors (Judy et al., 1973; Schooley et al.,
1973; Pratt and Tobe, 1974; Tobe and Pratt, 1974a,b; Jennings
et al., 1975). It was demonstrated that corpora allata in organ
culture were capable of de novo synthesis of the hormones from
small precursors such as acetate, propionate and methionine. Thus,
the construction of the carbon chain, as well as its final modification into the epoxy sesquiterpene ester are metabolic properties
of this gland.

Our own investigations were undertaken to define the individual
enzymatic steps in hormone formation. We started with the experimental techniques successfully employed by Judy et al. (1973), for
Manduca sexta corpora allata culture, and then disrupted the glands
to see if we could move back stepwise from the hormone product. In
our first paper (Reibstein and Law, 1973) we described the synthesis
of hormone from labeled S-adenosylmethionine (SAM) in these gland
homogenates. The present paper is an account of our further experiments that have clarified the pathway from farnesyl pyrophosphate
to hormone in homogenates of M. sexta corpora allata.

131

MATERIALS AND METHODS

Materials. Methyl trans, trans-farnesenate was prepared essentially by the method of Anderson et al. (1972), except that a mixture of tetrahydrofuran and dimethylformamide was used as the reaction solvent. JH III, the 10,11-epoxide of methyl farnesenate (methyl juvenate) was also prepared by the method of these workers, but on a smaller scale. The crude product was distilled, and a fraction of boiling point 135°C (0.25 mm) was purified by preparative TLC (silica gel, 8% ether in petroleum ether, 2 developments). The pure material was eluted from a band of R_f 0.4. These two products had spectroscopic properties that agreed well with those reported by Anderson et al. (1972).

For the preparation of 10,11-epoxy farnesenic acid (JH III acid), methyl trans,trans-farnesenate (1.15 g, 4.2 mmol) was dissolved in 25 ml of 80% aqueous tetrahydrofuran, and the solution was treated with N-bromosuccinimide (0.9 g, 5.0 mmol). The solution was stirred for 3 hr at 0°, and the reaction was quenched in brine. The mixture was extracted with three portions of ether; the combined organic layers were dried (MgSO$_4$) and the solvent was removed.

The oil obtained was dissolved in 50 ml of ethanol and the solution was treated with 50 ml of 2.5% aqueous sodium hydroxide. After stirring for 20 hr at room temperature, brine was added and the mixture was acidified with 6 N H$_2$SO$_4$. Isolation of the product in ether afforded about 500 mg of an oil. The crude product was chromatographed on silica gel plates which had been predeveloped with 10% triethylamine in ether and activated at 65°C for 90 min. A broad band was recovered and rechromatographed in the same way. Three developments with 2% methanol in dichloromethane gave 0.113 g (9%) of an oil, R_f 0.2. This product had spectroscopic properties expected for the epoxy acid, and a small amount, on treatment with diazomethane gave an ester with the properties of JH III.

trans,trans-Farnesenic acid was prepared from farnesol (Givaudan) and purified as the S-benzyl thiopseudouronium salt (Law et al., 1966). JH I (methyl dihomojuvenate) was purchased from Eco Control. It was homogeneous, as judged by GLC, and gave a mass spectrum identical to that published by Bieber (1973). Oleic acid was from Applied Science and 9-keto-2-decenoic acid was kindly provided by Dr. K. Eiter, Bayerwerk-Leverkusen. 2-Hexadecenoic acid was prepared by the method of Myers (1951), while cis-9,10-epoxystearic acid was synthesized by the procedure of Findley et al. (1943). Unlabeled farnesyl pyrophosphate was a gift of Dr. Hans Rilling.

[Methyl-^{14}C]-S-adenosylmethionine, 57 mCi/mmol; [methyl-^{3}H]-S-adenosylmethionine, 4.5-8.5 Ci/mmol; [5-^{3}H]-mevalonic acid, dibenzylethylenediamine salt, 6.7 Ci/mmol; and methyl [7-ethyl-1,2-^{3}H]-JH I, 14 Ci/mmol, were obtained from New England Nuclear. In each case the materials were examined by TLC, GLC or paper chromatography and found to be at least 90% pure. Minor impurities corresponded to thiomethyladenosine, in the case of SAM, and the diol ester, in the case of JH I. Nonisotopic S-adenosyl-L-methionine was purchased from Boehringer, while ATP, NAD and NADPH were products of P-L Laboratories; all other chemicals were commercial products of reagent grade.

[^{3}H]-JH I acid was prepared by base hydrolysis of [7-ethyl-1,2-^{3}H] JH I. The labeled compound (2 - 10 µCi) was mixed with 10 µg of unlabeled carrier. The solvent was removed and 1 ml of water was added. Brief sonication was employed to aid dissolution. One ml of 0.2 N NaOH was added and the mixture was incubated at room temperature for 4 hr. After neutralization, the acid product was extracted and purified by TLC on silica (hexane: ethyl acetate, 7:3) and the acid (R_f 0.4-0.5) was eluted with methanol (radiochemical yield 60%). As a check on the identity and purity of the acid, a portion of the labeled compound was treated with diazomethane which converted it quantitatively to authentic hormone, as judged by TLC and GLC.

[Methyl-^{3}H] methyl _trans,trans_,-farnesenate was prepared by exchange with labeled methanol (Amersham-Searle, 128 mCi/mmol) by the method of Sanburg _et al._ (1975). The product was purified by elution from a 2.5 ml column of Florisil with 20% ether in pentane. The product was judged homogeneous by TLC and GLC and had a specific activity of 26 mCi/mmol.

[1,5,9,-^{3}H] _trans,trans_-Farnesyl pyrophosphate was prepared by the method of Popják (1969). The product was purified by TLC (Brinkmann Polygram Sil G with isopropanol: 30% ammonium hydroxide: water, 6:3:1). The labeled product (20 Ci/mmol) was eluted with a small volume of 10% aqueous ammonium hydroxide and stored at -20°C. The identity of the product was confirmed by cleavage to farnesol (characterized by GLC and TLC) by the action of calf intestinal alkaline phosphatase (Boehringer) and by enzymatic conversion to squalene (Rilling and Bloch, 1959), which was identified by TLC and GLC after admixture with authentic squalene (Eastman Distillation Products).

M. sexta eggs were a gift of Dr. R.A. Bell, USDA, Fargo, North Dakota. Larvae were reared at 27°C in a 15-hr light, 9-hr dark photoperiod, using the Yamamoto diet (Yamamoto, 1969), as modified by Dr. Bell. Only female adults were used as a source of glands.

Methods. Adult female moths, 0-2 days post-emergence, were chilled, the legs and wings were removed and scales were scraped

from the head. Each animal was placed in a clay mold under cold
saline and two diagonal cuts meeting at the level of the antennae
were made through the cervical membrane and head capsule. A trans-
verse cut of the cervical membrane with dissecting scissors allowed
removal of a triangle of cuticle. The tracheae and muscles thus
exposed were carefully cut away revealing the corpora allata –
corpora cardiaca complexes lying next to the aorta behind the brain.
These were excised, washed thoroughly (Judy et al., 1973) and
transferred to a microhomogenizer (Hall et al., 1970) containing
10 μl of chilled Grace's medium (Gibco) supplemented with penicillin,
streptomycin, and 1% w/v bovine serum albumin. Following homogenization, the disrupted tissue mixture was centrifuged at 10,000 x g
for 4 min to remove whole cells and cell debris. Aliquots of
supernatant, each equivalent to one gland complex, were transferred to vials containing substrates for incubation.

All substrates for incubation (except the tissue homogenate)
were placed in a Reacti-Vial (Pierce Chemical Co.) 0.3 ml capacity,
and evaporated to dryness under a stream of nitrogen. In the case
of acidic materials (labeled SAM, for example, is supplied in
H_2SO_4, pH 2) sufficient phosphate buffer (0.1 M, pH 7.5) was added
to raise the pH above 5 before removal of solvent. After addition
of the aliquot of homogenate, the pH was checked by means of a
micro electrode (Bolab, Derry, N.H.) to ensure that the value was
5.0 – 5.2, except where otherwise noted. All homogenizers and
incubation vials were sterilized before use and every attempt was
made to achieve essentially sterile conditions during incubation.
The Reacti-Vials were sealed tightly and maintained at 27°C in a
sterile, water-saturated desiccator for 8 – 18 hr.

At the end of the incubation period, the total incubation
mixture plus a benzene rinse was streaked at the origin of a TLC
plate (Brinkmann Polygram Sil N-HR, 0.2 mm thickness or Sil G,
0.25 mm thickness) and carrier JH was applied to the same streak
or adjacent channels. The plate was developed with hexane:ethyl
acetate (7:3), unless otherwise specified. The JH region (R_f 0.6 –
0.7) was located by brief exposure to iodine vapor. The adsorbant
from this region was scraped into a sintered glass funnel, and
the hormone was quantitatively eluted with methanol. This procedure could also be used to recover methyl farnesenate (R_f 0.75 –
0.85) which was distinctly separated from hormone.

When farnesenic acid and JH III acid were labeled products
(as in the case of incubation of labeled farnesyl pyrophosphate),
the TLC plate was developed with hexane:acetone (9:1). The two
acids were not separated from each other, but were separated from
the two esters. The acid zone (R_f 0.04 – 0.20) was eluted and the
recovered acids were resolved by TLC in hexane:ethyl acetate, 7:3,
in which farnesenic acid has an R_f 0.5 – 0.6, and JH III acid
0.04 – 0.18.

After elution from silica, hormone samples were subjected to GLC. 3% OV-17 on Gas Chrom Q, 100 - 200 mesh (Applied Science laboratories) was packed in a 6-foot, glass column 3 mm I.D. We used an F and M 402 instrument operated isothermally at 150°C and with a helium flow rate of 37 ml/min. The effluent stream was split between the detector and the collection outlet at a ratio of approximately 1:10. The retention times of methyl trans,trans-farnesenate, JH III and JH I were 6, 12 and 15 min, respectively. Samples were trapped at the exit port in 7" lengths of Teflon tubings, from which they were washed with ethyl ether (collection efficiency nearly 80%). Enzymatic hormone products collected from the GLC column were degraded by the methods of Judy et al. (1973) to 10-hydroxy acids by ring opening followed by periodate cleavage and reduction. Non-hormone labeled impurities constituted as much as 50% of the products separated by the initial TLC step, but after collection of the hormone fraction from GLC, at least 80% of the labeled material behaved as authentic hormone through the two degradation steps.

Other enzymatic products were characterized as follows: JH III acid was converted to JH III by treatment with diazomethane, subjected to TLC, GLC and degradation as in the case of hormone produced biosynthetically. Farnesenic acid was crystallized to constant specific activity as the 2-benzyl thiopseudouronium salt. It was also converted to the methyl ester by treatment with diazomethane and subjected to TLC and GLC, where it behaved as the authentic compound. The methyl farnesenate prepared enzymatically with labeled SAM was converted to hormone by epoxidation with m-chloroperbenzoic acid, and the hormone was characterized by TLC and GLC.

RESULTS

Stimulation of Synthesis by Epoxy Acids

In an earlier report (Reibstein and Law, 1973) we showed that [methyl-^{3}H]-S-adenosylmethionine (SAM) could be converted to labeled juvenile hormones in a 12,000 x g supernatant fraction from homogenates of the corpus allatum - corpus cardiacum complex of M. sexta. The yields of labeled hormone were small and highly variable. We suspected that this was due to a lack of appropriate endogenous methyl group acceptors in the gland homogenates, and we therefore undertook experiments in which exogenous acids related in structure to the hormone products were added to the homogenates.

Since it had been shown (Metzler et al., 1972; more recently confirmed by Rodé-Gowal et al., 1975) that trans,trans,cis-3,11 dimethyl-7-ethyl-10,11 epoxy-2,6-tridecadienoic acid (JH I acid) could be converted efficiently by the H. cecropia moth to JH I, it seemed likely that this acid or its lower homolog could serve as a methyl acceptor. Table 1 (lines 1 and 2) shows that this is indeed the case. Addition of JH III acid caused at least a ten-fold increase in incorporation of SAM into JH III. Furthermore, incubation of JH I acid with homogenate led to the synthesis of JH I exclusively (Table 1, line 3). Other acids with double bonds (oleic acid, 2-hexadecenoic acid), epoxide groups (9,10-epoxystearic acid) or an α,β double bond and a keto group (9-keto-2-decenoic acid, the "queen substance" of Apis mellifera) were not alkylated by the gland homogenates (Table 1, lines 7 - 10).

No carboxyl alkylating activity could be demonstrated in extracts of the corpus cardiacum, brain or fat body, even when exogenous JH III acid was added along with labeled SAM.

In order to investigate the stoichiometry of the aklylation reaction, we incubated [^{14}C methyl] SAM and [^{3}H] JH I acid (prepared by alkaline hydrolysis of the commercial methyl [^{3}H 7-ethyl] JH I) with a gland homogenate. Nearly equal amounts of each iso-tope were incorporated into JH I (line 5, Table 1). When the labeled acid was incubated alone with homogenate, none of it was converted to hormone (line 6). These experiments demonstrate that neither endogenous epoxy acids nor endogenous SAM were present in any significant amount in the gland homogenates.

Table 2 shows that the rate of hormone production by gland homogenates was nearly saturated with SAM at 5×10^{-5} M, but epoxy acid did not reach saturation even at 10^{-3} M. The rate of the carboxyl alkylation reaction increased about three-fold when the pH was raised from 5 to 7.

Carboxyl Alkylation of Farnesenic Acid

When unlabeled farnesenic acid and labeled SAM were incubated together with gland homogenate, labeled methyl farnesenate was found in reasonable amounts, but only a small amount of labeled hormone was produced (Table 3). On the other hand, if NADPH was also included, good yields of labeled JH III were produced with no accumulation of methyl farnesenate. To probe the specificity of the epoxidizing system, oleic acid was added to a homogenate with NADPH and labeled SAM. No methyl epoxystearate was formed. These results indicated the presence of a mixed-function oxidase that could introduce the epoxide group, but they did not allow a conclu-sion concerning the substrate undergoing epoxidation. As shown in Fig. 1, farnesenic acid could be epoxidized to JH III acid, followed by alkylation to JH III; or it could be first alkylated to methyl farnesenate which could then be converted to the hormone by the operation of an oxidase.

TABLE 1. SUBSTRATES FOR JUVENILE HORMONE SYNTHESIS[a].

Labeled Substrate	Nonisotopic Additions	Labeled Products Formed (pmol)[b]		
		JH III	JH I	Other
[methyl-^{3}H]-S-adenosylmethionine (5.5 x 10^{-5} M)	none	0.093 ± 0.12	0	--
"	JH III acid (10^{-5} M)	1.25	0	--
[^{3}H] JH I acid (4.1 x 10^{-5} M)	S-adenosylmethionine (5.5 x 10^{-5} M)	0	4.4	--
"	" (no homogenate)	0	0	--
a. [^{3}H]-JH I acid (10^{-5} M)	none	0	[^{3}H] 2.4	--
b. [methyl-^{14}C] S-adenosylmethionine (5.5 x 10^{-5} M)		0	[^{14}C] 2.1	
[^{3}H] JH I acid (10^{-5} M)	none	0	0	--

TABLE 1 (Cont.)

Labeled Substrate	Nonisotopic Additions	Labeled Products Formed (pmol)[b]		
		JH III	JH I	Other
[methyl-^{3}H]-S-adenosyl-methionine (5.5 x 10^{-5} M)	oleic acid (10^{-5} M)	0.10	0	methyl oleate 0
"	9,10-epoxystearic acid (10^{-5} M)	0	0	methyl epoxy-stearate 0
"	2-hexadecenoic acid (10^{-5} M)	0.08	0	methyl 2-hexa-decenoate 0
"	9-keto-2-decenoic acid (10^{-5} M)	0	0	methyl 9-keto-2-decenoate 0

a Incubations with gland supernatant at 27°C for 18 hr.

b Products identified and quantitated by GLC.

TABLE 2. DEPENDENCE OF RATE OF METHYLATION ON SUBSTRATE CONCEN-
TRATIONS[a].

JH III M	[S-adenosyl-methionine], M[b]	Incorporation pmol[c]
(Homogenate 1)		
0	5.5×10^{-5}	0.27
1.0×10^{-6}	5.5×10^{-5}	0.21
1.0×10^{-5}	5.5×10^{-5}	3.3
1.0×10^{-5}	4.5×10^{-4}	7.0
1.0×10^{-5}	1.2×10^{-3}	7.9
(Homogenate 2)		
1.0×10^{-5}	1.2×10^{-3}	28.0
1.0×10^{-4}	1.2×10^{-3}	73.5
1.0×10^{-3}	1.2×10^{-3}	141

[a] Incubation with gland homogenate for 8 hr at 27°C.

[b] [methyl-^{3}H]-SAM, 4.52 Ci/mmol, was mixed with the appropriate amount of unlabeled SAM to give the final concentration shown.

[c] Labeled material in the JH zone of TLC.

We had demonstrated reactions 2 and 5 (Fig. 1), but in order to decide between the alternative pathways shown in Fig. 1, we needed an additional experiment, the direct assessment of reaction 4. Therefore, we prepared labeled methyl farnesenate and incubated it with gland homogenate and NADPH. Another portion of the same homogenate was incubated with unlabeled farnesenic acid, NADPH and labeled SAM. While farnesenic acid was efficiently converted to hormone, no label from methyl farnesenate appeared in hormone (Table 4). The specific activity of the methyl farnesenate was

TABLE 3. EFFECT OF FARNESENIC ACID AND NADPH ON ENZYMATIC JUVENILE
HORMONE SYNTHESIS.

Additions	Labeled Products Formed (pmol)		
	JH III	methyl farnesenate	methyl epoxystearate
Epoxyfarnesenic acid	1.25	--	--
Farnesenic acid	0.22	2.86	--
Farnesenic acid + NADPH	3.53	0	--
Oleic acid + NADPH	0.001	--	--

All incubations contained [methyl-^{3}H]-SAM, 5.5 x 10^{-5} M.
Concentration of added fatty acids 10^{-5} M, and of NADPH, where
added, 10^{-3} M. Incubation with gland supernatant at 27°C for
18 hr. Products identified and quantitated by GLC.

sufficiently high that its conversion to JH III could have been
detected even if the yield had been only 10% that produced from
farnesenic acid in the parallel experiment. Furthermore, although
methyl farnesenate appeared to be hydrolyzed during the course of
the incubation, the amount of unhydrolyzed ester recovered after
incubation was still far in excess of the hormone produced in the
parallel experiment.

Conversion of [^{3}H] Farnesyl Pyrophosphate to Hormone

In an effort to trace the pathway for hormone synthesis
backward from farnesenic acid, we prepared [^{3}H]-farnesyl pyrophosphate from mevalonic acid by an enzymatic method. The purified
[^{3}H] farnesyl pyrophosphate was then incubated in homogenates of
corpora allata along with unlabeled SAM and pyridine nucleotides.
NAD was added in addition to NADPH, because we reasoned that
cleavage of the pyrophosphate to farnesol would be followed by
alcohol and aldehyde dehydrogenase reactions, most likely employing NAD as a cofactor. From Table 5 it can be seen that farnesyl
pyrophosphate was converted to moderate amounts of farnesenic acid

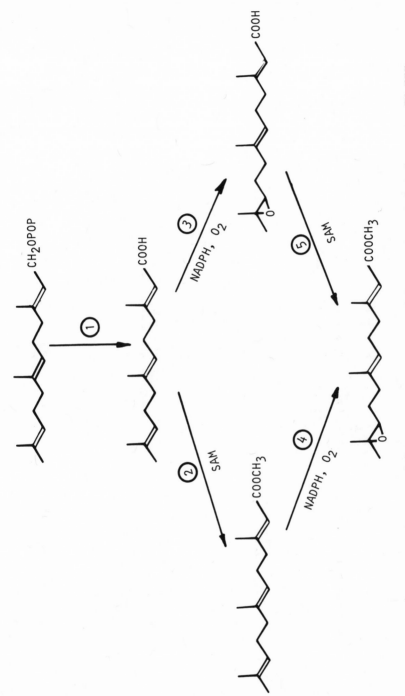

Fig. 1. Enzymatic pathways for the conversion of farnesyl pyrophosphate to JH III.

TABLE 4. COMPARISON OF METHYL FARNESENATE AND FARNESENIC ACID AS JH PRECURSORS[a]

Additions				Labeled products recovered (pmol)	
Farnesenic acid[b] (3.0 mM)	[³H]-methyl farnesenate (3.2 mM)	[methyl-³H]-SAM (1.24 mM)	NADPH (5 mM)	JH III	Methyl farnesenate
+	–	+	+	48.1	94.9
–	+	–	+	0	6×10^3

a Products quantitated by GLC.

b Frinton Labs; predominantly trans,trans.

TABLE 5. CONVERSION OF FARNESYL PYROPHOSPHATE TO JUVENILE HORMONE AND RELATED PRODUCTS.

	Additions			Labeled Products Formed (pmol)			
[^{3}H]-farnesyl pyrophosphate	10^{-3} M NAD	10^{-3} M NADPH	5.5×10^{-5} SAM	Farnesenic acid	Epoxyfarnesenic acid	methyl farnesenate	JH III
+	+	+	+	1.43	1.70	0.014	0.14
+	+	+	-	1.02	0.92	--	--

[^{3}H]-Farnesyl pyrophosphate: 20.2 ci/mmol; final concentration 8.62×10^{-6} M. Incubations with gland homogenate at 27°C for 20 hr. Products identified and quantitated by GLC.

and JH III acid, while smaller amounts of hormone and minute amounts of methyl farnesenate were formed. If SAM was omitted from the incubation mixture, no methyl esters were formed.

DISCUSSION

The enzymatic synthesis of juvenile hormones can readily be demonstrated in homogenates of the corpora allata of M. sexta, provided substrates of relatively high specific radioactivity are available. Exogenous precursors must be added because only minute amounts of hormone precursors are present in the homogenates. No further fractionation of the 10,000 x g supernatant has yet been attempted, and enzyme components responsible for the late stages of hormone biosynthesis may well be localized in subcellular organelles. No attempt has been made to assess the level of esterolytic or other hydrolytic enzymes present in the homogenates nor have inhibitors of esterases been employed. If esterolytic enzymes were present, they may have been responsible for the large variations in synthetic capacity from one homogenate to the next. Because of this variability it is imperative to perform parallel experiments with homogenates from pooled glands if any quantitative comparison is to be made.

The results of experiments with several putative hormone precursors, starting with farnesyl pyrophosphate, are summarized in Fig. 1. Farnesol pyrophosphate is presumably cleaved to farnesol, and then oxidized to farnesenic acid, which is subsequently enzymatically epoxidized to JH III acid in the presence of NADPH and molecular oxygen. We have been unable to demonstrate this epoxidation step directly because we have not had labeled farnesenic acid of sufficiently high specific activity. However, this sequence from farnesenic acid to JH III acid is clearly implicated in the experiments with labeled farnesyl pyrophosphate.

The conversion of epoxy acid to hormone in an enzymatic carboxyl alkylation reaction similar to that demonstrated earlier in bacterial extracts (Akamatsu and Law, 1970), has been unequivocally established. The enzymatic synthesis of methyl farnesenate has also been shown, but not its conversion to hormone. Indeed, no such conversion could be detected, even under conditions demonstrably favorable for hormone synthesis from farnesenic acid. The accumulation of methyl farnesenate occurs only when NADPH (and probably oxygen) is in short supply. It is conceivable that methyl farnesenate is merely a storage form for farnesenic acid when some metabolic control is exerted to decrease NADPH production. It is also possible, however, that in corpora allata of M. sexta farnesenic acid can be converted to hormone through both branches of the pathway, and that under the particular conditions we have used for our incubations only one branch of the pathway operates.

Pratt and Tobe (1974; Tobe and Pratt, 1974a) have also clearly demonstrated the accumulation of methyl farnesenate in the corpora allata of _Schistocerca_ maintained in organ culture and provided with labeled farnesenic acid. Although the accumulation of a compound in the whole organ does not constitute good evidence that it is a biosynthetic intermediate, as claimed by these authors, the circumstantial evidence is compelling that this is the case. Furthermore, Hammock (1975) has recently demonstrated conversion of methyl farnesenate, labeled in the acid chain, to labeled JH III by microsomes of _Blaberus_ corpora allata in the presence of oxygen and NADPH. While it is still conceivable that the ester is first cleaved to farnesenic acid, which is epoxidized and then alkylated, this seems unlikely in a microsomal system without added SAM. Ultimately, of course, the ideal experiment would be one with methyl farnesenate containing one isotope in the chain and another in the methyl group. If the isotope ratio remains unchanged during hormone synthesis, direct epoxidation would be established.

The difference between our own experiments and those of other investigators could also be explained by different pathways in Lepidoptera as opposed to other insect orders. Further investigations with cell-free extracts of corpora allata from a variety of species will be necessary to establish the relative contributions of the two pathways from farnesenic acid to hormone.

It is of some interest that the carboxyl alkylating enzyme, although it shows a high specificity for the branched sesquiterpene chain, readily converts both JH III acid and JH I acid to the corresponding hormones. Thus, the homosesquiterpene pyrophosphate produced from homomevalonic acid can presumably be cleaved, oxidized and handled readily by the epoxidizing enzymes and the carboxyl alkylation system.

It is hardly necessary to point out that every enzyme in the late stages of hormone biosynthesis offers a new target for screening insect control agents. These processes are very likely unique to JH biosynthesis, and any agent that can effectively inhibit them will have antihormonal activity. The potential practical use for antihormones should be at least as great as for compounds with hormonal activity.

ACKNOWLEDGEMENTS

This research was supported in part by grants from The National Science Foundation, BMS-74-21379 and MPS-73-08691 and from the National Institutes of General Medical Sciences, GM13863.

REFERENCES

Akamatsu, Y., and Law, J.H., 1970, J. Biol. Chem.245:709.

Anderson, R.J., Henrick, C.A., Siddall, J.B., and Zurflüh, R., 1972, J. Am. Chem. Soc. 94:5379.

Bieber, M.A., 1973, Thesis, Michigan State University.

Findley, T.W., Swern, D., and Scanlan, J.T., 1945, J. Am. Chem. Soc. 67:412.

Hall, Z.W., Bownds, M.D., and Kravitz, E.A., 1970, J. Cell Biol. 46:290.

Hammock, B.D., 1975, Life Sci. 17:323.

Jennings, R.C., Judy, K.J., and Schooley, D.A., J. Chem. Soc. Chem. Communs. 1975:21.

Judy, K.J., Schooley, D.A., Dunham, L.L., Hall, M.S., Bergot, B.J., and Siddall, J.B., 1973, Proc. Nat. Acad. Sci. USA 70:1509.

Law, J.H., Yuan, C., and Williams, C.M., 1966, Proc. Nat. Acad. Sci. USA 55:576.

Metzler, M., Meyer, D., Dahm, K.H., Röller, H., and Siddall, J.B., 1972, Z. Naturforsch. 276:321.

Myers, G.S., 1951, J. Am. Chem. Soc. 73:2100.

Popják, G., 1969, in "Methods in Enzymology" (R.B. Clayton, ed.) 15:393.

Pratt, G.E., and Tobe, S.S., 1974, Life Sci. 15:575.

Reibstein, D., and Law, J., 1973, Biochem. Biophys. Res. Communs. 55:266.

Rilling, H., and Bloch, K., 1959, J. Biol. Chem. 234:1424.

Rodé-Gowal, H., Abbott, S., Meyer, D., Röller, H., and Dahm, K.H., 1975, Z. Naturforsch 30c:392.

Röller, H., Dahm, K.H., Sweeley, C.C., and Trost, B.M., 1967, Angew. Chem. Int. Ed. Engl. 6:179.

Sanburg, L.L., Kramer, K.J., Kézdy, F.J., and Law, J.H., 1975, J. Insect Physiol. 21:873.

Schooley, D.A., Judy, K.J., Bergot, B.J., Hall, M.S., and Siddall, J.B., 1973, Proc. Nat. Acad. Sci. USA 70:2921.

Tobe, S.S., and Pratt, G.E., 1974a, Biochem. J. 144:107.

Tobe, S.S., and Pratt, G.E., 1974b, Nature 252:474.

Yamamoto, R.T., 1969, J. Econ. Entomol. 62:1427.

FARNESENIC ACID STIMULATION OF JUVENILE HORMONE BIOSYNTHESIS AS AN EXPERIMENTAL PROBE IN CORPUS ALLATUM PHYSIOLOGY

S.S. Tobe and G.E. Pratt

Department of Zoology, University of Toronto, Toronto
Ontario, Canada and Agricultural Research Council,
University of Sussex, Falmer, Brighton, England

INTRODUCTION

Recent publications from our laboratories have indicated some experimental applications of short-term radiochemical procedures in vitro to the investigation of the temporal patterns of endocrine activity in the corpus allatum (CA) of adult female insects from two different species (Tobe and Pratt, 1975a; Weaver et al., 1975). These assays have relied upon the specific incorporation from the S-methyl group of (methyl-^{14}C)-methionine into the methyl ester of known juvenile hormones, which lent itself to the estimation of rates of synthesis and release of C_{16}JH (JH III) by isolated glands incubated in defined tissue culture medium for 3 hr followed by chromatography and radiochemical quantitation. It has been previously reported that the addition of 2E,6E farnesenic acid to the incubation medium frequently stimulates the production of JH III by corpora allata (CA) from adult females of both Schistocerca gregaria and Periplaneta americana (Pratt and Tobe, 1974; Pratt and Weaver, 1974). It was proposed that farnesenic acid was esterified to methyl farnesoate and subsequently epoxidized at the 10,11 double bond to yield JH III (Pratt and Tobe, 1974). Detailed investigations with glands from S. gregaria showed that at optimal concentrations of farnesenic acid (20 µM), the rates of JH synthesis could be increased by up to 200-fold, and that these stimulated rates were constant over a 4 hr incubation period (Tobe and Pratt, 1974a).

Further investigations on spontaneous rates of JH biosynthesis, namely those which are observed in defined tissue culture medium lacking farnesenic acid or its homologues, showed that these also were constant for at least 3 hr after removal of glands from adult

female S. gregaria (Pratt et al., 1975a). CA from S. gregaria
which exhibit naturally low rates of spontaneous JH biosynthesis,
maintain these low rates for periods exceeding 3 hr whereas glands
exhibiting high initial rates of JH biosynthesis rapidly lose
activity after 3 hr; this may suggest the acute dependence of high
rates of spontaneous JH biosynthesis upon the ongoing stimulation
of the CA by neuroendocrine factors which are absent from the
isolated preparation. These studies also revealed that in glands
exhibiting a wide range of rates of spontaneous JH biosynthesis,
the rate of release of JH was directly proportional to the rate
of synthesis in each pair of glands, suggesting that there may be
no direct control over the rate of release of JH from the CA.
This hypothesis was confirmed, at least in the case of glands from
S. gregaria, by Tobe and Pratt (1974b) who showed that high rates
of JH release could be demonstrated in glands of widely differing
overt activities when the rates of JH biosynthesis were maintained
at a high level by the appropriate addition of farnesenic acid to
the incubation medium. Thus, the glands appear to have an
unvarying capacity to release newly synthesized JH which can be
defined on the basis of a simple physical diffusion model as being
governed by a "release coefficient" of 2.7 hr^{-1}. In accordance
with this model, and in view of the rapidity with which newly
synthesized JH appears in the incubation medium (Tobe and Pratt,
1974a), we further concluded that the release of JH from the CA
was not effected by processes which manifest themselves as changes
in cytomorphology; rather, we attributed ultrastructural differences between the cells of the CA to developmental changes in their
JH synthetic capacity.

     We routinely have employed short-term radiochemical assays
in vitro to investigate the changes in spontaneous JH synthetic
activity (Pratt et al., 1975a) and in the farnesenic acid stimulated activity (Tobe and Pratt, 1974a) of the CA during post-
ecdysial maturation and ensuing gonotrophic cycles in adult female
S. gregaria (Tobe and Pratt, 1975a). These studies have revealed
rapid changes in both spontaneous and stimulated JH biosynthetic
rates and interesting differences between the two patterns of
activity in the glands. In particular, stimulated rates of biosynthesis were always higher than the corresponding spontaneous
rates to an extent which varied throughout the gonotrophic cycles,
and the range of observed spontaneous activities, expressed as a
percentile of the observed maximum, was much greater than that
observed under conditions of stimulated biosynthesis. It was
concluded that the physiological rate-limiting step in JH biosynthesis is always prior to the point of entry of exogenous farnesenic
acid into the biosynthetic pathway in CA from this species (Tobe
and Pratt, 1975a) (see Fig. 1). At present, there is no information on the precise point of entry of exogenous farnesenic acid
into the biosynthetic pathway; in particular, it is not known if
farnesenic acid is an obligatory intermediate in the spontaneous

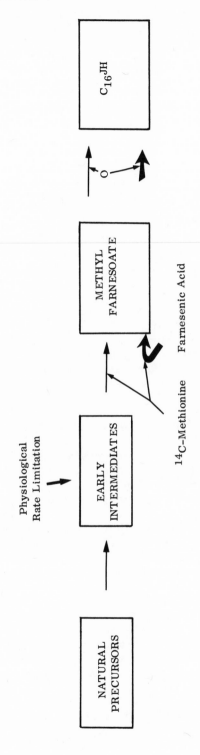

Fig. 1. Generalized flow sheet for JH III ($C_{16}JH$) biosynthesis by locust CA under both spontaneous and stimulated conditions. Physiological rate-limitation is prior to the entry of [^{14}C]-methionine and [^3H]-farnesenic acid into the biosynthetic pathway.

pathway. This area merits further investigation since such information would further support the use of exogenous farnesenic acid
as an exploratory tool.

In addition to the above studies revealing quantitatively
smaller changes in the stimulated as opposed to the spontaneous
JH synthetic capacity of the CA during the reproductive cycles,
we have also related these rapid and quantitatively large changes
in the rate of spontaneous JH biosynthesis to the slow and relatively small magnitude changes in the total volume of the CA
(Tobe and Pratt, 1975b). We reported developmental changes in the
rate of spontaneous JH biosynthesis per unit of CA volume, hereafter referred to as "volume specific activity" of up to 200-fold,
whereas under conditions of optimal farnesenic acid stimulation,
only small and presumably insignificant changes in volume specific
activity were observed. We speculated that there might be a
causal relationship between the total cytoplasmic volume of the
gland and its ability to utilize exogenous farnesenic acid, i.e.
to effect the last two stages in JH biosynthesis.

In the present report, we shall further explore the relationship between spontaneous and stimulated JH biosynthesis and shall
relate these findings to an autoradiographic identification of the
glandular compartments involved in the last two stages of JH biosynthesis as revealed by the discriminating use of [³H]-farnesenic
acid as a cytoradiochemical probe.

MATERIALS AND METHODS

Desert locusts, S. gregaria were reared after the method of
Tobe and Pratt (1975a). Standard incubation procedures for the
estimation of spontaneous and stimulated rates of JH III biosynthesis by individual pairs of isolated corpora allata were as
described by Tobe and Pratt (1975a). Glands were incubated in
tissue culture medium TC199 plus Ficoll (20 mg/ml) (20 mM HEPES
buffered at pH 7.2) at 30°C. Rates of synthesis of JH III were
calculated from the incorporation of radioactivity from (methyl-
¹⁴C)-methionine of known specific activity. For autoradiography,
[³H]-farnesenic acid (specific activity 25 mCi/mmol) was added to
the TC199 (final concentration 20 μM) and glands from female
locusts 2, 4, 6, 8 and 11 days after fledging were incubated for
30 min at 30°C. At the end of this period, glands were subjected
to one of the following three treatments: a) the medium was
removed by aspiration and the CA were fixed immediately; b) the
medium was removed by aspiration, the CA were washed in "cold"
medium and the glands were chased for 20 min in 0.1 ml "cold"
medium under air; c) the medium was removed by aspiration, the CA
were washed in "cold" medium and the glands were chased for 20 min

in 0.1 ml "cold" medium under <u>nitrogen</u> ($\leq$5 ppm oxygen). At the end of the chase periods (b and c), the medium was removed by aspiration and the glands immediately fixed. Control glands were incubated as above but the incubation medium contained no [^{3}H]-farnesenic acid, and were fixed after the appropriate treatments. Fixation and embedding procedures were as previously described (Tobe <u>et al</u>., 1976). To determine total radioisotope content at the end of the embedding procedure, representative CA were solubilized as previously described (Tobe <u>et al</u>., 1976) and assayed for radioactivity by conventional liquid scintillation counting (LSC). In parallel experiments, CA were not fixed but were subjected to radiochemical analysis for their content of [^{3}H]-farnesenic acid, [^{3}H]-methyl farnesoate and [^{3}H]-JH III by extraction, chromatography and LSC (Pratt and Tobe, 1974).

All fixation and embedding solutions were routinely collected and assayed for radioactivity by LSC in order to determine the washout of radioactivity from the CA during each step in the embedding procedure.

For light microscopic autoradiography, embedded CA were sectioned at 0.5-1 µm, mounted on glass slides and coated with either L-4 emulsion (Ilford Ltd., Ilford, England) or NTB-2 emulsion (Eastman Kodak, Rochester, New York) diluted 1:1 with distilled water, after the method of Kopriwa and Leblond (1962). After 2-3 weeks exposure, slides were developed with Kodak D-19 developer, fixed and stained with Kingsley stain (Bogoroch, 1972). For evaluation of autoradiographs, the number of grains per 100 µm^2 was determined for each glandular compartment, e.g. cytoplasm, nuclei, extracellular space, connective tissue sheath, tracheae, and axons on representative sections of CA from each age.

RESULTS

We employed two different experimental protocols in order to explore the relation between spontaneous and stimulated JH bio-synthesis during sexual maturation of <u>S</u>. <u>gregaria</u>, based on the standard assay procedures. In the first series, the CA from four animals of the same age were bilaterally randomized into two groups of four CA each. Three hour radiolabel incubations were employed to estimate the rate of spontaneous JH biosynthesis by one group of CA and the rate of stimulated JH biosynthesis by the other group of CA. In the second series, CA from four animals on each experimental day were individually subjected to consecutive 2 hr estimations of their rates of spontaneous JH biosynthesis and then of their rates of stimulated JH biosynthesis (see Tobe and Pratt, 1975a). THus using these procedures, we were able to com-pare spontaneous and stimulated JH biosynthesis on each experimen-tal day. Data from both series of experiments afforded average

estimates of the synthetic capacity of the CA in these two respects
while the second series indicated the range of capacities of individual pairs of CA.

On any one day or in any one pair of CA, the rate of spontaneous biosynthesis divided by the rate of stimulated biosynthesis
is a fraction (equal to or less than 1.0) which represents the
extent to which the rate of spontaneous biosynthesis approaches
that of stimulated biosynthesis. We refer to this fraction as the
"Fractional Endocrine Activity Ratio" (FEAR). Figure 2 shows the
developmental pattern of FEAR using the two experimental protocols,
during sexual maturation and the first two gonotrophic cycles.

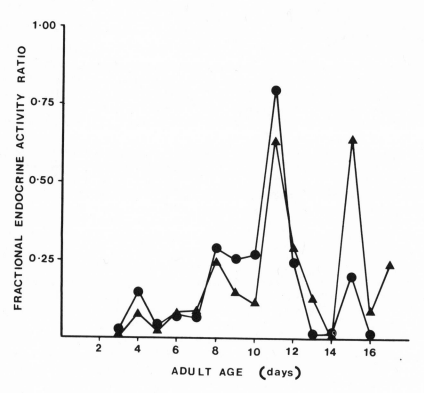

Fig. 2. Developmental changes during sexual maturation and early
reproductive life in female S. gregaria in the "Fractional Endocrine Activity Ratio" (FEAR) (see text) of the CA, showing the
cyclic controlled release of rate-limitation in spontaneous JH
biosynthesis. Closed circles and triangles represent the first
and second experimental series respectively.

It can be seen that there is good agreement between the results
from the two experimental series and in particular, there is an
exact coincidence of the cyclic peaks of CA activation as reflected
by elevated values of FEAR. It is noteworthy that this represen-
tation of the level of activation of the CA provides a clearer indication of the changing states of physiological activation of the
CA during reproductive life than does simple estimation of the
overt rates of spontaneous biosynthesis (see Tobe and Pratt,
1975a). This is because much of the between-animal variation in
the observed activities of the CA from animals of a given age is
associated with parallel variability in rates of both spontaneous
and stimulated biosynthesis so that the between-animal variation
is reduced when the level of activation of the CA is represented
by a ratio of the two. Because of the fact that the developmental changes in the spontaneous biosynthesis are quantitatively
much larger than those changes in stimulated biosynthesis (Tobe
and Pratt, 1975a), the normalization of spontaneous rates of JH
biosynthesis into a FEAR value does not alter the timing of the
derived cyclical peaks of activation of the CA. Thus, the same
temporal correlation is observed between peaks of CA activation
and the induction of growth in the oocytes, as has been previously
reported (Tobe and Pratt, 1975a). Thus, two quantitatively small
peaks of CA activation are observed on days 4 and 8, followed by
much larger peaks on days 11 and 15; during these cycles of activation, average values of FEAR range from 0.003 to 0.750, dramatically indicating the significance of changing control of the early
rate-limiting step(s) in JH biosynthesis.

The developmental pattern of changing values of FEAR reported
here closely parallels the corresponding pattern of spontaneous
JH biosynthesis per unit volume CA already reported (Tobe and
Pratt, 1975b). Because we had previously speculated that there
might be a functional relationship between the total cytoplasmic
volume of the CA and the ability to effect the last two stages
in JH biosynthesis, we were interested to investigate directly
which parts of the gland and its component cells were so involved.
To this end, we have used [^{3}H]-farnesenic acid as an autoradiographic probe, because it has been previously demonstrated that
this compound is incorporated into only two major products, [^{3}H]-
methyl farnesoate and its product, [^{3}H]-JH III (see Fig. 1)
(Pratt and Tobe, 1974).

The fixation procedures employed in the present study were
based on preliminary chemical observations which demonstrated that
osmium tetroxide-glutaraldehyde solution (2:1) reacts very rapidly
with farnesenic acid, methyl farnesoate and JH III to yield black,
insoluble products. No specific problems were encountered in the
use of this combined fixative in contrast to conventional glutaraldehyde fixative, which is unreactive toward, and therefore

cannot adequately fix, the particular radiolabeled compounds under study (unpublished findings). The actual losses observed during the fixation, dehydrating and embedding procedures were as described elsewhere (Tobe et al., 1976) and in general accounted for 40% of the total radioactivity observed in the glands immediately prior to fixation. A large proportion of the loss of the radiolabel occurs during the first two fixation steps and probably represents the wash-out of radiolabeled compounds on the surface of the glands and in the peripheral extracellular spaces. Processing losses of this magnitude appear to be unavoidable using these procedures and must be regarded as satisfactory in as much as they compare favorably with those reported by other authors for the fixation of cholesterol and its precursors in other types of tissue (Darrah et al., 1971; Mak and Trier, 1972). For obvious reasons, we were unable to show that the percentage composition of the radioactivity remaining in the glands after processing was identical to that immediately prior to fixation and therefore cannot exclude the possibility that there were differential losses of the three major radiolabeled compounds during tissue processing.

The three different radiolabeling treatments employed in the present study yielded glands which differed both in their total radiolabel content and also in the proportions of the three major metabolites (see Fig. 3). We have already reported that shortly after 20 min from the addition of [^{3}H]-farnesenic acid to the incubation medium, steady state intraglandular concentrations of [^{3}H]-JH III are achieved (Tobe and Pratt, 1974a). It therefore seemed appropriate to select a pulse duration of 30 min for optimum labeling of the metabolites in JH biosynthesis. Figure 3 (treatment A) shows that over 50% of the radioactivity in the gland at that time can be associated with [^{3}H]-JH III. In treatment B, the 30 min pulse was followed by a 20 min aerobic chase in medium lacking [^{3}H]-farnesenic acid, and from Fig. 3, it can be seen that this results in an appreciable loss of radioactivity from the glands, with a majority of the residual radioactivity being present in the JH III fraction, as would be predicted from the rapid kinetics of the system. When steady-state radiolabeled CA are chased with [^{3}H]-farnesenic acid-free medium under anaerobic conditions for 20 min (treatment C), most of the [^{3}H]-JH III is lost from the glands and the complete inhibition of the final stage in JH III biosynthesis results in [^{3}H]-methyl farnesoate being the principal radioactive metabolite present in the glands at the end of the anaerobic chase. The total amount of radioactivity remaining in these glands after treatment C is intermediate between that observed after treatments A and B, as might be expected.

The general validity of the autoradiographic procedure for indicating the levels of radioactivity in the sections is borne out by the fact that the average grain densities from CA subjected to the three different treatments were similarly proportional to

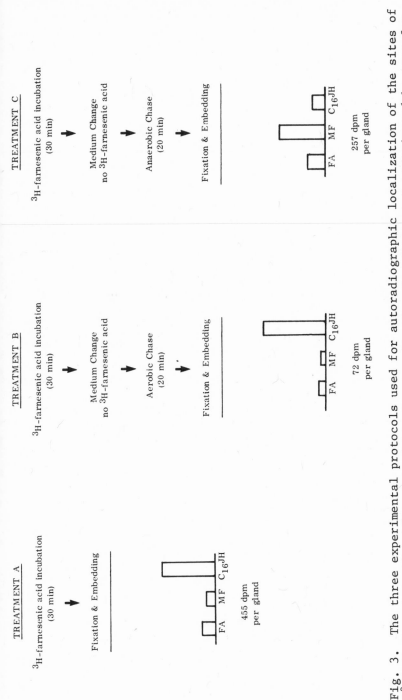

Fig. 3. The three experimental protocols used for autoradiographic localization of the sites of [3H]-farnesenic acid utilization within S. gregaria CA. Treatments were terminated by transfer of glands to freshly prepared glutaraldehyde-osmium tetroxide fixative at 0°C. The histograms show the percentage composition of the radiolabel in the CA at these times as determined on parallel glands which were subjected to radiochemical analysis instead of fixation; the total radioactivity in selected glands is also indicated. FA = [3H]-farnesenic acid; MF = [3H]-methyl farnesoate; C$_{16}$JH = [3H]-C$_{16}$JH (JH III).

the known total radioactivities in CA subjected to the same respective treatments (see also Tobe et al., 1976). However, this
quantitative difference apart, we have not observed any difference
in the distribution of radiolabel in the tissues subjected to the
three different treatments (see Fig. 4). In view of the fact that
[^{3}H]-JH III is the principal radioactive component in glands subjected to treatment A whereas [^{3}H]-methyl farnesoate is the principal radioactive component in glands subjected to treatment C, we
interpret this uniform distribution of label throughout the gland
under all treatments as indicating that both the sites of esterification and epoxidation stages in JH biosynthesis occur uniformly
throughout the cytoplasmic mass of the CA. Similarly, we have not
observed any age-dependent difference between CA subjected to the
three treatments. We have previously shown that over the range of
experimental ages selected for the present study, the rate of
stimulated JH biosynthesis changes by a factor of 2 to 3 (Tobe and
Pratt, 1975a) which could suggest that a varying percentage of the
cells of the CA are involved in stimulated JH biosynthesis. However, our present studies indicate that such is not the case and
suggest that all cells of the CA are actively involved in stimulated JH biosynthesis at all times during sexual maturation, which
we might associate with the fact that there are no observable
differences between the cells of the CA, at the level of the light
microscope, using the histological procedures employed in the
present study.

It is noteworthy that after all treatments, the great majority
of label (more than 90%) in the sections was associated with the
cytoplasmic compartment of the CA cells and only few grains were
located over the nuclei (see Fig. 4). In addition, little label
could be found associated with the contiguous connective tissue
sheath or nerves. This allows us to exclude the possibility that
the nuclei of the cells of the CA are in any way involved in the
final two stages of JH biosynthesis. Similarly, it appears
unlikely that contiguous tissues associated with CA play any part
in JH biosynthesis. No mitotic figures were observed in the adult
CA at any of the ages examined; this is, in this respect, in
agreement with the observations of Odhiambo (1966) on the adult
male CA of S. gregaria.

DISCUSSION

The results presented here and elsewhere (Tobe and Pratt,
1975a,b) enable us to correlate the developmental changes during
sexual maturation of adult female S. gregaria as revealed by three
distinct experimental criteria. Thus we have monitored a) the
changing ability of CA to synthesize and release JH III spontaneously; b) the changing ability of the CA to utilize optimal concentrations of farnesenic acid to support stimulated biosynthesis;

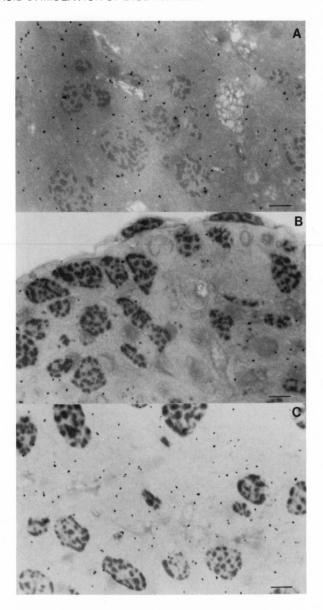

Fig. 4. Typical autoradiographs of S. gregaria CA showing grain
distribution after the three treatments described in the text.
A) CA of 4 day old S. gregaria after treatment A, i.e. no chase
and fixation immediately after incubation; B) CA of 11 day old
S. gregaria after treatment B, i.e. aerobic chase after incubation;
C) CA of 4 day old S. gregaria after treatment C, i.e. anaerobic
chase after incubation. Note uniform distribution of label in
cytoplasm with little label in nuclei. Scale = 7 μm.

and c) changes in the total volume of the CA. Each of these experimental criteria have been previously employed to assess the state
of activity of the CA in this and other insect species (a)-
Tobe and Pratt, 1975a; Weaver et al., 1975; b) Tobe and Pratt,
1975a; c) Johnson and Hill, 1973, 1975; Tobe and Pratt, 1975b; see
also Engelmann, 1970). One of the purposes of the present paper
is to synthesize the different types of information which derive
from these three classes of observation, as they apply to the
desert locust, into a unified hypothesis which describes at least
some of the developmental changes in the CA during adult reproductive life. However, such an hypothesis must necessarily be
preliminary since there are many aspects of CA biochemistry which
are beyond direct experimental observation at this time.

The utilization of exogenous [^{3}H]-farnesenic acid by CA in
vitro, which results in stimulated rates of JH biosynthesis,
bypasses the earlier rate-limiting steps in JH biosynthesis de
novo and is believed not to occur in vivo (Tobe and Pratt, 1974a).
Rather it permits an assessment of the enzymic competence of the
CA to affect the esterification and epoxidation of the farnesenic
acid. We do not know what constitutes the rate-limiting step
under conditions of stimulated JH biosynthesis, although the
exogenous farnesenic acid concentration is an important determinant of the stimulated rate (Tobe and Pratt, 1974a). In view
of the fact that the final epoxidation step is almost certainly
not rate-limiting in CA of this species (Pratt et al., 1975b),
it would appear that it is the activity of the enzymes associated
with the esterification of farnesenic acid by the S-methyl group
of methionine which determines the measured rates of stimulated
biosynthesis. To date, there is no definitive evidence of the
subcellular compartment in which this esterification occurs
although the experiments of Reibstein and Law (1973), using homogenates of the CA of adult Manduca sexta, suggest that it does
not take place in the nucleus. This is evidenced by their report
that the ability to incorporate radiolabel from (methyl) [^{3}H]-SAM
into known JHs is restricted to a supernatant obtained after centrifugation of CA homogenates at 12,000 x g for 5 min; no evidence
was presented for the precise cytoplasmic localization of the
enzymes involved.

It has been reported by Hammock (1975) that the final stage
in JH biosynthesis, namely the epoxidation of the triene ester,
is largely confined to the microsomal fraction obtained by differential centrifugation of CA homogenates from Blaberus giganteus.
This finding satisfactorily confirms earlier speculation that the
site of epoxidation is closely associated with the place of release
of JH from the cells of the CA (Tobe and Pratt, 1974b).

We have previously reported that during the period of sexual maturation of adult female S. gregaria, there are parallel changes in the total volume of the CA and in the overt enzymic activities of the final two stages in JH biosynthesis as revealed by rates of stimulated JH biosynthesis (Tobe and Pratt, 1975b). This led us to speculate that there might be a functional relationship between the final two stages in JH biosynthesis and the total cytoplasmic volume. Thus the reported works of Reibstein and Law (1973), Hammock (1975) and Tobe and Pratt (1974b, 1975b), taken together, strongly suggest exclusive involvement of cytoplasmic components to effect the last two stages in JH biosynthesis. The present study confirms these suggestions by way of autoradiographic procedures, which demonstrate the exclusively cytoplasmic distribution of the substrates and products of the enzymes involved in these reactions. Unfortunately, light microscopic autoradiography does not reveal whether the pools of these metabolites are contained in any specific cytoplasmic compartments such as mitochondria, smooth endoplasmic reticulum or Golgi complexes. The subcellular distribution of these pools is presently under investigation using electron microscopic autoradiography.

Our present finding that all the cells of the CA are involved in the conversion of farnesenic acid to JH III provides the basis for a direct comparison between stimulated JH biosynthesis and total cytoplasmic volume. Thus, our previous failure to demonstrate significant changes in the stimulated volume specific activity of the CA during sexual maturation of S. gregaria is easily understood (Tobe and Pratt, 1975b). In view of the fact that physiological changes in spontaneous JH biosynthesis are virtually independent of changes in the volume of the gland (Tobe and Pratt, 1975b), it would be of interest to determine whether all cells of the CA are involved equally in spontaneous JH biosynthesis, or whether changes in the overall spontaneous rate should be attributed to changes in the percentage of spontaneously active cells in the CA. This problem has not yet been investigated and an autoradiographic approach to it awaits the identification of a suitably specific physiological precursor. Until this question is resolved, it is not possible to devise a unique mechanistic model which describes the changing values of FEAR at the cellular level, because we cannot distinguish between the two extreme alternatives that: a) the cellular response to activation is all or nothing, in which case FEAR simply indicates the fraction of the cells in the CA which are spontaneously synthesizing JH; and b) all cells in the CA are equally capable of graded rates of spontaneous JH biosynthesis, in which case FEAR indicates the extent of their activation. Similar questions could be asked in the case of CA from other species which biosynthesize different relative quantities of the three known JHs in vitro (Röller and Dahm, 1970; Judy et al., 1973; Jennings et al., 1975). The advantage of using

FEAR as an experimental parameter resides in the fact that it allows us to evaluate the control of spontaneous JH biosynthesis independently of changes in the total volume of the glands which have been demonstrated to be poor indicators of the endocrine activity of the CA in this species (Tobe and Pratt, 1975b). That is not to say, however, that we attach no significance to changes in the volume of the CA, since as noted above, the volume correlates well with the ability of the glands to carry out at least two of the enzymic stages in JH biosynthesis which are "downstream" from the physiological rate-limiting step(s).

According to our interpretation, the rate of stimulated JH biosynthesis dictates the maximum possible rate of spontaneous JH biosynthesis (no rate limitation at the physiological control step; FEAR = 1.0). Our experiments have demonstrated that this condition is never realized, at least under our incubation conditions. However, it is possible that the progressive increase in stimulated JH biosynthesis and parallel increase in total glandular volume observed during the first 10 days of adult life reflect preparatory changes in the CA which are manifested in the high rates of spontaneous JH biosynthesis observed on days 11 and 15. Although we have no direct evidence that these changes in glandular volume and in stimulated JH biosynthesis are controlled by mechanisms different from those regulating spontaneous JH biosynthesis, the physiological significance which we attach to these parameters is sufficiently different that we have classified them into two groups: fast responses, changes in the rate of spontaneous JH biosynthesis; and slow responses, changes in the enzymic activity of the physiologically non-rate-limiting stages in JH biosynthesis and associated changes in cytoplasmic volume (see Fig. 5). The application of the above concepts, as they apply to developmental changes in the CA of adult female S. gregaria are diagrammatically summarized in Fig. 5.

The scheme outlined in Fig. 5 is consistent with all known data on the CA of S. gregaria but should not be assumed to be applicable to the functioning of the CA of other insects or for that matter to larval S. gregaria. In particular, the close parallel between total glandular volume and stimulated JH biosynthesis may not apply, perhaps because of large and independent changes in the volume of the extracellular spaces within the CA or large and independent changes in cytoplasmic components which are not involved in stimulated JH biosynthesis but which represent a significant proportion of the total cellular volume. In addition, it is possible that other types of CA do not always possess an excess enzymic competence to effect the last two stages in JH biosynthesis; in this case, the measured value of FEAR would equal 1.0 which would indicate that the physiological site of rate-limitation in the biosynthetic pathway was at a later step than is the case in adult female S. gregaria. If such differences between

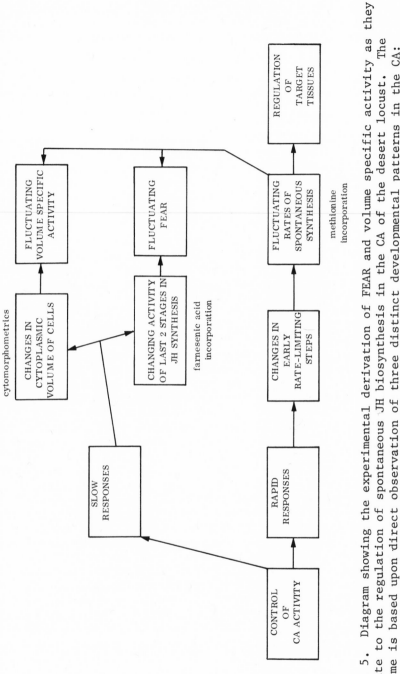

Fig. 5. Diagram showing the experimental derivation of FEAR and volume specific activity as they relate to the regulation of spontaneous JH biosynthesis in the CA of the desert locust. The scheme is based upon direct observation of three distinct developmental patterns in the CA: 1) spontaneous JH biosynthesis as shown by [^{14}C]-methionine incorporation into JH III; 2) stimulated JH biosynthesis as shown by both [^{3}H]-farnesenic acid and [^{14}C]-methionine incorporation into JH III; 3) total volume of the CA as computed from direct measurement of isolated glands. The functional distinction between slow and fast responses of the CA to presumed regulatory signals is described in the text.

CA from other sources exist, they would alter the relationship
between spontaneous JH biosynthesis and either CA volume or stimulated JH biosynthesis which could influence the physiological
significance of changes in the two latter parameters.

The rate of spontaneous biosynthesis is a direct estimate of
the overall activity of the CA and will clearly be the most relevant parameter to correlate with either hemolymph titers of JH or
with the manifest response of JH target tissues. By contrast,
the fractional endocrine activity ratio (FEAR) relates more
directly to the mechanism of control of CA activity in S. gregaria,
as it provides an opportunity to explore the extent of rate-limitation in consecutive parts of the JH biosynthetic pathway. Thus
FEAR indicates the relative importance of rate-limitation at early
and as yet undefined steps in JH biosynthesis, notwithstanding the
changes in the enzymic competence of the glands to affect the last
two stages. In our experiments, exogenous farnesenic acid is
efficiently utilized by the CA so as to nullify the effect of
rate-limitation at an early step(s) in JH biosynthesis. The discovery of other compounds capable of efficient entry into the intracellular JH biosynthetic pathway, at stages earlier than that of
the entry of farnesenic acid, might facilitate the more precise
identification of the site of physiological rate-limitation.

SUMMARY

The stimulation of $C_{16}JH$ (JH III) biosynthesis in freshly
isolated corpora allata of adult female Schistocerca gregaria by
optimum concentrations of [^{3}H]-farnesenic acid has been investigated using short-term radiochemical incubations and autoradiography. In any individual pair of corpora allata or in corpora
allata taken from animals of the same physiological age, the rate
of spontaneous JH biosynthesis divided by the rate of stimulated
JH biosynthesis is defined as the "Fractional Endocrine Activity
Ratio" (FEAR). Thus, FEAR describes the extent of rate limitation
at early step(s) in JH biosynthesis relative to the enzymic
activities associated with the final two stages in the pathway.
Determination of FEAR on each day during sexual maturation and
the first two gonotrophic cycles has revealed a developmental
series of peaks having reproducibly different magnitudes and which
can be associated with selected parameters of ovarian growth.
Incubation with [^{3}H]-farnesenic acid and pulse-chase techniques
have been used to obtain glands which contain different relative
amounts of [^{3}H]-farnesenic acid, [^{3}H]-methyl farnesoate and [^{3}H]-
JH III for autoradiographic investigation of the cellular distribution of these compounds. In general, these late intermediates
in JH biosynthesis were present in the cytoplasm of all cells of
the glands in approximately equal amount at all ages studied but

were largely absent from the nuclei. It is concluded that stimulated JH biosynthesis is carried out by all cells of the corpora
allata; this explains the correlation between stimulated JH biosynthesis and total glandular volume observed in this species.
A general scheme is presented for exploring the relationship
between spontaneous JH biosynthesis and other parameters of corpus
allatum activity.

ACKNOWLEDGEMENTS

We wish to thank Dr. A.F. White for supplying authentic
hormones and related compounds. SST acknowledges financial support
from the National Research Council of Canada.

REFERENCES

Bogoroch, R., 1972, in "Autoradiography for Biologists" (P.B.
    Gahan, ed.), pp. 66–94, Academic Press, New York.
Darrah, H.K., Hedley-Whyte, J., and Hedley-Whyte, E.T., 1971,
    J. Cell Biol. 49:345.
Engelmann, F., 1970, "The Physiology of Insect Reproduction,"
    Pergamon Press, Oxford.
Hammock, B.D., 1975, Life Sciences 17:323.
Jennings, R.C., Judy, K.J., Schooley, D.A., Hall, M.S., and Siddall,
    J.B., 1975, Life Sciences 16:1033.
Johnson, R.A., and Hill, L., 1973, J. Insect Physiol. 19:2459.
Johnson, R.A., and Hill, L., 1975, J. Insect Physiol. 21:1517.
Judy, K.J., Schooley, D.A., Dunham, L.L., Hall, M.S., Bergot, B.J.,
    and Siddall, J.B., 1973, Proc. Nat. Acad. Sci. USA 70:1509.
Kopriwa, B.M., and Leblond, C.P., 1962, J. Histochem. Cytochem.
    10:269.
Mak, K.M., and Trier, J.S., 1972, Biochim. Biophys. Acta 280:316.
Odhiambo, T.R., 1966, J. Insect Physiol. 12:655.
Pratt, G.E., and Tobe, S.S., 1974, Life Sciences 14:575.
Pratt, G.E., and Weaver, R.J., 1974, J. Endocrinol. 64:67P.
Pratt, G.E., Tobe, S.S., Weaver, R.J., and Finney, J.R., 1975a, Gen.
    Comp. Endocrinol. 26:478.
Pratt, G.E., Tobe, S.S., and Weaver, R.J., 1975b, Experientia 31:120.
Reibstein, D., and Law, J.H., 1973, Biochem. Biophys. Res. Commun.
    55:266.
Röller, H., and Dahm, K.H., 1970, Naturwissenschaften 57:454.
Tobe, S.S., and Pratt, G.E., 1974a, Biochem. J. 144:107.
Tobe, S.S., and Pratt, G.E., 1974b, Nature (London) 252:474.
Tobe, S.S., and Pratt, G.E., 1975a, J. Exp. Biol. 62:611.
Tobe, S.S., and Pratt, G.E., 1975b, Life Sciences 17:417.
Tobe, S.S., Pratt, G.E., and Saleuddin, A.S.M., 1976, C.N.R.S.
    Colloque N° 251: in press.
Weaver, R.J., Pratt, G.E., and Finney, J.R., 1975, Experientia 31:
    597.

CONTINUOUS MONITORING OF JUVENILE HORMONE RELEASE BY SUPERFUSED

CORPORA ALLATA OF PERIPLANETA AMERICANA

G.E. Pratt, R.J. Weaver and A.F. Hamnett

Agricultural Research Council, University of Sussex
Brighton, England

INTRODUCTION

The discovery by Röller and Dahm (1970) that significant
quantities of juvenile hormone could be isolated from cultures in
vitro of corpora allata from the adult moth Hyalophora cecropia
heralded the approach of a new era of experimental investigations
of the physiology of the insect corpus allatum, since it showed
that the glands continued to make and release the hormone after
removal from the body and therefore might be amenable to direct
observation in this way. Standard organ culture procedures for
the production of relatively pure preparations of juvenile hormones
biosynthesized in vitro, were then employed in conjunction with
specially developed analytical procedures by workers in two other
laboratories, so as to identify which of the three known juvenile
hormone homologs were produced by the corpora allata of the adult
female of several insect species (Judy et al., 1973a,b; 1975;
Jennings et al., 1975; Muller et al., 1974). K.J. Judy and his
co-workers also employed to their advantage in identifying the
homologs, the efficient and relatively specific incorporation of
radioactivity from the S-methyl group of methionine into the methyl
ester of juvenile hormones, previously discovered during experiments on H. cecropia in vivo by Metzler et al. (1971). Such
positive identifications are clearly a sine qua non for biochemical
investigations of the endocrine function of the corpus allatum.
The above investigations were aimed primarily at synthesizing sufficient hormone material for identification purposes, and in some
cases (Schooley et al., 1973; Jennings et al., 1975) for demonstration of the contribution made by simple precursors towards
the carbon skeleton of the hormones: the use of relatively long-
term organ culture techniques (involving the replacement of culture

medium every few days or weeks) was not favorable towards calculation of the actual rates of synthesis or release of hormones in
vitro. Furthermore, with the possible exception of the early work
on corpora allata from H. cecropia (Röller and Dahm, 1970), the
overall yields of hormone from these classic organ culture experiments were not sufficiently high to encourage the prospect that
it might be possible to measure the rates of hormone biosynthesis
over relatively short periods of time in vitro.

However, the situation changed when exploratory work by Pratt
and Tobe (1974) revealed that corpora allata from adult female
Schistocerca gregaria could utilize exogenous sesquiterpenoid acids
such as farnesenic acid to produce juvenile hormones at "rates"
which were two to three orders of magnitude higher than those
obtained from conventional organ cultures. With these much higher
rates of incorporation of radioactivity from methyl-labeled methionine, it proved possible to demonstrate linear rates of synthesis
of JH III over periods of just a few hours after removal from the
animal, and to examine the effects of incubation conditions, such
as the concentration of methionine, farnesenic acid, pH and gaseous
oxygen tension on these stimulated rates of hormone biosynthesis
(Tobe and Pratt, 1974a). The use of different radiolabels in
farnesenic acid and methionine, together with glands exhibiting
low spontaneous rates of hormone biosynthesis, indicated that
methionine was an almost, if not entirely, unique source of the
methyl ester in the double-labeled hormone thus synthesized, and
could therefore be used as a mass marker of known specific activity.
Similarly, Judy et al. (1975) used mass fragmentographic analysis
of highly purified JH III which had been biosynthesized by isolated retro-cerebral complexes of adult female Tenebrio molitor,
to demonstrate only a slight, albeit measurable, dilution of the
radiolabel from [methyl-^{14}C]-methionine; whilst it is possible
that the observed dilution by natural radionucleid was due to the
ongoing utilization by the glands in vitro of an alternative donor
of methyl groups (but one which was less preferred to methionine),
it could also have been due to the presence of unlabeled hormone,
or an intermediate such as S-adenosyl methionine (Reibstein and
Law, 1973) or methyl farnesoate (Pratt et al., 1975a), in the
glands at the time of extirpation. The usefulness of methyl-
radiolabeled methionine as a quantitative marker for biosynthesized
juvenile hormones led to the development of standard short-term
radiochemical assay procedures for comparing the rates of spontaneous synthesis and release of JH III by freshly isolated corpora
allata of adult female Periplaneta americana and S. gregaria (Pratt
et al., 1975b).

The recent use of these short-term radiochemical assay procedures for direct estimation of the changing patterns of juvenile
hormone biosynthesis and secretion by the corpora allata of female
adult P. americana and S. gregaria has already served to confirm

and considerably extend several earlier concepts on the endocrine
role of the corpus allatum in the adult female insect[*]. A particularly illuminating discovery is that of the rapidity with which
the activity of the glands changes during sexual maturation and
the subsequent ongoing gonotrophic cycles in these two primitive
species of insect. In maturing females of both species a quantitatively small pulse of JH synthetic activity shortly after molting is followed by a series of major peaks of glandular activity,
each of which is approximately coincident with the onset of vitellogenesis in the successive (and in these spp., highly synchronized) batches of oocytes (Tobe and Pratt, 1975; Weaver et al.,
1975; Weaver and Pratt, 1975). During these gonotrophic cycles
of the corpus allatum, the rates of JH biosynthesis can increase
from their minimum to maximum observed values in little more than
a day, representing increases of approximately 25-fold and 100-
fold in P. americana and S. gregaria, respectively. It is of
considerable interest that the reproductively primitive P. americana, in which the ovaries exhibit vitellogenesis continuously
once sexual maturity has been attained, nevertheless has pronounced and rapid cycles of corpus allatum activity; it will be
of interest to see whether the mechanisms responsible for this
rapidly rising and falling activity of the corpus allatum in spp.
such as these, are the fore-runners of the mechanisms which are
believed to operate in those spp. with more elaborate reproductive
systems [for example, see Engelmann (1970)].

The manifestly dynamic role of the corpus allatum in adult
females of P. americana and S. gregaria carries obvious implications for the design of several sorts of experiments, including
the production in vitro of sufficient quantities of hormone for
chemical identification. For example, the long-term organ cultures of corpora allata employed by Muller et al. (1974) for
their chromatographic and mass fragmentographic identification of
JH III as the principal juvenile hormone of adult female P. americana, were taken from animals at a stage in the oviposition cycle
when the rate of synthesis of JH III is apparently at its lowest
(Weaver et al., 1975). Judy et al. (1975) have recently noted
the importance of using correctly aged females of T. molitor for
the production by their retro-cerebral complexes in vitro of
identifiable quantities of JH III. Even so, the overall yield
of JH III obtained by Muller and his co-workers from their long-
term (6-10 weeks) organ cultures, calculates out to a maximum
estimated "rate" of JH biosynthesis which is at least 10 times
lower than the average minimum rates of biosynthesis observed by
Pratt and Weaver (1974), Weaver and Pratt (1975) and Weaver et al.
(1975) during the course of 3 hr incubations. It is already known
that the rates of biosynthesis of JH III by individual pairs of

---

[*] For a general discussion of classical corpus allatum physiology
in the adult female insect see Engelmann (1970).

corpora allata from adult female P. americana remain constant for
a period of 5 hr after removal from the animal, irrespective of
the spontaneous rate exhibited (Pratt et al., 1975b). A simple
hypothesis which reconciles these apparent quantitative disparities
would be that after an unknown period of time in vitro the glands
begin to lose synthetic activity. This idea receives some support
from the observations of Judy et al. (1973a) on JH III production
by corpora allata from M. sexta over a period of 15 days, and of
Pratt (unpublished observations) on the production of JH III by
corpora allata from S. gregaria over a period of 5 days, but it is
clearly unwise to place much more than trivial reliance upon the
outcome of comparisons of the results obtained with these very
different experimental procedures. There is clear need for observations on the long-term hormone productivity of corpora allata
in vitro which are more directly comparable with the results
obtained from short-term incubations. Advantages of several kinds
might accrue from an accurate description of the changing performance of the glands in vitro, by laying the basis for further experiments on the mechanisms of control of the synthetic activity of
the glands. Although conventional short-term incubation procedures
have been employed successfully to assess the activity of acute
chemical inhibitors of JH biosynthesis in corpora allata from adult
female P. americana (Pratt and Finney, 1975), it is probable that
elucidation of the activity and mode of action of natural regulators of corpus allatum activity will require observations to be
made over much longer periods of time. We shall describe here,
results obtained with corpora allata from adult female P. americana
which have been continuously superfused with radiolabeled chemically
defined tissue culture medium, for long periods of time, in a
system which automatically provides hourly samples of the released
juvenile hormone.

MATERIALS AND METHODS

The rearing of experimental animals was as described by
Weaver et al. (1975) and the standard 3 hr incubation procedure
for estimation of the spontaneous rates of $C_{16}JH$ (JH III) biosynthesis, by individual pairs of freshly isolated corpora allata
was as described by Pratt et al. (1975b). The culture medium
employed was TC199 medium (with glutamine) (Flow Laboratories,
Edinburgh) plus 20 mg/ml Ficoll and 20 mM HEPES buffer (pH 7.1-
7.2) at 30°C, other compounds being added as indicated in the text.
As in the case of the superfusion analyses described below, rates
of synthesis of JH III were calculated from the incorporation of
radioactivity from [methyl-^{14}C]-methionine of known specific radioactivity in the medium. The females employed were young, but of
unknown age, and physiological synchronization was achieved by
selecting animals with recently extruded and still mid-brown
ootheca, and the animal then aged for the required number of hours

before sacrifice. The efficiency of the synchronization procedure
was checked by measurement of the sizes of the two most terminal
groups of oocytes, and as before (Weaver et al., 1975) a small
percentage of animals was rejected because the temporal coordination between the growth of T and T-1 oocytes deviated markedly
from the norm.

   Continuous superfusion of isolated corpora allata was carried
out in a water-jacketed and modified 1 cm diameter chromatography
column fitted with adjustable column supports. The column and its
jacket were shortened so that both column supports could meet in
the middle, thereby enclosing a gap of approximately 0.7 mm depth
in which the corpora allata were held. The original sintered
polythene support discs had an extremely high adsorption affinity
for JH in aqueous solution, and were replaced with 0.02 mm pore
nylon mesh held in place with a small quantity of epoxy-resin
adhesive. Radiolabeled culture medium was pumped from a refriger-
ated reservoir via a silicone tubing occlusion pump, usually at a
rate of 0.5 ml/min, into the bottom of the superfusion chamber,
which was held at 30°C. The effluent from the top of the chamber
was immediately joined by an equal stream of methanol which swept
the products of the superfusion to the fraction collector. All
tubing connections were in 1/16 inch Teflon. The entire assembly,
including dissection facilities but excluding the fraction collector, was located within a five feet wide horizontal laminar flow
sterile air-bench. Sterile procedures were employed throughout
the setting-up and operation of the superfusion apparatus; the
apparatus itself was sterilized by washing through for 1 hr with
1% sodium hypochlorite solution, followed by an overnight stand
filled with 1% glutaraldehyde. The glutaraldehyde was removed by
repeated pumping through with sterile water, and finally with
sterile non-radiolabeled culture medium. We routinely employed
10 pairs of corpora allata, which were passaged through five
changes of sterile culture medium before individual transfer on a
small stainless steel loop to the open superfusion chamber: the
corpora allata sink to the bottom of the chamber and come to rest
on the lower support net. The pre-sterilized upper support was
then lowered into the corpora allata and 2.5 ml of radiolabeled
medium pumped through at high speed to establish the correct level
of radioactivity in the effluent from the chamber. The operational
flow rate was then established and the effluent delivered into
tubes pre-loaded with 2.0 ml chloroform which were changed every
hour. The final specific activity of [methyl-^{14}C]-methionine in
the medium was set at 10-12.5 mCi/mmol in these experiments.
Extraction, thin-layer chromatography and quantitation of the JH III
released was done as described by Pratt and Tobe (1974) excepting
that the extraction volumes were increased in proportion to the
aqueous volume, and only the sections of the plate corresponding
to JH III and methyl farnesoate (as appropriate) were taken for
scintillation counting. In some superfusions, every tenth tube

was submitted to micro-preparative thin layer chromatography using
JH I and JH III as lateral markers, prior to radio-gas chromatography.

Gas chromatography was conducted on 20 M glass capillary
columns coated with UCON HB5100 (Jaeggi, Switzerland) via an all-
glass solvent-free injector (Phase Separations, England). Authentic samples of the three known JH homologs and some of their
unnatural geometrical isomers, either non-radioactive or [^{3}H]-
labeled, were added to the sample on the injection probe in 50-
200 ng quantities as reference markers. Double-labeled radioactivity in the effluent was determined by means of an effluent
collector developed by Dr. C.M. Wells, Department of Chemistry,
University of Nottingham, England, in which the products of total
combustion from the flame ionization detector are passed through
heated glass-lined steel tubing to a scrubbing tower containing a
continuously flowing annular curtain of 5% 2-phenyl ethylamine in
ethyl cellosolve which is collected every 12 sec with the help of
an automatic valve. Actual recoveries of radioactivity were
determined with standard double-labeled methyl palmitate solutions
during each group of analyses and were typically 60% for [^{14}C] and
90% for [^{3}H]. The effluent fractions were counted in a Wallac
8100 liquid scintillation spectrometer using a napthalene/butyl-
PBD/dioxan scintillant, calibrated with standard [^{14}C,^{3}H]-n
hexadecanes.

RESULTS

Early experiments showed that it was not possible to measure
the output of JH from a single pair of glands in this system without the use of large quantities of radiolabeled precursor. The
majority of observations were conducted with 10 or 12 pairs of
glands in the chamber and lower levels of labeled methionine. A
detection limit in the scintillation counter equivalent to 0.1-
0.2 pmol, hr^{-1} per pair was realized. Figure 1 shows a selection
of four superfusion profiles obtained with glands taken from
females at various times during the oviposition cycle, which illus-
trate the two patterns of activity observed thusfar. The degree
of radiochemical purity of the thin-layer chromatographic cut
corresponding to JH III which was employed for these measurements,
is attested by Fig. 2 which shows a radio-gas chromatogram of the
plate area corresponding to JH III through JH I: essentially all
the [^{14}C] radioactivity cochromatographs with authentic 2E,6E
JH III. Glands removed from animals at their time of minimum
activity, corresponding approximately to the ovulation of chorio-
nated T oocytes, continue to exhibit low rates of hormone release
when cultured for over two days. Glands removed from animals
after the onset of vitellogenesis in the T-1 oocyte, which exhibit
either high or declining spontaneous activity (Weaver et al., 1975),

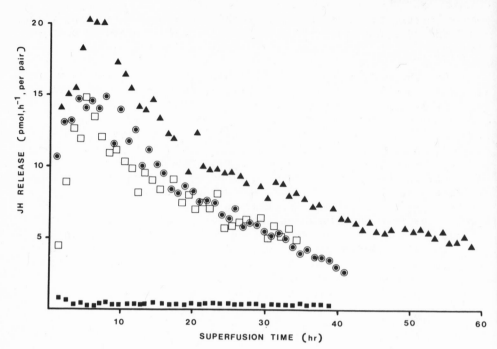

Fig. 1. Time-courses of the rates of release of JH from groups of 10 pairs of corpora allata removed from females at different times after formation of an ootheca, during continuous superfusion in vitro. Solid squares, 0 hr; circles and squares, 30 hr; triangles, 36 hr. Glands were superfused in augmented TC199 medium containing radiolabeled methionine, and the rates of release of newly synthesized JH were calculated from the measured incorporation of radioactivity into JH isolated by thin-layer chromatography.

exhibit similar rates of release of newly-synthesized JH during superfusion for up to 8-10 hr; thereafter the activity declines, and reaches the "resting level" of 1-3 pmol, hr^{-1} per pair observed in short term incubations, after periods of from 50 to 80 or more hours. Continuation of the superfusion for several more days reveals a continued very slow decline to levels of less than 0.5 pmol, hr^{-1} per pair, which we do not regard as physiological in glands from reproductively mature animals of this species. It is noteworthy that the higher the initial activity of the glands at the time of extirpation, the longer it takes to reach the "resting level." We also made the detrimental observation that in most of the superfusions carried out with glands exhibiting medium to high activities it took 3 or 4 hr for the labeled JH III content of the effluent to reach its short-lived plateau value. The simple

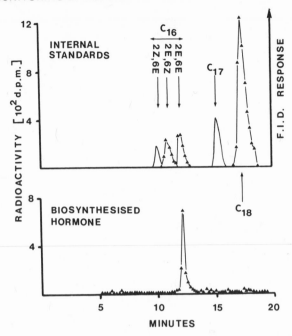

Fig. 2. Radio-gas chromatograph of the "JH" zone on thin-layer chromatography of the product of corpora allata after more than 30 hr superfusion in medium containing [^{14}C]-methionine. Non-radioactive or [^{3}H]-labeled authentic hormones were added as internal reference compounds. Carrier gas: Argon at 15 p.s.i., temperature program 160°–180°C at 2°/min, scrubbing flow-rate 11 ml/min.

analysis of short-term test-tube incubations has already shown that there is a radiochemical lag period of half to one hr in the corpora allata of this species representing the time taken for the radiolabel in exogenous methionine to reach isotopic equilibrium in the newly-synthesized JH leaving the gland (Pratt et al., 1975b) and that following this lag the rate of release of labeled JH remains constant for at least a further 4 hr. Further, the super-fusion chamber itself and the downstream Teflon plumbing add a further lag of approximately one half hr, at the superfusion rates employed here. These predicted lags are insufficient to account for the progressive increase during the first few hr of superfusion, which must therefore be regarded as artificial. The source of the artifact has not been identified (for example it might be due to adsorption of hormone onto the walls of the superfusion chamber), and unless it can be eliminated, simple test-tube incubations must be the method of choice for observing the synthetic activity of the glands during the first 4 or 5 hr after isolation.

The failure of inactive glands to become activated, and the
eventual decline in activity of glands which were active at the
time of removal, have to be interpreted within the framework of
the physiological time-scale of the insect. In this respect, the
relationship between oocyte length and glandular activity established by Weaver et al. (1975) was not immediately helpful, but
subsequent work on maturing females of known age (Weaver and Pratt,
1975; and unpublished data) did indicate that the in vivo periodi-
city of the glands in sexually mature female P. americana was
probably shorter than would be predicted from rates of change
observed during superfusion in vitro. In order to compare directly
the changes taking place in vivo with those obtained in vitro,
groups of synchronized animals were divided into two sub-sets:
(a) animals which provided corpora allata at a particular time in
the oviposition cycle for superfusion analysis; and (b) animals
providing corpora allata for short-term incubation to assess the
rate of change of spontaneous activity in vivo. The temporal
pattern of corpus allatum activity during the oviposition cycle is
shown in Fig. 3; the development of the T, T-1 and T-2 oocytes
during this period are shown diagrammatically for comparison.

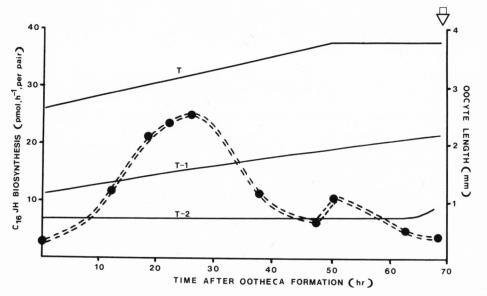

Fig. 3. Changing rates of biosynthesis of JH III in isolated corpora allata through the oviposition cycle. Each point is the mean
of 3 or 4 determinations made on individual glands incubated for
3 hr in medium containing [^{14}C]-methionine; the points indicate
the beginning of each group of incubations. Continuous lines
indicate the respective increases in length of the three most terminal oocytes during the same period.

The observed relationship between glandular activity and ovarian
development is the same as that observed previously (Weaver et al.,
1975) except that activation of the T-1 oocytes occurs sooner
after chorionation of the T oocytes in the present animals,
perhaps due to increased vigor in the colony (see also Pratt and
Finney, 1975). These observations show rapid activation of the
corpora allata shortly after ovulation of the T oocytes, reaching
maximum activity approximately 24 hr after extrusion of the
ootheca. The complete failure of glands isolated at the time of
ootheca formation spontaneously to reactivate in vitro during this
period confirms many earlier proposals (Engelmann, 1970) that high
activity of the corpus allatum in adult female insects requires
the action of fairly specific trophic stimuli, which would not
exist for the denervated glands in the chemically defined superfusion medium employed. The subsequent decline of glandular
activity in vivo is more drawn out than the activation, and here
the difference between the effects of ageing in vivo and in vitro
is less marked. Glands isolated at 30 and 36 hr after ootheca
formation do appear to lose activity more slowly than their
counterparts in vivo, such that by 60 or 70 hr after ootheca formation their rate of release of JH III is approximately twice
that of freshly isolated glands of the same experimental age.
Whilst this difference could suggest that the environment and
chemical circumstances of the superfusion system employed here are
conducive to the maintenance of high synthetic activity in the
corpora allata, we do not feel that it is sufficient to justify
a claim to discovery of in vivo inactivation of the corpora allata.
Further work will be necessary to test this interesting possibility.

These experiments all call to question the general suitability
and cellular compatability of the culture medium employed for superfusion. As indicated above, the fact that active glands aged in
vitro for up to 40 hr do not have lower synthetic rates than
contemporary glands "aged" in vivo, provides only incomplete support for the belief that the medium is physiologically suitable.
Therefore, we were interested to obtain more direct indices of the
biochemical normality of the glands cultured under these conditions.
Unfortunately, very few biochemical parameters of the metabolic
status of either the intact or recently isolated corpus allatum
have been investigated thusfar, which considerably restricts the
choice. We elected to examine the ability of aged glands to respond to exogenous 2E,6E farnesenic acid, since this response has
been examined in some detail with freshly isolated glands from
both P. americana and S. gregaria (Pratt and Weaver, 1974; Tobe
and Pratt, 1974a; Pratt et al., 1975a; Tobe and Pratt, 1975a).

In the present experiments, active glands were superfused for
just over 40 hr until their rate of release of JH had dropped to
less than 50% the initial value, and the culture medium was then
replaced at a high pumping rate with medium of identical composition

and specific radioactivity to which had been added a nominal concentration of 20 μM farnesenic acid. Superfusion with this forti-
fied medium was then continued at the standard flow rate for a
further 11.5 hr, after which the farnesenic acid was flushed out
and chased with unfortified medium. The results of two such experiments are shown in Fig. 4, which clearly reveals the immediate
stimulation of JH III biosynthesis by the added farnesenic acid.

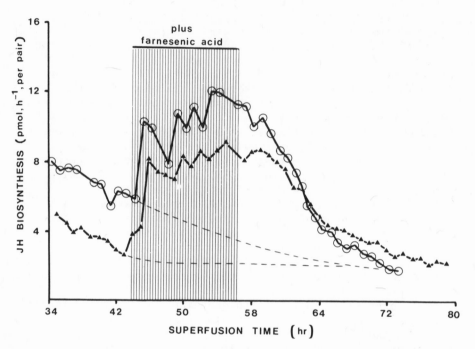

Fig. 4. Stimulation of the rate of release of newly synthesized
JH from corpora allata aged in vitro, by added farnesenic acid.
Two separate experiments are shown. The narrow dashed lines are
interpolations showing the continued decline of spontaneous synthetic activity during and after the farnesenic acid challenge.
Each point represents a single determination of the release of
radiolabeled JH from 10 pairs of glands during the course of a
1 hr period of superfusion in medium containing [^{14}C]-methionine.

This shows that these aged glands, like these freshly isolated
from the animal, are capable of utilizing exogenous farnesenic
acid to promote enhanced rates of JH biosynthesis, indicating that
the rate-limiting step in JH biosynthesis is still prior to the
level of entry of farnesenic acid into the biosynthetic pathway
(Pratt and Tobe, 1974; Pratt and Weaver, 1974), which serves as an

encouraging index of the metabolic "health" of the glands. It is noteworthy that at no time during or after these exposures to farnesenic acid, was any radioactive methyl farnesoate detected in the effluent, although it is known that large quantities of this intermediate accumulate within the gland cells during exposure to farnesenic acid (Pratt et al., 1975a): this provides indirect evidence of the continued integrity of the cytomembranes of the corpus allatum cells during this period of waning biosynthetic activity.

DISCUSSION

The repeatable patterns of changing synthetic activity in vitro of corpora allata taken from adult female P. americana at various times during the oviposition cycle, suggest that continuous superfusion may be of reliable use for appropriate further investigations on the natural control of JH biosynthesis. In addition to the ease of automation, continuous superfusion eliminates the possibility inherent in long-term static cultures that previously synthesized hormone remaining in contact with the glands, may be broken down by unknown mechanisms. However, an inherent disadvantage of the system in its present form is that there is no provision for a gas phase in the vicinity of the glands, so that the oxygen supply to the glands is limited to that dissolved in the perfusing medium, which is equivalent to a supply of 2-3 µl (N.T.P.)/hr at the superfusion rate employed, when in equilibrium with atmospheric oxygen. In view of the facts that isolated muscle from P. americana having the same weight as the 10 pairs of corpora allata in the superfusion chamber would only consume oxygen at the rate of 0.5-1 µl (N.T.P.)/hr (Barron and Tahmisian, 1948), and that the poor tracheation of the corpora allata in P. americana suggests that these glands have a relatively low oxygen demand (Pratt et al., 1975a), it seems reasonable to conclude that oxygen is not seriously rate-limiting under superfusion conditions. However, it seems likely that the provision of a more efficient oxygen supply would make the method more generally useful, particularly in view of the highly aerobic appearance of corpora allata from S. gregaria (Pratt et al., 1975a) and the observation by Chino et al. (1974) that the highly tracheolated prothoracic glands of Bombyx produced larger quantities of α-ecdysone in long-term hanging drop cultures when the oxygen tension in the gas phase was increased above atmospheric.

Pratt et al. (1975b) and Tobe and Pratt (1974b) have demonstrated with short-term experiments on freshly isolated corpora allata from reproductively active females of both P. americana and S. gregaria that the rate of release of newly synthesized JH is invariably proportional to the rate of synthesis, and these authors concluded that the release of hormone must be a rapid and relatively simple process. Although we have not been able to measure both the rate of release of JH and the simultaneous intra-glandular content of JH in the present experiments, we have no reason to believe that this simple relationship between the two will change during superfusion, and that is the basis upon which we have compared the rates of release of JH from superfused glands with the rates of synthesis and release from glands during short-term incubations. The fact that addition of farnesenic acid to the medium superfusing "aged" glands effects an immediate increase in the rate of release of newly synthesized hormone, further supports the conclusion that the decline in the rate of release of JH during long-term culture is not because the glands have switched from a "synthesize and release" mode to a "synthesize and store" mode, but is because of a progressive decrease in the anabolic flow through some early rate-limiting step in the biosyn-thetic pathway.

The choice of a culture medium is clearly of considerable importance in experiments such as these. The medium should be one which: (a) permits the glands to express fully their natural range of spontaneous activities; (b) affords reproducible rates of JH synthesis by glands in any particular physiological state; and (c) does not contain any specific activators or inhibitors of JH biosynthesis which might themselves be of physiological inter-est. Previous work in this laboratory has suggested that the augmented TC199 medium employed in the present and other experi-ments comes reasonably close to having the desired "physiological indifference" towards corpora allata from both P. americana and S. gregaria. Thus, the observed activities of the glands can be reliably associated with particular physiological states of the gland, and the medium permits the expression in vitro of a wide range of spontaneous synthetic activities by freshly isolated glands (Weaver et al., 1975; Tobe and Pratt, 1975). However, it may be more difficult to decide upon the suitability of the medium for long-term cultures, partly because what one considers to be a "physiologically indifferent" medium depends upon the particular experimental question one is asking. For example, if the declin-ing phase of corpus allatum activity observed here during long-term superfusion could be attributed to the absence from the medium of specific metabolic substrates or humoral factors which are always present in the hemolymph, then the medium we have employed would be unphysiological and far from indifferent. Alter-natively, if such factors did exist in vivo, but it was their cyclical appearance and disappearance which was in part or whole responsible for the observed cycles of corpus allatum activity during the oviposition cycle, then the medium employed here would be "physiologically indifferent" and might provide a satisfactory system for the demonstration of such trophic factors. Such matters can only be solved by further exploratory research. In view of the almost certain involvement of both nervous connections to the corpora allata and the composition of the bathing hemolymph in determining the synthetic activity of the glands (Engelmann, 1970), the experimental solutions to these problems are unlikely to be simple.

SUMMARY

It is already known that in adult females of two reproductively simple insects, Schistocerca gregaria and Periplaneta americana, there are wide and rapid fluctuations in the rates of synthesis and release of C_{16}JH (JH III) associated with successive waves of vitellogenesis in the ovaries, which imply potent natural mechanisms for regulating the activity of the glands. A system is described in which corpora allata are continuously superfused under sterile conditions with defined culture medium containing methyl-labeled methionine, and hourly fractions collected. Biosynthesized JH is isolated and identified by chromatography, and quantitated by liquid scintillation counting. The rates of release of hormone during the first 8-10 hr of superfusion are in reasonable agreement with the rates observed during short-term incubation of glands from females of the same physiological age. Glands which exhibited low rates of synthesis at the time of explantation remain relatively inactive during superfusions of up to 54 hr: during this period the glands would have been maximally activated had they been allowed to remain within the animal. Active glands begin to lose activity after 8-10 hr superfusion, and their rate of hormone release falls to the resting level after 50 to 80 hr superfusion. Glands which have lost over 50% of their release activity still respond to the addition of farnesenic acid to the medium by increased rates of hormone synthesis, indicating that the rate limiting step in biosynthesis is still upstream prior to the entry of farnesenic acid. The "programmed" decline of activity is reproducible, provided animals of similar physiological age with respect to the ovarian maturation cycles are employed as donors, and may serve as the basis for experimental investigation of slow-acting activators and inhibitors of JH biosynthesis.

ACKNOWLEDGEMENTS

We are grateful to Dr. A.F. White and Dr. R.C. Jennings for supplying most of the authentic reference compounds, to Mr. K.G. Pike for manufacture of the superfusion chamber and the chromatographic effluent gas scrubbing tower, to Dr. C.M. Wells for helpful advice on construction and operation of the latter, and to Mrs. Kathy Stott for technical assistance. We also recognize the generous encouragement of Prof. A.W. Johnson FRS, and Mr. S.E. Lewis.

REFERENCES

Barron E.S.G., Tahmisian, T.N., 1948, J. Cell. comp. Physiol.
    32:57.
Chino, H., Sakurai, S., Ohtaki, T., Ikekawa, N., Miyazaki, H.,
    Ishibashi, M., and Abuki, H., 1974, Science 183:529.
Engelmann, F., 1970, in "The Physiology of Insect Reproduction"
    pp. 143-192, Pergamon Press, Oxford.
Jennings, R.C., Judy, K.J., Schooley, D.A., Hall, M.S., and
    Siddall, J.B., 1975, Life Sciences 16:1033.
Judy, K.J., Schooley, D.A., Dunham, L.L., Hall, M.S., Bergot, B.J.,
    and Siddall, J.B., 1973a, Proc. Nat. Acad. Sci. USA 70:1509.
Judy, K.J., Schooley, D.A., Hall, M.S., Bergot, B.J., and Siddall,
    J.B., 1973b, Life Sciences 13:1511.
Judy, K.J., Schooley, D.A., Troetschler, R.G., Jennings, R.C.,
    Bergot, B.J., and Hall, M.S., 1975, Life Sciences 16:1059.
Metzler, M., Dahm, K.M., Meyer, D., and Röller, H., 1971, Z.
    Naturforsch. 26b:1270.
Muller, P.J., Masner, P., Trautmann, K.H., Suchý, M., and Wipf,
    H.K., 1974, Life Sciences 15:915.
Pratt, G.E., and Finney, J.R., 1975, "Chemical inhibitors of
    juvenile hormone biosynthesis", in "The Evaluation of Biological
    Activity: The Future of Pesticides" Academic Press, New York,
    in press.
Pratt, G.E., and Tobe, S.S., 1974, Life Sciences 14:575.
Pratt, G.E., and Weaver, R.J., 1974, J. Endocrinol. 64:67p.
Pratt, G.E., Tobe, S.S., and Weaver, R.J., 1975a, Experientia 31:
    120.
Pratt, G.E., Tobe, S.S., Weaver, R.J., and Finney, J.R., 1975b,
    Gen. Comp. Endocrinol. 26:478.
Reibstein, D., and Law, J.H., 1973, Biochem. biophys. Res. Commun.
    55:266.
Röller, H., and Dahm, K.H., 1970, Naturwissenschaften 57:454.
Schooley, D.A., Judy, K.J., Bergot, B.J., Hall, M.S., and Siddall,
    J.B., 1973, Proc. Nat. Acad. Sci. USA 70:2921.
Tobe, S.S., and Pratt, G.E., 1974a, Biochem J. 144:107.
Tobe, S.S., and Pratt, G.E., 1974b, Nature 252:474.
Tobe, S.S., and Pratt, G.E., 1975, J. Exp. Biol. 62:611.
Weaver, R.J., and Pratt, G.E., 1975, Gen. Comp. Endocrinol., in
    press.
Weaver, R.J., Pratt, G.E., and Finney, J.R., 1975, Experientia 31:
    597.

SUMMARY OF SESSION III:

JUVENILE HORMONE EFFECTS AT THE CELLULAR LEVEL

N.A. Granger and H.A. Schneiderman

Department of Developmental & Cell Biology
University of California, Irvine, California 92664

Most of the papers in this next section focus on the interaction between juvenile hormone and ecdysone. They deal with a
number of questions, but three are recurrent themes. When and how
do epidermal cells change their commitment from secreting larval
cuticle to synthesizing pupal cuticle? Is mitosis or DNA replication necessary for this reprogramming? What is the nature of the
interactions between juvenile hormone (JH) and ecdysone at the
cellular level and at the level of the whole organism?

The reader will be struck by the fact that in addition to
analyzing the results on whole organisms, most of the papers report
results obtained from in vitro studies: cell cultures, epidermal
vesicles, fragments of skin, imaginal discs, salivary glands. The
obvious asset of these in vitro systems is that they circumvent
such problems as endogenous hormone titers, the complexities of
hormone metabolism, and the interaction of exogenous hormones with
the neuroendocrine system. However, most of the papers also point
out the difficulty of obtaining in vitro a morphological and biochemical response to physiological levels of JH and ecdysone,
similar to that found in vivo. Apparently the effective concentrations of hormones in vivo are regulated by factors which have not
yet been duplicated in vitro; in addition, in vitro studies utilizing JH, as pointed out in the report of Oberlander and Silhacek,
must take into account degradation, nonspecific binding, and
sequestration of the added hormone.

Notwithstanding these problems, the results reported from the
in vitro studies are enormously compelling. For example, Riddiford
shows that brief in vitro exposure to low concentrations of ecdysone
changes the commitment of larval epidermal cells to pupal syntheses

179

and that JH in vitro blocks this change in commitment. She also
shows that the actual synthesis of cuticle depends on a second,
larger dose of ecdysone, but that cuticle synthesis itself is not
affected by juvenile hormone. There seems little doubt that in
vitro studies of this kind will soon dominate efforts to analyze
hormonal interactions in insects.

     Several of the studies provide a clear answer to the question
of when during the last larval instar JH is able to block metamorphosis. In four separate cases it is shown that the critical period
during which juvenile hormone is effective is the first half of
the last instar. Furthermore, Riddiford can duplicate the critical
period in vitro by manipulating the titers of JH and ecdysone in
cultures of fragments of larval epidermis. Her study and the
others in this section identify the time in the developmental
history of an epidermal cell at which JH acts, both in relation to
the time of DNA synthesis in that cell and to the time of acquisition by the cell of the competence to secrete a new kind of cuticle
(i.e. "reprogramming").

     These same studies also lay to rest the notion that DNA
synthesis or mitosis is a necessary prerequisite for a change in
cellular commitment from the synthesis of larval cuticle to the
synthesis of pupal cuticle, and raise questions about the usefulness of the concept of quantal mitosis, i.e., that cell division
is an essential component of differentiation. In the case of
insect metamorphosis, the concept of quantal mitosis proposes that
JH acts at the time of ecdysone-induced DNA synthesis and that
ecdysone-stimulated DNA replication must occur in the absence of
JH if new information (e.g. pupal versus larval) is to be expressed.
Although initial evidence supported this idea, it may not be correct.
The experiments of Caveney (1970) in Tenebrio provide convincing
evidence that during cuticle deposition, one epidermal cell can
secrete first larval and then adult structures, in the absence of
an intervening ecdysis and presumably in the absence of DNA synthesis. Evidence presented by Riddiford in her report strongly
supports the idea that for larval epidermis, the JH titer at the
time of the cell divisions preceding the metamorphic molt is irrel-
evant. Instead, the acquisition of competence to respond to ecdysone with the subsequent expression of pupal characteristics depends
on the absence of JH during a critical period preceding the occurrence of DNA synthesis. However, DNA synthesis may be an essential
component in the expression of that commitment.

     Oberlander and Silhacek reach a similar conclusion about the
processes involved in the acquisition of competence from their
in vitro studies of the response of imaginal discs to hormones.
Unlike other parts of the integument, the imaginal discs do not
secrete cuticle until the time of pupation, although they become
competent to do so during larval life. Oberlander and Silhacek

show that the acquisition by imaginal discs of the competence to respond to β-ecdysone in vitro and secrete cuticle is associated with the growth of the imaginal discs and that JH prevents the discs from developing to a state in which they are receptive to ecdysone.

It will become apparent to the reader of these reports that there is now evidence from several in vitro systems that JH exercises its effect prior to overt differentiation - that the change in a cell's developmental commitment is a separate event from the ecdysone-induced expression of that commitment in overt differentiation.

The other recurrent theme in several of the papers is the nature of the interaction between juvenile hormone and ecdysone, their possible antagonism, synergism, or mimicry. The reader will soon discern that many of the apparent differences among the authors are semantic and for the most part do not reflect real differences in interpretation. Wigglesworth dealt with the question of JH-ecdysone interaction on a number of occasions and concluded that the hormones did two quite different things: "Ecdysone brings about the activation of epidermal cells that is necessary for growth and molting. Once growth has been set going in this way, the production of larval or adult characters is determined by the presence or absence of the juvenile hormone" (Wigglesworth, 1970). According to this view, at least as far as cellular metamorphosis is concerned, JH is the handmaiden to ecdysone. JH has no action in cellular metamorphosis except when it is present along with ecdysone. Several of the results presented, especially those of Willis and Hallowell on "hyperecdysonism", provide support for this idea of separate actions of ecdysone and juvenile hormone. They show that high levels of ecdysone elicit prompt cuticle deposition and molting, causing isolated pupal abdomens to molt into second pupal abdomens. Superficially, it appears that these high ecdysone titers are mimicking the action of JH, but the authors explain the results on the basis of residual JH persisting in these abdomens or of cuticle deposition occurring so rapidly that insufficient time is available for full development. The conclusion is made that there is no convincing evidence in the case of epidermal cells that JH and ecdysone mimic or antagonize each other at the target cell level.

However, when the experimenter leaves the target cell level, various interactions between juvenile hormone and ecdysone may be possible, including inhibition. Lezzi and Wyss review the extensive literature on JH-ecdysone interactions both in vivo and in vitro, where JH is said to antagonize, inhibit, or prevent the action of ecdysone. They emphasize the difficulty of deciding from in vivo experiments at what level JH-ecdysone interactions occur, and propose that ecdysone and JH may interact at any one of

several levels from neuroendocrine to target cell and that these
interactions may appear as antagonisms, mimicry, or synergisms in
the response of the target cells. They also discuss their own
experiments on the effects of JH and ecdysone on puffing patterns
of salivary gland chromosomes and on cell divisions of cultured
Drosophila cells, interpreting their results in terms of an interaction between JH and ecdysone at the target cell level. On the
other hand, the JH-ecdysone antagonism demonstrated by Masner et al.
is a clear example of hormonal interactions which occur at a level
other than the target cell. They show that exogenously supplied
JH suppresses the synthesis of ecdysone and prolongs the last larval
instar, causing an artificial diapause-like state in the larva.
They conclude with the attractive suggestion that this JH suppression of ecdysone synthesis may be an endocrine fail-safe mechanism
which operates under conditions of hormonal derangement, as well
as during normal physiological development (see also Williams, this
volume).

A theoretical consideration of the role of JH in regulating
the expression of larval, pupal, and adult characteristics can be
found in the contribution of Krishna Kumaran. Krishna Kumaran
proposes a model for the role of JH in the development of holometabolous insects, traditionally viewed as a three choice mechanism that depends on declining JH titers. According to his model,
at each of two developmental steps, two alternatives are possible,
based on the presence or absence of JH. Thus, larval cells have a
choice of expressing larval or pupal genes, and pupal cells, pupal
or adult genes, in the presence or absence of JH, respectively.

The final two papers in this section by Yagi and Sehnal,
which were not presented at the symposium, point out the increasing
number of roles JH has been found to play in insect life and analyze
the application of JH analogs in pest control. The role of JH in
the larval diapause of lepidopterans was first proposed by Fukaya
and Mitsuhashi in 1957. Yagi has extended this finding and demonstrated that JH regulates not only the diapause but also the phase
variations of larval lepidopterans.

Sehnal's definitive contribution represents the first cohesive
analysis of the effects of JH analogs on the class Insecta as a
whole. From a careful examination of results obtained from a
decade of research, he concludes that JH analogs can affect insect
development at all stages, from egg to adult. He also calls attention to a remarkable regularity in the order- and family-specific
variations in response to analogs, which makes it possible to
predict whether a species will respond to a JH analog and what the
response will be at different developmental stages.

The authors of the reports in this section have presented us with a number of <u>in vitro</u> systems in which the morphogenetic action of JH can be clearly demonstrated. Among them surely is at least one which will enable us within the next decade to understand the mechanism of JH action at the molecular level.

RELATIONSHIP BETWEEN DNA SYNTHESIS, JUVENILE HORMONE AND

METAMORPHOSIS IN GALLERIA LARVAE

A. Krishna Kumaran[*]

Department of Biology, Marquette University
Milwaukee, Wisconsin 53233

INTRODUCTION

Epidermal cells in holometabolous insects exhibit progressive differentiative changes. At hatching from the egg, the larval epi-cells have already secreted a characteristic larval cuticle. In Galleria mellonella, the greater wax moth, the larval cuticle is untanned, soft and pliable with hexagonal plaques on the surface. In insects, each epidermal cell secretes the cuticle that is over-lying it. Thus, each hexagonal plaque represents the boundary lines of the cuticle secreted by the epidermal cell underlying the plaque. The larva passes through several morphologically and ana-tomically similar instars. At each of these succeeding larval molts the epidermal cells secrete a morphologically and chemically similar cuticle. After a definite number of larval molts, usually seven in Galleria, the larva metamorphoses into a pupa. At this molt the epidermal cells secrete a pupal cuticle characterized by its tanned exocuticle and knob-like elevations on the outer surface. Acquisition of the ability to secrete a pupal cuticle by the larval epidermal cells has been regarded as a differentiative event.

The metamorphosis in insects has been shown to be under the control of two hormones, ecdysone and juvenile hormone (JH) (see Gilbert and King, 1973). α-Ecdysone is a steroid hormone secreted by the prothoracic glands upon stimulation by a prothoracicotropic hormone from the brain. Ecdysones have been shown to promote

---

[*] Formerly known as A. Krishnakumaran

metamorphosis. On the other hand juvenile hormone, a terpenoid
derivative that is secreted by the corpora allata, also possibly
under brain hormone stimulation, prevents metamorphosis. Because
metamorphosis in insects can be controlled by manipulating the
levels of the two hormones, because of the availability of the
two hormones in a pure state and because of the ease with which
insects can be surgically manipulated to extirpate the endogenous
sources of hormones, differentiation of insect epidermal cells at
metamorphosis has been studied extensively as a model for differentiation (Schneiderman, 1969).

Some of these studies have been directed at an analysis of
the relationship between DNA synthesis and differentiation. These
studies suggested that insect epidermal cells have to replicate
their genome before they can express genes not hitherto functional
and led to the so called "clean genes" hypothesis and the concept
of quantal mitosis (Holtzer et al., 1972). The data that led to
these concepts are the close chronological correlation between DNA
synthesis in epidermal cells and the assumed time of release of
ecdysone (e.g. Bowers and Williams, 1964), absence of DNA synthesis
in developmental stages of insects that are normally lacking
ecdysones (Krishnakumaran et al., 1967) and absence of cell divisions in the epidermal cells of insects that are surgically deprived
of prothoracic glands (Strich-Halbwachs, 1959). Later it was also
shown that α-ecdysone promotes DNA synthesis in imaginal wing discs
of larval Lepidoptera in vitro (Oberlander, 1972).

Since ecdysone is needed for metamorphosis, the correlation
between occurrence of DNA synthesis and presence of ecdysone has
been interpreted as a causal connection. However, these are mere
correlations and a causal connection between DNA synthesis and
differentiation (metamorphosis) of larval epidermis has not been
unequivocally proven.

The time of action of JH is also related to this question.
Since JH inhibits metamorphosis and because of the assumption that
metamorphosis is always preceded by DNA synthesis and cell division,
it has been widely suggested that JH acts at the time of DNA synthesis and/or cell division. In support of this contention are the
facts that injured cells are more sensitive to JH than normal cells
(Schneiderman et al., 1965) and that JH is most effective, when
injected into last instar Galleria, before the period of active DNA
synthesis (see Sehnal and Meyer, 1968).

The data to be reported here has a bearing on these two questions, the relationship between DNA synthesis and differentiation
of larval epidermal cells and the time of action of JH in relation
to cell division. The results show that DNA synthesis and differentiation in Galleria larval epidermal cells are not causally
related. Epidermal cells that secreted a larval cuticle at the

preceding molt can secrete a pupal cuticle without undergoing a
round of DNA synthesis. The results also suggest that differentiation of larval epidermis is dependent on the effective removal
of JH before the action of ecdysone and is unrelated to the JH
being absent at the time the cell division preceding the metamorphic
molt.

MATERIALS AND METHODS

Larvae of the greater wax moth, Galleria mellonella were raised
in the laboratory according to procedures described earlier
(Krishnakumaran, 1972). Last instar larvae (designated in the text
as last larvae) and penultimate instar larvae were used in these
experiments.

Autoradiographic procedures are similar to those used earlier
(Krishnakumaran et al., 1967). Labeled thymidine of specific activity 2 Ci/mmol obtained from New England Nuclear or of specific
activity 5 Ci/mmol obtained from Amersham/Searle Co. was used.

Surgical procedures used in these experiments followed the
same general precautions suggested in the literature (Schneiderman,
1967).

RESULTS

Time of Acquisition of Pupal State by Larval Epidermal Cells During the Last Instar

In the first series of experiments we examined the time at
which the larval cells acquire the ability to synthesize a pupal
cuticle. The last larval stadium in Galleria spans about 6 to
6 1/2 days. At the end of this period the larva empties its gut,
spins a cocoon and enters a pharate pupal stage during which it
deposits a pupal cuticle. What is the earliest stage at which an
epidermal cell is capable of secreting a pupal cuticle? The
developmental state of an insect epidermal cell can be determined
by injecting a relatively high dose of molting hormone, since epidermal cells under these conditions promptly secrete a cuticle and
this cuticle represents the state of determination of the cells at
the time of injection. For this purpose last larvae of different
ages were injected with 6 µg of inokosterone or ecdysterone per
larva. Forty eight to 60 hr later such larvae deposit a new cuticle
whose morphological features reflect the developmental state of the
epidermal cell at the time of injection of the hormone.

The results show that all 3-day old larvae and most 4-day old larvae under these conditions secrete a larval cuticle (Table 1). However 5-day old larvae invariably secrete a pupal-type cuticle in the mid-dorsal region of all the abdominal tergites. Although this region of the tergite exhibits a distinctly pupal, well tanned cuticle, it is interesting to note that the mid-dorsal ridge normally present in this region is not prominent. The extent of the pupal cuticle secreted by day 6 larvae is greater. In several larvae most of the tergite and parts of the sternite are tanned and are clearly pupal in appearance. In two larvae only the anterior and posterior margins, in addition to the mid-dorsal region, of the tergite are tanned. In these larvae the wing discs were also everted and the prolegs and anal opening were absent.

TABLE 1. TYPE OF CUTICLE SECRETED BY <u>GALLERIA</u> LARVAE WHEN INJECTED WITH 6 μg OF MOLTING HORMONE ON DIFFERENT DAYS OF THE LAST LARVAL INSTAR.

Age of Larva	No. of Larvae Injected	No. of Larvae that Survived and Deposited a Cuticle	Type of Cuticle[*]
3	8	8	Larval (8)
4	8	7	Larval (6) Pupal[**] (1)
5	8	7	Pupal (7)
6	8	8	Pupal (8)

Day of ecdysis into last instar is regarded as day 1.

* Number of insects that deposited this type of cuticle is given in parentheses.

** Presence of any pupal tanned cuticle, everted wing discs, absence of prolegs and absence of an anal opening is regarded as pupal.

From this observation it is concluded that even though the
larval epidermal cells do not normally secrete a pupal cuticle
until much later, they may have acquired the ability to secrete
a pupal cuticle much earlier. By day 5 the last larval epidermal
cells in several regions of the body acquired the ability to deposit
a pupal cuticle.

Developmental State of Ligated Last Larvae

In order to determine the factors responsible for the acquisition of the ability to secrete a pupal cuticle the following experiments were carried out with ligated larvae. For this purpose
last larvae of diverse ages were ligated behind the prothoracic
segment and the ligated preparations were injected with 6 µg of inokosterone either immediately after ligation or 5 days after ligation.

The results reveal that both day 3 and day 4 ligated larvae,
if stimulated to secrete a cuticle soon after ligation, invariably
secrete a larval cuticle (Table 2). In these preparations the wing
discs did not evert, the prolegs were persistent and the tip of the
abdomen was characteristically larval. Day 5 and day 6 ligated
larvae, on the other hand, deposited cuticles which exhibited pupal
features and the wing discs were everted.

When day 3 and day 4 ligated larvae were injected with 6 µg
of molting hormone 5 days after ligation, not all preparations
deposited a cuticle. However all the preparations that deposited
cuticle (3 among day 3 larvae and 4 among day 4 larvae) expressed
distinctly pupal characters. The wing discs were everted, the
prolegs were absent and the tip of the abdomen was pupal. However,
the cuticle did not show signs of pupal tanning. These data
suggest that epidermal cells from a day 3 larva can acquire the
ability to express pupal genes after a period of time in an apparently hormone-free environment. Hence the ability to express
pupal characteristics by day 3 last larval ectodermal cells is not
dependent on the positive effects of one or the other of the insect
hormones. A mere lapse of time seems to bestow this ability on
the epidermal cells.

Nature of the Factors That Influence the Acquisition of Pupal State

Since ligated day 3 last larvae acquire the ability to express
pupal characteristics in the apparent absence of hormones, the
nature of the factors involved in this transformation is examined.
For this purpose pieces of integument from day 1 penultimate instar
larvae were implanted into day 4 or day 7 last larvae. When the
host metamorphosed the implant deposited a new cuticle whose type
varied depending on the age of the host.

TABLE 2. TYPE OF CUTICLE SECRETED BY LIGATED LARVAE AFTER INJECTION OF ECDYSONE

Age at Ligation	Time of Injection	No. of Larvae that Survived and Deposited a Cuticle	Type of Cuticle
3	Immediately after ligation	6	Larval
	5 days after ligation	3	Pupal
4	Immediately after ligation	6	Larval
	5 days after ligation	4	Pupal
5	Immediately after ligation	6	Pupal
	5 days after ligation	2	Pupal
6	Immediately after ligation	6	Pupal

Larvae were ligated behind the prothoracic segment and injected with 6 µg of inokosterone at the indicated time.

The data show that penultimate larval integument implanted into day 4 last larvae frequently secreted a pupal cuticle while the same integument implanted into day 7 host secreted a larval cuticle (Table 3). This experiment suggests that day 4 last larvae have a factor that causes the implanted epidermal cells to metamorphose. The difference between the two types of hosts may also merely be the time that elapses from implantation to pupation. Day 4 last larvae take a longer time than day 7 hosts to pupate. That it is not merely the passing of time is shown by culturing the implant in different primary hosts prior to implantation into day 7 last larvae. When cultured first in day 2 pharate adults, the implant secreted a larval cuticle. In contrast to the above experiment, when penultimate larval integument was first cultured for 24 hr in normal or ligated day 4 larvae prior to implantation into day 7 last larval host, the implanted epidermal cells frequently deposited a pupal cuticle. This observation suggests that day 4 larvae have a factor that influences the expression of pupal genes. Since the ligated larvae also possess this factor, it is concluded that the factor may not be hormonal.

TABLE 3.   NATURE OF CUTICLE DEPOSITED BY DAY 1 PENULTIMATE LARVAL
GALLERIA INTEGUMENT FOLLOWING IMPLANTATION INTO LAST INSTAR LARVAE.

Age of Host	Primary Host for Preliminary In Vivo Culture of Implant	Number of Successful Implants	Type of Cuticle Secreted by Implant	
			Larval	Pupal
4	None	12	2	10
7	None	16	16	0
7	None	6	0	6*
7	Day 2 pharate adult	5	5	0
7	Day 4 last larva	14	2	12
7	Day 4 ligated larva**	14	2	12

* Day 1 last larval integument was implanted in this experiment.

** Larvae were ligated on day 4 and were used for in vivo culture
24 hours later.

        Similar implantation studies to determine the developmental
potential of epidermal cells from last larvae were conducted. When
implanted into day 7 hosts, integument from day 1, 2, 3, 4 or 5
last larvae deposited a pupal cuticle (data for day 1 implants
only are included in Table 3).

    Cell Divisions in Tergal Epidermal Cells of Last Instar Larvae

        In order to determine whether the acquisition of the ability
to express pupal genes coincides with cell divisions in the epidermal cells, the increase in the number of larval epidermal cells was
investigated. In an earlier investigation it was noted that mitotic
figures occur in tergal epidermal cells of last larval Galleria
between day 3 and 6 (R. Granger, A. Krishnakumaran and H. Schneiderman unpublished data). We examined the ratio of epidermal cells
to plaques in the 2nd thoracic and 2nd abdominal tergites from
days 1, 3 and 5, last larvae (Table 4).

TABLE 4. RATIO OF EPIDERMAL CELLS TO CUTICULAR PLAQUES IN LAST INSTAR <u>GALLERIA</u> LARVAE.

	Segment	
Age of Larva	2nd Thoracic	2nd Abdominal
1	1.0	1.0
3	1.2	1.6
5	1.8	1.1

Since each plaque arises from the secretory product of the epidermal cell underlying it, the ratio of cells to plaques is 1 on day 1. On day 3 the ratio in the 2nd abdominal tergite is 1.6 while that of 2nd thoracic tergite is 1.2. On day 5 the ratios are 1.1 and 1.8 respectively. Since the ratio of cells to plaques increases following cell division these data suggest that in day 3 larvae the epidermal cells have already divided. The decrease noted in the ratio of cells to plaques in the abdominal tergite from day 5 larvae may be in part the result of cell death. In preliminary studies we examined whether the distribution of mitotic figures bears any relation to the areas that express pupal characters on day 5, but these studies have not yielded a positive correlation.

Is There DNA Synthesis Prior to Secretion of Pupal Cuticle by
Larval Implants?

Since the penultimate larval integument can be programmed to synthesize either a larval cuticle or pupal cuticle by implantation into the appropriate host, it is possible to monitor incorporation of [^{3}H]thymidine into epidermal cells under both experimental conditions. For this purpose, following implantation of day 1 penultimate larval integument, the hosts were injected with [^{3}H]thymidine. Three schedules of isotope injection were used. In one series different groups of host larvae were injected with isotope on each of the days from implantation to pupation. The larvae were sacrificed 24 hr after injection of label. The implants were retrieved and processed for autoradiography. In the second series the host larvae received repeated injections of label on each of the successive days during the entire period from implantation to pupation. Following pupation of the hosts the implants were retrieved and

processed. In the third series a different batch of larvae were
injected with [^{3}H]thymidine at intervals of 3 hr commencing from
implantation to pupation. The hosts were sacrificed 6 hr after
injection of isotope and the implants were examined by autoradiography for thymidine incorporation.

All these studies demonstrated that penultimate larval epidermal cells do not engage in DNA synthesis before they secrete a
pupal cuticle. When the implants secrete a larval cuticle as well,
these cells do not incorporate thymidine. These observations
suggest that DNA synthesis does not occur prior to reprogramming
of larval epidermal cells into pupal cells.

DISCUSSION

DNA Synthesis, Metamorphosis and Reprogramming

The results reported here clearly show that larval epidermal
cells in Galleria can reprogram to synthesize and deposit a pupal
cuticle in the apparent absence of an intervening round of DNA
synthesis and presumably, cell division. In other words, reprogramming in these insect cells is unrelated to cell division and
in this respect differs from some of the vertebrate cells that
require a quantal mitosis prior to reprogramming (see Holtzer
et al., 1972).

This conclusion is supported by two lines of evidence. The
penultimate larval integument can deposit a pupal cuticle when
implanted into a last larva without an intervening cell division.
These larval epidermal cells presumably underwent a round of DNA
synthesis prior to depositing the larval cuticle of the penultimate
stadium. The same cells having once secreted a larval cuticle
go on to secrete a pupal cuticle without an intervening round of
DNA synthesis. Presumably no cell divisions occurred either.
Hence, differentiation of larval lepidopteran epidermal cells
into pupal cells can occur without an intervening cell division.

The second line of evidence in support of this view comes
from the study of ligated day 3 last larvae. Ligated day 3 last
larvae when injected with molting hormone soon after ligation
secreted a larval cuticle. However if similar ligated preparations
were injected with molting hormone 5 days after ligation their
epidermal cells express pupal characteristics. Since epidermal
cells in ligated larvae do not engage in DNA synthesis (Krishnakumaran et al., 1967) it suggests that epidermal cells from day 3
last larvae can reprogram without cell division. Several investigators reported earlier a strong correlation between the time
of ecdysone action and DNA synthesis. What are the possible causes

for the lack of similar correlation in the current studies?   In
these earlier studies, the normal pattern of DNA synthesis in
intact insects was examined.   In intact larvae stretching of the
integument that invariably accompanies growth may stimulate cell
divisions.   Since the growth of last larvae also occurs before the
acquisition of the pupal state on day 5, it is likely that cell
divisions that occur on days 3 and 4 of the last larval stadium
may be associated with growth.   Yet we noted DNA synthesis in epidermal cells of adult saturnid moths parabiosed to pupae (Krishnakumaran et al., 1967).   Similarly Oberlander (1972) observed DNA
synthesis in isolated lepidopteran wing discs that are exposed to
α-ecdysone.   In these two instances stretching related to growth
is not the apparent cause for initiation of DNA synthesis.   Hence
the role of ecdysone, if any, in initiation of DNA synthesis in
epidermal cells is obscure.

Time of JH Action

These studies have a bearing on the time and mode of action
of JH.   It is generally assumed that JH may act at the time of
cell division (Schneiderman et al., 1969).   The fact that injured
cells are more sensitive to JH certainly supports the hypothesis.
More importantly this hypothesis is strongly supported by the
assumption that reprogramming of the genome occurs during cell
division.   The results reported here show that cell division is
not necessary for reprogramming.   Similarly, Lawrence (1975)
reviewed the literature and concluded that reprogramming in insect
epidermal cells does not require cell division.   Hence, the question of time of JH action should be reexamined.

Day 3 last larvae lose the JH influence within five days
following ligation.   Similarly, penultimate larval integument
cultured for 24 hr in day 4 last larvae loses the influence of JH.
These results suggest that in order for JH to be effective, it
should be present at the time when molting hormone acts and is
not related to DNA synthesis and the cell cycle.   If this assumption is valid, then JH must influence the quality of transcription
and/or translational events initiated by molting hormone.   In the
presence of JH, ecdysones stimulate transcription, processing
and/or translation of genes that specify larval features.   On the
other hand, in the absence of JH, ecdysone permits expression of
pupal genes.   This control may be achieved by JH acting as a co-
repressor as hypothesized by Williams and Kafatos (1971).

One difficulty in explaining the role of JH in the control of
metamorphosis in holometabolous insects is that it occurs in two
stages.   According to the traditional view the three choices, i.e.
larval, pupal and adult are potentially available for realization
each time a cuticle is made.   The realization of one of these

morphological states is dependent on JH levels. In the presence of
high titers the cells express larval genes whereas in its absence
they express adult genes. In the presence of intermediate levels
of JH, pupal genes are expressed. However there is as yet no
proof that intermediate levels of JH occur at pupation. In fact,
many studies show that JH is not detectable even at pupation.
There are also difficulties in visualizing the possible mode of JH
action if one accepts this model. On the other hand, if one views
development in holometabolous insects as two sequential sets of
two alternatives one can explain all the observed developmental
phenomena.

    According to this hypothesis (Fig. 1) a larval epidermal cell
has only two choices: larval or pupal. Similarly a pupal cell can
express either pupal or adult genes. In the presence of JH, larval
cells express larval genes and pupal cells express pupal genes.
In the absence of JH these cells express pupal and adult genes
respectively. This hypothesis differs from the traditional view
in that at each metamorphic molt the cell has to choose from among
two alternatives rather than from among three choices. The hypothesis also assumes that the set of alternatives available is
dependent on the past history and developmental state of the cell.

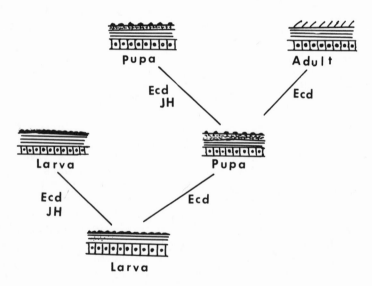

Fig. 1. Sequential choice model to explain juvenile hormone
control of metamorphosis. Ecd: Ecdysone; JH: Juvenile Hormone.

This hypothesis can explain many observations such as our inability to find intermediate levels of JH at pupation and the fact that insect larvae do not ordinarily skip a pupal stage under normal and under many experimental conditions (Kato, 1973; unpublished observations). The few examples that suggest omission of the pupal stage pertain to eyes, antennae, legs and wings of lepidopteran larvae. This is not surprising because these are imaginal structures which remain in a determined state during the larval stage and can attain only two overtly differentiated states, i.e. pupal and adult. At a pupal molt their two choices are to respond to ecdysone by secreting a cuticle or not to respond. Once they acquire the ability to respond to ecdysone they have two choices; either to make a pupal or adult cuticle. These choices are dependent on the presence or absence of JH. In experiments on imaginal structures in which the pupal state was omitted the explanation may be the delay in pupation that often accompanies these experimental conditions. Because of this delay, the imaginal structures acquired the competence to respond to ecdysone but were not exposed to ecdysone. After this delay, when they were exposed to ecdysone they directly expressed the adult genes.

The sequential choice hypothesis can also explain the reversal of metamorphosis documented in the literature. Normally the pupal epidermis cannot be caused to secrete a larval cuticle by increasing the concentration of JH. This fact speaks against the multiple choice theory. However when cultured in vivo in the presence of JH, epidermal cells occasionally reverted to a juvenile state. An analogous situation is found in many animals. For example, cells in the limb stump of an amphibian or of an amputated planarian dedifferentiate and revert to a more juvenile state. Similarly, many cell types dedifferentiate when placed in culture or after grafting to an ectopic region and under certain conditions these dedifferentiated cells can redifferentiate. The insect epidermal cells as well appear to have dedifferentiated and later when the dedifferentiated cells are forced to synthesize a cuticle in the presence of JH, they express juvenile characters.

The great advantage of this model is that it eliminates the need to hypothesize a novel mechanism by which juvenile hormone controls the choice of one among three alternative paths of development. JH like other hormones merely controls the "on" or "off" switch and the choice as to which genes are accessible for regulation is determined by the developmental history of the target cell. Examples of sequential choices that become manifest at successive developmental stages are common in vertebrate embryology. The canalization of neural retina cell in an amphibian embryo illustrates this point. In its developmental history it must first choose between general ectoderm and neural ectoderm and later between neural tube and optic vesicle. Subsequently, the choice between neural retina and pigment retina is made.

SUMMARY

The relationship between DNA synthesis, JH and metamorphosis in Galleria is examined. The results show that last instar larval epidermal cells acquire the ability to express pupal genes towards the latter half of the stadium. DNA synthesis and cell division do not appear to be the critical events that permit expression of pupal genes. Pupal genes seem to be expressed upon inactivation of endogenous JH. The same is true of penultimate larval stage epidermal cells. A model is presented to explain the role of juvenile hormone in regulating the expression of larval, pupal and adult genes at successive stages.

ACKNOWLEDGEMENTS

This research was aided by grants from the National Institute of Child Health and Human Development N.I.H. (HD08089) and Marquette University. The author wishes to thank Mr. Kenneth Condon and Miss Ann Brennan for technical help and Mr. William H. Kastern for collaboration during a portion of these studies.

REFERENCES

Gilbert, L.I., and King, D.S., 1973, in "Physiology of Insecta" (M. Rockstein, ed.) Vol. 1, 2nd Edition, Academic Press, New York.
Holtzer, H., Weintraub, H., Mayne, R., and Mochan, B., 1972, in "Current Topics in Developmental Biology" (A.A. Moscona and A. Monroy, eds.) Vol. 7, Academic Press, New York.
Kato, Y., 1973, J. Insect Physiol. 19:495.
Krishnakumaran, A., Berry, S.J., Oberlander, H., and Schneiderman, H.A., 1967, J. Insect Physiol. 13:1.
Lawrence, P.A., 1975, in "Results and Problems in Cell Differentiation" (J. Reinert and H. Holtzer, eds.) Vol. 7, Springer-Verlag, New York.
Oberlander, H., 1972, J. Insect Physiol. 18:223.
Schneiderman, H.A., 1967, in "Methods in Developmental Biology" (F.H. Wilt and N.K. Wessells, eds.) Thomas Y. Crowell Co., New York.
Schneiderman, H.A., 1969, in "Biology and Physical Sciences" (S. Devons, ed.) Columbia University Press, New York.
Schneiderman, H.A., Krishnakumaran, A., Kulkarni, V.G., and Friedman, L., 1965, J. Insect Physiol. 11:1641.
Schneiderman, H.A., Krishnakumaran, A., Bryant, P.J., and Sehnal, F., 1969, in "Proc. Symp. Potentials in Crop Protection" New York State Agricultural Experiment Station, Cornell University, Ithaca, New York.
Sehnal, F., and Meyer, A.S., 1968, Science N.Y. 159:981.

Strich-Halbwachs, M.-C., 1959, "Controle de la mue chez _Locusta_
    migratoria" Thesis, Universite de Strasbourg.
Williams, C.M., and Kafatos, F.C., 1971, Mitteil. Schweiz. Entomol.
    Gesellschaft. 44:151.

JUVENILE HORMONE CONTROL OF EPIDERMAL COMMITMENT IN VIVO AND

IN VITRO

L.M. Riddiford

Department of Zoology, University of Washington
Seattle, Washington 98195

INTRODUCTION

The study of the cellular and molecular basis of the morphogenetic "status quo" action of juvenile hormone (Williams, 1961)
has been hampered by the lack of a good system in which this action
of juvenile hormone (JH) is completely separated from the events
of cuticle synthesis itself. When we first found that there were
two releases of ecdysone necessary for the larval-pupal transformation in the tobacco hornworm Manduca sexta (Truman and Riddiford,
1974) and that JH only affected the type of cuticle made if given
before the first release (Truman et al., 1974), it seemed that
these epidermal cells were ideal for a study of JH action. This
paper will summarize our current state of knowledge about these
cells and their response to hormones both in vivo and in vitro.

During the last few years the endocrine physiology of both
the larval-larval molt and the larval-pupal transformation in
Manduca has been worked out in some detail. Since some facets
are crucial for an understanding of the in vitro system we have
developed, a brief summary follows and is depicted in Fig. 1. In
the fourth (penultimate) larval instar the JH titer fluctuates
somewhat but remains high (between about 5 and 15 ng JH I equiva-
lents/ml) through the time of prothoracicotropic hormone (PTTH)
and ecdysone release (Fain and Riddiford, 1975). PTTH release
occurs during a well-defined portion of the photoperiod and is
rapidly followed by ecdysone release to initiate the molting process (Truman, 1972; Fain and Riddiford, 1976a). After this time
the JH titer declines, but then abruptly rises again to about
6 ng/ml after ecdysis to the fifth instar followed by a decline to
an 0.8 ng/ml plateau level (Fain and Riddiford, 1975). When the

198

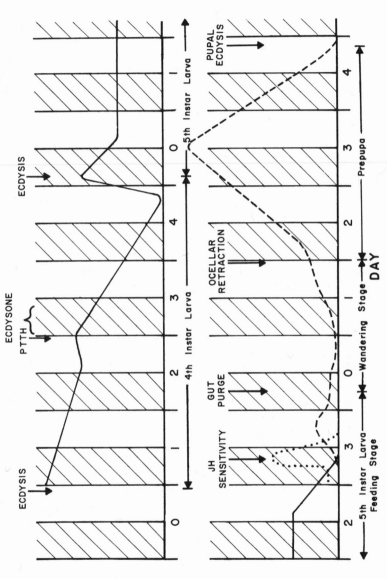

Fig. 1. Timing of endocrine events during the fourth and fifth larval stages and the larval-pupal transformation of *Manduca sexta* in a 12L:12D photoperiod, 25°C. Juvenile hormone titer (solid curve) from Fain and Riddiford (1975) (4th and early 5th) and Nijhout and Williams (1974b) (late 5th); ecdysone titer (dashed curve) from Bollenbacher et al. (1975); JH sensitivity (dotted curve) from Truman (1972) (4th) and Truman and Riddiford (1974) (end of 5th).

larva attains a weight of 5 grams, the corpora allata decline in
activity and the JH titer in the hemolymph becomes undetectable by
the time the animal weighs about 7 grams (Nijhout and Williams,
1974a,b; Nijhout, 1975; Williams, this volume). Then in the absence of detectable JH in the hemolymph, PTTH and ecdysone are
released and initiate metamorphosis with its associated behavioral
changes – the switch from the feeding to the wandering stage
(Truman and Riddiford, 1974; Nijhout and Williams, 1974b). Although this small release of ecdysone (Bollenbacher et al., 1975)
does not elicit pupal cuticle synthesis, it is sufficient to evoke
a change in commitment of the epidermal cells. Consequently, they
can no longer make larval cuticle when exposed to ecdysone in the
presence of JH but can only make pupal cuticle irrespective of the
presence or absence of JH (Truman et al., 1974; Riddiford, 1975).
The actual production of pupal cuticle does not begin until a second release of PTTH followed by ecdysone that occurs two days
later (Truman and Riddiford, 1974). This much larger amount of
ecdysone remains elevated for about 3 days while the cuticle is
being formed, then falls just before pupal ecdysis (Bollenbacher
et al., 1975).

Since the switchover in the commitment of the epidermal cells
occurs two days before the beginning of the expression of that
commitment (Truman et al., 1974), it seemed to be an excellent
system in which to look at the cellular events involved in the
change in commitment. Also, the action of JH in preventing this
change could be studied unhindered by the events leading directly
to cuticle synthesis.

MATERIALS AND METHODS

Animals. Manduca sexta were reared on an artificial diet as
previously described (Truman, 1972) under a 12L:12D photoperiod
at 25°C. They were staged carefully by weight or by day of en-
trance into the wandering phase (Truman and Riddiford, 1974;
Nijhout and Williams, 1974a).

Histological Procedures. The tissue was dissected free and
immediately fixed either in Bouin's or in buffered glutaraldehyde
and then embedded in paraffin or in epon and sectioned at 5μ or 1μ
respectively. The autoradiography after a 2 hr exposure to 5 μCi/
gm [^{3}H]-thymidine was performed by standard techniques as described
in Riddiford (1975).

Cuticular Protein Synthesis. The dorsal cuticle from the
third to fifth abdominal segment down to the spiracles was removed,
scraped clean of epidermis, and dried overnight at 95°C to a constant dry weight. The piece was then suspended in 1 N KOH and
incubated at 37-45°C for 18-24 hr. After centrifugation the

protein content of the supernatant was determined by the method
of Lowry et al. (1951).

Culture Methods. After surface sterilization the dorsal
integument of segments 3 through 6 of a 6.5 gm fifth instar (day
2 in Fig. 1) larva was removed, dissected free of muscle and fat
body, then placed in Grace's medium (Gibco) as previously described
(Riddiford, 1976). The tissue was incubated at room temperature
under a 95% O_2 - 5% CO_2 atmosphere for a given length of time with
or without hormones. Then it was implanted into a fourth instar
larva before the initiation of the molt to the fifth larval stage
to assay its commitment (hereafter referred to as the "implantation assay") (Riddiford, 1975; 1976). Alternatively, the hormones
were removed from the culture media and the tissue allowed to form
cuticle in vitro under various conditions defined in the Results
section (Mitsui and Riddiford, 1976).

Hormones. β-Ecdysone (Rhoto Pharmaceutical Co.) was dissolved
in 10% isopropanol sterilized through a Millipore filter, then its
concentration determined by absorption at 240 nm (ε_{240} = 12,400).
It was premixed with Grace's medium at the desired concentration.
$C_{18}JH$ (JH I) (EcoControl; 95% the natural trans,trans,cis isomer)
and the JH mimic epoxygeranylsesamole (EGS) was dissolved in sterile
Grace's medium by sonication using siliconized glassware. These
solutions were stored at -20°C and diluted when necessary.

Macromolecular Synthesis. Stock solutions of the inhibitors
cycloheximide, puromycin, cordycepin, and actinomycin D (all from
Sigma) were prepared in distilled water and aliquots added to the
Grace's medium to obtain the desired concentration. For labeling
experiments 5 μCi of [^{3}H]-uridine (Schwartz-Mann; specific activity 22 Ci/mmol) in 2% ethanol, pH 7.5, or 5 μCi of [^{3}H]-leucine
(Schwartz-Mann; specific activity 48 Ci/mmol) was added to 0.5 ml
Grace's medium. Integument (epidermis plus cuticle) from the
dorsal midline to the spiracle of one abdominal segment was incubated in the medium with or without hormones for a specified time.
Incorporation of labeled precursor into RNA and protein of the
epidermis was determined by the method of Fristrom et al. (1973).

RESULTS

In Vivo Cellular Events During Time of Commitment

The epidermis in response to the first release of ecdysone
detached from the overlying larval cuticle as seen in Fig. 2 and
the cells changed in shape from cuboidal to columnar (Curtis and
Riddiford, unpublished information). These cells remained detached

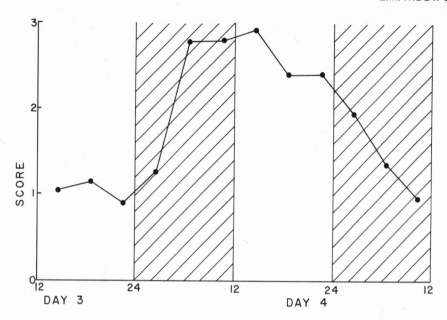

Fig. 2. Timing of detachment of epidermis from overlying cuticle
during the late fifth instar of Manduca. Scoring as follows:
0 = firm attachment; 1 = beginning detachment of tiny strips of
epidermal cells; basement membrane can be stripped off separately;
cells cuboidal. 2 = detachment of small sheets of epidermal cells.
3 = complete detachment and melding with basement membrane; cells
columnar. Ten larvae were observed for each time point.

throughout the time of the switchover in commitment and then
reattached to the cuticle (see Fig. 2) when the ecdysone titer
decreased (Bollenbacher et al., 1975) (see Fig. 1). As seen in
Fig. 3, during the time of detachment (between 7 and 9 gms), the
epidermal cells continued to secrete endocuticle at the same rate
as before detachment (Logan, unpublished information). Our 1μ
sections of epidermis at this time confirmed the presence of
cuticle on the apical border of these detached cells (Curtis and
Riddiford, unpublished information). Since the animals are still
feeding and growing, this continued deposition of endocuticle is
not surprising.

DNA Synthesis During the Larval-Pupal Transformation

It has been often stated by insect endocrinologists that in
order for cells to change their program from larval to pupal DNA
synthesis was required and that to be effective in blocking the
change, JH must act before this round of DNA synthesis (see Willis,

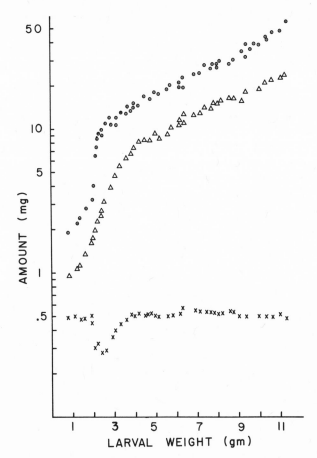

Fig. 3. Cuticle production by the dorsal epidermis of the third
to fifth abdominal segment during the fifth larval instar of
<u>Manduca</u>. The circles (o) indicate dry weight of the cuticle; the
triangles ($\triangle$) indicate the amount of protein; the x's indicate
the percentage of cuticle that is protein.

1974). The larval-pupal transformation of the epidermis in <u>Manduca</u>
seemed to be an ideal system in which to test this notion since the
expression of the differentiated state was so far removed from the
determination of the commitment. Autoradiographic studies of
various areas of epidermis after 2 hr pulses of 5 μCi/gm [^{3}H]-
thymidine to feeding and wandering larvae as well as to prepupae
revealed (Fig. 4) that the only round of DNA synthesis in the
dorsal abdominal epidermis occurred on day 1 of the wandering stage
when the second release of ecdysone was beginning (Riddiford, 1975).
There was no detectable incorporation by the [^{3}H]-thymidine either
at the end of the third day of the instar or during the fourth day

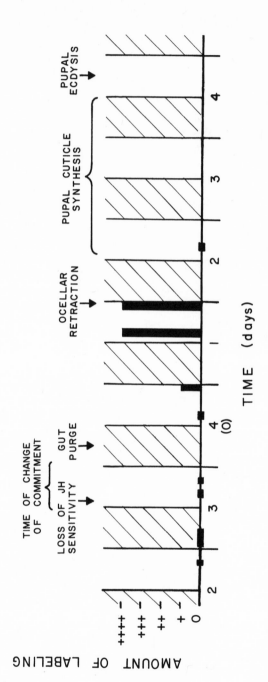

Fig. 4. The timing of DNA synthesis as detected by autoradiography in the dorsal part of the fifth abdominal segment of Manduca during the larval-pupal transformation. 0: no label; +: 10 cells labeled per section; ++: 10-30 cells; +++: 30-50 cells; ++++: >50 cells labeled per section. Labeling on day 0 of wandering stage only in spiracular region.

during the time that ecdysone was causing the changeover in commitment. Thus, DNA synthesis is not associated with the ecdysone-
induced switchover in commitment but rather occurs immediately
before pupal cuticle synthesis in response to the larger second
release of ecdysone.

Yet juvenile hormone is only effective in preventing pupal
cuticle synthesis if given before the first exposure to ecdysone
(Truman et al., 1974). It is most effective at the time that the
epidermis becomes fully detached (compare Figs. 1 and 2) and then
rapidly loses its potency. By 2 hr after lights-on on the fourth
day of the instar, topically applied JH had no detectable effect
on later pupal cuticle synthesis (Truman et al., 1974). Therefore,
its action must have nothing directly to do with DNA synthesis.

Commitment of Epidermis in Response to Hormones In Vitro

To study the cellular and molecular events occurring in the
epidermis at this time in response to both ecdysone and JH, an
in vitro system was needed. To this end I began culturing epidermis from 6.5 gm feeding larvae. The epidermis at this time has not
yet seen any ecdysone. Although the JH titer in the hemolymph is
nearly undetectable, the cells have retained their commitment to
form larval cuticle when challenged with ecdysone in the presence
of JH during an in vivo larval-larval molt (Riddiford, 1975). When
this epidermis was explanted to Grace's medium for 24 hr in the
absence of hormones, it also retained its larval commitment
(Riddiford, 1975; 1976) as seen in Table 1. If, however, 1 μg/ml
β-ecdysone was added to the culture medium and the tissue incubated
for 24 hr, then implanted into a fourth instar larva, it always
formed pupal cuticle when the host molted. After incubation with
1 μg/ml β-ecdysone in the presence of either 5 μg/ml JH I (Ro6-
9550 from Hoffman-LaRoche) or 10 μg/ml epoxygeranylsesamole (EGS)
for 24 hr, the epidermis always formed larval cuticle as tested by
the implantation assay (Riddiford, 1975). The switchover in
commitment in response to β-ecdysone alone was dose-dependent with
a 50% response occurring at 0.1 μg/ml (Riddiford, 1976). This
concentration of β-ecdysone of about 2×10^{-7} M is similar to that
found necessary in other in vitro systems, i.e. evagination of
Drosophila imaginal discs (Fristrom et al., 1973), puffing in
Drosophila salivary gland chromosomes (Ashburner, 1973), and
crochet formation of Manduca crochet epidermis (Fain and Riddiford,
1976b).

Similarly, the dose-response curve for the inhibition of the
change in commitment by the natural trans,trans,cis isomer of JH I
indicated that 3×10^{-7} M JH was sufficient for 50% of the cysts
to form larval cuticle (Riddiford, 1976; see Table 1). An order
of magnitude lower dose (2×10^{-8} M) was sufficient for 50% of the

TABLE 1. IMPLANTATION ASSAY OF COMMITMENT OF EPIDERMIS FROM 6.5 gm MANDUCA LARVAE AFTER 24 HR IN VITRO EXPOSURE TO HORMONES.

Hormonal Treatment (µg/ml)		Number	Type of Cuticle (%)		
Ecdysone	JH		Larval	Larval + Pupal	Pupal
0	0	268	92	8	0
1	0	150	0	7	93
1	5[a]	6	100	0	0
1	10[b]	10	100	0	0
1	3[c]	22	77	23	0
1	0.1[c]	12	50	50	0

[a] Ro6-9550 (mixture of isomers of JH I; data from Riddiford, 1975).

[b] epoxygeranylsesamole.

[c] >95% t,t,c JH I (data from Riddiford, 1976).

recovered cysts to show >50% larval cuticle. This latter concentration is the same as that found to be present in the hemolymph
at the time of the initiation of the molt to the fifth instar
(Fain and Riddiford, 1975). Thus, both ecdysone and juvenile
hormone at physiological concentrations seem to affect commitment
in these epidermal cells in vitro.

In Vitro Test of Epidermal Commitment

When the commitment of the epidermis was tested by the implantation assay, it was exposed to the high JH environment of the
host during the time that ecdysone initiated the molt (Fain and
Riddiford, 1975). Therefore, it was of interest to show that the
effects of both ecdysone and JH in the epidermis in vitro were not
dependent in any way on the method of assay. Therefore, we developed an in vitro system in which wandering stage epidermis would
form pupal cuticle (Mitsui and Riddiford, 1976). To obtain tanned
pupal cuticle from the dorsal epidermis of the fifth abdominal
segment required an optimum of 4 days exposure to 5 µg/ml β-ecdysone.
After this time the ecdysone was removed and the tissue incubated
in hormone-free Grace's medium for another 3 to 8 days to obtain
tan pupal cuticle.

In the in vitro assay of commitment, dorsal epidermis from a
6.5 gm larva was first exposed to 1 µg/ml β-ecdysone in Grace's
medium for 24 hr, then rinsed and placed in ecdysone-free medium
for 24 hr. Upon a second exposure, this time to 5 µg/ml β-ecdysone
for 4 days, followed by hormone-free medium, 72% of the pieces
formed pupal cuticle (Table 2). None formed larval cuticle, thereby indicating that 24 hr exposure to 1 µg/ml ecdysone caused the
change in commitment just as had been shown by the in vivo assay.
When both ecdysone and 3 µg/ml of the JH mimic EGS were present
during the first 24 hr period, the epidermis did not change its
commitment as seen in Table 2. The dose response curve for the
effectiveness of this JH mimic in the inhibition of the ecdysone-
induced switchover is shown in Fig. 5. Fifty percent of the pieces
formed all larval cuticle after exposure to 0.2 µg/ml EGS. The
decline in effectiveness of this mimic is very similar to that seen
in the in vivo implantation assay (Riddiford, 1975) again confirming that the in vitro change in commitment is permanent and is not
affected by the method of assay.

Furthermore, the in vitro assay allows us to test whether the
epidermis exposed to β-ecdysone and JH together can produce larval
cuticle. As seen in Table 2, when exposed to 1 µg/ml β-ecdysone
in the presence of EGS for 24 hr, followed by no hormone, only 50%
of the pieces formed larval cuticle. But when exposed to 5 µg/ml
β-ecdysone in the presence of EGS, 100% formed larval cuticle.

TABLE 2. IN VITRO FORMATION OF CUTICLE BY EPIDERMIS FROM 6.5 gm MANDUCA LARVAE.

| Hormone Exposure (μg/ml) | | | Number | Type of Cuticle (%) | | | | |
1st 24 hr Ecdysone	JH[a]	Days 3-6 Ecdysone		Larval	Larval + Pupal	Pupal	Indeter- minate	None
.1	0	5	47	0	6	72	19	2
1	3	5	20	80	10	0	10	0
1	3	0	10	50	0	0	20	30
5	3	0	5	100	0	0	0	0

a Epoxygeranylsesamole.

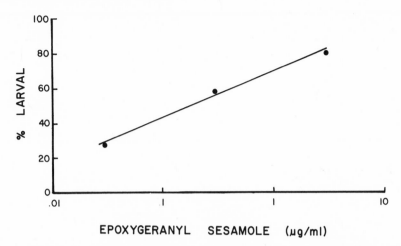

EPOXYGERANYL   SESAMOLE   (µg/ml)

Fig. 5. Dose-response curve for inhibition by the JH mimic EGS
of the switchover in cellular commitment of dorsal abdominal epidermis of the fifth segment induced by 1 µg/ml β-ecdysone for 24 hr
in vitro. One day after washout of the hormones the commitment was
tested in vitro by exposure to 5 µg/ml β-ecdysone for 4 days followed by incubation in Grace's medium. Twenty pieces were scored
for 3 µg/ml; 12 for the other two doses.

Thus, this latter condition can effectively mimic the hormonal
environment of a larval-larval molt.

Time Course of Ecdysone-Induced Switchover in Commitment

        Previous studies (Riddiford, 1975; 1976) had shown that 1 µg/
ml β-ecdysone for 24 hr was sufficient to effect a complete changeover in commitment in the epidermal cells such that when tested
by the implantation assay they formed pupal cuticle. This dosage
appeared to be in the physiological range (Riddiford, 1976) and was
consequently used to investigate the time course of ecdysone action
in this system. The epidermis was incubated in Grace's medium containing 1 µg/ml β-ecdysone for varying lengths of time after which
the tissue was removed from the ecdysone-containing medium, rinsed
through three rinses of Grace's medium and incubated for the remainder of the 24 hr period in hormone-free Grace's medium. It
was then implanted into the fourth instar hosts. Figure 6 shows
that after 3 hr exposure a very few cells had seen sufficient ecdysone to affect a change in their commitment. By 12.5 hr half of
the recovered cysts had more than 50% pupal cuticle, and by 16 hr
half of the cells had irreversibly changed their commitment from
larval to pupal. A similar time course is seen when epidermal
commitment is assayed in vitro (Mitsui and Riddiford, unpublished
information).

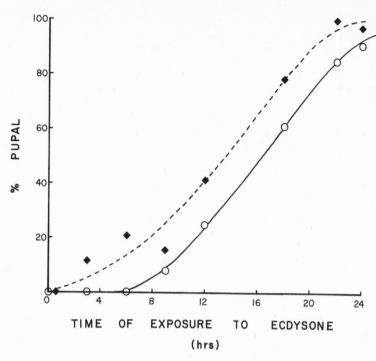

Fig. 6. The switchover in cellular commitment of dorsal abdominal and gin trap epidermis as a function of time of exposure to 1 µg/ml β-ecdysone in vitro. The circles (o) indicate the percentage of recovered cysts from the implantation assay which showed >90% pupal cuticle; the diamonds indicate those cysts having >50% pupal cuticle. Nineteen to 24 cysts were scored per point except 12 at 3 hr and 38 at 9 hr.

Time Course of JH Sensitivity

To obtain the dose-response curve for the inhibitory effect of JH I, the epidermis had always been pretreated in the JH-containing Grace's medium for 3 hr before the ecdysone was added (Riddiford, 1976). To find if this pretreatment was necessary, the epidermis was placed in Grace's medium containing 1 µg β-ecdysone/ml, then at various times thereafter the tissue was transferred to Grace's medium containing 0.2 µg/ml JH I and 1 µg/ml β-ecdysone and cultured for the remainder of the 24 hr. Then it was assayed for commitment by the implantation assay. This concentration of JH in the medium inhibited 50% of the epidermis from changing its commitment, irrespective of whether it was given 3 hr before the ecdysone or concurrently as seen in Fig. 7. But when the tissue was exposed to ecdysone first, then later to JH, the effect of JH rapidly declined so that after 6 hr of exposure to ecdysone before addition of JH, none of the epidermal cysts formed all larval cuticle

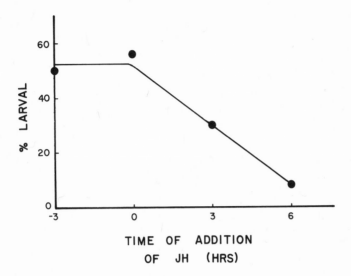

**TIME OF ADDITION
OF JH (HRS)**

Fig. 7. The complete inhibition of the switchover in cellular
commitment in dorsal abdominal and gin trap epidermis induced by
1 µg/ml β-ecdysone <u>in vitro</u> by 0.2 µg/ml <u>t</u>,<u>t</u>,<u>c</u> JH I as a function
of time of addition of JH relative to the beginning of the 24 hr
exposure to ecdysone.   Ten to 15 cysts were scored per point.

although 76% of the recovered cysts formed some larval cuticle.
Thus, in the absence of JH ecdysone acts rather rapidly on the
epidermal cells to make them aloof to JH.   By 9 to 12 hr after the
initiation of ecdysone exposure, the addition of JH cannot reverse
this new commitment.

Macromolecular Events During the Change in Commitment

From the autoradiographic experiments described above, DNA
synthesis did not seem to be necessary to commit the epidermis to
make pupal cuticle.   Therefore, attention has been focused on
protein and RNA synthesis during <u>in vitro</u> exposure to hormones
(Logan, Graves, and Riddiford, unpublished information).   The
approach has been two-fold:
    (1) to look at the effects of various inhibitors on the
ecdysone-induced switchover;
    (2) to look at the kinds of syntheses occurring during the
24 hr exposure to β-ecdysone.

Protein Synthesis. The epidermis from 6.5 gm larvae was incubated with 1 µg/ml β-ecdysone and various concentrations of the protein synthesis inhibitors cycloheximide and puromycin for 24 to 25 hr, then its commitment assayed by the implantation assay. As seen in Fig. 8, not even the highest dose of 100 µg/ml seriously affected the ability of the epidermis to change its commitment in vitro nor subsequently to make pupal cuticle in the implantation assay. Yet these higher concentrations of inhibitors (>0.1 µg/ml) blocked increasing amounts of protein synthesis as measured by [^{3}H]-leucine incorporation into total protein (see Fig. 8).

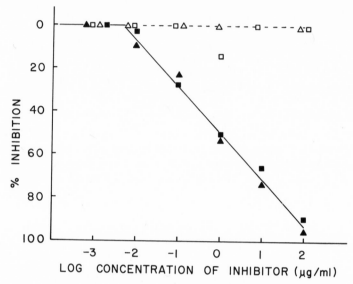

Fig. 8. The inhibition of the 1 µg/ml β-ecdysone-induced change in cellular commitment in vitro (open symbols) and of [^{3}H]-leucine incorporation into total protein (closed symbols) of dorsal abdominal epidermis as a function of concentration of the inhibitors cycloheximide (circles) and puromycin (triangles). After exposure to cycloheximide 8 to 13 recovered cysts were scored for each dosage except for 1 µg/ml where 22 were scored. For puromycin only 4 cysts per point were scored. The inhibition of leucine incorporation is an average of 2 replicates of duplicate samples.

The second approach was to study protein synthesis as a function of time of exposure to ecdysone. But Table 3 shows that the incorporation of [^{3}H]-leucine into protein by the epidermis in vitro over a 24 hr period does not depend on the hormone in the medium. There is no significant difference between the epidermis exposed to Grace's medium, JH (EGS) alone, β-ecdysone alone, or JH and β-ecdysone together. Thus, from these two separate lines of

evidence we conclude that protein synthesis is probably not
necessary for the changeover in commitment.

TABLE 3. PROTEIN SYNTHESIS IN LARVAL EPIDERMIS DURING IN VITRO
EXPOSURE TO HORMONES.

Treatment	Incorporation of $[^3H]$-leucine cpm/A_{260} (± 2 S.D.)
No hormone	10,422 ($\pm$ 2409)
3 µg/ml epoxygeranylsesamole	12,424 ($\pm$ 2773)
1 µg/ml β-ecdysone	10,409 ($\pm$ 1879)
1 µg/ml β-ecdysone + 3 µg/ml epoxygeranylsesamole	10,308 ($\pm$ 3297)

RNA Synthesis. When incubated with 1 µg/ml ecdysone and
either of the two RNA synthesis inhibitors, actinomycin D and
cordycepin, at concentrations of 0.1 µg/ml or lower, the epidermis
still formed pupal cuticle in the implantation assay (Fig. 9).
But as the concentration of either inhibitor was further increased,
the switchover in commitment was progressively inhibited. Figure 9
shows that after exposure to 100 µg/ml actinomycin D or cordycepin
simultaneously with ecdysone, nearly all of the epidermis forms
larval cuticle indicating that these inhibitors have prevented the
changeover in commitment but have not interfered with the ability
of the epidermal cells to produce cuticle after the inhibitor is
removed. That these inhibitors have their expected effect on RNA
synthesis is shown by the incorporation of $[^3H]$-uridine into total
RNA as indicated in Fig. 9. Both actinomycin D and cordycepin had
no effect on total RNA synthesis at 0.1 µg/ml but at higher concentrations showed the expected concentration-dependent inhibition.
Cordycepin, even at higher concentrations, only inhibits about 30%
of total RNA synthesis as expected from its role in preventing
poly A synthesis (Brawerman, 1974) and thus blocking only messenger
RNA processing. Hence, it appears that RNA synthesis, possibly
mRNA synthesis, is important for the switchover in commitment of
the epidermal cells.

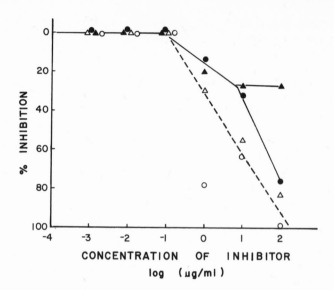

Fig. 9.   The inhibition of the 1 µg/ml β-ecdysone induced change
in cellular commitment in vitro (open symbols) and of [³H]-uridine
incorporation into total RNA (closed symbols) of dorsal abdominal
epidermis as a function of the concentration of the inhibitors
actinomycin D (squares) and cordycepin (triangles). After exposure
to actinomycin D 23 to 32 cysts were scored per point at 0.01, 0.1
and 1 µg/ml, 8 at 10 µg/ml, and 3 at 0.001 µg/ml. After exposure
to cordycepin 4 cysts were scored per point. The inhibition of
uridine incorporation represents an average of 2 replicates of
duplicate samples.

        To confirm that RNA synthesis was indeed occurring during the
time of exposure of the epidermis to ecdysone, [³H]-uridine was
added to the Grace's medium for varying periods of time. As seen
in Fig. 10, when the epidermis was incubated without hormones there
was a very low level of incorporation which did not change significantly over the 25 hr period. But when either 1 µg/ml β-ecdysone
or 3 µg/ml EGS was present in the Grace's medium, total RNA synthesis increased linearly showing about a 5-fold increase in 25 hr.
When both hormones were present in the medium, the rate of RNA
synthesis increased more dramatically and by 25 hr. there was a
14-fold increase in [³H]-uridine incorporation. Thus, RNA synthesis
occurs during the ecdysone-induced switchover in commitment but it
also occurs in the two hormonal situations in which there is no
change in commitment. Obviously, there must be differences in the
RNA synthesized in these various situations and work is in progress
to try to identify these differences.

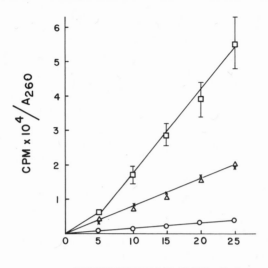

DURATION OF CULTURE (hr)

Fig. 10. The incorporation of [³H]-uridine into total RNA as a
function of time of exposure of dorsal abdominal epidermis to
hormones in vitro. Circles (o) indicate no hormones in the
Grace's medium; triangles (Δ) 1 μg/ml β-ecdysone; x's 3 μg/ml
epoxygeranylsesamole; squares 1 μg/ml β-ecdysone and 3 μg/ml
epoxygeranylsesamole. The points are averages of 6 to 12 replicates. The bars indicate ± two standard deviations when these
were larger than the symbols.

DISCUSSION

The results outlined above show that the Manduca larval
epidermis is an excellent system in which to study the action of
juvenile hormone as well as of ecdysone. When exposed to ecdysone
either in vivo or in vitro in the presence of JH, it forms larval
cuticle (see Fig. 1 and Tables 1 and 2). But when exposed to
ecdysone in the absence of JH, it changes its commitment and can
then only make pupal cuticle when given a second exposure to
ecdysone. Thus the action of ecdysone in causing a change in cellular commitment so that pupal cuticle can be synthesized is well-
separated in time from the expression of that cellular commitment.
Furthermore, it is this event that can be blocked by JH, rather
than the events of cuticle synthesis themselves.

The question now is, "What is the cellular event which can be
induced by ecdysone and prevented by juvenile hormone?" From these
studies thus far one can make a few solid conclusions and a few
speculations. The autoradiographic evidence on the timing of DNA

synthesis during the larval-pupal transformation (Riddiford, 1975; Fig. 4) strongly indicates that there is no DNA synthesis during the time of the switchover in commitment but only just before the expression of that commitment two days later. Thus, although there may well be a "quantal mitosis" (Holtzer et al., 1972) before the cell can express its new differentiated state, DNA synthesis is not the cellular event which occurs during the irreversible change in commitment. Furthermore, our in vitro studies with protein synthesis inhibitors as well as measurement of total protein synthesis over the 24 hr commitment period suggest that protein synthesis is not essential for the change.

Similar experiments with the RNA synthesis inhibitors actinomycin D and cordycepin indicate that RNA synthesis is required. Although at the higher concentrations used, actinomycin D should block both ribosomal and messenger RNA synthesis (Brawerman, 1974), the effect of cordycepin suggests that mRNA synthesis may be necessary for the switch in commitment. Cordycepin is known to inhibit poly A synthesis, thus blocking the proper processing and transport of newly synthesized mRNA to the cytoplasm except for the mRNA which has no poly A such as for histone (Brawerman, 1974). Further evidence on the role of mRNA synthesis in the switchover will be afforded by our studies in progress on the α-amanitin sensitivity of the system. The identity of this messenger(s) is unknown, but it seems unlikely that they are messenger(s) for the pupal cuticle protein since the epidermal cells have yet to undergo a cell division before synthesizing the pupal cuticle. Rather it seems more probable that they are necessary for the synthesis of regulatory protein(s) which "set the stage" for the epidermal cell. Consequently, when it next sees ecdysone, it immediately undergoes DNA synthesis, then begins producing pupal cuticle. These proteins show few qualitative differences but some quantitative differences from larval cuticle proteins (Logan and Riddiford, unpublished studies).

Our studies of total RNA synthesis over the 24 hr commitment period (see Fig. 10) show that β-ecdysone induces a 5-fold increase but so also does JH (EGS) alone. Moreover, the two hormones together cause a 14-fold increase. This latter increase is most probably due to synthesis of all types of RNA necessary for larval cuticle protein synthesis since exposure of larval epidermis to ecdysone in the presence of JH promotes a larval molt either in vivo (Fig. 1) or in vitro (see Table 2). The effect of JH alone on promoting RNA synthesis is more perplexing, especially since the kinetics of the increase are identical with that of β-ecdysone. Yet in the presence of ecdysone the cells are changing their commitment, but in JH alone as in medium with no hormones they are maintaining a "status quo" at least as far as commitment is concerned. After the last larval molt of Manduca the JH titer rises and remains at a readily detectable level through about 75% of the

feeding stage. Therefore, it apparently has some function in
maintaining the integrity of the larval tissues (e.g. the crochet
epidermis, Fain and Riddiford, 1976b). These effects therefore
are presumably reflected in the JH-stimulated increase in RNA
synthesis in Fig. 10. Obviously there are differences in the RNAs
synthesized under these two hormonal conditions and one is tempted
to speculate that the difference lies in the type of mRNAs synthesized. Hopefully we will be able to pinpoint these differences
in further studies.

Another intriguing aspect of ecdysone's action in changing
the commitment of the epidermal cells is that it renders them
aloof to the action of JH at least until after they have made
pupal cuticle. No longer can that cell produce larval cuticle.
In their theory of JH action, Williams and Kafatos (1972) have
postulated a change in the RNA polymerase which transcribes the
master regulatory gene for the three gene sets - larval, pupal,
and adult. Then one part of the pupal gene set is to make a
repressor protein which is insensitive to JH and which permanently
turns off the larval gene set. This hypothesis has not been satisfactorily tested and possibly could be studied in this system once
we know more about what the change in commitment is on a molecular
basis.

But it also seems possible that the absence of any JH effect
on the formation of pupal cuticle once the switchover has been
made could simply be due to a lack of JH receptors. In vertebrate
systems such as the rat uterus continued synthesis of estrogen
cytoplasmic protein receptors is necessary for the whole spectrum
of estrogen effects seen _in vivo_ (Clark, 1973). Therefore,
assuming that JH does indeed act similarly to a steroid hormone
in that it enters the cell, combines with a cytoplasmic receptor
protein and thence moves to the nucleus, it would require a cytoplasmic receptor protein in order to have its effect. Thus there
is a possibility that when exposed to ecdysone in the absence of
JH, the larval epidermal cell as part of its switchover in commitment ceases production of JH-receptor protein. Consequently, the
cell would become insensitive to JH. Since it regains sensitivity
to JH after pupal ecdysis, one would have to suppose that this
protein is again made during or just after the time pupal cuticle
is made. When we have a labeled JH of very high specific activity
we should be able to test the hypothesis that there is such a cytoplasmic protein and to look for the reasons for this ecdysone-
induced turning-off of sensitivity to JH.

In summary, the larval epidermis of _Manduca_ promises to be an
extremely good system in which we can now study the cellular and
molecular basis of the morphogenetic action of JH.

SUMMARY

In <u>Manduca</u> <u>sexta</u> the larval-pupal transformation requires two releases of ecdysone: the first causes a change in the commitment of the epidermal cells so that they can no longer make larval cuticle; the second induces these cells to make pupal cuticle. During the period of the change in commitment the epidermal cells were detached from the overlying cuticle although they continued to produce larval cuticle. No DNA synthesis was detectable by autoradiography after 2 hr pulses of [^{3}H]-thymidine until the second release of ecdysone. The switchover in commitment was affected <u>in vitro</u> by 24 hr exposure to 1 µg/ml β-ecdysone in Grace's medium irrespective of whether assayed by implantation into a penultimate instar larva before the initiation of the molt or by a further 4-day exposure to 5 µg/ml β-ecdysone <u>in vitro</u>. Thirteen to 16 hr exposure to 1 µg/ml β-ecdysone or 24 hr exposure to the physiological concentration of about 10^{-7} M ecdysone was sufficient for half of the cells to change their commitment. When juvenile hormone was added before or with the ecdysone, it prevented the change in commitment so that the cells continued to make larval cuticle. A physiological concentration of 2 x 10^{-8} M t,t,c JH I was sufficient to inhibit the change in 50% of the cells as was 0.2 µg/ ml of the JH mimic epoxygeranylsesamole (EGS). The protein synthesis inhibitors cycloheximide and puromycin had no effect on the ecdysone-induced switchover <u>in vitro</u>, but the RNA synthesis inhibitors actinomycin D and cordycepin at concentrations above 1 µg/ml prevented the change in commitment. Total RNA synthesis increased 5-fold during the 24 hr exposure to either ecdysone or EGS alone and increased 14-fold in the presence of both hormones (as a prelude to larval cuticle synthesis). Thus, in exerting its "status quo" action on epidermal commitment, JH promotes general RNA synthesis, while blocking the ecdysone-induced specific, possibly messenger, RNA synthesis.

ACKNOWLEDGEMENTS

I thank Ms. Anna Curtis for technical assistance; Ms. Barbara Graves, Dr. W.R. Logan, and Dr. J.W. Truman for many helpful discussions; and the latter two for a critical reading of the manuscript. The work was supported by grants from NSF (BMS 7516360), NIH (AI 12459-01) and the Rockefeller Foundation.

REFERENCES

Ashburner, M., 1973, <u>Develop. Biol</u>. 35:47.
Bollenbacher, W.E., Vedeckis, W.V., Gilbert, L.I., and O'Connor,
    J.D., 1975, <u>Develop. Biol</u>. 44:46.
Brawerman, G., 1974, <u>Ann. Rev. Biochem</u>. 43:621.

Clark, J.H., 1973, in "Receptors for Reproductive Hormones"
    (B.W. O'Malley and A.R. Means, eds.) p. 15, Plenum Press,
    New York.
Fain, M.J., and Riddiford, L.M., 1975, Biol. Bull. 149:506.
Fain, M.J., and Riddiford, L.M., 1976a, Gen. Comp. Endocrinol.,
    in press.
Fain, M.J., and Riddiford, L.M., 1976b, Develop. Biol., submitted
    for publication.
Fristrom, J.W., Logan, W.R., and Murphy, C., 1973, Develop. Biol.
    33:441.
Holtzer, H., Weintraub, K., Mayne, R., and Mochan, B., 1972,
    Curr. Topics Devel. Biol. 7:229.
Lowry, O.H., Rosebrough, N.J., Farr, A.L., and Randall, R.J.,
    1951, J. Biol. Chem. 193:265.
Mitsui, T., and Riddiford, L.M., 1976, Develop. Biol., submitted
    for publication.
Nijhout, H.F., 1975, Biol. Bull. 49:568.
Nijhout, H.F., and Williams, C.M., 1974a, J. Exp. Biol. 61:481.
Nijhout, H.F., and Williams, C.M., 1974b, J. Exp. Biol. 61:493.
Riddiford, L.M., 1975, in "Invertebrate Tissue Culture: Applications
    in Medicine, Biology, and Agriculture." (K. Maramorosch and
    E. Kurstak, eds.) Academic Press, in press.
Riddiford, L.M., 1976, Nature, 259:115.
Truman, J.W., 1972, J. Exp. Biol. 57:805.
Truman, J.W., and Riddiford, L.M., 1974, J. Exp. Biol. 60:371.
Truman, J.W., Riddiford, L.M., and Safranek, L., 1974, Develop.
    Biol. 39:247.
Williams, C.M., 1961, Biol. Bull. 121:572.
Williams, C.M., and Kafatos, F.C., 1972, in "Insect Juvenile
    Hormones, Chemistry and Action" (J.J. Menn and M. Beroza, eds.)
    pp. 29-41, Academic Press, New York.
Willis, J.H., 1974, Ann. Rev. Entomol. 19:97.

ACTION OF JUVENILE HORMONE ON IMAGINAL DISCS OF THE INDIAN MEAL

MOTH

H. Oberlander and D.L. Silhacek

Insect Attractants, Behavior, and Basic Biology Research
Laboratory, Agric. Res. Serv., USDA, Gainesville,
Florida 32604

INTRODUCTION

The action of the corpora allata, which secrete juvenile
hormone (JH), on the development of imaginal discs has received
limited attention during the three decades that followed Vogt's
pioneering work with Diptera. Vogt (1943) reported that removal
of the medial cells (corpora allata) of the ring gland in larvae
resulted in accelerated development of the eye discs. Precocious
development of imaginal discs was also obtained by transplanting
imaginal discs into a JH-free milieu. In these experiments
Williams (1961) transplanted wing discs from Cynthia into chilled
pupae. The discs metamorphosed with the host into pupal, pupal-
adult, or directly into adult structures. Not only could the
imaginal discs be induced to metamorphose prematurely by reducing
their exposure to JH, but they could also be maintained in an
undifferentiated state by extending the JH period. These early
studies demonstrated that the corpora allata prevented the differentiation of imaginal discs in both dipteran (Vogt, 1946; Possompes,
1953) and lepidopteran larvae (Sehnal, 1968). Other investigations
have shown that imaginal discs are targets of the ecdysones
(Oberlander, 1972).

The interactions of imaginal discs with JH and β-ecdysone could
not be satisfactorily explored prior to the availability of pure
hormones, as well as the development of sensitive in vivo and in
vitro test systems. Therefore, we undertook a detailed examination
of the ability of wing discs from the Indian meal moth (Plodia
interpunctella Hübner) to respond to JH I and β-ecdysone. We are
now in a position to address the following questions: (1) Are the
wing discs direct targets of JH? (2) does JH prevent differentiation

of the imaginal discs by directly opposing β-ecdysone (antagonism),
or by preventing the acquisition of competence to respond to β-
ecdysone?

MATERIALS AND METHODS

The Indian meal moths used in these experiments were reared
by the standardized procedures of Silhacek and Miller (1972).
Unless otherwise noted, in vivo and in vitro hormone treatments
of wing discs were conducted as previously reported (Silhacek and
Oberlander, 1975; Oberlander et al., 1973). For the in vivo
experiments newly ecdysed last-instar larvae were taken from
standard cultures and transferred to test diets treated with either
hormone solution or solvent. The duration of treatment was varied,
but in all cases the wing discs were examined seven days after the
beginning of the instar. The ability of the discs to differentiate was assessed by culturing the tissue with fat body and β-
ecdysone in Grace-Yunker's medium.

Four stages in the development of the wing discs during the
last larval instar are illustrated in Fig. 1. The wing discs
grow and develop lacunae on the second day of the instar. On the
third day (wandering larvae) the tracheoles migrate into the lacunae,
and on the fourth day, the discs evaginate as the larvae commence
cocoon spinning. Pupal cuticle is deposited on the fifth to sixth
day. These developmental stages are also elicited in vitro after
incubation with β-ecdysone (Oberlander and Tomblin, 1972). Incubation of lacunar discs with ecdysone induced tracheole migration,
evagination, and deposition of a cuticle that was tanned, lacked
scales or hairs, and was indistinguishable from pupal cuticle. In
the present experiments the discs were examined daily with a dissecting microscope and the presence of tanned cuticle noted.

In the JH metabolism experiments the tissues were cultured in
200 µl of medium in glass micro dishes coated with 1% polyethylene
glycol (20,000) to prevent the hormone from adhering to the glass.
The hormone was extracted and chromatographed on thin layer plates
according to the method of Slade and Zibitt (1972).

RESULTS

Response of Wing Discs to β-Ecdysone In Vitro.

Before investigating the action of JH we determined the optimal conditions in vitro for stimulation of wing disc development
by β-ecdysone. The best response to β-ecdysone was obtained by
culturing wing discs from mid-last-instar larvae (weighing more

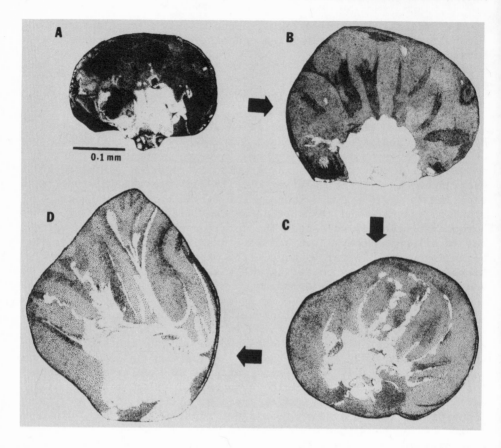

Fig. 1. Development of wing discs during the last larval instar
of Plodia moths. A, prelacunar disk, first day; b, lacunar disk,
second day; C, tracheolated disk, third day; D, evaginated disk,
fourth day (Silhacek and Oberlander, 1975).

than 12 mg) in medium that contained larval fat body (Oberlander
and Tomblin, 1972; Dutkowski and Oberlander, 1973; 1974; see
Figs. 2 and 3). Wing discs from larvae during the first two days
of the last instar and weighing less than 11 mg responded poorly
or not at all when incubated with β-ecdysone. This transition in
capacity to develop in vitro corresponded to the development of
lacunae in discs from day-two larvae.

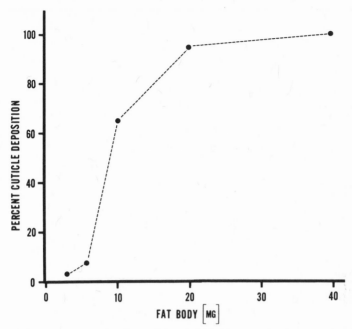

Fig. 2. The effect of different amounts of fat body on cuticle
deposition in three-day wing discs (15–18 mg larvae). The wing
discs were cultured in 1 ml Grace-Yunker's medium in which 2 μg
β-ecdysone and fat body had been incubated for 24 hr (Dutkowski and
Oberlander, 1973).

Response of Wing Discs to JH In Vitro

The conditions that are optimal for β-ecdysone action are not
necessarily the most favorable for JH action. For example, the
stability of JH in vitro is a matter of concern since both imaginal
discs (Drosophila melanogaster, Chihara et al., 1972; Manduca sexta,
Hammock et al., 1975) and fat body (Manduca sexta, Hammock et al.,
1975) degrade the hormone. Therefore, we examined the fate of
[^{3}H]-JH in our cultures.

The isotope was incubated with 2.1 X 10^{-4} M unlabeled JH,
10 mg fat body and ten wing discs. After 9.5 hr of incubation,
only 9% of the radioactivity recovered from the medium was undegraded JH (as determined by thin-layer chromatography). However,
when JH binding protein isolated from the hemolymph of Manduca
sexta (Kramer et al., 1974) was added to the medium, the percent
of radioactivity recovered as undegraded JH after 9.5 hr was 94%,
and after 24 hr was 90% (Table 1). The radioactivity in extracts
of the discs from individual cultures was negligible.

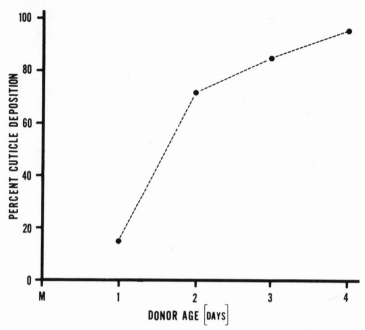

Fig. 3. The effect of donor age on cuticle deposition in cultured wing discs. The conditions are the same as in Fig. 2. Regardless of the age of the donor of the discs, the fat body was taken from three-day larvae (Dutkowski and Oberlander, 1974).

The percentage of radioactivity in the fat body recovered as JH was at least 90% whether or not carrier protein was present (see also Hammock et al., 1975). In the absence of carrier protein, approximately three-quarters of the total radioactivity recovered from the cultures as unmetabolized JH was within the fat body (Table 2). In vitro the fat body may rapidly sequester juvenile hormone from the medium and may serve as a reservoir for its slow release. This could explain our observation that fat body was required to prevent the toxic effects of JH on the discs in medium.

The degradation of JH by tissues in vitro, as well as non-specific binding to serum proteins in the medium (Ferkovich and Oberlander, unpublished observations) probably necessitates the use of high levels of hormone. However, when JH-carrier protein was used, 10^{-4} M hormone was still required (Sanburg et al., 1975). Because of these considerations the addition of 10^{-4} M JH to the culture medium does not necessarily mean that the disc cells are actually exposed to this concentration; in fact, they are probably exposed to considerably less.

TABLE 1. METABOLISM OF JUVENILE HORMONE IN TISSUE CULTURE.[a]

Incubation time (hr)	% radioactivity recovered from medium as JH		% radioactivity recovered from fat body as JH	
	No carrier protein	With carrier protein[b]	No carrier protein	With carrier protein
24 (no tissue)	94	98	---	---
2	79	---	96	---
4	73	---	97	---
9.5	9	94	95	91
16	7	91	90	94
24	6	94	90	92

[a] Ten wing discs from larvae that weighed 12-15 mg were incubated with fat body (10 mg) in 200 µl of Grace-Yunker's tissue culture medium. The medium contained 2.1×10^{-4} M unlabeled JH and 8 µCi of [^{3}H]-JH (14.1 Ci/mmol). The wing discs, fat body, and medium were separately extracted. Negligible label was recovered from the discs. After extraction and chromatography of the medium and tissue, the thin layer plates were scraped in 5 mm sections and counted in a Packard Tricarb Scintillation Spectrometer.

[b] Sixty µg of homogeneous binding protein isolated from the hemolymph of Manduca sexta.

TABLE 2.   PERCENT OF TOTAL RADIOACTIVITY RECOVERED AS JUVENILE
HORMONE FROM THE MEDIUM AND TISSUES THAT IS FOUND IN THE FAT BODY.[a]

Incubation time (hr)	No carrier protein	Carrier protein
2	8	--
4	24	--
9.5	76	5
16	77	7
24	73	1

[a] Conditions are the same as in Table 1.

     The effect of JH on the capacity of wing discs to differen-
tiate in vitro was assessed by incubating the discs with fat body
and 3.3×10^{-4} M JH for 24 hr, and then adding β-ecdysone. We
found a correlation of donor age and effectiveness of JH that was
the reverse of the relationship for β-ecdysone. JH inhibited
cuticle deposition in discs from one and two-day last-instar larvae, but had little effect on discs from four-day larvae (Oberlander and Tomblin, 1972). These in vitro experiments demonstrated
that in last-instar larvae of the Indian meal moth, wing discs
become responsive to β-ecdysone at the same time that they are
losing their sensitivity to JH.

Response of Wing Discs to JH In Vivo

     We never observed the formation of lacunae by day-one discs,
and were unable to investigate the acquisition of competence by
the discs to respond to β-ecdysone in vitro. Therefore, we turned
to a combination of in vivo and in vitro techniques to explore the
acquisition of competence. We administered JH to Indian meal moth
larvae through their diet (Firstenberg and Silhacek, 1975) and
examined the response of wing discs to β-ecdysone in vitro.

     Last-instar larvae that are reared on a diet containing 100 ppm
JH do not metamorphose unless they are transferred to hormone-free
diet. This reversibility is observed even after treatments as long
as four weeks. However, when hormone treatment began on day three

or four of the last instar, many (40% and 85%, respectively) of
the test insects metamorphosed (Silhacek and Oberlander, 1975).
Thus, JH treatment in vivo must begin in the early part of the last
instar to be effective.

We monitored the morphology of the imaginal discs in the
treated larvae to determine whether their development was arrested
or reversed. Discs from larvae that were kept on JH treated diet
from day zero of the last instar to day seven did not develop
beyond the prelacunar stage. However, if treatment began on day
one when the discs were still prelacunar, only one-half remained
in that state by the seventh day. The results of this treatment
were striking because 25% of the wing discs grew substantially and
deposited cuticle even though none of these larvae had pupated.
Although cuticular discs were found only in treated larvae that
had undergone a supernumerary molt, not all wing discs from these
larvae deposited cuticle. When JH treatment began on day two of
the last instar most of the larvae pupated and the wing discs
formed normal pupal structures (Silhacek and Oberlander, 1975).
Similar results were obtained by exposing the larvae to a JH analog
(piperonyl 6,7-epoxy-3-ethyl-7-methyl-2-nonenyl ether) first synthesized by Bowers et al. (1965) (Table 3).

The morphology of the discs from only those larvae that
molted while on JH diet is shown separately in Table 4. An examination of the wing discs in the supernumerary larval instar
revealed the following. When larvae were placed on hormone-
treated diet from day zero to day seven nearly all of the wing
discs in the supernumerary larval instar remained small and prelacunar. In contrast, if JH treatment was begun on day one, all
of the discs in these larvae deposited cuticle. Finally, if
the larvae were placed on JH-treated diet on day zero and transferred to hormone-free diet on either day one or two, about half
the discs in the larvae that had molted developed lacunae while
the other half grew substantially. Thus, an additional larval
ecdysis as a result of JH treatment resulted in the stimulation
of moderate growth with lacunae formation; extensive growth without lacunae formation; or extensive growth with cuticle deposition.

Finally, we tested the capacity of wing discs from larvae
treated with JH in vivo to respond to β-ecdysone in vitro. Wing
discs were dissected from the JH-treated larvae after seven days
and cultured in vitro with fat body and β-ecdysone (Silhacek and
Oberlander, 1975). The chronological age of the larval donors had
no effect on the response of the discs to β-ecdysone. The most
critical factor was associated with the morphology of the discs.
Few of the small prelacunar discs made cuticle in vitro, while
lacunar discs responded well to β-ecdysone whatever the nature of
their in vivo JH exposure (Table 5). One exception was that most
of the discs from larvae that had an additional larval ecdysis and

TABLE 3. EFFECTS OF JUVENILE HORMONE ON WING DISC DEVELOPMENT
IN VIVO.

Exposure	Disc morphology		
(days of last instar)	% prelacunar	% lacunar	% cuticle
JH I[a] (100 ppm)			
0–1	––	33	67
0–2	11	45	45
0–3	––	78	22
0–7	100	––	––
1–7	50	25	25[c]
2–7	––	11	89
3–7	––	––	100
JH analog[b] (100 ppm)			
0–7	90	10	––
1–7	20	10	70 (30[c])
2–7	––	40	60 (20[c])
3–7	––	10	90

[a] Methyl-10,11-epoxy-7-ethyl-3,11-dimethyl-2,6-tridecadienoate
(Silhacek and Oberlander, 1975).

[b] Piperonyl 6,7-epoxy-3-ethyl-7-methyl-2-nonenyl ether.

[c] Percentage of discs that made cuticle in larvae that did not
pupate.

TABLE 4. MORPHOLOGY OF WING DISCS IN THE SUPERNUMERARY LARVAL
INSTAR.

Exposure of JH (days of "last" instar)	% prelacunar (normal)	% prelacunar (enlarged)	% lacunar	% cuticle
0-1	--	50	50	--
0-2	--	40	60	--
0-7	91	9	--	--
1-7	--	--	--	100

TABLE 5. WING DISC DEVELOPMENT IN VITRO FOLLOWING JUVENILE
HORMONE[a] TREATMENT IN VIVO.

Initial morphology	% cuticle	
	2 µg β-ecdysone	Ethanol
Prelacunar (normal)	3	0
Prelacunar (enlarged)	60	0
Lacunar	94	0

[a] Silhacek and Oberlander (1975).

grew without forming lacunae were able to make cuticle in response
to β-ecdysone treatment in vitro. Hence, lacunae formation is a
convenient marker of the acquisition of sensitivity to β-ecdysone
but is not essential for the discs' response in vitro.

DISCUSSION

The Imaginal Discs as Targets of JH.

Because we examined the effects of JH on imaginal discs from the Indian meal moth in medium that contained fat body, it is not certain that JH acted directly on the discs. Although fat body-conditioned medium may be substituted for fat body-containing medium, small groups of fat body cells may remain in the medium (Oberlander, unpublished observations). Mass-isolated imaginal discs from Drosophila appear to respond directly to JH, but even here small amounts (4%) of contaminating tissue are present (Logan et al., 1975). Moreover, the high concentration of JH required for inhibition of imaginal discs cultured in vitro raises some question about the physiological nature of the interaction. Imaginal discs may be direct targets of JH but we conclude that further refinements in the in vitro techniques are necessary to demonstrate this unequivocally.

Competence of Imaginal Discs to Respond to JH and Ecdysone

Our experiments conducted with the Indian meal moth show that β-ecdysone-induced development of cultured imaginal discs is prevented if sufficient JH is present in the medium. According to Willis (1974) "the inhibition of cuticle synthesis in imaginal discs is not due to any antagonistic action," but rather to the normal role of JH in preventing maturation of imaginal discs. On the other hand, Fristrom (1972) concluded from his investigation of JH inhibition of ecdysone-induced protein synthesis that "juvenile hormone and ecdysone act directly and antagonistically on the development of discs." Subsequently, Fristrom's group showed that JH also inhibited ecdysone-induced RNA and DNA synthesis (Chihara and Fristrom, 1973; Logan et al., 1975). Our position is that the terminology used to describe the interactions between ecdysone and JH is unimportant as long as the restricted frame of reference is clearly defined. In our papers on imaginal disc development we have not used the term "antagonism" because we believe our findings can best be understood in relation to the development of competence of the discs to respond to hormones.

Our results with imaginal discs from the Indian meal moth point to an increased competence to respond to β-ecdysone as the larvae mature. These observations are in accord with earlier results for Drosophila (Bodenstein, 1939), Chilo supressalis (Agui et al., 1969) and Galleria mellonella (Oberlander, 1969).

One requirement for the development of competence of wing discs
to respond to β-ecdysone is extensive growth. In our experiments
we demonstrated that growth of the wing discs in the absence of
discernible morphological change led to their acquisition of competence to respond to β-ecdysone. These results with the Indian
meal moth are compatible with those of Mindek and Nothiger (1973)
with Drosophila. They showed that transplanted early second-instar
eye-antennal discs become competent to metamorphose when cultured
in flies fed on standard food. In contrast, imaginal discs did not
develop competence when cultured in starved flies in which cell
multiplication in the discs was suppressed. They concluded that
"the acquisition of competence requires a minimum number of cell
divisions...."

Other examples in which growth is correlated with the acquisition of competence may be found in the literature. For example,
Bodenstein (1943) transplanted both ring glands and leg discs from
various aged Drosophila larvae into adult flies. Under these conditions both young and old discs metamorphosed. An examination of
the discs led Bodenstein to conclude that "the young discs grow
to a certain size before their differentiation leading to imaginal
completion begins." This growth period probably accounts for the
development of imaginal discs transplanted by Williams (1961) from
fourth instar Cynthia larvae into pupae. Indeed imaginal disc
cells derived from embryos can develop a capacity for metamorphosis after a period of growth in adult flies (Hadorn, 1968) or
under in vitro conditions (Dubendorfer, 1975).

It is evident from the results with last-instar larvae of
the Indian meal moth that JH prevents the wing discs from developing to an ecdysone-receptive state. In our experiments JH was
most effective as an inhibitor when it was added to the culture
medium simultaneously with the discs, allowing no lag period. We
found that once the wing discs acquired full competence to respond
to ecdysone they were insensitive to JH.

Because the titer of β-ecdysone rises in last-instar larvae
as JH titer falls (Patel and Madhaven, 1969) there is little
opportunity in vivo for direct repression of β-ecdysone effects
on imaginal discs by JH. Instead, JH acts by preventing the development of the wing discs to an ecdysone-receptive state. JH may
prevent the development of ecdysone-sensitivity by controlling the
cell cycle (Lobbecke, 1969), because growth of the discs influences
the acquisition of competence. Although JH may sometimes act in
physiological opposition to ecdysone, the suppression by JH of the
development of ecdysone-sensitivity in imaginal discs is the most
likely explanation of our results.

SUMMARY

The interactions of imaginal discs of the Indian meal moth, Plodia interpunctella Hubner with JH and β-ecdysone were investigated under in vivo and in vitro conditions. We considered whether JH controls metamorphosis by preventing the acquisition of competence to respond to β-ecdysone. We found that wing discs during the last larval instar became responsive to β-ecdysone as they lost their sensitivity to JH. Growth of the discs was associated with the development of ecdysone-sensitivity. We conclude that although JH may sometimes act in physiological opposition to β-ecdysone, the suppression of competence to respond to this hormone is the most critical action of JH on imaginal discs.

ACKNOWLEDGEMENTS

We are grateful to Hoffmann-LaRoche for providing us with the synthetic JH I and the JH analog used in our experiments. Professor John Law (University of Chicago) generously provided us with Manduca sexta JH binding protein. Dr. Karl Kramer suggested the use of polyethylene glycol coated glassware as well as the extraction and chromatographic methods used in the JH metabolism studies. Ms. Karen Kohl and Mr. C.E. Leach provided excellent technical assistance. Figure 1 is reprinted with permission of Pergamon Press.

REFERENCES

Agui, N., Yagi, S., and Fukaya, M., 1969, Appl. entomol. Zool. 4:158.

Bodenstein, D., 1939, J. exp. Zool. 82:1.

Bowers, W.S., Thompson, M.J., and Uebel, E.C., 1965, Life Sci. 4:2323.

Chihara, C.J., and Fristrom, J.W., 1973, Devel. Biol. 35:36.

Chihara, C.J., Petri, W., Fristrom, J., and King, D., 1972, J. Insect Physiol. 18:1115.

Dubendorfer, A., 1975, "Metamorphosis of imaginal disc tissue grown in vitro from dissociated embryos of Drosophila." In Proc. 4th International Conference on Invertebrate Tissue Culture, Montreal, in press.

Dutkowski, A.B., and Oberlander, H., 1973, J. Insect Physiol. 19:2155.

Dutkowski, A.B., and Oberlander, H., 1974, J. Insect Physiol. 20:743.

Firstenberg, D., and Silhacek, D.L., 1975, "Food consumption by Plodia interpunctella feeding on diets containing insect growth regulators" J. Ga. Entomol. Soc., in press.

Fristrom, J.W., 1972, in "The Biology of Imaginal Disks" (H.
    Ursprung and R. Nothiger, eds.) pp. 109-154, Springer-Verlag,
    New York.
Hadorn, E., 1966, Devel. Biol. 13:424.
Hammock, B., Nowock, J., Goodman, W., Stamoudis, V., and Gilbert,
    L.I., 1975, Mol. Cell. Endocrinol. 3:167.
Kramer, K.J., Sanburg, L.L., Kézdy, F.J., and Law, J.H., 1974,
    Proc. Nat. Acad. Sci. USA 71:493.
Lobbecke, E.A., 1969, Roux' Arch. Entw. Mech. Org. 162:1.
Logan, R.W., Fristrom, D., and Fristrom, J.W., 1975, J. Insect
    Physiol. 21:1343.
Mindek, G., and Nothiger, R., 1973, J. Insect Physiol. 19:1711.
Oberlander, H., 1969, J. Insect Physiol. 15:297.
Oberlander, H., 1972, in "The Biology of Imaginal Disks" (H.
    Ursprung and R. Nothiger, eds.), pp. 155-172, Springer-Verlag,
    New York.
Oberlander, H., and Tomblin, C., 1972, Science (Wash.) 177:441.
Oberlander, H., Leach, C.E., and Tomblin, C., 1973, J. Insect
    Physiol. 19:993.
Patel, N., and Madhavan, K., 1969, J. Insect Physiol. 15:2141.
Possompes, B., 1953, Arch. Zool. exp. gen. 89:203.
Sanburg, L.L., Kramer, K.J., Kézdy, F.J., Law, J.H., and
    Oberlander, H., 1975, Nature (Lond.) 253:266.
Sehnal, F., 1969, J. Insect Physiol. 14:73.
Silhacek, D.L., and Miller, G.L., 1972, Ann. Entomol. Soc. Amer.
    65:1084.
Silhacek, D.L., and Oberlander, H., 1975, J. Insect Physiol.
    21:153.
Slade, M., and Zibitt, C.H., 1972, in "Insect Juvenile Hormones"
    (J.J. Menn and M. Beroza, eds.) pp. 156-176, Academic Press,
    New York.
Vogt, M., 1943, Biol. Zbl. 63:395.
Williams, C.M., 1961, Biol. Bull. 121:572.
Willis, J.H., 1974, Ann. Rev. Entomol. 19:97.

ECDYSONE DEFICIENCY IN JUVENILE HORMONE TREATED LARVAE OF THE

GERMAN COCKROACH, BLATTELLA GERMANICA (L.)

P. Masner, W. Hangartner, and M. Suchý

Biological Laboratory, Dr. R. Maag Ltd., 8157 Dielsdorf,
Switzerland and Chemical Laboratory, Socar Ltd., 8600
Dübendorf, Switzerland

INTRODUCTION

According to Wigglesworth (1957), juvenile hormone (JH) acts
in opposition to the second principal insect hormone, ecdysone,
but this relationship is not construed as a simple inhibitory
effect. JH was given a key role in determining the outcome of
ecdysis induced by ecdysone by either suppressing new genetic
information (Williams, 1961; Schneiderman and Gilbert, 1964), or
activating specific genes of a conservative character (Wigglesworth, 1961). Both theories are based on the varying titer of
JH during postembryonic development and both neglect the simultaneous variations in the amount of ecdysone. In other words,
ecdysone is supposed to be always present in a quantity high enough
to stimulate morphogenesis but the given amount of JH limits its
activating capacity.

The titer of ecdysone also varies considerably during postembryonic development, as has been found in many insect species.
Shaaya and Levinson (1966) proposed the presence of either ecdysone or JH in the hemolymph of blowfly larvae and discussed the
antagonistic relationship of both hormones. The inhibitory effect
of synthetic JH and other insect growth regulators (IGRs) with JH
activity on morphogenesis and ecdysis as reported by Hangartner
and Masner (1973) and Madhavan (1973) could thus be explained by
the deficiency of ecdysone in the treated insects. The results
presented here demonstrate that the effect of JH and corresponding
IGRs is causally connected to an ecdysone deficiency syndrome.

MATERIALS AND METHODS

Ten freshly hatched fifth or sixth (last) stage larvae of the German cockroach, Blattella germanica (L.), were confined in 200 ml plastic cups with two folded 10 cm^2 paper discs previously treated with an acetone solution of JH or IGR at doses of 100, 10, 3.2 and 1 $\mu g/cm^2$, or with pure acetone as a control. Dog food and drinking water were supplied, and incubation took place at 24°C, 45% relative humidity, and 18 hr light/day. JH Ro 6-9550 (methyl 10,11-epoxy-7-ethyl-3,11-dimethyl-2,6-tridecadienoate, cis/trans mixture) and JH-active IGR Ro 8-9801 (6,7-epoxy-1-[p-ethylphenoxy]-3,7-dimethyl-2-octene, cis/trans mixture) were used for these studies. For injection, IGR was also used in combination with ecdysone as an emulsifiable concentrate (ACR-2022-A).

Treated and untreated insects were collected at equal time intervals; the last sample was taken when 20% of the population already ecdysed into the next stage. The insect material was stored at -20°C and ecdysone was extracted using the water-butanol procedure described by Karlson and Shaaya (1964). The ecdysone titer was established using the Musca domestica bioassay of Kaplanis et al. (1966) as modified by Staal (1967). The extracts as well as the α-ecdysone standard were diluted with small amounts of absolute ethanol and further with twice distilled water in order to get 2% solutions of the extracts and a 16 x 10^{-5}% solution of the ecdysone standard as the highest concentrations. 5 µl of test solution were injected through the pupated head into the larval abdomen of the ligated fully grown fly larvae. A scoring system of 0.5, 0.75 and 1 was used to determine ecdysone activity. Under our test conditions, one Musca Unit (MU), causing a 50% positive response, corresponds to 4 ng of either α- or β-ecdysone. Results of the ecdysone titrations were calculated in ng equivalents of ecdysone/g fresh weight.

RESULTS

Development of Normal Larvae

The postembryonic development of the German cockroach takes about nine weeks and passes through six larval stages. The younger instars including the penultimate, last ten days and the metamorphic sixth stage lasts 15 days (Table 1). The next ecdysis is being prepared within a few hr of the preceding molt, when the separation of the existing cuticle from the hypodermis starts. Apolysis is complete by the sixth day.

TABLE 1. COURSE OF DEVELOPMENT OF NORMAL AND JH OR IGR TREATED LARVAE OF THE TWO LAST LARVAL STAGES*.

Larval stage	Substance Ro	Dose µg/cm²	Apolysis (day)	Ecdysis (day)	Result of the next ecdysis
Penultimate	Untreated	0	5.8 ± 0.40	10.9 ± 0.40	6th instar larvae
	JH (6-9550)	100	6.9 ± 0.76	12.4 ± 0.54	6th instar larvae
		10	6.2 ± 0.55	12.0 ± 0.62	6th instar larvae
	IGR (8-9801)	100	10.5 ± 3.15	12.5 ± 1.29	dead in ecdysis
		10	7.4 ± 1.34	13.4 ± 2.09	6th instar larvae
		1	6.4 ± 0.99	11.4 ± 1.04	6th instar larvae
Last	Untreated	0	6.1 ± 0.26	14.5 ± 0.27	Adult
	JH (6-9550)	100	7.5 ± 0.82	15.9 ± 1.22	Supernumerary larvae
		10	6.8 ± 0.40	15.5 ± 0.64	Supernumerary larvae
	IGR (8-9801)	100	21.4 ± 4.00	>29	Permanent larvae
		10	12.6 ± 1.39	22.6 ± 2.14	Supernumerary larvae
		3.2	7.1 ± 0.85	15.6 ± 1.05	Supernumerary larvae
		1	6.3 ± 0.24	15.4 ± 0.76	Supernumerary larvae or deformed adults

* The figures are means of 10-300 insects. Not included in the means are the permanent larvae, i.e. those which required more than twice as long for ecdysis than did the controls.

Development of JH or IGR Treated Larvae of Younger Instars

The development of larvae exposed to JH and IGR is limited.
The two main functions of ecdysone, i.e. metamorphosis and induction and coordination of ecdysis, appear to be inhibited. This
second effect of the JH is not necessarily connected with morphogenetic activity, which is clearly demonstrated by the experiment
in which younger instar larvae are exposed to the highest dose
of IGR (100 µg/cm^2). They undergo delayed development starting
with apolysis and ending with synthesis of the new cuticle. The
treated larvae fail, however, to hatch and die during ecdysis,
which is connected with no, or very limited, morphogenetic transformation (Table 1, Fig. 1). They are trapped inside the old
cuticle, the endocuticle of which has not been digested although
the new exocuticle was already sclerotized (Figs. 2-4). The body
of those larvae cannot expand and the small exuvial space is
blocked by the coarse wrinkled new cuticle. The limited pressure
of the body and high friction of both cuticles are the main
reasons for the failure to ecdyse (Hangartner and Masner, 1973).
The development of larvae of the penultimate stage exposed to
the lower doses of IGR or to the highest dose of JH is slightly
delayed but continuous, and the insects molt into the next instar
(Table 1).

Development of JH or IGR Treated Last Stage Larvae

In last stage larvae exposed to the highest dose of IGR,
apolysis proceeds very slowly and the process of cuticle synthesis
is more or less completely inhibited as long as the insects remain
in contact with the treated paper (Masner and Hangartner, 1973;
Das and Gupta, 1974). The block in the postembryonic development
of those permanent larvae represents the greatest effect of the JH
active substance. Treatment with lower doses of either JH or IGR
suppresses metamorphosis and the larvae molt into supernumerary
instars characterized by dark pigmentation (Table 1).

JH Sensitive Period of Larvae

The larvae are most sensitive to elevated JH levels during
the first half of the instar. This was revealed by experiments
which combined topical and continuous contact treatments. The
insects were treated at the beginning of the experiment by a single
topical application of 100 µg IGR/larva and then exposed to the
paper impregnated with 100 µg/cm^2 of the same substance. This
treatment ensured an immediate as well as a continuous and long
lasting effect of the development-limiting dose of the JH-active
compound.

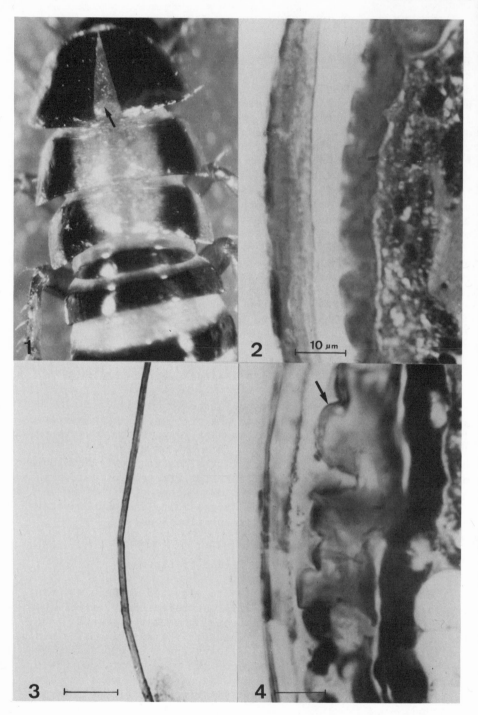

Figs. 1-4

Nearly all penultimate stage larvae treated up to the fourth day ceased development and died during ecdysis (Table 2). Later treatment had only a limited effect. Development of most of the last stage larvae treated on the sixth day or earlier was blocked and larval "diapause" was induced. Larvae treated after this critical period commenced ecdysis but were unable to complete it and died during the process. The inability to cast the old cuticle was very similar to the effect in younger instars treated before the critical period. More than 50% of the last stage larvae treated two to three days before the final ecdysis hatched into adults, most of which were sterile and exhibited slightly deformed wings (Table 3).

Selective Inhibition of Ecdysone Functions in IGR Treated Larvae

The amount of substance retained by larvae exposed for eight days to the dose of 100 $\mu g/cm^2$ has been established using tritium labeled IGR which is closely related chemically to the IGR used here. Preliminary results obtained by radioassay revealed 1.3 $\mu g/$ larvae or 22 $\mu g/g$ fresh weight of the labeled material. About one half of this material represents unchanged active substance (Masner, unpublished data).

Fig. 1. <u>Effect of JH on larval ecdysis</u>. The larva was reared in contact with paper impregnated with 100 $\mu g/cm^2$ of JH and died during ecdysis at the end of the first stage. New coriaceous cuticle is visible between the two parts of the split pronotal sclerite of the old cuticle (arrow).

Fig. 2. <u>Section through the tibia of an untreated penultimate stage larva shortly before ecdysis</u>. The broad exuvial space separates both cuticles. The old endocuticle is not yet digested; the new exocuticle is not sclerotized. Epoxy-embedding, basic fuchsin and alkalinized methylene blue stain.

Fig. 3. <u>Section through the tibia of the cast exuvia from an untreated larva of the same stage as in Fig. 2</u>. Endocuticle completely digested; only sclerotized exocuticle remains.

Fig. 4. <u>Section through the tibia of an IGR-treated penultimate stage larva dying during ecdysis</u>. The old endocuticle is not digested; the new exocuticle is fully sclerotized (arrow); the exuvial space is narrow.

TABLE 2. COURSE OF DEVELOPMENT OF PENULTIMATE STAGE LARVAE EXPOSED TO IGR* DURING DIFFERENT PERIODS OF TIME.

Days between last ecdysis and onset of IGR treatment	Morphogenetic effect (number of insects)			Length of period (in days) between two ecdyses
	Next larval stage	Dead during ecdysis	Permanent larvae	
2	0	28	3	13.8
4	0	29	1	12.1
6	24	15	0	11
8	25	5	0	10.2

* Single topical treatment (100 µg/larva) and contact with paper (100 µg/cm^2).

TABLE 3. COURSE OF DEVELOPMENT OF LAST STAGE LARVAE EXPOSED TO IGR* DURING DIFFERENT PERIODS OF TIME.

Days between last ecdysis and onset of IGR treatment	Morphogenetic effect (number of insects)					Length of period (in days) between two ecdyses
	Perfect adults	Deformed adults	Supernumerary larvae	Dead during ecdysis	Permanent larvae	
5				5	25	26
6				6	23	19.3
8	1		1	53	4	16.5
10	4	15	9	20	1	15.5
12	8	11	2	9		15

* Single topical treatment (100 µg/larva) and contact with paper (100 µg/cm^2).

This dose is neither toxic nor harmful to the basal metabolism of the insect as documented by the similar course of the growth curve for body weight with both treated and control insects (Fig. 5), normal life span and functionality of the female gonads. The growing oocytes are naturally resorbed during the advanced vitellogenic period as the route to oviposition is not differentiated. The functions of the ecdysone are, however, selectively limited. The permanent larvae are activated by a single injection of 10 µg of ecdysone; the inhibitory effect of JH is lifted and the larvae initiate ecdysis within ten days (Table 4).

Ecdysone is known to be an activator not only of synthesis but also of lysis of the cuticle during ecdysis since it activates the enzyme chitinase responsible for the digestion of the endo-cuticle (Kimura, 1973a,b). Thus, the main reason for the difficulties of younger IGR-treated larvae to ecdyse is the fact that the endocuticle of the old cuticle is undigested. This effect can also be attributed to a deficiency in ecdysone.

The Titer of Ecdysone During Larval Development

The key to understanding the mechanism of action of the IGR with all the reported morphogenetic and inhibitory effects, appeared to be connected with the changes in the ecdysone regime of the treated insects. We therefore decided to titer ecdysone under different experimental conditions during the two last larval stages.

The Amount of Ecdysone Permitting Apolysis, Ecdysis and Metamorphosis. Traces of ecdysone are present in 0–24 hr old larvae (Fig. 6). This amount, remaining from the preceding ecdysis, promotes development at the beginning of each stage when apolysis starts. The hormone disappears within the next 24 hr, and during the following two days it is either absent or at a titer below the biodetectable limit. It is at this time that immature insects exhibit the highest sensitivity to administered JH-active IGR as reported by many authors (Krishnakumaran, 1974; Lanzrein, 1974; Silhacek and Oberlander, 1975). The active substance applied during this period apparently has a good opportunity to prevent or postpone the normal increase in ecdysone titer that occurs later. This JH sensitive period is, in the German cockroach, terminated at the sixth day, which corresponds exactly with the re-appearance of ecdysone. In eight day old larvae there are 300 ng equivalents of ecdysone/g of fresh weight in the penultimate and 690 ng in the last instar. This amount seems to be indispensable for complete metamorphosis, whereas ecdysis takes place when the ecdysone titer drops to about one-fifth of the peak value.

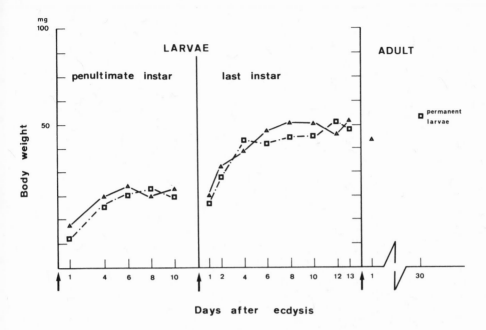

Fig. 5. <u>Effect of IGR on body weight</u>. Body weight was measured
during the course of development of the penultimate and last
larval stages in untreated larvae (Δ—Δ) and larvae exposed to
IGR at a dose of 100 μg/cm² (squares). The values are expressed
in mg/larva and are means of 20 insects. The arrows indicate
the time of the molt.

<u>Deficiency and Absence of Ecdysone in JH or IGR-Treated</u>
<u>Larvae</u>. The appearance of ecdysone in larvae treated with JH or
lower doses of IGR is delayed, and no ecdysone is detectable at
the time that the controls undergo metamorphosis. The ecdysone
titer, however, increases towards the end of the stage but obviously
cannot reach the level necessary for metamorphosis because ecdysis
at that time terminates development when the titer rises to the
proper concentration. In other words, the epidermis of larvae in
the middle of the instar is not responsive to an ecdysis-inducing
titer of ecdysone which increases in untreated larvae to the concentration ensuring metamorphosis. However, in full grown larvae
exposed to JH the epidermis and all other tissues involved are
activated by the ecdysis-inducing amount of ecdysone, and the molt
follows. The resulting supernumerary larval stage is the unavoidable consequence of the lack of ecdysone during the preceding
developmental period.

TABLE -4. EFFECT OF ECDYSONE INJECTED ALONE, AND IN COMBINATION WITH IGR, ON THE DEVELOPMENT OF NEWLY EMERGED LAST STAGE LARVAE.

Treatment		Number of dead last stage larvae	Number of insects hatching to the next stage			Length of the last stage (days)	Number of permanent larvae
IGR (μg/cm^2 paper)	Ecdysone (μg)		Adults	Extra-larvae	Dead in old cuticle		
	5		20			16	
	10	2		20		11.8	
100	5				18	11	20
100	10				20	10.1	20
100	50	1			19	8.5	

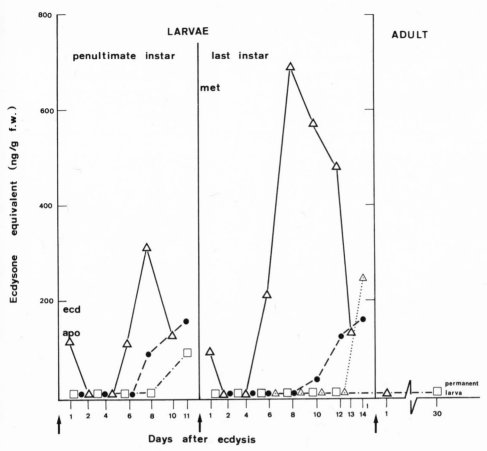

Fig. 6. Effect of JH on the ecdysone titer. Untreated larvae
(Δ——Δ); larvae exposed to JH at a dose of 100 µg/cm^2 (o---o) or
IGR at a dose of 3.2 µg/cm (Δ····Δ) and 100 µg/cm (squares).
The values are expressed in ng ecdysone equivalents/g fresh weight.
Hypothetical levels for apolysis (apo), ecdysis (ecd) and metamor-
phosis (met) are indicated.

No trace of ecdysone has ever been found in the permanent last
stage larva exposed to the highest dose of IGR. They remain in a
state of developmental arrest similar to JH-induced diapause
reported for several insect species (see Yagi, this volume).

Besides the inhibition of metamorphosis and cuticle synthesis,
JH-active IGR can limit development by preventing ecdysis in
younger larvae exposed to the highest dose of IGR. The amount of
ecdysone found in the extracts of penultimate larvae shortly before
the abortive attempt to molt does not reach the value present in
ecdysing control larvae.

Effect of IGR on Ecdysone Bioassay

Theoretically, the water-butanol ecdysone extracts from the treated insects could contain some JH-active material. The presence of this material could obscure the results obtained from the Musca bioassay. To exclude this possibility, we diluted the ecdysone standard with a 0.2% aqueous emulsion of IGR. The mixture (10 µg IGR/larva) was then injected into Musca which pupated 15 hr after treatment. In contrast to the results of Shaaya and Levinson (1966) who used unligated Calliphora larvae 40 hr before pupation, we found similar activities for both the pure standard and the IGR/ecdysone mixture (Fig. 7).

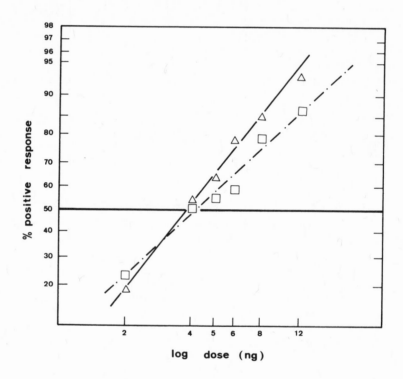

Fig. 7. Effect of JH on the response of ligated Musca larvae to ecdysone. Ligated larvae were injected with various doses of α-ecdysone diluted either in water (Δ——Δ) or in a 0.2% water emulsion of IGR giving 10 µg/larva (squares). The results are average values of two replicates with 20 insects at each dose. Probit/log scale.

DISCUSSION

Developmental Arrest in the Metamorphic Instar Treated
with JH-Active IGR

The inhibition of all ecdysone dependent processes, observed nearly exclusively in the last metamorphic stage treated with JH-active IGR, demonstrates the presence of a fail-safe mechanism which makes further development in this instar impossible under such conditions. A JH-induced diapause-like stage of developmental arrest has been reported for many insect species (Highnam, 1958; Fukaya and Mitsuhashi, 1961; Fukaya and Kobayashi, 1966; Sehnal and Meyer, 1968; Metwally and Sehnal, 1969; Benz, 1973; Karnavar, 1973; Yin and Chippendale, 1974; Hoppe, 1975; Sláma, 1975; Silhacek and Oberlander, 1975). The time and amount of ecdysone synthesized under the conditions of elevated titers of JH may vary in different species and therefore some insects respond to this "diapause" either by stationary molts or by becoming permanent larvae. However, this mechanism does fail in last stage larvae exposed to the highest dose of IGR after the critical period when ecdysone synthesis has already commenced. These larvae can synthesize new cuticle, due to the low ecdysone titer present before IGR application, but are generally unable to ecdyse. They die during ecdysis as do the larvae of all younger instars exposed to the highest dose of IGR from the beginning of the instar. The younger larvae seem to lack the JH fail-safe mechanism and produce, even under unfavorable conditions, a limited amount of ecdysone which permits synthesis of the new cuticle but precludes ecdysis.

The shortening of the metamorphic stage, occasionally observed in JH-treated insects, does not contradict the general ecdysone-inhibitive concept for this hormone. When the substance and/or dose applied is not completely suppressive, the ecdysone titer rises to the ecdysis-inducing level. The insect omits the last third of the metamorphic stage and molts into the supernumerary instar at the time of the larval-larval molt of the younger stages.

The permanent larvae may continue development into fertile adults when the titer of JH decreases, as seen in Blattella and Galleria (Masner, unpublished results). The reversibility of this JH effect has to be carefully considered in projects aimed at developing JH based pesticides. The younger larval stages which do not possess the fail-safe mechanism described above, are in this respect better targets for JH based IGRs which are to be developed as inhibitors of ecdysis.

Physiological and Phylogenetic Basis of JH—Ecdysone Antagonism

JH and IGR appear to inhibit the two main functions of ecdysone, i.e. differentiation during embryonic and postembryonic
development and the whole process of ecdysis in Blattella germanica
(Hangartner and Masner, 1973). Both effects can be explained by
the antagonistic relationship between the two hormones proposed by
Shaaya and Levinson (1966) and Benz (1973). This interpretation
is supported by many in vitro results which demonstrate the failure
of ecdysone in the presence of JH to stimulate RNA synthesis in
fat body and imaginal discs (Congote, 1969; Patel and Madhavan,
1969; Sekeris, 1972; Chihara and Fristrom, 1973), synthesis of
cuticle in imaginal discs (Oberlander and Tomblin, 1972), adult
differentiation of ovaries (Laverdure, 1971) or the puffing
activity of chromosomes (Lezzi, 1974). Direct evidence for such
an antagonistic relationship in vivo was shown when the inhibition
of development in the JH induced permanent larvae of Blattella
was broken by injection of ecdysone (Masner and Hangartner, 1973).

We do not know how this antagonistic relationship is mediated,
but the involvement of the neurosecretory system is a possibility.
This suggestion is supported by the results of Nijhout (1973) who
demonstrated an inhibition of the release of a prothoracicotropic
neurosecretory factor in Manduca when the JH titer is high as a
result of either endogenous of applied JH. Siew and Gilbert
(1971), Nasar (1974) and Grzelak et al. (1975) found, on the other
hand, that JH stimulated development in diapausing pupae. Perhaps
JH affects the prothoracic glands differently in larvae and pupae
(see Wilson and Larsen, 1974 for a discussion of diapause break).

The limiting of ecdysone activity by JH has to be complemented
by the negative effect of ecdysone on JH functions. The gonadotropic effect of JH is inhibited by the injection of ecdysone into the
females of Leucophaea (Englemann, 1959). Ecdysone appears to block
both specific protein synthesis as well as corpus allatum activity,
and Engelmann (1971) has explained the periodic alternation of egg
maturation and molting observed in Thermobia domestica by Watson
(1964) on the basis of mutual inhibition between the two insect
hormones.

The concept of antagonism between JH and ecdysone of insects
parallels the situation existing in phylogenetically close rela-
tives – the Crustacea. JH of insects inhibits differentiation and
ecdysis in larvae and stimulates the function of gonads in adults
in a similar fashion to the action of the molt-inhibiting hormone
(MIH) in crabs (Rao et al., 1973). The development and reproduction of crabs are based on the dynamic antagonism between ecdysone
and MIH (Lowe et al., 1968). It is important to note that extracts
of the MIH producing eyestalk gland of crabs was found to possess
the highest JH activity of the many non-insect extracts tested by

Schneiderman and Gilbert (1958) on the saturniid moth <u>Antheraea</u>
<u>polyphemus</u>. One of the most primitive insects, <u>Thermobia</u>, with the
reproduction and ecdysis cycles mentioned, may represent an important milestone on the path of development of the arthropod hormones.
The mutually exclusive presence of either gonadotropin or molting
hormone is the basic prerequisite for successful growth and reproduction.

The results presented in this paper show a causal connection
between the application of JH and IGR with consequent morphogenetic,
inhibitory and lethal effects on the one hand and the deficiency
or absence of ecdysone on the other. This observation may contribute to our understanding of the mechanism of the effect of IGRs
on insects. The relationship between these two hormones is of
principal importance for insect development and reproduction. The
simple fact of the absence of ecdysone in the JH treated larvae
implies unequivocally, that this relation is, at least in the
German cockroach, an antagonistic interplay rather than any kind
of synergism. It is difficult to speculate about the direct morphogenetic effect of the applied JH-active substance, when the
inductor of this process, ecdysone, is absent.

SUMMARY

Changes in the ecdysone titer of larvae of the German cockroach, <u>Blattella germanica</u>, exposed continuously to the juvenile
hormone (JH) or to the insect growth regulator (IGR) with JH activity, can be correlated with the nature of the substance applied,
its dose and the time of application. The younger larvae exposed
to the high dose of IGR die in the next ecdysis, whereas the same
treatment induces a diapause-like stage of developmental arrest in
the last larval stage. The affected larvae have very little, or
no ecdysone, the synthesis of which takes place in the second part
of the instar. The same treatment after this period has a lesser
effect. The extent of the effect is correlated with the amount of
ecdysone synthesized before the application of IGR. Last instar
larvae exposed to the lower dose of IGR or JH lack the peak of
ecdysone normally found in the controls at the end of the second
third of the instar when metamorphosis takes place. In these
insects, a first rise of the ecdysone titer begins towards the end
of the instar, and ecdysis into the supernumerary larval stage is
initiated when the ecdysone titer reached a level permitting ecdysis.

A direct or indirect antagonism between these hormones, both
fundamental to insect development, can explain the morphogenetic,
inhibitory, and lethal effects observed in insects treated with JH
or IGR with JH activity.

ACKNOWLEDGEMENTS

We wish to express our thanks to Dr. W. Vogel for stimulating discussions and for his help with the statistical analyses, to Prof. Dr. H.A. Schneiderman, Dr. W. Mordue and Mr. M. Gruber for helpful criticisms and correction of the manuscript and to Dr. V. Hromadkova, Mrs. T. Mühle, Mrs. B. Mühlemann and Mr. K-H. Trautmann for technical assistance.

REFERENCES

Benz, G., 1973, Experientia 29:1437.

Chihara, C.J., and Fristrom, J.W., 1973, Devel. Biol. 35:36.

Congote, L.F., Sekeris, C.E., and Karlson, P., 1969, Expl. Cell Res. 56:338.

Das, Y.T., and Gupta, A.P., 1974, Experientia 30:1093.

Engelmann, F., 1959, Z. Vergleich. Physiol. 41:456.

Engelmann, F., 1971, Acta Phytopathol. Acad. Sci. Hungaricae 6:211.

Fukaya, M., and Kobayashi, M., 1966, Appl. Ent. Zool. 1:125.

Fukaya, M., and Mitsuhashi, J., 1961, Bull. natn. Inst. agric. Sci. C 13:1.

Gilbert, L.I., and Schneiderman, H.A., 1961, Trans. Am. microsc. Soc. 79:38.

Grzelak, K., Szysko, M., and Lassota, Z., 1975, Insect Biochem. 5:409.

Hangartner, W., and Masner, P., 1973, Experientia 29:1358.

Highnam, K.C., 1958, Quart. J. microsc. Sci. 99:171.

Hoppe, T., 1974, J. Econ. Entomol. 67:789.

Kaplanis, J.N., Tabor, L.A., Thompson, M.J., Robbins, W.E., and Shortino, T.J., 1966, Steroids 8:625.

Karlson, P., and Shaaya, E., 1964, J. Insect Physiol. 10:797.

Karnavar, K.G., 1973, Indian J. exp. Biol. 11:138.

Kimura, S., 1973a, J. Insect Physiol. 19:115.

Kimura, S., 1973b, J. Insect Physiol. 19:2177.

Krishnakumaran, A., 1974, J. Insect Physiol. 20:975.

Lanzrein, B., 1974, J. Insect Physiol. 20:1871.

Laverdure, A., 1971, Gen. Comp. Endocrinol. 17:467.

Lezzi, M., 1974, Molec. Cell.Endocrinol. 1:189.

Lowe, M.E., Horn, D.H.S., and Galbraith, M.N., 1968, Experientia 24:518.

Madhavan, K., 1973, J. Insect Physiol. 19:441.

Masner, P., and Hangartner, W., 1973, Experientia 29:1550.

Masner, P., Hangartner, W., and Suchý, M., 1975, J. Insect Physiol. 21:1755.

Metwally, M.M., and Sehnal, F., 1973, Biol. Bull. mar. biol. Lab., Woods Hole 144:368.

Nasar, H.A., 1974, Z. Ang. Entomol. 76:137.

Nijhout, H.F., 1973, Am. Zool. 13:1272.

Oberlander, H., and Tomblin, C., 1972, Science 177:441.

Patel, N., and Madhaven, K., 1969, J. Insect Physiol. 15:2141.

Rao, K.R., Fingerman, S.W., and Fingerman, M., 1973, J. Insect
  Physiol. 19:1105.

Schneiderman, H.A., and Gilbert, L.I., 1958, Biol. Bull. 115:530.

Schneiderman, H.A., and Gilbert, L.I., 1964, Science 143:325.

Sehnal, F., and Meyer, A.S., 1968, Science 159:981.

Sekeris, C.E., 1972, Gen. Comp. Endocrinol. Suppl. 3:149.

Shaaya, E., and Levinson, H.Z., 1966, Riv. Parasitol. 27:211.

Siew, Y.C., and Gilbert, L.I., 1971, J. Insect Physiol. 17:2095.

Silhacek, D.L., and Oberlander, H., 1975, J. Insect Physiol. 21:
  153.

Sláma, K., 1975, Colloque Internat. C.N.R.S. Endocrinol. Inverte-
  bres (in press).

Staal, G.B., 1967, Proc. Roy. Neth. Acad. Sci. Let. C 70:409.

Watson, J.A.L., 1964, J. Insect Physiol. 10:399.

Wigglesworth, V.B., 1957, Symp. Soc. exp. Biol. 11:204.

Wigglesworth, V.B., 1961, Symp. R. Entomol. Soc. Lond. 1:104.

Williams, C.M., 1961, Biol. Bull. 121:572.

Wilson, G.R., and Larsen, J.R., 1974, J. Insect Physiol. 20:2459.

Yin, C.M., and Chippendale, G.M., 1974, J. Insect Physiol. 20:1833.

THE ANTAGONISM BETWEEN JUVENILE HORMONE AND ECDYSONE

M. Lezzi and C. Wyss

Institute for Cell Biology, Swiss Federal Institute
of Technology, Zurich, Switzerland

INTRODUCTION

The aim of this paper is to discuss the following three
questions: 1) is there an antagonism between juvenile hormone (JH)
and ecdysone (Ec); 2) if so, what is the nature of the antagonism;
3) what is its role in insect development? However, it is not the
aim of this paper to disprove the notions of those people (Doane,
1973; Laufer and Calvet, 1972; Wyatt, 1972; Willis, 1974) who use
other terms to describe the interaction between JH and Ec. We
emphasize this point for three reasons: 1) whether or not one
designates the effect of JH in relation to that of Ec "antagonistic",
"modificatory" or "in concert" is, in our opinion, mainly a matter
of semantics; 2) we have never claimed that antagonism is the only
mode of JH-Ec interaction as we have always agreed that the JH-Ec
antagonism is not strict (Lezzi and Frigg, 1971) because there are
conditions under which JH does not antagonize the effect of ecdysone (Lezzi and Gilbert, 1969); 3) we ourselves shall present
some evidence that JH and Ec are not likely to bind to each other
or to a common receptor. Thus, the interaction between JH and Ec
is not an antagonism in the strict molecular or pharmacological
sense. However, in a general sense we may speak of an antagonism
whenever one hormone interferes with a step previous or subsequent
to the interaction of another hormone with a receptor.

In Table 1, examples of experiments are given in which JH,
when given in vivo (i.e. to intact or operated animals), was said
to antagonize, inhibit, or prevent the action of Ec. (Examples of
experiments which demonstrate a synergistic or independent JH
action are not listed, though they might exist.) Since a thorough

252

TABLE 1. EXPERIMENTS INDICATING AN ANTAGONISM BETWEEN JH AND EC IN VIVO.

Authors	Species	Developmental stage	Hormone treatment	Effect	Interpretation by authors
Patel & Madhavan, 1969	Samia cynthia ricini	Last larval instar	JHA: injected Ec: injected	JHA, Ec: stimulation of RNA synthesis in wing discs. JHA & Ec: no effect.	Antagonistic behavior.
Srivastava & Gilbert, 1969	Sarcophaga bullata	Last larval instar	JH: injected EC: injected	JH: delay of molt. JH & Ec: prevention of delay.	JH may inhibit synthesis or release of Ec.
Lezzi & Gilbert, 1969; Lezzi & Frigg, 1971	Chironomus tentans, C. thummi	Last larval instar	JH: externally or injected Ec: injected	JH: regression of Ec-specific puffs; Ec: regression of JH-specific puffs in salivary glands.	JH and Ec may act antagonistically.
Frey, 1973	Pieris brassicae	Last larval instar, ligated abdomens	JH, JHA: externally Ec: injected	JH & Ec: delay of molt (larval-pupal); JHA & Ec: delay of molt (L5-L6).	Antagonism.

TABLE 1 (Cont.)

Authors	Species	Developmental stage	Hormone treatment	Effect	Interpretation by authors
Masner & Hangartner, 1973	Blattella germanica	Last larval instar	JH: externally Ec: injected	JH: failure to synthesize new cuticle; JH & Ec: synthesis of new cuticle.	Antagonism.
Socha & Sehnal, 1973	Tenebrio molitor	Pupa	JHA: injected Ec: injected	JHA, Ec: cuticle deposition; JHA & Ec: no cuticle deposition.	Mutually antagonistic action.
Bryan et al., 1974	Oncopeltus fasciatus	Last larval instar	JH: externally Ec: injected	JH & Ec: less supernumary larvae, more adults than with JH alone.	Antagonistic action.
Silhacek & Oberlander, 1975	Plodia interpunctella	Last larval instar	JH in diet	JH: prevention or delay of development, larval-pupal molt, eclosion.	JH prevents development to Ec-sensitive state.

TABLE 1 (Cont.)

Authors	Species	Developmental stage	Hormone treatment	Effect	Interpretation by authors
Sláma, 1975	Pyrrhocoris anterus, Galleria mellonella	Last larval instar	JH: externally or locally with injury Ec: injected or locally with injury	JH locally & Ec injected: metathetelic patches of epidermis; Ec locally & JH externally: prothetelic patches in pupae.	JH is "red signal" for morphogenesis promoted (among others) by Ec.
Srihari et al., 1975	Acheta domestica	Young adults	JH: externally Ec: injected	Ec: retardation of flight muscle degeneration; JH: enhancement of above.	Ec interferes with JH action.

JH = juvenile hormone

JHA = juvenile hormone analog

Ec = α- or β-ecdysone

appraisal of every experiment cannot be given here we only wish to
draw attention to the great diversity of biological systems in
which an antagonism has been observed and, particularly, to those
cases in which by the administration of JH not only the type of
molt is changed but the molting process as such is inhibited. It
is most important to note that this inhibition need not be due to
an unspecific abolition of the insect's functioning since the molts
which may occur after a certain lag period or upon the injection of
Ec proceed normally, leading to pupae or supernumerary larvae.

     With in vivo experiments it is difficult to know at which
level the antagonism between JH and Ec operates. The following
possibilities may be conceived: 1) at the endocrine level: JH may
inhibit the synthesis and/or release of Ec (see e.g. Wanyonyi,
1974); 2) at the circulatory level: JH may stimulate Ec degradation, excretion or inactivation; 3) at the target cell level:
a) JH may disturb the interaction between Ec and its presumed
receptor; b) JH may interfere with one of the consecutive steps
which eventually lead to the realization of the Ec effect. Of
course, JH is not necessarily assumed to act directly at the above
mentioned levels or to act exclusively at one of these levels.
Furthermore, what is said for JH may hold for Ec, correspondingly.
At least in one case listed in Table 1, i.e. with the experiments
performed by Frey (1973), an antagonism at the endocrine level
seems to be excluded since the body region analyzed was separated
from the major endocrine organs by ligation.

     The in vitro studies listed in Table 2 strongly suggest the
existence of antagonisms at the cellular level, although hormone
inactivation (level 2) cannot always be excluded. Most of the
examples given here reveal an inhibitory action of JH in that JH
prevents Ec from exerting its stimulatory effect. Considering
the facts that the JH concentrations required are usually rather
high (above the solubility of JH?), that the stereochemical specificity is not always ascertained, and that inhibition is a "negative accomplishment", which can be achieved by many other more or
less specific inhibitors (e.g. actinomycin D), it is difficult to
decide whether these JH effects are pharmacological effects or
indeed reflect a natural function of JH which is, for example, the
prevention of metamorphosis. (Note that many experiments of Table 2
deal with imaginal discs which naturally do not react to ecdysone
prior to metamorphosis in terms of the features mentioned in
Table 2.) For these reasons we deem it justified to concentrate
the following discussion on the two instances which demonstrate a
positive accomplishment by JH: these are 1) the puff induction in
Chironomus tentans salivary gland chromosomes and 2) the stimulation of growth in a cell line of Drosophila melanogaster.

TABLE 2. IN VITRO EXPERIMENTS INDICATING AN ANTAGONISM BETWEEN JH AND EC.

Authors	Species	Tissues, cells	Developmental stage	Hormone active substances	Concentration (μg/ml)	Duration of treatment	Effect
Laverdure, 1971	Tenebrio molitor	Ovary	Pupa	Farnesyl-methyl-ester Ec (α ?)	5 3	5-12 d	Ec: growth and differentiation Ec & JH: no effect
Oberlander & Tomblin, 1972	Plodia interpunctella	Imaginal discs	Last larval instar	JH I β-Ec	100 2	21-28 d	Ec: cuticle deposition Ec & JH: no effect
Chihara & Fristrom, 1973	Drosophila melanogaster	Imaginal discs	Last larval instar	JH I or JH II β-Ec	20-60 0.001-100	3 hr or 9 hr	Ec: stimulation of RNA synthesis and evagination Ec & JH: no effect
Logan et al., 1975	Drosophila melanogaster	Imaginal discs	Last larval instar	JH III β-Ec	20-60 10	20 hr	Ec: stimulation of DNA synthesis EC & JH: no effect

TABLE 2 (Cont.)

Authors	Species	Tissues, cells	Developmental stage	Hormone active substances	Concentration (μg/ml)	Duration of treatment	Effect
Lezzi, 1974	Chironomus tentans	Salivary gland	Last larval instar (pre-pupa)	JH I β-Ec	1-10 50	4 hr	Ec: maintenance of Ec-specific puff Ec & JH: regression of Ec-specific puff, induction of JH-specific puff
Wiebusch, 1975; Rensing, 1975	Drosophila melanogaster	Salivary gland	Last larval instar	Farnesyl-methyl-ether α-Ec	0.1-1 1	2-3 hr	Ec: induction of puff 2B JH: induction of puff 3C Ec & JH: no effect
Mitsuhashi & Grace, 1970	Antheraea eucalypti	Cell line, derived from ovariols of pupae		Farnesol β-Ec	0.1 0.1	4 d	Ec: stimulation of multiplication

TABLE 2 (Cont.)

Authors	Species	Tissues, cells	Developmental stage	Hormone active substances	Concentration (μg/ml)	Duration of treatment	Effect
Mitsuhashi & Grace, 1970 (cont.)							Ec & JH: no effect
Courgeon, 1975	Drosophila melanogaster	Cell line, derived from embryos		JH I β-Ec	0.003-30 0.005	1-6 d	Ec: morphological change; Ec & JH: delayed or no effect
Wyss, 1975	Drosophila melanogaster	Cell line, derived from embryos		Ethyl-dichloro-farnesoate β-Ec α-Ec	0.2-20 0.001-0.01 0.01-10	3 d	Low Ec: stimulation of multiplication; High Ec: inhibition; JH: inhibition; Low Ec & JH: inhibition; High Ec & JH: stimulation

MATERIALS AND METHODS

In vitro Incubation of Salivary Glands of Chironomus tentans.
The methods of incubation, preparation and evaluation are fully
described by Lezzi (1974). Here, it only should be emphasized
that: 1) a synthetic medium and synthetic hormones (JH I and β-Ec)
were used; 2) only unwounded glands, by the criterion of trypan
blue exclusion, were considered for the analyses reported in this
paper; 3) each incubated gland was compared with its unincubated
sister gland; 4) the quantitative and statistical evaluation was
based on chromosome measurements (approximately 100 chromosomes
per point) rather than morphological staging. This last point
is particularly important to keep in mind because, for the sake
of clarity attempted in this paper, the following symbols are used
(for numerical data, see Lezzi, 1974).

$\dagger$ = no puff $\dagger \leftarrow$ puff site I- 19 - A₁

ϕ = small puff $\dagger \leftarrow$ puff site I- 18 - C

ϕ = large puff

Cell Line KcC7 of Drosophila melanogaster. The cell line
and the methods of cultivation and hormone treatment are described
by Wyss (1975). It is important to note that a clonal subline
designated, KcC7, of the original line Kc was used and that the
incubation medium contained a defined amino acid mixture instead
of lactoalbumine hydrolysate. Ethyl-dichloro-farnesoate (EDCF,
gift from Hoffmann-La Roche Company, Basel) served as a JH analog
in these studies.

RESULTS

Chromosomal Puffing in Isolated Salivary Glands

Since we worked with salivary glands of prepupae it was
important to establish in vitro cultivation conditions under which
the prepupal pattern was obtained. This was realized by the administration of high doses of β-Ec (final concentration 10^{-4} M) to
the medium:

$$\text{↓ chromosome} \xrightarrow[\text{4 hr}]{Ec} \text{↓ chromosome}$$

When in addition to Ec a high concentration of JH (10^{-4} M) is included in the medium the Ec-specific puff is suppressed, clearly demonstrating the antagonistic action of JH:

$$\xrightarrow[\text{4 hr}]{Ec,\ JH\ (10^{-4})}$$

This antagonism appears to be mutual in that Ec apparently prevents the formation of what we call the JH-specific puff:

$$\xrightarrow[\text{4 hr}]{JH\ (10^{-4})}$$

However, this antagonism is obviously rather paradoxical in that Ec, though being capable of suppressing the JH-specific puff at very low (10^{-7} M) and very high (10^{-4} M) JH concentrations, cannot antagonize the action of intermediate JH concentrations:

$$\xrightarrow[\text{4 hr}]{JH\ (10^{-5}),\ Ec}
\qquad
\xrightarrow[\text{4 hr}]{JH\ (10^{-6}),\ Ec}$$

In passing we note that, with these intermediate JH concentrations (specifically with 10^{-6} M) in combination with 10^{-4} M Ec, a condition is met at which the Ec-specific as well as the JH-specific puffs may co-exist. The occurrence of such puff combinations was the reason that even in the first papers on JH and puffing (Lezzi and Gilbert, 1969; Lezzi and Frigg, 1971), we designated the JH-Ec

antagonism as being not strict. Laufer and coworkers (Laufer and
Holt, 1970; Laufer and Calvet, 1972) in their in vivo experiments
apparently also found such a coexistence of Ec- and what they
believed to be JH-specific puffs (for criticism of Laufer's puff
classification see Lezzi and Gilbert, 1972; Lezzi, 1975a) which
led them and others (Wyatt, 1972; Doane, 1973; Willis, 1974) to
conclude that in general there is no JH-Ec antagonism at the puffing level. The reduction in the size of the Ec-specific puffs
upon administration of high doses of JH, which was also observed
by Laufer and Calvet (1973), was interpreted by these authors as
"a spurious or otherwise inexplicable result" which would be
caused by the generally deleterious effects of toxic JH concentrations. The reason that Laufer favored this rather than our inter-
pretion (which is the existence of a limited JH-Ec antagonism)
resides in the observation by several groups of a natural coexistence of Ec- and JH-specific puffs during the larval molt (Clever,
1963; Kroeger, 1973; Lezzi, 1974) when the Ec- as well as JH-
titers are assumed to be highest. The justification for this
assumption shall be discussed subsequently. For now let us follow
this argument and agree with Laufer and others that JH and Ec
independently induce their puffs. We may then ask: what would
one predict would happen to the respective chromosome regions in
the absence of both hormones, e.g. the non-appearance of both,
the Ec- and JH-specific puffs? This is not what we observed:

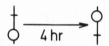

Puff I-19-A$_1$, hitherto regarded as JH-dependent, develops spontaneously, i.e. without the administration of JH or Ec. One
therefore could visualize this puff as being responsive to low
Ec concentrations rather than being JH-specific. The idea of JH-
Ec antagonism could then be rescued by assuming that an increase
in Ec concentration promotes the development of a sequence of
qualitatively different puffing patterns while JH somehow diminishes the effectiveness of Ec in a dose dependent manner, thereby
antagonizing the Ec action:

It is obvious from the results already reported that this interpretation is also too simple. Besides, it is not compatible with:
1) the outcome of experiments involving the effect on Ec and JH
action of an artificial increase in cell membrane permeability,
2) the kinetics of these actions and 3) findings on non-prepupal,
i.e. intermolt salivary glands. While the first two points are
discussed elsewhere (Lezzi, 1974, 1975a,b) the third shall be
presented here, in a preliminary form:

As intermolt larvae are assumed to have a low Ec titer it was
expected that their salivary glands, when kept in an Ec-free
medium, would either maintain the original puffing condition or
reduce puff I-19-A$_1$. Instead, this puff increases to a supersize, a
behavior which is inconsistent with the above stated tentative
hypothesis that this puff is responsive to Ec of a limited concentration. In agreement with the original suggestion, this puff
could be sensitive to JH so that removal of the slightly suppressive, low ecdysone concentrations (assumed to be present during
intermolt) would allow the full expansion of this puff to an
extent which is determined by a previous (and lasting?) stimulus
of JH. Then, however, it is difficult to understand why incubation with excess Ec does not suppress this puff completely. Thus,
this interpretation is unsatisfactory as well.

    There is no reason to further discuss the interaction of JH
and Ec in puff regulation. It suffices to state that: 1) such an
interaction must exist since, in general, JH and Ec influence
each other in their effects; and 2) this interaction has many
features of an antagonism which, however, is not of a simple
balance type. The following experiments on cell cultures will
support these two statements.

Cell-Multiplication in a Cell Line of D. melanogaster

    Ecdysone influences the growth of cultures of D. melanogaster
cells in two contrasting ways: at low concentrations it stimulates
while at higher concentrations it inhibits cell-multiplication
(Fig. 1). This bimodal effect is observed not only with α-Ec but
also with β-Ec, though at 100x lower concentrations. The JH
analog, when added alone, suppresses cell growth in a dose dependent manner. It is therefore not surprising that it diminishes

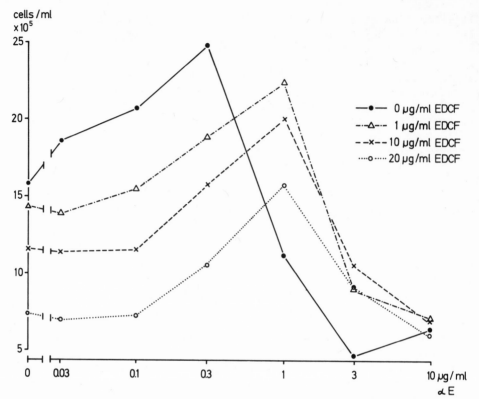

Fig. 1.   Effect of hormonal composition of culture medium on
multiplication of cell line KcC7.   Abscissa: concentration of
α-ecdysone.   Ordinate: cell density after 3 days culturing in a
medium of the respective hormonal composition.   The initial cell
density was 2.3 x 10^5 cells per ml.   The curves are based on data
to be published with statistical analysis by Wyss (1975).

the stimulatory effect of low Ec on cell multiplication.   However,
at intermediate or higher Ec concentrations, instead of enforcing
the inhibition by Ec, it reverses this inhibition such that a
stimulation of cell growth may result.   This is a paradoxical
phenomenon which above a certain Ec-concentration limit (approx.
0.6 µg/ml for α-Ec) resembles that observed with puffing.
In the case of cell multiplication, the higher the Ec concentration the better it is antagonized by low JH, while in the case of
puffing, the higher the JH the better it is antagonized by Ec.
However, the situation becomes clearer when the entire dose effect
curves are compared with one another.   The effect of JH (e.g. 1 µg/
ml) can then be described as a shifting of the Ec-curve towards
higher Ec concentrations, an effect which would be in keeping with

a simple antagonism of the balance type. JH somehow reduces the
effectiveness of Ec making higher Ec concentrations necessary for
obtaining the same results as in the absence of JH. Yet, even
this interpretation is unsatisfactory since an increase in the
JH concentration does not cause a further shift to higher Ec concentrations but rather appears to depress the shifted Ec curve
along its whole length. With regard to this depression, the
nature of JH could be envisaged as generally inhibitory. This
would be in accordance with numerous observations on in vivo
systems in which high JH concentrations were used.

In conclusion one may state that the studies with cultured
cells also point to a complexity of the partially antagonistic
interaction between JH and Ec which, in this case, includes at
least two tendencies: 1) one which reduces the effectiveness of
Ec and 2) one which is generally inhibitory.

DISCUSSION

Both types of experiments described above reinforce our
belief that JH and Ec may act antagonistically at the cellular
level in a complex way. With regard to the problem of revealing
the nature of this antagonism one should be aware of the differences between the two experimental systems. While the puff
response of salivary chromosomes occurs relatively early after
hormone administration, the growth response of the cell cultures
has a considerable lag (about two days). This is reasonable
since puffing is regarded as a manifestation of genetic activities by which cellular processes like cell division may be second-
arily controlled. Thus, the puffing response appears to be closer
to the initial interaction between the hormone and the cell,
therefore being more relevant to investigations on the primary
action and interaction of hormones. Karlson (1963) postulated a
direct interaction of Ec (possibly being bound to a receptor)
with the Ec-controlled gene. It is intriguing to note that this
hypothesis evolved primarily from in vivo studies with exactly the
same Ec-specific puff as that discussed here, i.e. I-18-C whose
behavior, however, led us to quite different conclusions (Lezzi,
1975a,b). Subsequent work by Congote et al. (1969) on total RNA
synthesis in isolated fat body nuclei seemed to support Karlson's
hypothesis and, in addition, suggested a mutual neutralization of
Ec and JH or their respective receptors. According to Karlson
et al. (1971), this would result in a simple balance type of JH-Ec
antagonism which would be consistent neither with the observation
described in this paper nor with those by others (Chihara and
Fristrom, 1973). In recent reviews (Lezzi, 1975a,b) several other
models of the JH-Ec interaction have been discussed extensively.
An inherent feature of all these models is that they are very
sophisticated and yet rather unsatisfactory. On the basis of

studies (see Lezzi, 1974) involving permeability increases of the
cell membranes, the response of other puff-sites, and, primarily,
the time course of puff reactions, we were inclined to believe
that: 1) Ec and JH influence gene activities indirectly rather than
directly; 2) there exist at the cellular level regulatory circuits
controlling gene activities; 3) permeability changes of cell membranes play a key role in the JH and possibly Ec action.

In leaving our discussion of the cellular mechanism of JH
and Ec action and interaction, we wish to turn attention to the
problem of the so-called "physiological significance" of the in
vivo and in vitro experiments involving exogenous JH and Ec.
Borst et al. (1974) determined by biochemical methods the nature
and titer of the molting hormone in D. melanogaster and found that
0.1 µg/ml of β-Ec is present in the hemolymph during metamorphosis.
This concentration, when tested in cell cultures of D. melanogaster,
is strongly inhibitory to cell multiplication (see Wyss, 1975),
corresponding to an α-Ec concentration of 10 µg/ml. The nature
and titer of the molting hormone of Chironomus has not yet been
determined. The β-Ec concentration used for in vitro cultivation
of Chironomus salivary glands is 50 µg/ml which, clearly, is
extremely high. This concentration was arrived at by purely
empirical means. It is ten times the concentration necessary for
maintaining the prepupal puffing pattern in vitro. (The concentration was increased by a factor of ten in order to preclude Ec
shortage as an explanation for the regression of the Ec-specific
puff under certain experimental conditions.) However, even 5 µg/
ml is a very high Ec concentration and it remains to be elucidated
why such high concentrations are required. (Note: Clever et al.,
1973, also used 1-10 µg/ml α- or β-Ec for their in vitro experiments.) It may be that β-Ec is not the natural molting hormone
of Chironomus, a possibility which is further suggested by the
discrepancy concerning the effects of α- and β-Ec upon the induction of puff I-18-C in isolated C. tentans salivary glands
(compare with Clever et al., 1973). The JH has not been chemically
characterized for either Chironomus or Drosophila. It is of
course, not EDCF and whether it is JH I is also uncertain. Besides,
the possibility exists that, as in other insects, more than one
JH form may exist in Chironomus and Drosophila. DeWilde et al.
(1971) determined the JH-titer in penultimate larvae of Philosamia
cynthia to be about 10^{-6} M. In our in vitro system with Chironomus
salivary glands, this is just the concentration which significantly
increases the JH-specific puff and starts decreasing the Ec-specific puff. As good as this correlation may appear, it may be
premature to take it as proof for the "physiological significance"
of the hormonal effects observed in our in vitro system. The
point is, that in the insect the effective hormone concentration
is regulated by parameters which are absent in vitro, e.g. JH
binding proteins (see Nowock et al., this volume).

However, one must consider not only the absolute, but the relative concentration of JH in relation to that of Ec. Whether one claims that these two hormones act "in concert" or "antagonistically", it is an indispensible requirement that they work together, i.e. that they are present at the same time. Unfortunately, there are very few data available on titer variations of both hormones during normal development. Krishnakumaran and Schneiderman (cited by Patel and Madhavan, 1969) showed that the JH titer is very low at that time (e.g. immediately before the molt to the fifth instar) when the Ec titer is very high, and the converse. If that is true, the whole discussion about JH-Ec antagonism becomes moot. When these two hormones are normally not present at the same time they clearly cannot antagonize each other nor work in concert. Their interrelationship would, then, be confined to consecutive control of the same developmental program. The role of JH would consist of determining the nature of the subsequent steps in that program, which would later be realized by the action of Ec. This does not exclude a decisive morphogenetic role for Ec itself when acting for an extended period of time before inducing molting (e.g. during the prepupal phase). Perhaps one should think along the lines of newer ideas formulated for example by Truman et al. (1974), Sláma (1975) and Silhacek and Oberlander (1975) who perceive insect development as the realization of a program in which each hormone has its specific time (critical period?) for coming into play. The antagonistic effects observed with experimental hormone treatment, whether in vivo or in vitro, could then be interpreted as being the result of the action of built-in safety mechanisms correcting for the disturbed hormonal situation caused by an untimely (e.g. experimental) increase in titer of one or the other hormones. Interestingly enough, such correcting mechanisms appear to function at the endocrine and circulatory levels as well as at the target cell. However, before all these more or less far-reaching conclusions are drawn, one must clearly know how the titers of JH and Ec change with relation to one another during normal development.

ACKNOWLEDGEMENTS

We thank Prof. G. Benz (Entomological Institute, Swiss Federal Institute of Technology, Zürich) and Prof. L. Rensing (Zoological Institute, University of Göttingen) for permission to use the data of H. Frey's and D. Wiebusch' Diploma Theses. The unpublished experiments reported in the present paper were assisted by Miss G. Bachmann and E. Hansson whose assistance is gratefully acknowledged. We are grateful to Mrs. R. Räber and Dr. K. Downing for drawings and help with the English manuscript, respectively. This work was supported by grant Nr. 3.2280.74 and grant Nr. 3.8640. 72 of the Swiss National Science Foundation.

REFERENCES

Borst, D.W., Bollenbacher, W.E., O'Connor, J., King, D.S., and Fristrom, J.W., 1974, Develop. Biol. 39:308.

Bryan, M.D., Brown, T.M., and Monroe, R.E., 1974, J. Insect Physiol. 20:1057.

Chihara, C.J., and Fristrom, J.W., 1973, Develop. Biol. 35:36.

Clever, U., 1963, Chromosoma (Berl.) 14:651.

Clever, U., Clever, I., Storbeck, I., and Young, N.L., 1973, Develop. Biol. 31:47.

Congote, L.F., Sekeris, C.E., and Karlson, P., 1969, Exp. Cell Res. 56:338.

Courgeon, A.-M., 1975, C.R. Acad. Sc. Paris, Serie D 280:2563.

Doane, W.W., 1973, in "Developmental Systems: Insects" (S.J. Counce and C.H. Waddington, eds.) Vol. 2, pp. 291-497, Academic Press, London and New York.

Frey, H., 1973, Diploma Thesis, Swiss Federal Institute of Technology, Zurich, pp. 1-31 (unpublished).

Karlson, P., 1963, Perspect. Biol. Med. 6:203.

Karlson, P., Congote, L.F., and Sekeris, C.E., 1971, Mitteilungen Schweiz. Entomol. Ges. 44:171.

Kroeger, H., 1973, Z. Morph. Tiere 74:65.

Laufer, H., and Calvet, J.P., 1972, Gen. Comp. Endocrinol. Suppl. 3:137.

Laufer, H., and Holt, T.K.H., 1970, J. Exp. Zool. 173:341.

Laverdure, A.-M., 1971, Gen. Comp. Endocrinol. 17:467.

Lezzi, M., 1974, Molec. Cell. Endocrinol. 1:189.

Lezzi, M., 1975a, Habilitation Thesis, Swiss Federal Institute of Technology, Zurich, pp. 1-82 (unpublished, publication in preparation).

Lezzi, M., 1975b, Nachrichten der Akademie der Wissenschaften in Gottingen, II. Mathematisch-physikalische Klasse, Jhg. 1975, Nr. 11:26.

Lezzi, M., and Frigg, M., 1971, Mitteilungen Schweiz. Entomol. Ges. 44:163.

Lezzi, M., and Gilbert, L.I., 1969, Proc. Nat. Acad. Sci. USA 64:498.

Lezzi, M., and Gilbert, L.I., 1972, Gen. Comp. Endocrinol. Suppl. 3:159.

Logan, W.R., Fristrom, D., and Fristrom, J.W., 1975, J. Insect Physiol. 21:1343.

Masner, P., and Hangartner, W., 1973, Experientia 29:1550.

Mitsuhashi, J., and Grace, T.D.C., 1970, Appl. Ent. Zool. 5:182.

Oberlander, H., and Tomblin, C., 1972, Science 177:441.

Patel, N., and Madhaven, K., 1969, J. Insect Physiol. 15:2141.

Rensing, L., 1975, Nachrichten der Akademie der Wissenschaften in Gottingen, II. Mathematisch-physikalische Klasse, Jhg. 1975, Nr. 11:38.

Silhacek, D.L., and Oberlander, H., 1975, J. Insect Physiol. 21:153.

Sláma, K., 1975, J. Insect Physiol. 21:921.
Socha, R., and Sehnal, F., 1973, J. Insect Physiol. 19:1449.
Srihari, T., Gutmann, E., and Novak, V.J.A., 1975, J. Insect
    Physiol. 21:1.
Srivastava, U.S., and Gilbert, L.I., 1969, J. Insect Physiol.
    15:177.
Truman, J.W., Riddiford, L.M., and Safranek, L., 1974, Develop.
    Biol. 39:247.
Wanyonyi, K., 1974, Insectes Sociaux, Paris 21:35.
Wiebusch, D., 1975, Diploma Thesis, University of Gottingen
    (unpublished, publication in preparation).
Wilde, J., De, Kort, C.A.D., De, and Loof, A., De, 1971,
    Mitteilungen Schweiz. Entomol. Ges. 44:79.
Willis, J.H., 1974, Annual. Rev. Entomol. 19:97.
Wyatt, G.R., 1972, in "Biochemical Actions of Hormones" (G.
    Litwack, ed.) Vol. 2, pp. 385-490, Academic Press, New York
    and London.
Wyss, 1975, in preparation.

THE INTERACTION OF JUVENILE HORMONE AND ECDYSONE: ANTAGONISTIC, SYNERGISTIC, OR PERMISSIVE?

J.H. Willis and M.P. Hollowell

Department of Entomology and Provisional Department of
Genetics and Development, University of Illinois
Urbana, Illinois 61801

INTRODUCTION

A central issue in insect endocrinology is the nature of the interaction of ecdysone and juvenile hormone (JH) in controlling metamorphosis (Willis, 1974). While we await determination of their interaction at the level of cellular receptors or their ability to affect each other's metabolism or uptake into target tissue, it is possible to obtain other significant information by more classical means.

Much of our work has centered around the question of whether high levels of ecdysone can mimic JH. One of the symptoms of hyperecdysonism in Samia cynthia and Tenebrio molitor is the formation of second pupae (Williams, 1968; Socha and Sehnal, 1972). Bowers (1968) found that some synergists of insecticides mimicked JH by inhibiting metamorphosis in Tenebrio and Oncopeltus fasciatus, and suggested that JH might act by synergizing the activity of ecdysone. Such a hypothesis is in accord with the well known observation that, in the presence of JH, the period between apolysis and ecdysis is accelerated relative to a metamorphic molt which occurs at low or null JH concentrations. If JH's action was merely to accelerate deposition of cuticle and fixation of the state of differentiation of the internal organs, it might well be trivial at the molecular level -- a mere potentiation or synergism of ecdysone's action. Low levels of ecdysone would keep the developmental program running while effectively higher levels (JH assisted in larvae) would cause the program to be expressed. A critical test of such a scheme is to see whether ecdysone truly mimics JH. It is possible that the ecdysone-induced second pupae referred to above were produced in the presence of endogenous levels of JH,

possibly secreted in response to the high levels of ecdysone (see
Siew and Gilbert, 1971).

Accordingly, our first set of experiments dealt with the
effects of high levels of ecdysone in isolated abdomens of the
Cecropia silkmoth. Secondly, we extended the study of hyperecdysonism to a hemimetabolous insect, Oncopeltus. The results
from these two studies refute the contention that ecdysone is a
genuine mimic of JH. The final series of experiments with tissue
cultures of Cecropia wings yielded data which though preliminary
show the potential of such an in vitro system for clarifying the
interaction of these hormones.

MATERIALS AND METHODS

Cecropia Abdomens. Isolated abdomens (weighing about 1.5 g)
were prepared from diapausing male pupae of Hyalophora cecropia
and stored at 25°C. Hormones were injected five days after the
abdomens were prepared. Ecdysterone (Rhoto Pharmaceutical) was
dissolved in 10% isopropanol and pure synthetic Cecropia JH I
(methyl trans, trans, cis 3,11-dimethyl-7-ethyl-10,11-epoxytrideca-
2,6-dienoate; Eco Control) was diluted with light weight mineral
oil. The abdomens were dissected three weeks after receiving
hormones and frozen at -70°C. Cuticles were processed according
to the methods of Ruh and Willis (1974) except that electrophoresis
was carried out in 10% acrylamide gels containing 0.1% sodium
dodecyl sulphate (SDS) and 8 M urea.

Oncopeltus Development. We used white body mutants of Oncopeltus fasciatus (cb, wb,re; Lawrence, 1970) synthesized from stocks
provided by Dr. Peter Lawrence. Fresh (pale) fifth instar larvae
were collected, maintained at 29°C, and injected at the appropriate
age with 1 µl of the hormone solutions described above. Animals
were checked at approximately 3 hr intervals during their 18 hr
photophase. None of the animals injected with hormones actually
underwent ecdysis and the time of ecdysis given in Table 1 represents the time at which the cuticle of the pronotum split or the
insect lay on its back unable to right itself. Such animals had
formed a new cuticle and digested the inner layers of the old one.
Solvent injected insects ecdysed normally. Animals were fixed
in 70% ethanol one to two days after "ecdysis"; analyses were
made on whole mounts of cuticles which had been boiled in saturated KOH.

Tissue Cultures. The epidermal sac of the fore-wing from
diapausing Cecropia pupae was freed from the overlying cuticle
under aseptic conditions, placed in Grace's medium (Grand Island
Biologicals) containing a few crystals of phenylthiourea (PTU)
for 1-2 hr, transferred to fresh medium with PTU, and minced

with fine scissors. After an additional hour of incubation to
remove blood from the lumen, the fragments were divided between
culture dishes and 2 ml of fresh culture medium added. Ecdysterone was dissolved directly in Grace's medium, and an aliquot
of a concentrated stock solution added to the cultures. All
glassware, including culture dishes, which were used in experiments
with JH were coated with polyethylene glycol to prevent adsorption
of JH to the glass (J. Law, personal communication). JH was
dissolved by sonication to a final concentration of 5 x 10^{-5} M
(Kramer et al., 1974). We did not replace the medium during the
culture period. Culture dishes were maintained in sealed desica-
tors at 25°C.

Morphological analyses were carried out on tissues processed
with Bouin's fixative and Mallory's stain. Isotope incorporation
was measured in cultures from which 1 ml of medium had been removed
prior to the addition of 50 µl of labeled leucine (4,5-[^{3}H] and
[^{14}C], New England Nuclear Corp.). The cultures were kept at 25°C
for an additional 4 hr after the addition of isotope, then rinsed
three times with cold Grace's medium, and homogenized in 0.01 M
sodium phosphate buffer containing 0.1% SDS. They were then lyophilized and resuspended in a smaller volume of buffer. Appropriate samples were mixed and run on 10% SDS-urea gels, fixed,
stained, scanned, sliced, and counted (Ruh and Willis, 1974).
Labeled cuticular proteins were prepared from fresh pupae and
pharate adults after an 18 hr pulse of labeled leucine, and
processed as described by Ruh and Willis (1974).

RESULTS

Action of Ecdysterone and JH in Pupal Abdomens

In order to see whether ecdysterone could mimic JH, we tested
four hormonal treatments on isolated abdomens. Two groups were
treated with high levels of ecdysterone (150 µg), while two groups
received one tenth that amount. One group receiving each dose was
also injected with 20 µg JH. The low dose of ecdysterone we used
(about 10 µg/g) appeared to be the minimum necessary to sustain
development of isolated abdomens. On two occasions, abdomens
receiving this dose of ecdysterone molted into second pupae. The
more typical results (Willis et al., 1974) described below and
based on over 100 abdomens were quite different.

Cuticular Morphology. Abdomens injected with 15 µg ecdysterone formed a wide spectrum of adult abdomens ranging from some
which were covered with pigmented scales arranged in a normal
pattern to others without scales but which had a thin, amber adult-
like cuticle. Histological sections of cuticles from these naked

adults had surface sculpturing characteristic of adults. All of
the abdomens which received low amounts of ecdysterone plus JH
formed second pupal abdomens. The development of the external
genitalia was suppressed, the cuticle was hard and brown, and
polygonal fields, characteristic of pupal intersegmental membranes
(Willis, 1969), were present.

High doses of ecdysterone caused the formation of naked adult
abdomens or occasionally abdomens which bore a few patches of
unpigmented scales. Some of the naked abdomens had dark brown
cuticle in the mid-dorsal region. Histological sections of these
animals revealed three to five new cuticles. The outermost had
typical adult sculpturing, and the innermost had identical staining properties to the outer ones, each having well defined endo-
and exocuticle. Animals molting in the presence of high levels
of ecdysterone plus JH formed brown, mottled cuticles. Only
rarely were multiple cuticles formed; the surface sulpturing of
the polygonal fields was absent, and the histological sections
could not be classified as either pupal or adult.

Analysis of Cuticular Proteins. A comparison of the proteins
extracted from the cuticles secreted by these isolated abdomens
made it possible to do what we could not do by morphological criteria alone, namely classify the animals which had received high
levels of ecdysterone. The cuticular proteins from isolated
abdomens were not identical in all respects to those extracted
from intact animals. The adult cuticles, in particular, had some
additional bands, but there were no differences between cuticles
from abdomens treated only with ecdysterone, irrespective of the
dose (Fig. 1 C,D); they were clearly of the adult type. We could
also detect no differences in proteins isolated from the brown and
pale regions of the abdomens which received 150 µg of ecdysterone.
The banding patterns of the cuticular proteins taken from abdomens
treated with JH were also very similar, once again irrespective
of the level of ecdysterone (Fig. 1 A,B). The rapidly migrating
bands which are the most conspicuous unique property of pupal
cuticular proteins (Fig. 4B) were present.

Action of Ecdysterone and JH in Oncopeltus Larvae

Oncopeltus is singularly appropriate for studying the interaction of ecdysterone and JH because several distinct characteristics can be used to differentiate larval and adult cuticles
(Lawrence, 1969) (Table 1). Injection of either ecdysterone
(20 µg) or JH (10 µg) into last instar larvae (weighing about
50 mg) at either 12 or 36 hr after ecdysis shortened the inter-
ecdysial period significantly (Table 1). Normally, cuticle
deposition does not begin until 96 hr (Lawrence, 1969), which is
about the time some of our experimental animals were ready for

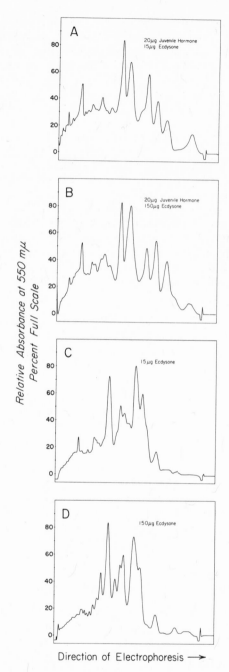

Fig. 1. Densitometric scans of SDS–polyacrylamide gels of cuticular proteins made by abdomens isolated from diapausing Cecropia pupae in response to the hormonal treatment indicated on each panel. The dye front is the blip on the right of each scan.

TABLE 1.   COMPARISON OF THE ACTION OF JH AND ECDYSTERONE IN <u>ONCOPELTUS</u>.

Cuticular Character	Normal Form[a]		Form in Hormone Treated			
			at 12 hr		at 36 hr	
	Larva	Adult	JH	Ecd.	JH	Ecd.
Cuticle around SG	P	NP	L	A	L	A
Dorsal sculpturing	-	+	L	A	L	A
Third sternite	NP	P	L	A	A	A
Fourth sternite	NP	P	L	L	A	L+A
Hairs	-	+	L	L	A	L
Hr to ecdysis[b]	141±3		102±2	95±3	115±1	114±1

[a] Lawrence (1969) and present study.

[b] Hr from fresh fifth instar to ecdysis (control) or other morphological sign of ecdysis (see Text). Based on animals where time was known to within 4 hr. Mean±S.E.

Abbreviations: P(pigmented), NP(not pigmented), +(present), -(absent), L(larval form), A(adult form).

ecdysis. The morphology of the new cuticles was distinctive for each hormone and time of application (Table 1). Animals injected with either hormone at 12 hr produced cuticles which shared two larval features - lack of hairs, and lack of pigmentation on the fourth abdominal sternite, but for all other features only JH had a juvenilizing action. Thus, the conspicuous larval pigmentation around the opening of the stink glands was present on cuticles made by JH treated animals but not by animals treated with ecdysterone, although the latter had formed dark cuticle on other regions of their bodies. Injection of ecdysterone at 36 hr still had a "juvenilizing" action on hair formation, but significantly by then it was too late for JH to influence this character. Most other larval features were still preserved by JH, but not by ecdysterone (Table 1). The tergites were smooth in JH treated animals and outlines of pore canals were visible. Ecdysterone treated animals had formed small tubercles on the tergal cuticle, which while frequently smaller or more irregular in shape than the normal adult tubercles never had the flat-toothed form Lawrence (1969) found in animals treated with JH at later times.

Interaction of JH and Ecdysterone in Wing Cultures

Morphological Features. When minced wing tissue from diapausing pupae was cultured in Grace's medium without ecdysterone it formed large numbers of hollow vesicles within 2 or 3 days (Fig. 5A). The wall of each fluid-filled vesicle was composed of a single layer of epidermal cells, and the cells appeared healthy even after three weeks in the same medium (Fig. 5B). If ecdysterone (0.05-5 µg/ml) was added to such cultures after initial isolation, or to freshly prepared cultures, the epidermis responded by making cuticle (Fig. 5E). Cuticle formation was frequently preceeded by outgrowth of long projections from the cells which gave the tissues a fuzzy appearance (Fig. 5C). In favorable sections (Fig. 5D), it could be seen that these outgrowths arose from giant cells, similar to the tricogen cells of normal wings, and were surrounded by distinct sockets. The scale processes never acquired the flattened, ridged shape of normal scales, and from the outset were narrower than the outgrowths seen during scale formation in vivo. Cuticle generally covered one surface of all of the epidermal cells in a fragment, and endo- and exocuticle were well differentiated with Mallory's stain. When wing tissue was cultured with ecdysterone and JH it made cuticle, which by morphological criteria was pupal (Fig. 5F). There was a very thick endocuticle and the exocuticle was brown. Such cuticles were never encountered in wings treated only with ecdysterone.

Protein Synthesis. Our experiments with leucine incorporation have been aimed at measuring the synthetic capacities of cultured tissues under different hormonal conditions, and at comparing

proteins made in vitro with isolated cuticular proteins made in
vivo.

Wings cultured with ecdysterone had different banding patterns
(Fig. 2) and different patterns of incorporation (Figs. 3, 4A) than
tissues cultured without this hormone. These differences were
reproducible, and we could detect no essential differences in
patterns of incorporation in wings cultured with ecdysterone 7, 10,
or 14 days, and all were distinctly different from their hormone-
free controls.

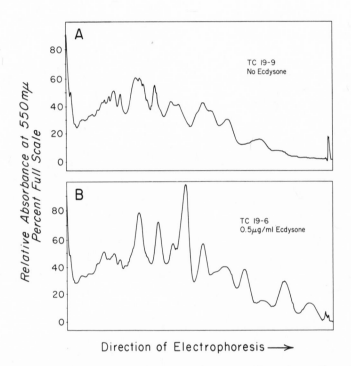

Fig. 2. Densitometric scans of SDS-polyacrylamide gels of proteins
synthesized by wing tissues taken from diapausing Cecropia pupae
and maintained in vitro for 14 days under the hormonal conditions
indicated on each panel.

The most sensitive comparison of synthetic activities under
two different sets of conditions is in doubly labeled gels. We
have found it useful to plot the percentage of counts each slice is
of the total number of counts recovered from a gel, thus making it
possible to assess the emphasis the tissue is placing on the synthesis of a particular protein(s) which migrates to a specific
region of the gel.

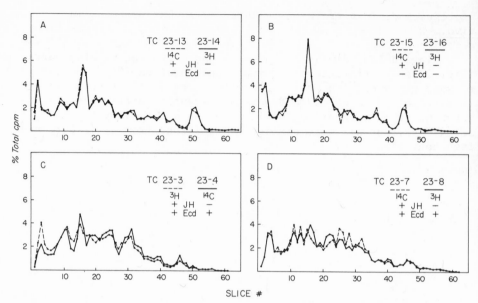

Fig. 3. Electrophoretic distribution of proteins synthesized in cultures of wing epidermis maintained in vitro for 10 days under the hormonal condition indicated in each panel. Isotopically labeled leucine was present during the final 4 hr of culturing. The incorporation of leucine per gel slice is plotted as the percentage of the total number of counts of the respective isotope on the entire gel. TC-# indicates the number of the culture. Aliquots from several of these cultures were also used for some preparations in Fig. 4.

        In one experiment (Figs. 3,4), pairs of wings from one animal were cultured separately for 10 days with only one member of the pair being exposed to JH. Half of the pairs also had ecdysterone. The patterns of incorporation of leucine were virtually identical in wings cultured without ecdysterone, irrespective of whether JH was present (Fig. 3A,B). In contrast to this finding were the results when ecdysterone was present. Then there were distinct differences (primarily quantitative) in the synthetic activites of tissues from the same animal in the presence and absence of JH (Fig. 3C,D). Unfortunately, in this experiment we did not detect morphological differences between cuticles made in the presence and absence of JH in sections of fragments removed just prior to the addition of the isotope.

        We also inquired whether the proteins synthesized in vitro bore any relationship to cuticular proteins. There are conspicuous differences between the proteins which can be extracted from pupae and pharate adults after an 18 hr pulse of isotope (Fig. 4B). We

compared such cuticular proteins standards with proteins made
following several treatments in vitro. Tissues cultured without
ecdysterone neither make proteins resembling their last secretory
product, pupal cuticle, nor the proteins they are ultimately destined to secrete for the adult cuticle (Fig. 4C,D). There are
differences in proteins synthesized by cultures grown with and
without ecdysterone (Fig. 4A), but even the ecdysterone primed
cultures, which were known to have made cuticle, were not producing as their major product proteins which resembled either the
major pupal or adult cuticular proteins (Fig. 4E,F).

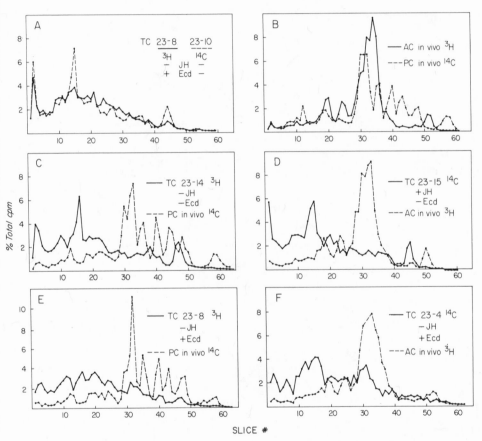

Fig. 4. Electrophoretic distribution of proteins synthesized in
vitro under the conditions indicated for Fig. 3, and of pupal (PC)
and adult (AC) cuticular proteins synthesized in vivo during the
18 hr following the administration of isotopically labeled leucine.
Other designations are as described in Fig. 3.

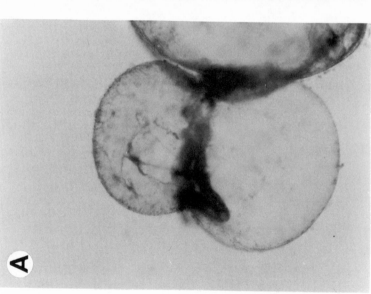

Fig. 5A,B. Morphological responses of wings, taken from diapausing Cecropia pupae, to different conditions in vitro.

  A. Phase contrast micrograph of living vesicles, photographed after 5 days in vitro without ecdysterone.

  B. Section cut through a vesicle fixed after 20 days in vitro without ecdysterone.

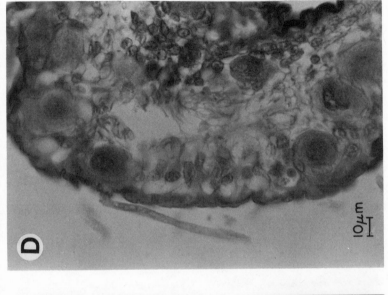

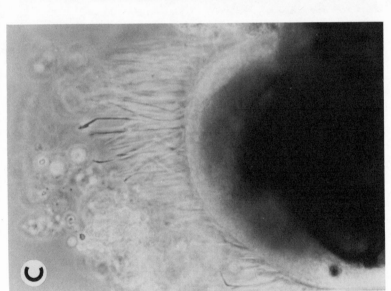

Fig. 5C,D. Morphological responses of wings, taken from diapausing Cecropia pupae, to different conditions in vitro.

C. Phase contrast micrograph of living culture with scales, formed after 7 days in vitro in the presence of 0.1 μg/ml ecdysterone.

D. Section showing scales, sockets, and cuticle from a culture maintained for 18 days with 2 μg/ml ecdysterone.

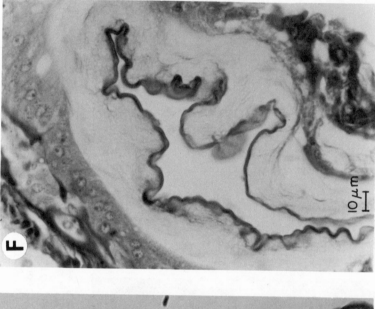

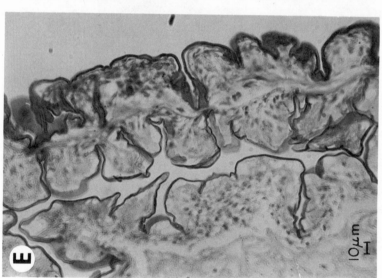

Fig. 5E,F. Morphological responses of wings, taken from diapausing Cecropia pupae, to different conditions in vitro.

E. Section depicting typical cuticle formed in response to ecdysterone. Fixed after 21 days in culture with 0.1 μg/ml ecdysterone.

F. Section with pupal-type cuticle formed in response to ecdysterone (0.1 μg/ml) plus JH (15 μg/ml). Fixed after 14 days in vitro.

Cultures maintained without ecdysterone always have one band (R_f = 0.24) which accounts for a significant fraction of the isotope incorporation (about 15% of all counts on the 60-slice gel are in 3 slices in this region). One of the striking responses to ecdysterone is the diminished significance of this region to the overall pattern of incorporation. It has fewer absolute counts and becomes one of the minor bands on a relative scale (Fig. 4A).

DISCUSSION

Ecdysone Cannot Substitute for JH

The results from these studies clearly demonstrate that even massive levels of ecdysterone never fully mimic the action of JH. Abdomens isolated from diapausing pupae of Cecropia are fully competent to respond to JH and secrete a second pupal cuticle. When injected with high levels of ecdysterone alone they only form naked adult cuticles which have a spectrum of cuticular proteins indistinguishable from those made by such abdomens receiving low levels of ecdysterone (Fig. 1C,D). Furthermore, high levels of ecdysterone do not antagonize the action of JH. Analyses of cuticular proteins revealed that the secretory products of abdomens molting in the presence of JH were identical irrespective of the level of ecdysterone used to induce the molt (Fig. 1A,B).

How can we explain the instances where high levels of ecdysones cause the formation of second pupae, namely in a few isolated pupal abdomens of Cecropia (this paper), isolated abdomens of Tenebrio (Socha and Sehnal, 1972), and intact Cecropia pupae (Williams, 1968)? We suspect that in all of these cases low but effective levels of JH were present. JH has been detected in diapausing Cecropia pupae by bioassay (Gilbert and Schneiderman, 1961), and it is reasonable to assume some is still present in Tenebrio on the first day after pupal ecdysis. Ecdysone may activate the corpora allata in intact animals (Siew and Gilbert, 1971).

Fortunately our results with hyperecdysonism in Oncopeltus require no assumption in their interpretation. Here is was clear that ecdysterone mimicked JH's action on only two individual cuticular characteristics (pigmentation on the fourth sternite, hair formation), but never caused an extra larval instar (Table 1). The sensitive period for inhibition of hair formation lasts even longer for ecdysterone than it does for JH, thus showing that they have different modes of action. Surface sculpturing and stink gland pigmentation were always adult following administration of ecdysterone, yet these characters show larval properties when JH is applied as late as 72 hr after larval ecdysis (Lawrence, 1969; unpublished observations).

Two other critical interpretations are possible as a result of this study: 1) JH does not work merely by accelerating molting processes. 2) Its step-wise action on a single cell (Wigglesworth, 1940; Lawrence, 1969) is also direct in that it is not related to premature cuticle deposition.

In Oncopeltus the inter-ecdysial period was shortened identically following injections of either ecdysterone or JH at 36 hr, and ecdysterone had even a greater effect than JH at 12 hr. Since only two larval characters were preserved in the ecdysterone treated larvae, it is apparent that the form of the new cuticle depends primarily on factors unrelated to the timing of cuticle deposition.

Lawrence (1969) has shown, and we have confirmed, that if JH is administered later than the first day of the instar, areas of cuticle are secreted by single cells which are mosaic for adult and larval structures. Thus, one can find cuticle around the stink glands (a larval character) which has teeth (modified tubercles, an adult character). We had worried whether these intermediate cells might arise by the combination of a direct action of JH and an indirect action caused by an acceleration of the time of cuticle deposition. Our results show that this is not the case. Both pigmentation and cuticular sculpturing are adult in animals treated with high levels of ecdysterone at either 12 or 36 hr despite the fact that such animals are ready for ecdysis at the same time as larvae given JH.

Cultured Pupal Wings

Appropriate Model for Studying Hormone Interaction. Experiments with intact animals suffer from several uncertainties. We do not yet know the titers of endogenous hormones or how administration of one hormone will affect feedback relationships with its own production or degradation or that of other hormones. Our ability to interpret cuticle morphology from experimentally produced intermediate forms is fraught with subjectivity (see Sláma, 1975). For these reasons we have been directing our efforts towards developing a tissue culture system which might circumvent these difficulties.

We have chosen not to use larval imaginal discs because, despite their impressive reaction to ecdysone, their response to ecdysone plus JH is a negative one, that is they do nothing. Indeed, we suspect that one could not always distinguish between the action of JH and a low dose of cyanide. Furthermore, imaginal discs of flies are destined to give rise to a large array of different tissues, including muscles, nerves, and a variety of types of cuticle. By contrast, the pupal wings of Lepidoptera are composed of a homogeneous epidermis with only a few blood cells and tracheae. This epidermis gives rise to but three cell types - scale cells, socket cells, and general epithelial cells. Most significantly, it makes cuticle in the presence as well as the absence of JH, and these cuticles differ in morphology and protein composition. Finally, we have been successful in maintaining pupal wing epidermis in a fully defined medium, without the derivatives from calves and chickens which appear to be required for many larval discs.

Response to Hormones. Minced pupal wings respond to ecdysterone by forming scales and sockets and supporting cuticle. The scales, while not of the proper adult dimensions, are produced by giant epidermal cells. In the presence of JH and ecdysterone, the tissues make a thick, tanned cuticle, which by morphological appearance is pupal (Fig. 5).

Studies on the incorporation of leucine into wing fragments also demonstrate their responsiveness to hormones. Cultures maintained without ecdysterone always have one electrophoretic band which accounts for a significant fraction of leucine incorporation. Ecdysterone diminishes the contribution of this band to the overall pattern of incorporation, also causes the appearance of new bands (Fig. 2), and differences in the entire pattern of incorporation (Fig. 4A). In cultures primed with effective concentrations of ecdysterone, addition of JH caused only quantitative differences in the incorporation of leucine into individual bands (Fig. 3C,D). The absence of morphological differences between the cultures with and without JH in this experiment precludes further interpretation.

The one drawback of the system encountered so far is that the bulk of the labeled proteins produced in cultures do not co-migrate with the major cuticular proteins synthesized in vivo, even when the cultures are treated for 7 to 14 days with ecdysterone (Fig. 4C-F). Certainly some of the cuticular proteins may be made by tissues other than epidermis, for cuticular and blood proteins are homologous by several criteria (Koeppe and Gilbert, 1973; Ruh and Willis, 1974). We did not expect, however, that the major cuticular proteins might be in this class. Equally likely is the possibility that the proteins being synthesized in the cultured wings include precursors of cuticular proteins. It is worth noting, in this regard, that labeled proteins from wing extracts were of higher molecular weights than the predominant cuticular proteins.

CONCLUSIONS

The evidence we have presented is consistent with the notion that ecdysone does not mimic the action of JH nor is it antagonistic to it. JH alone has no influence on protein synthesis in the epidermis, and ecdysone thus is a permissive agent for JH activity. Certainly JH is antagonistic to ecdysone when its action prolongs larval life (Sláma et al., 1974), but when JH directs the course of metamorphosis neither antagonism nor synergism are its mode of action. JH, therefore, is doing something more than notifying a cell that the time has come to make a cuticle and act upon the present state of the developmental program.

SUMMARY

Abdomens isolated from diapausing pupae of Hyalophora cecropia were injected with two levels of ecdysterone (150 and 15 µg) in the presence and absence of JH (20 µg). Electrophoretic analyses of proteins isolated from their new cuticles revealed that pupal cuticles were formed in the presence of JH and adult cuticles secreted in its absence, irrespective of the level of ecdysterone.

Ecdysis can be accelerated in final instar larvae of Oncopeltus fasciatus to the same extent with either JH (10 µg) or ecdysterone (20 µg), but only JH causes an extra larval instar.

Wing epidermis from diapausing Cecropia pupae, cultured in Grace's medium without macromolecular additives, responded to ecdysterone by making scales and adult-like cuticle; when JH was also present pupal cuticle was formed. Studies on the incorporation of labeled leucine revealed altered patterns of protein synthesis in response to ecdysterone; JH in the absence of ecdysterone had no effect. The major proteins synthesized in vitro failed to co-migrate with the predominant cuticular proteins synthesized in vivo.

These results have been interpreted as showing that ecdysterone, a permissive agent for JH's action, neither mimics the activity of JH nor antagonizes it. JH's prime action is more than to accelerate the onset of cuticle deposition.

ACKNOWLEDGEMENTS

This work was supported by grants HD-06195 and AG-00248 from the National Institutes of Health. The manuscript was prepared while J.H.W. was working in Prague as an Exchange Scientist of the U.S. NAS and CSAV. She thanks her colleagues in Prague, Drs. F. Sehnal, K. Sláma, R. Socha, and J. Žďárek for their critical

discussions of the issues of this paper. We gratefully acknowledge
Prof. John S. Willis' critical comments on the manuscript.

REFERENCES

Bowers, W.S., 1968, Science 161:895.
Gilbert, L.I., and Schneiderman, H.A., 1961, Gen. Comp. Endocrinol.
    1:453.
Koeppe, J.K., and Gilbert, L.I., 1973, J. Insect Physiol. 19:615.
Kramer, K.J., Sanburg, L.L., Kézdy, F.J., and Law, J.H., 1974,
    Proc. Nat. Acad. Sci. USA 71:493.
Lawrence, P.A., 1969, Devel. Biol. 19:12.
Lawrence, P.A., 1970, Gen. Res. Camb. 15:347.
Ruh, M.F., and Willis, J.H., 1974, J. Insect Physiol. 20:1277.
Siew, Y.C., and Gilbert, L.I., 1971, J. Insect Physiol. 17:2095.
Sláma, K., 1975, J. Insect Physiol. 21:921.
Sláma, K., Romaňuk, M., and Šorm, F., 1974, "Insect Hormones and
    Bioanalogues," Springer, New York.
Socha, R., and Sehnal, F., 1972, J. Insect Physiol. 18:317.
Wigglesworth, V.B., 1940, J. Exp. Biol. 17:201.
Williams, C.M., 1968, Biol. Bull. 134:344.
Willis, J.H., 1969, J. Embryol. Exp. Morphol. 22:27.
Willis, J.H., 1974, Ann. Rev. Entom. 19:97.
Willis, J.H., Friedman, L., and Swanson, S.C., 1974, Am. Zool.
    14:1289.

THE ROLE OF JUVENILE HORMONE IN DIAPAUSE AND PHASE VARIATION

IN SOME LEPIDOPTEROUS INSECTS

S. Yagi

Laboratory of Applied Zoology, Faculty of Agriculture
Tokyo University of Education, Komaba, Meguro-ku
Tokyo 153, Japan

INTRODUCTION

It is well known that various insects respond to environmental stimuli by entering diapause or exhibiting phase variation. In lepidopterous insects, for example, mature larval diapause of both the rice stem borer, Chilo suppressalis and European corn borer, Ostrinia nubilalis is mainly induced by a short day photoperiod (SD), while under a long day photoperiod (LD) almost all larvae fail to enter diapause (Inouye and Kamano, 1957; Beck, 1962). Similar photoperiodic responses were observed in the cabbage armyworm, Mamestra brassicae. In this insect, pupal diapause was mainly induced by SD conditions during the larval period (Matsumoto et al., 1953). On the other hand, it was shown in both the common armyworm, Leucania separata and the tobacco cutworm, Spodoptera litura that larval characters are varied with the density of population (Iwao, 1962).

It was proposed from a series of experiments that the larval diapause of Chilo suppressalis is induced and maintained by a hormone secreted from the corpus allatum (Fukaya and Mitsuhashi, 1957, 1958, 1961). In 1962, Fukaya showed that farnesol caused the persistence of larval diapause of this borer. Recently, it was proven that juvenile hormone (JH) is the key factor which regulates the larval diapause of the borer (Yagi and Fukaya, 1974). Two remarkable characteristics were well recognized in the diapausing larvae of this insect: 1) the prolongation of the larval period, especially of the last larval instar; 2) the development of gonads and germ cells was retarded (Fukaya and Mitsuhashi, 1961; Yagi and Fukaya, 1974). The Chilo-type diapause widely exists in other insect species which have larval diapause. It was

proposed that in the southwestern corn borer, <u>Diatraea</u> <u>grandiosella</u>
and the khapra beetle, <u>Trogoderma</u> <u>granarium</u>, JH is the principal
agent for diapause (Chippendale and Yin, 1973; Yin and Chippendale,
1973; Nair, 1974).

The present studies were undertaken to further investigate
the role of JH in diapause and phase variation in several lepidopterous insects.

RESULTS AND DISCUSSION

Larval Diapause of <u>Chilo</u> <u>suppressalis</u>

<u>Application of JH to Non-Diapause Larvae</u>. When larvae were
reared on an artificial diet containing a high concentration of
JH analog (JHA; Chang and Tamura, 1971) under LD conditions, the
following results were obtained: 1) prolongation of the larval
period, especially of the last larval instar, 2) retardation of
gonad and germ cell development of mature larvae as in diapause
insects and, 3) induction of immediate pupation by reincubation
after chilling. Thus, the larval diapause of this borer could
be induced by JH which was applied to diets even under LD conditions. In addition to this, the intensity of JH-induced diapause
larvae was found to depend upon both the concentrations and
activities of JH analogs contained in the diets (Yagi and
Fukaya, 1974).

As shown in Table 1, when 2 µg of JH I was topically applied
at daily intervals for 30 days to 30 day-old (early) larvae, all
surviving larvae entered diapause within 40 days after the onset
of the treatment, while the application of 0.2 µg JH I increased
the number of larvae which became intermediates. However, the
same doses of JH applied to 36 day-old (late) larvae resulted in
the pupation of many larvae, although some became intermediates of
various grades depending upon the hormone dose. Therefore, the
early stage of the last larval instar seems to be critical for JH
induction of diapause.

<u>JH Titers in the Hemolymph of Non-Diapause and Diapause</u>
<u>Larvae</u>. JH titers in the hemolymph of non-diapause and diapause
larvae of the last larval instar were examined by means of the
<u>Galleria</u> wax test (Schneiderman <u>et al</u>., 1965; de Wilde <u>et al</u>.,
1968). As indicated in Fig. 1, the hemolymph titers of non-
diapause and diapause larvae were 2400 and 3700 G.U./ml respectively,
soon after ecdysis. In non-diapause larvae, the titer rapidly
decreased thereafter. In diapause larvae, however, the high hormone
titer was maintained and only gradually decreased during diapause.
Therefore, the maintenance of the high titer of JH during the

TABLE 1. EFFECTS OF JH I ON NON-DIAPAUSE LARVAE OF C. SUPPRESSALIS.

Larval age (days)	Dose (μg)	No. of larvae used[a]	Results 40 days after onset of treatment		
			No. of larvae in diapause[b]	No. of intermediates[b,c]	No. of pupae[b]
30 (Early)	control	25	0 (0)	0 (0)	19 (76)
	0.2	25	12 (48)	5 (20)	8 (32)
	2	25	19 (76)	0 (0)	0 (0)
36 (Late)	control	25	0 (0)	0 (0)	23 (92)
	0.2	25	0 (0)	3 (12)	21 (84)
	2	29	1 (3)	15 (52)	11 (38)

[a] JH I dissolved in 1 μl of acetone was topically applied daily to early (30 day-old) and late (36 day-old) stage larvae for 30 days.

[b] Figures in parentheses represent percentage.

[c] Intermediates obtained were slightly progressive (intermediate +) or heavily progressive (intermediate ++) (see Yagi and Fukaya, 1974).

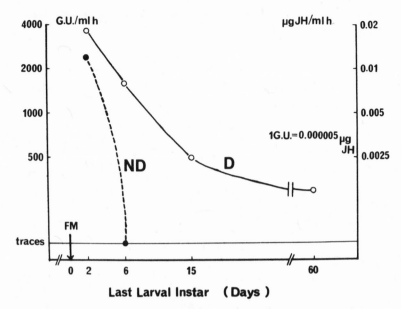

Fig. 1. JH titers in the hemolymph of non-diapause (ND) and
diapause (D) larvae of C. suppressalis. Titers lower than
30 G.U./ml of blood were not evaluated quantitatively and were
indicated as "traces". F.M.: final larval ecdysis.

early stage of the last larval instar may be related to the
induction of larval diapause.

Effects of Allatectomy on Diapause Larvae.

   Termination of larval diapause by allatectomy: After allatectomy of mature diapause larvae at an early stage (35 day-old),
they terminated diapause and pupated at a high rate within 30 days
of the operation (Table 2). The effects of allatectomy were diminished when the operation was conducted at a later stage (60 day-
old); i.e. only 10% of larvae pupated and 14% ecdysed to intermediates. The diapause larvae were indicative of high corpora allata
activity during the early stage of the last larval instar (Table 2).
In addition, the brain-prothoracic gland system was considered to
be still active at this stage, because stationary larval ecdysis
sometimes occurred spontaneously as was described by Chippendale
and Yin (1973) for Diatraea. Therefore, it was assumed that the
persistence of a high titer JH inactivated this system at the later
stages. A similar mechanism was suggested for last instar larvae
of Manduca sexta (Nijhout and Williams, 1974).

TABLE 2.   EFFECTS OF ALLATECTOMY ON DIAPAUSE LARVAE OF
C. SUPPRESSALIS.

Larval age (days)	Operation	No. of larvae	Results after 30 days	
			No. of pupae[a]	No. of intermediates[a,b]
35 (Early)	allatectomy	40	30 (75)	1 (3)
	sham	39	0 (0)	0 (0)
60 (Late)	allatectomy	59	6 (10)	8 (14)
	sham	42	0 (0)	0 (0)

[a] Figures in parentheses represent percentage.

[b] Intermediates obtained were slightly progressive (intermediate +).

Effects of JH on allatectomized diapause larvae: As much as
0.2 or 2 µg of JH I was applied daily to either allatectomized or
sham operated diapause larvae at an early stage (35 day-old) in
order to examine if JH is able to sustain diapause in allatectomized larvae. All of the larvae which had received JH remained
in diapause for as long as 40 days after onset of the treatment,
but in the control 56% of the larvae pupated even under SD conditions (Table 3). This result clearly showed that JH is a key
factor which regulates larval diapause of the rice stem borer and
causes stationary larval ecdysis.

Larval Diapause of Ostrinia nubilalis
(Ostrinia furnacalis; see Mutuura and Munroe, 1970)

Application of JH to Non-Diapause Larvae. When larvae were
continuously reared on an artificial diet containing JHA or
received daily topical applications of JH I for 20 days from the
third larval ecdysis, the following phenomena were observed:
1) prolongation of the larval period; and 2) retardation in the

TABLE 3. EFFECTS OF JH I ON ALLATECTOMIZED DIAPAUSE LARVAE OF
C. SUPPRESSALIS.

| Operation | Dose (μg) | No. of larvae[a] | Results 40 days after onset of JH treatment | |
			No. of pupae[b]	Frequency of stationary larval ecdysis
Allatectomy	control	18	10 (56)	1
	0.2	12	0 (0)	12
	2	9	0 (0)	9
Sham	control	8	0 (0)	7
	0.2	11	0 (0)	12
	2	7	0 (0)	16

[a] Larvae were 35 days old (early stage), and were continually
reared under SD conditions. Soon after the operation, JH I
dissolved in 1 μl of acetone was topically applied daily to each
larva for 30 days.

[b] Figures in parentheses represent percentage of pupae.

development of gonads as occurs in diapause larvae (Yagi and
Akaike, 1976). In this case, however, larvae of a state similar
to diapause larvae were not obtained. Therefore, it was assumed
that the Ostrinia larvae are more sensitive to the photoperiodic
stimulus than Chilo larvae. This was confirmed by experiments
in which the termination of diapause could be rapidly induced by
changing the photoperiod from SD to LD during the early stage of
diapause (Beck and Apple, 1961).

Effects of Allatectomy on Diapause Larvae. Pupation was
greatly accelerated by the extirpation of the corpora allata from
diapause larvae, although both allatectomized and sham operated
larvae pupated within 35 days if the photoperiod was changed from
SD to LD just after the operation (Table 4).

TABLE 4.   EFFECTS OF ALLATECTOMY ON DIAPAUSE LARVAE OF O. NUBILALIS.

			Days required for pupation after operation		
Operation[a]	No. of larvae	No. of pupae obtained[b]	Minimum	Maximum	Average
Allatectomy					
a pair	27	4 (23)	8	20	12.0
a half	38	11 (27)	13	31	22.3
Sham	45	12 (33)	17	35	22.5

[a] Three or four day-old larvae of the last instar were allatectomized or sham operated. After the operation, the photoperiod was
changed from SD to LD.

[b] Figures in parentheses show the number of dead larvae.

From these results, it is obvious that JH also regulates the
induction and maintenance of diapause in this borer. The JH titer
is probably rather high at the early stage of the last larval
instar since almost all diapause larvae at this stage ecdysed to
larvae when 5 µg of β-ecdysone was injected (Akaike and Yagi,
unpublished). Mitsuhashi (1963) confirmed by histological observations that the corpora allata of Ostrinia nubilalis were active
during diapause. Furthermore, the "proctodone hypothesis" (Beck
and Alexander, 1964a,b; Beck, 1968) was not supported by the experiments with either Ostrinia or Diatraea (Chippendale and Yin,
1975). Thus, the Chilo-type diapause, as previously mentioned,
seems to occur in Ostrinia.

Pupal Diapause of Mamestra brassicae

Application of JH to Larvae Destined to Non-Diapause.   In
the cabbage armyworm, Mamestra brassicae, it was reported that the

larval period in larvae destined to diapause was a little longer
than in those destined to non-diapause (Uchida and Masaki, 1952),
and that spermatogenesis was accelerated in larvae destined to
non-diapause when compared to those destined to diapause (Santa
and Otuka, 1955). Recently, these results were conclusively
proven (Yagi, 1975). The prolongation of the larval period occurred
when JHA was administered orally to larvae destined to non-diapause
under LD conditions, although they did not enter diapause after
pupation.

JH Titer in the Hemolymph of Larvae Destined for Non-Diapause
or Diapause. As shown in Fig. 2, the JH titer at the early stage
of the last larval instar was high in larvae destined to diapause
compared with those destined to non-diapause. The titer decreased
towards the pre-pupal stage in both types of larvae. An interesting fact was that the titer rapidly increased again at the late
pre-pupal stage in both larval types, and thereafter it suddenly
decreased just after pupation.

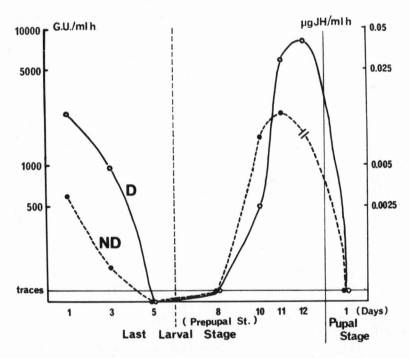

Fig. 2. JH titers in the hemolymph of M. brassicae destined to
non-diapause (ND) or diapause (D). Titers lower than 20 G.U./ml
of blood were not evaluated quantitatively and were indicated as
"traces".

Effects of Allatectomy on the Larvae Destined to Non-
Diapause or Diapause. If the corpora allata were removed from
1 day-old larvae of the last instar, the duration of the last
larval period was scarcely affected in larvae destined to non-
diapause. However, the duration was considerably shortened in
those destined to diapause (Table 5). If JHA was topically applied
daily for 3 days to 1 day-old larvae of the last instar destined
to non-diapause, the last larval period was prolonged proportionately to the amount of hormone applied (Yagi, 1975). By allatectomy of 1 day-old larvae of the last instar destined to diapause,
however, the development of spermatocysts was greatly accelerated
and many elongated spermatocysts were observed in the diapause
pupae just after pupation (Yagi, 1975).

TABLE 5. LENGTH OF THE LAST LARVAL PERIOD OF ALLATECTOMIZED
M. BRASSICAE DESTINED TO NON-DIAPAUSE OR DIAPAUSE.

	Destined to non-diapause		Destined to diapause	
	allatectomy	sham	allatectomy	sham
No. of larvae[a]	8	22	21	45
Days last larval period (Mean ± S.D.)	12.0 ± 0.8	11.7 ± 0.8	12.5 ± 1.1	15.4 ± 1.0

[a] One day-old larvae of the last instar were allatectomized or sham
operated.

From these results, it is suggested that the secretory activity
of the corpora allata is high in early stage larvae destined to
diapause. It was proven that in larvae destined to diapause, the
high JH titer at this particular stage caused the inhibition of
spermatogenesis, especially of spermiogenesis, and prolongation of
the last larval period as in the Chilo-type larval diapause.

Phase Variation of Spodoptera litura

Larval Period of Larvae Reared under Isolated and Crowded Conditions. It has been shown that in both the common armyworm, Leucania separata and tobacco cutworm, Spodoptera litura, the duration of the larval stage was prolonged in those larvae reared under isolated conditions when compared to those reared under crowded conditions (Iwao, 1962). This observation was confirmed by the present experiments. In isolation, 19.5 days on an average were required for pupation, whereas 17.5 days were required under crowded conditions. It is noteworthy that the isolated larvae showed about a 1 day prolongation of the last larval instar when compared with crowded ones.

JH Titers in the Hemolymph of Isolated and Crowded Larvae. As indicated in Fig. 3, just after the fifth larval ecdysis (0 day-old), the JH titers in both isolated and crowded larvae were about 5000 G.U./ml hemolymph. In crowded larvae, the titer rapidly decreased on the next day (1 day-old). In isolated larvae, however, the hormone level of 1 day-old larvae was higher than that of 0 day-old and the titer rapidly decreased by the third day. As in the case of Mamestra, the titers in both isolated and crowded larvae suddenly increased at the pre-pupal stage, and rapidly decreased again just after pupation.

Spermiogenesis of Isolated and Crowded Larvae. The progress of spermiogenesis was examined daily in both isolated and crowded insects. It was considerably promoted in 1 day-old crowded larvae of the last instar, while in the isolated ones the elongation of spermatocysts was delayed for 2 days (Fig. 4). The result clearly showed that the acceleration of spermiogenesis was closely related with the sudden decrease in JH titer during the last larval instar.

These data indicate similar fluctuations in JH titer in Mamestra and Spodoptera larvae and perhaps a physiological similarity between diapause in Mamestra and the solitary phase of Spodoptera. In the latter species, the corpora allata seem to be greatly affected by the population density during the larval period. Similar results were obtained in Leucania separata larvae (Kuramochi and Yagi, unpublished). Moreover, it was observed that pale coloration was characteristic of mature larvae which had been continuously reared with high doses of JH analog (ZR-515) under crowded conditions, as is normally seen with isolated larvae (Kuramochi et al., unpublished). These phenomena are analogous to phase variation in the migratory locusts where JH plays an important role (Joly and Meyer, 1970; Joly and Joly, 1974).

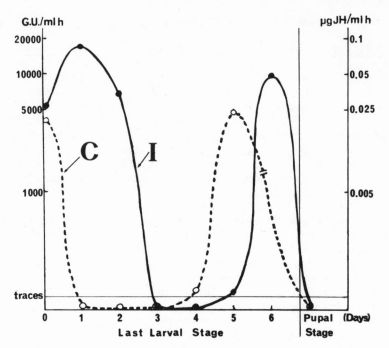

Fig. 3. JH titers in the hemolymph of S. litura reared under
isolated (I) and crowded (C) conditions. Titers lower than 30
G.U./ml of blood were not evaluated quantitatively and were indicated as "traces".

CONCLUSIONS

The present studies dealt with several aspects of the role
of JH in diapause and phase variation in some lepidopterous insects.
In the species studied, it was concluded that the secretory activity
of the corpora allata is greatly affected by environmental conditions, such as photoperiod, temperature or rearing density. As
the results of activation of corpora allata by environmental conditions, the larval period, especially the last larval period, was
prolonged and spermiogenesis was inhibited.

It seems reasonable to assume that JH, which has been known
to regulate the Chilo-type diapause, plays a very important role
in the physiological changes of various insect species including
larval and pupal diapause, and also phase variation. However,
further studies will be needed to clarify how the activity of the
corpora allata is controlled by both internal and external stimuli.

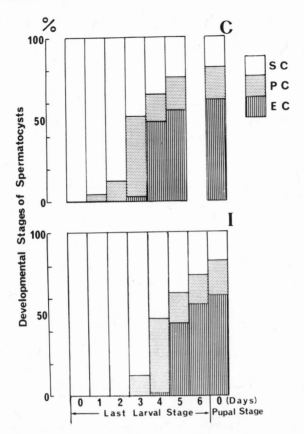

Fig. 4. Spermiogenesis of S. litura reared under isolated (I) and crowded (C) conditions. The developmental stages of spermatocysts were divided into 3 classes: SC, spherical spermatocysts; PC, pyriform spermatocysts; EC, elongated spermatocysts (see Yagi and Fukushima, 1975).

ACKNOWLEDGEMENTS

I wish to thank the late Prof. M. Fukaya of my laboratory, for his helpful suggestions and criticisms. Thanks are also due to Dr. N. Agui and the members of my laboratory, for their fruitful discussions and technical assistance throughout this study. Further I wish to express my thanks to Dr. J.B. Siddall of Zoecon Corporation and Prof. S. Tamura of Tokyo University, for supplying the hormones used in this study, to Dr. J. Mitsuhashi of the National Institute of Agricultural Sciences and Dr. Y. Takahashi of the Association of International Education in Japan, for their kindness in reading this manuscript. I am also grateful to Professor L. I. Gilbert for editing my English text.

REFERENCES

Beck, S.D., 1962, Biol. Bull. 122:1.
Beck, S.D., 1968, "Insect Photoperiodism," Academic Press, New York and London.
Beck, S.D., and Alexander, N., 1964a, Science 143:478.
Beck, S.D., and Alexander, N., 1964b, Biol. Bull. 126:185.
Beck, S.D., and Apple, J.W., 1961, J. Econ. Entomol. 54:550.
Chang, C.-F., and Tamura, S., 1971, Agr. Biol. Chem. 35:1307.
Chippendale, G.M., and Yin, C.-M., 1973, Nature 246:511.
Chippendale, G.M., and Yin, C.-M., 1975, Biol. Bull. 149:151.
Fukaya, M., 1962, Jap. Jour. Appl. Ent. Zool. 6:298.
Fukaya, M., and Mitsuhashi, J., 1957, Jap. Jour. Appl. Ent. Zool. 1:145.
Fukaya, M., and Mitsuhashi, J., 1958, Jap. Jour. Appl. Ent. Zool. 2:50.
Fukaya, M., and Mitsuhashi, J., 1961, Bull. Nat. Inst. Agr. Sci. ser. C 13:1.
Inouye, H., and Kamano, S., 1957, Jap. Jour. Appl. Ent. Zool. 1:100.
Iwao, S., 1962, Mem. Coll. Agric., Kyoto Univ. 84:1.
Joly, L., and Joly, P., 1974, Compt. Rend. Acad. Sci. Paris 279: 1007.
Joly, P., and Meyer, A.S., 1970, Arch. Zool. exp. gén. 111:51.
Matsumoto, S., Santa, H., and Otuka, M., 1953, Oyo-Kontyu 9:45.
Mitsuhashi, J., 1963, Bull. Nat. Inst. Agr. Sci. ser. C 16:67.
Mutuura, A., and Munroe, E., 1970, Mem. ent. Soc. Canada 71:1.
Nair, K.S.S., 1974, J. Insect Physiol. 20:231.
Nijhout, H.F., and Williams, C.M., 1974, J. exp. Biol. 61:493.
Santa, H., and Otuka, M., 1955, Bull. Nat. Inst. Agr. Sci. ser. C 5:57.
Schneiderman, H.A., Krishnakumaran, A., Kulkarni, V.G., and Friedman, L., 1965, J. Insect Physiol. 11:1641.
Uchida, T., and Masaki, S., 1952, Oyo-Kontyu 8:129.
de Wilde, J., Staal, G.B., de Kort, C.A.D., de Loof, A., and Baard, G., 1968, Proc. K. ned. Akad. Wet. (C) 71:321.
Yagi, S., 1975, Mem. Fac. Agr., Tokyo Univ. Education 21:1.
Yagi, S., and Akaike, N., 1976, J. Insect Physiol. 22: in press.
Yagi, S., and Fukaya, M., 1974, Appl. Ent. Zool. 9:247.
Yagi, S., and Fukushima, T., 1975, Appl. Ent. Zool. 10:77.
Yin, C.-M., and Chippendale, G.M., 1973, J. Insect Physiol. 19: 2403.

ACTION OF JUVENOIDS ON DIFFERENT GROUPS OF INSECTS

F. Sehnal

Institute of Entomology, Czechoslovak Academy of
Sciences, Praha, Czechoslovakia

INTRODUCTION

Extensive research on juvenoids in the last decade has involved
investigations of their effects on diverse insects. Most of these
comparative studies, however, suffered from the following short-
comings: the effects observed in various species were superficially
compared with those known to occur in a few well studied insects
such as Pyrrhocoris and Tenebrio; results obtained in a single
species were often regarded as typical for the respective order;
different investigators used different compounds which made comparison of activities on different species impossible. Thus,
despite the large body of information in the literature, no sound
generalization on the action of juvenoids on all insects can be
drawn. This situation provided a stimulus for this study.

The present paper examines selected effects and compares
activities of 27 juvenoids on forty species representing eleven
orders. The compounds include a majority of the most potent juvenoids so far described. We set as our prime aims the elucidation
of similarities and differences in the reaction to juvenoids of
different ontogenic stages and different species of insects, and
recognition of features which characterize these responses. The
results showed an orderly variation of responses to juvenoids
within the class Insecta and have important implications for
research aimed at the use of juvenoids in pest control.

Data presented in this article were selected from the results
obtained in the Entomological Institute of the Czechoslovak Academy
of Sciences during the last five years. The conclusions are primarily based on these results but all information available in the

literature was taken into account. Only a few papers of those
actually used can be cited here; the rest can be found in the
book by Sláma et al. (1974) and in the review by Staal (1975).

MATERIALS AND METHODS

Most juvenoids were prepared by the chemists of the Institute
of Organic Chemistry and Biochemistry, CSAV, Prague (Fig. 1).
Compounds II, III, VII, VIII, XVIII - XXI were provided by Zoecon
Corporation, Palo Alto, California; substance IX by Dr. W. Sobotka
of the Institute of Organic Chemistry, PAN, Warsaw, and substance
XV by Dr. R. Scheurer, Ciba-Geigy Corporation, Basel, Switzerland.

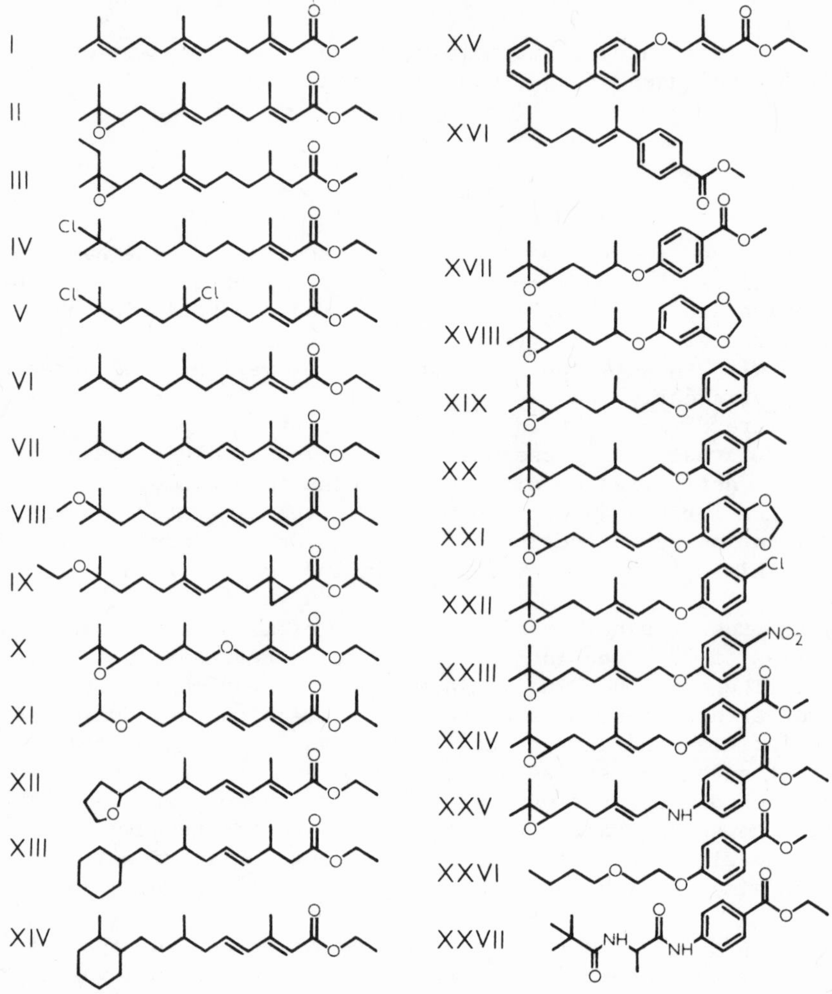

Fig. 1. Juvenoids utilized.

All compounds were racemic mixtures containing different amounts
of active isomers. Non-aromatic esters always contained 55-70%
of the most active isomer, i.e. 2E in I - VI and X, and 2E, 4E in
VII, VIII, XI - XIV. All tests on different species were performed
with the same batches of compounds which were stored at -20°C.

Juvenoids were applied topically in acetone to different
developmental stages of the species listed in Table 1. Applications were precisely timed in relation to the preceding ecdysis or
to a similar well defined point in the development. Some tests
were performed either by my colleagues or in close cooperation
with them. I am grateful for their contributions and cite their
names in Table 1.

RESULTS

Action of Juvenoids on Embryogenesis

Eggs treated with high doses of juvenoids often ceased to
develop due to suffocation of the embryo. This lethal effect can
be distinguished from a specific juvenilizing effect by tests with
olive or paraffin oils. We found that some eggs were sensitive to
10% solutions of oils in acetone but not to 1% solutions. Therefore, 1% solutions of juvenoids in acetone were in most cases the
highest concentrations tested. Insects which failed to hatch or
died shortly after hatching were considered to be affected. Failure to hatch was in a vast majority of cases accompanied by morphological abnormalities in the embryo. The deformities included
defective appendages, incomplete dorsal closure and cessation of
embryogenesis at an early stage. A similar range of morphological
effects was found in species as unrelated as Thermobia and Hylobius.

Eggs of all species studied were most sensitive to juvenoids
just after oviposition. The period of sensitivity to high doses
generally lasted for more than 50% of the incubation period: for
example, in Thermobia it encompassed 75% and in Carpocapsa and
Hylobius about 65% of the incubation period.

Activities of juvenoids were expressed in percent concentrations resulting in the failure to hatch in 50% of the eggs just
flooded with the respective solution of the compound in acetone
(EC-50 Ovicid., see Sláma et al., 1974). Selected results of
tests on seven species are summarized in Table 2, which shows that
virtually every species was most sensitive to a different compound.
Aromatic juvenoids, XVII - XXIV and XXVI showed some activity on
all species examined whereas the non-aromatic substances I - XIII
were virtually inactive on Hylobius and Ceratitis. Compounds which
are not included in Table 2 were tested on only some species and
showed only moderate or low activities.

TABLE 1. LIST OF SPECIES UTILIZED.

APTERYGOTA

Thysanura, Lepismatidae: <u>Thermobia</u> <u>domestica</u> Packard[+] - Rohdendorf
    and Sehnal (1973).

EXOPTERYGOTA

Blattodea, Blaberidae: <u>Nauphoeta</u> <u>cinerea</u> (Oliv.)[+] - Radwan and
    Sehnal (1974).
Orthoptera, Acrididae: <u>Schistocerca</u> <u>gregaria</u> Forsk.[+] - Němec <u>et al.</u>
    (1973).
Heteroptera, Pyrrhocoridae: <u>Pyrrhocoris</u> <u>apterus</u> L.[+] and <u>Dysdercus</u>
    <u>cingulatus</u> (F.)[+] - Sláma <u>et al.</u> (1974); Pentatomidae: <u>Grapho-</u>
    <u>soma</u> <u>lineatum</u> <u>italicum</u> Mull.[+] - Kontev <u>et al.</u> (1973); Scutel-
    leridae: <u>Eurygaster</u> <u>integriceps</u> Put. - Kontev <u>et al.</u> (1973).
Homoptera, Aphidae: <u>Acyrthosiphon</u> <u>pisum</u> (Harris)[+] - Holman
    (personal communication).
Phasmatodea, Lonchodidae: <u>Dixippus</u> (<u>Carausius</u>) <u>morosus</u> Br.[+] -
    Socha (1974).

ENDOPTERYGOTA

Neuroptera, Chrysopidae: <u>Chrysopa</u> sp. - Ružička and Sehnal
    (unpublished information).
Coleoptera, Tenebrionidae: <u>Tenebrio</u> <u>molitor</u> L.[+] - Sláma <u>et al.</u>
    (1974); Coccinellidae: <u>Semiadalia</u> <u>undecimnotata</u> (Schneider) -
    Hodek <u>et al.</u> (1973); Chrysomellidae: <u>Leptinotarsa</u> <u>decemlineata</u>
    (Say) - orig.; Dermestidae: <u>Trogoderma</u> <u>granarium</u> Everts[+] -
    Metwally and Sehnal (1973), and <u>Dermestes</u> <u>vulpinus</u> (F.)[+] -
    Sláma <u>et al.</u> (1974); Bruchidae: <u>Caryedon</u> <u>gonagra</u>[+] - Metwally
    and Sehnal (1973); Ipidae: <u>Ips</u> <u>typographus</u> (L.)[+] - Novák
    <u>et al.</u> (1976); <u>Hylobius</u> <u>abietis</u> (L.) - Novák <u>et al.</u> (1976).
Diptera (Nematocera), Chironomidae: <u>Chironomus</u> <u>thummi</u> Kiefer[+] -
    orig.; Cecidomyiidae: <u>Dasyneura</u> <u>laricis</u> (F. Loew) - Skuhravý
    and Sehnal (1973).
    (Orthorapha), Tabanidae: <u>Tabanus</u> <u>autumnalis</u> L. - orig.;
    (Cyclorrhapha), Syrphidae: <u>Syrphus</u> (<u>Metasyrphus</u>) <u>corollae</u>
    (F.)[+] - Ružička <u>et al.</u> (1974); Trypetidae: <u>Ceratitis</u> <u>capitata</u>
    (Wied.)[+] - Daoud and Sehnal (1975); Muscidae: <u>Musca</u> <u>domestica</u>
    L.[+] - Sehnal and Žďárek (1976); Calliphoridae: <u>Calliphora</u>
    <u>vomitoria</u> L.[+] - Sehnal and Žďárek (1976); Sarcophagidae:
    <u>Sarcophaga</u> <u>crassipalpus</u> Macq.[+] - Sehnal and Žďárek (1976).

TABLE 1 (Cont.)

___

ENDOPTERYGOTA (Cont.)

Lepidoptera, Galleriidae: <u>Galleria mellonella</u> L.[+] - Jarolím <u>et al</u>.
   (1969); Olethreutidae: <u>Carpocapsa</u> (<u>Cydia</u>) <u>pomonella</u> (L.)[+] -
   Gelbič and Sehnal (1973); Sphingidae: <u>Celerio euphorbiae</u> (L.)[+]-
   orig.; Hyponomeutidae; <u>Hyponomeuta malinella</u> Zell. - orig.;
   Lasiocampidae: <u>Malacosoma neustrium</u> (L.) - orig.; Lymantriidae:
   <u>Porthetria</u> (<u>Lymantria</u>) <u>dispar</u> (L.) - orig.; Noctuidae: <u>Plusia</u>
   (<u>Autographa</u>) <u>gamma</u> L.[+], <u>Agrotis ypsilon</u> Rott.[+], <u>Mamestra</u>
   (<u>Barathra</u>) <u>brassicae</u> L.[+] and <u>Spodoptera littoralis</u> (Boisd.)[+] -
   Sehnal <u>et al</u>. (1976); Pieridae: <u>Pieris brassicae</u> L. - orig.
Hymenoptera, Tenthredinidae: <u>Pteronus ribesii</u> Scop. - orig.;
   Cephidae: <u>Cephus pygmaeus</u> (L.) - orig.; Apidae: <u>Apis mellifera</u>
   L.[+] - Žďárek and Haragsim (1973).

___

Species marked with + were reared in the laboratory for at
least three generations; the remaining species were collected in
the field shortly before use. Results taken from the publications
cited were complemented with additional original data.

Action of Juvenoids on Larvae - Effects on Pigmentation

Juvenile hormone (JH) has long been known to play an important
role in the control of solitary and gregarious phases of locusts
which are characterized, among other criteria, by different body
coloration. We found that similar hormonally controlled phases also
occurred in caterpillars, and this led us to examine the effects
of juvenoids on pigmentation.

Application of juvenoids to larvae in the third quarter of
the penultimate instar caused a loss of dark pigments (presumably
melanins) in the final instar and thus induced the light pigmentation characteristic of the solitary phase in <u>Schistocerca</u>, <u>Celerio</u>,
<u>Mamestra</u> and <u>Spodoptera</u>. The effects were more variable in <u>Plusia</u>,
another species which shows phase variation. These results suggest
that the role of JH in the phase control of Orthoptera and Lepidoptera is similar or identical. Loss of the dark pigmentation also
occurred in pupae of <u>Celerio</u> and in puparia of <u>Syrphus</u> after application of juvenoids in the middle of the last larval instar, and
in adults of <u>Thermobia</u> after treatments in the preceding imaginal
instar. Since we do not know, however, if light and dark pupae,
puparia, and adults, respectively, are also formed in the field in
response to environmental conditions, we cannot interpret this
effect. In all instances, the loss of dark pigmentation was caused

with the same or lower doses as those inhibiting metamorphosis
when applied to the sensitive stage of the same species.

TABLE 2.   OVICIDAL ACTIVITIES OF JUVENOIDS.

| Compound | Species | | | | | | |
	Therm.	Euryg.	Carp.	Spodopt.	Leptin.	Hylobius	Cerat.
I	0.005	4	1	0.5	5	8	in.
II	0.005	2	2	5	in.	in.	in.
III	0.001	1	0.01	0.5	0.08	in.	in.
IV	0.04	7	0.4	0.5	5	in.	in.
V	0.08	10	3	in.	1	50	in.
VI	0.01	5	0.01	in.	10	in.	1
VII	0.01	0.4	0.01	0.01	1	in.	in.
VIII	1	0.5	0.08	1	1	in.	in.
X	0.05	5	0.5	0.5	0.8	in.	1
XIII	–	in.	–	5	in.	in.	in.
XVII	0.04	9	0.5	1	0.3	1	1
XVIII	0.05	1	0.5	in.	0.03	in.	0.1
XIX	0.01	6	0.5	0.5	0.5	50	0.1
XX	–	–	0.5	0.01	0.5	50	0.1
XXI	0.05	7	0.1	in.	0.3	10	0.1
XXII	–	0.5	–	0.1	0.5	8	0.1
XXIII	0.005	7	0.1	1	0.5	3	1
XXIV	0.008	0.6	0.5	5	4	50	–
XXV	0.1	6	1	1	5	in.	in.
XXVI	5	10	0.3	0.02	0.01	0.05	–

    Activities are expressed as concentrations (%) of juvenoids
affecting 50% of treated eggs; "in" indicates that 10% (Hylobius)
or 1% concentrations (other species) were inactive.

Inhibition of Metamorphosis

Juvenoids administered to last instar larvae or pupae impeded their transformation into adults. Each metamorphic stage was characterized by a period of sensitivity to juvenoids, the array of elicited effects, and the sensitivity to different compounds. In respect to the possible practical use of juvenoids we distinguished between reversible and irreversible effects (Table 3).

TABLE 3. EFFECTS OF JUVENOIDS ON METAMORPHOSIS.

Reversible effects	Irreversible effects
Prolongation of the last larval instar: - feeding continues (larvae grow); - no feeding (typically a larval diapause) Extra larval instar(s): - feeding and growth continue; - limited feeding (retrogression); - no feeding (diapause)	The insects develop into: - imperfect superlarvae (see text for details); - intermediates between preceding and successive developmental stage; - internally abnormal pupae and adults

In the case of reversible effects, the insects became adults after a delayed metamorphosis, whereas in the case of irreversible effects they perished as morphologically abnormal creatures.

Activities of juvenoids were assessed from the frequency and degree of morphological effects occurring after their administration at a time when minimal amounts were effective: the dose affecting the morphology of 50% of the treated insects (ED-50 Morph., see Sláma et al., 1974), expressed in µg per gram of body weight, was regularly taken as a measure of activity. Only those morphological effects which could clearly be distinguished from occasional developmental abnormalities were considered in the scoring. Juvenoids were not assayed on stages in which they produced only small morphological effects.

Effects on the Larval-Adult Transformation. Application of
juvenoids to the last instar larvae of exopterygotes listed in
Table 1 typically resulted in the preservation of larval features
after the next ecdysis, although in Dixippus we have occasionally
observed prolongation of the instar terminated with successful metamorphosis. The sensitive period to juvenoids extended for
a considerable portion of the last larval instar in all species
except Acyrthosiphon; in this aphid it ended within the penultimate
instar (Table 4). Treatment of Nauphoeta and Dixippus within
about the first third of the sensitive period elicited development
of perfect superlarvae which sometimes underwent one or two additional larval molts and eventually metamorphosed into adults.
This reversible effect, however, never occurred in other exoptery-
gots: the maximal effect produced with an early application of
juvenoids was typically formation of imperfect superlarvae. They
looked like supernumerary larvae but died, often due to ecdysial
failures, without becoming adults. Interestingly, the imperfect
superlarvae of Dixippus (they were produced with a slightly later
application than perfect superlarvae) contained mature eggs and
those of parthenogenetic Acyrthosiphon showed fully grown embryos
which normally are found only in the adults. Later treatments
within the sensitive period produced larval-adult intermediates
and adultoids in all species.

Activities of juvenoids were assayed in tests performed in
the middle of the penultimate instar in Acyrthosiphon, at the
start of the last larval instar in bugs and Schistocerca, and in
the middle of the last larval instar in Nauphoeta and Dixippus
(Table 4). Treatments at these times caused a dose-dependent
variety of effects ranging from imperfect superlarvae to adultoids.
The nymphs of Acyrthosiphon were allowed to walk for 3 hr on filter
paper treated with a known amount of juvenoid. The ED-50 doses
were expressed in $\mu g/cm^2$. The tests on other species were performed and evaluated in the standard manner.

As shown in Table 5, most exopterygotes responded to about
0.05 - 0.5 $\mu g/g$ of the most active compounds but Dysdercus and
Pyrrhocoris (not in the Table) were affected with as little as
0.0004 $\mu g/g$ of the peptidic compound XXVII. This compound was
inactive on all other species. Another obvious interspecific
difference was the low sensitivity of Nauphoeta and Acyrthosiphon
to aromatic juvenoids XVI - XXVI while Eurygaster and Graphosoma
were most sensitive to XXII - XXIV. Compounds IV, VI, and XXI
showed the highest activities out of the 17 juvenoids included
in preliminary tests on Dixippus.

TABLE 4.   REACTIONS OF SELECTED SPECIES TO JUVENOIDS.

Species	Instar length (days)	Sensitive period[a]	Response[b] Revers.	Response[b] Irrevers.	Time of application in assays[a]
Nauphoeta	27	85%	SL	IM	50%
Schistocerca	14	60%	–	IM	10%
Pyrrhocoris	7	50%	–	ISL, IM	0%
Acyrthosiphon	1	–	–	ISL, IM	–50%[c]
Dixippus	20	>75%	SL	ISL, IM	50%
Galleria	8	85%	L, SL	IM	75%
Carpocapsa	10	50%	L, (SL)	ISL, IM	30%
Plusia	9	75%	(L)	ISL, IM	50%
Spodoptera	7	55%	L, SL	ISL, IM	25%
Pieris	7	55%	L	ISL, IM	25%
Semiadalia	7	–	(L)	(IP)	–
Leptinotarsa	12	70%	L	(IP)	–
Trogoderma	13	85%	SL	IM	75%
Caryedon	17	85%	L	IM	75%
Hylobius	∿30	90%	L	(IP)	–
Chironomus	12	75%	(L)	IM	60%
Cyclorrhapha	4-6	–	(L)	–	–

[a] In % of the length of the last instar; [b] Reversible and irreversible effects on the larval-adult (first five species) and larval-pupal transformation (the remaining species) are recorded:
L - prolongation of the last larval instar; SL - superlarvae; IM - intermediates; IP - imperfect pupae; ISL - imperfect superlarvae. Effects in parentheses were rare. [c] Application in the middle of the penultimate instar.

TABLE 5. ACTIVITIES OF JUVENOIDS ON EXOPTERYGOTES.

Compound	Species					
	Nauph.	Schist.	Dysd.	Euryg.	Graph.	Acyrth.
I	400	600	500	1000	500	in.
II	150	–	10	200	100	50
III	1	–	4	40	10	100
IV	10	0.5	0.8	10	10	30
V	50	–	0.008	100	100	in.
VI	2	0.2	30	1000	1000	50
VII	1	0.3	50	10	100	3
VIII	0.1	4	0.5	–	–	10
X	200	–	0.05	5	1	50
XII	40	0.6	9	–	100	>100
XIII	0.1	0.08	10	–	100	100
XIV	0.2	0.8	5	–	500	8
XVII	50	–	3	10	3	–
XX	40	2	0.5	10	10	in.
XXI	80	0.08	40	50	50	in.
XXII	100	–	40	0.8	0.7	–
XXIII	–	–	10	2	0.7	in.
XXIV	40	–	0.5	0.5	0.1	in.
XXV	–	–	0.05	5	5	–
XXVII	–	–	0.0004	in.	in.	in.

Applied on the last or penultimate instar larvae (see Table 4) and expressed as ED-50 in $\mu g/g$ (most species) or in $\mu g/cm^2$ of filter paper on which the nymphs walked for 3 hr (Acyrthosiphon); "in" indicates compounds inactive at 10 $\mu g/spec.$ or 100 $\mu g/cm^2$. Body weights of species used: Schistocerca – 1250 mg; Nauphoeta – 500 mg; all bugs – 20 mg.

Effect on the Transformation from Larva to Eonympha. Larvae
of one or two last larval instars of some sawflies differ considerably from the feeding larvae of previous instars. They are
called eonymphs and pronymphs, respectively, and are regarded as
the first steps in metamorphosis. Compounds applied in the
middle of the last regular larval instar of Pteronus caused
preservation of larval integumental sculpture and of larval behavior in eonympha. (Pronympha does not exist in this species and
the effect of juvenoids on the transformation of eonympha to pupa
was not examined.)

Effects on the Larval-Pupal Transformation. The effects of
juvenoids on the larval-pupal transformation were diversified. A
reversible effect common virtually to all endopterygotes was
prolongation of the last larval instar. Length of the instar was
more than doubled in some moths (Carpocapsa, Spodoptera, Agrotis)
and beetles (Leptinotarsa, Dermestes, Hylobius) but only slightly
prolonged or unchanged in cyclorrhaphous flies. The growth of
affected larvae also varied. The larvae of Galleria, Spodoptera,
and some other moths continued feeding and often reached a body
weight twice that of the controls. Occasionally they produced
viable giant pupae. The larvae of Syrphus became just significantly
larger than the controls and for the most part developed into
fertile adults. The larvae of Leptinotarsa fed for several extra
days without any apparent increase of the body size. In Hylobius,
prolongation of the instar occurred in post-feeding larvae (this
case resembled the induction of larval diapause in the moth
Chilo suppresalis - see Yagi and Fukaya, 1974; Yagi, this volume).
A peculiar prolongation of the last larval instar characterized by
hypermetabolism was observed in Dermestes (Sláma and Hodková, 1975).

Occurrence of a delayed but successful metamorphosis indicated
that prolongation of the last larval instar provided time for the
removal of applied compounds. Experiments with Galleria revealed
that juvenoids cause no morphological effect unless they remain in
the body until shortly before secretion of the new cuticle (Sehnal
and Schneiderman, 1973). Our experience with other species also
suggested that persistence of applied compounds until about this
time is one of the prerequisites for induction of morphological
effects in Holometabola. The morphological effects observed in
the course of our study varied greatly (Table 4) but species of
the same order showed similarities in their responses.

In Lepidoptera, application of a sufficient dose of juvenoids
at the start of the last larval instar caused development of perfect (Carpocapsa, Galleria, Spodoptera and Agrotis) or imperfect
superlarvae (the remaining species listed in Table 1). The perfect superlarvae fed and grew, reaching up to three times the
weight of the controls. The superlarvae of Galleria occasionally
underwent another extra larval molt. Eventually, at least some

superlarvae of all species produced normal pupae and adults; the
remainder perished. The imperfect superlarvae also resembled
larvae but they usually possessed long antennae, and occasionally
also partially differentiated imaginal discs of legs, wings, and
compound eyes. They generally ecdysed well and lived for some time,
but eventually died; only a very few developed into pupae and
adults. Treatment in the second half or within the last third of
the sensitive period produced various larval-pupal intermediates
in all Lepidoptera. Intermediates of some species ecdysed well,
others remained trapped inside the last exuvia, and all of them
perished without reaching the adult stage.

Since juvenoids blocked the transformation to pupae in all
Lepidoptera, their application to the last instar larvae was used
for assaying their activities on this order. They were applied
at times recorded in Table 4. The maximum effect thus obtained was
formation of imperfect superlarvae or of larval-like intermediates,
and lower doses caused proportionally smaller effects. The results
of scoring are summarized in Table 6.

The unifying features of Lepidoptera were their high sensitivity to aliphatic dienoates VII and VIII and to cyclic juvenoids
XII - XIV, their moderate responsiveness to IV and XXI, and their
relative resistence to V, IX, XI, and most aromatic compounds
(XVI - XVIII, XXIII - XXVII). Substance XV was tested only on
Galleria and Spodoptera and showed moderate activity (5 µg/g and
10 µg/g, respectively). Interspecific differences among Lepidoptera included, for example, low sensitivity of noctuids to III
which was highly active on most other species. Carpocapsa,
Agrotis and Spodoptera responded only to 1 - 2 µg/g of the most
active compounds while the remaining species were affected with
0.01 - 0.1 µg/g of juvenoids.

Morphological effects of juvenoids on the larval-pupal transformation of Coleoptera were not uniform and often they were
negligible. Applications of juvenoids to Tenebrio and Trogoderma
within about two thirds of the sensitive period, however, caused
development of perfect superlarvae. The superlarvae of Tenebrio
fed and grew, whereas those of Trogoderma fed little and their
final body weight was the same as that of last instar larvae
although they lived for up to four months and underwent as many as
6 extra larval molts (normal larval development proceeds in 4 - 5
instars and lasts less than 2 months). Some of the superlarvae
of both species eventually developed into adults, others perished.
Juvenoids administered during the last third of the sensitive
period caused formation of larval-pupal intermediates in both
Tenebrio and Trogoderma and also in Caryedon and Ips. The maximal
but rare morphological effect of juvenoids on Semiadalia, Leptinotarsa, Dermestes and Hylobius was occurrence of slightly abnormal
pupae with remnants of larval appendages. Most treated larvae of

these species ecdysed as externally perfect pupae which often
developed into pupal-adult intermediates.

TABLE 6. ACTIVITIES OF JUVENOIDS ON LEPIDOPTERA.

Compound	Species							
	Gall.	Carp.	Hypon.	Porth.	Plus.	Agr.	Spod.	Pieris
I	in.	1000	80	–	–	in.	in.	–
II	30	10	50	20	–	10	200	25
III	0.2	1	0.02	1	150	200	500	7
IV	20	5	1	2	70	30	20	0.2
VI	50	500	–	–	300	in.	150	25
VII	0.5	5	2	10	30	100	12	0.4
VIII	4	–	0.01	0.1	0.3	50	2	0.04
X	10	50	30	in.	400	200	200	in.
XII	0.1	–	5	6	–	50	in.	–
XIII	0.1	–	0.05	0.2	0.6	50	6	0.04
XIV	0.2	–	–	–	0.07	1	2	0.2
XVII	3000	500	10	in.	in.	in.	500	in.
XIX	20	50	50	20	in.	200	200	20
XX	5	100	10	200	500	200	100	1
XXI	0.5	50	2	0.5	50	200	15	2
XXII	500	–	100	in.	700	in.	in.	in.
XXIII	3000	in.	80	–	in.	in.	in.	in.
XXIV	in.	1000	50	–	in.	in.	in.	200
X XV	in.	in.	150	–	in.	in.	–	200

Assayed on larval-pupal transformation and expressed as ED-50
in µg/g; "in" indicates that 100 µg/spec. was inactive. Compounds
V, IX, XVI, XVIII, XXVI and XXVII were inactive on all species
tested. Body weights of species used: Galleria, Plusia – 150 mg;
Carpocapsa, Hyponomeuta – 100 mg; Pieris – 400 mg; Porthetria,
Spodoptera – 500 mg; Agrotis – 1000 mg.

Larvae of most Diptera and Hymenoptera were only slightly sensitive to juvenoids. Application of juvenoids to water with the last instar larvae of Chironomus, however, caused formation of larval-pupal intermediates which died at ecdysis. The test was employed for assaying activities of juvenoids on this species (Table 7). Activities were expressed in ppm as EC-50 concentrations affecting morphologically 50% of the treated insects (they are discussed in the following section of this paper). Preliminary experiments with Tabanus indicated that juvenoids could elicit development of perfect superlarvae which did not feed and eventually pupated. No intermediates were produced. No morphological effect on the larval-pupal transformation was observed in cyclorrhaphous flies, although treatment of larvae often morphologically affected the pupal-adult transformation. Lack of morphological effect on the formation of pupae was also seen in Apis.

Effects on the Pupal-Adult Transformation. The imaginal differentiation of pupae was irreversibly impeded in all species examined but the morphological effects produced in different species varied greatly. The pupal instar was exceptionally prolonged in Caryedon and Ips but in most cases its length was either unchanged or shortened. The best morphological effects were always obtained by administering juvenoids to either pharate or newly ecdysed pupae. Later treatments were gradually less effective and those performed after the first third of the instar were mostly inactive. Second pupae, differing from the normal ones only by the malformation of some hairs and slight differentiation of compound eyes, were obtained in Galleria, Celerio, Tenebrio and Tabanus. Imperfect second pupae lacking some specialized pupal structures occurred, in addition to these four species, in Trogoderma, Dermestes, Apis, and most Lepidoptera. In the majority of beetles and in all cyclorrhaphous flies, however, the maximally affected specimens were about half pupal and half adult creatures. Chrysopa, Hylobius and Ceratitis seemed to be affected only internally; treated pupae developed into externally normal pharate adults which failed to escape from the pupal exuvia. Ecdysial failures also accompanied all other effects on pupae.

Application of juvenoids to postfeeding larvae of cyclorrhaphous flies, to cocoons with pharate pupae of Chrysopa, and to freshly ecdysed pupae of Dasyneura, Tabanus, beetles and Apis were employed as tests of juvenoid activities. Table 8 shows that the sensitivity of beetles to different compounds varied greatly. Most species responded readily to the derivatives of farnesoate III and IV, to dienoates VII, VIII, XI and XIII, and to aromatic ethers XIX - XXI. All beetles were either slightly sensitive or insensitive to compounds XVI - XVIII, XXV and XXVI, and none responded to XXVII. Compounds IX and XV affected Semiadalia and Leptinotarsa at doses of 5 - 15 μg/g (not included in Table 8). Most beetles responded to as little as a few ng/g of

the most potent juvenoids but Caryedon was affected with only
0.5 µg/g and Ips with 80 µg/g. Moderate sensitivity to juvenoids
was established in the lacewing Crysopa which responded to
0.7 µg/g of VII.

TABLE 7. ACTIVITIES OF JUVENOIDS ON DIPTERA.

Compound	Species							
	Chir.	Das.	Tab.	Syr.	Cer.	Musca	Call.	Sar.
I	>1	800	5000	in.	10	250	10	5
II	-	3000	800	300	0.8	250	100	5
III	1	2000	10	in.	0.1	250	50	5
IV	in.	2000	50	70	5	0.5	10	8
VI	in.	100	5000	300	5	4	6	8
VII	0.8	100	-	150	0.8	25	15	6
VIII	0.005	80	50	30	0.05	0.02	0.2	0.05
IX	0.003	-	-	30	0.8	0.0006	-	<0.1
X	0.5	500	500	2500	0.001	40	100	8
XI	0.3	-	-	-	0.005	0.4	0.01	0.01
XII	in.	-	-	in.	0.1	1500	100	100
XIII	0.5	100	-	in.	1000	35	in.	in.
XV	0.05	-	-	in.	in.	250	-	in.
XVII	1	-	500	3000	0.1	in.	1000	in.
XIX	0.1	800	80	1500	0.05	25	1	1
XX	0.1	800	-	3000	1000	100	0.1	6
XXI	0.1	100	100	2500	10	25	10	5
XXII	0.5	-	-	3000	500	50	in.	8

Tested on the larval-pupal transformation of Chironomus and on
the pupal-adult transformation of other species. Expressed as
EC-50 in ppm (Chir.) and as ED-50 in µg/g. Inactive compounds
caused no effect at 1 ppm (Chir.) or at 100 µg/spec. V, XIV, XVI,
XXIII - XXVII were slightly active or inactive. Average weights
of species used: Tabanus, Sarcophaga - 100 mg; Calliphora - 50 mg;
Syrphus - 30 mg; Musca - 20 mg; Ceratitis - 10 mg; Dasyneura -
1 mg.

TABLE 8.   ACTIVITIES OF JUVENOIDS IN PUPAL ASSAYS.

Compound	Species							
	Chrys.	Ten.	Sem.	Lept.	Trog.	Car.	Ips.	Apis
I	600	100	150	30	3	50	1000	350
II	1000	–	15	10	0.2	3	in.	3
III	10	4	1	0.003	0.01	25	in.	0.7
IV	40	10	2	0.8	0.2	0.5	in.	1
V	in.	500	15	40	30	5000	800	0.7
VI	1000	20	8	0.01	0.3	2500	500	0.4
VII	0.7	0.6	0.01	0.0002	0.3	0.5	100	0.7
VIII	40	0.01	1	0.01	–	–	in.	1
X	<1000	100	20	50	3	5	in.	0.007
XI	70	10	<0.2	8	–	–	in.	–
XII	<1000	10	25	8	–	–	200	–
XIII	700	5	0.1	0.08	–	–	80	0.7
XIV	50	1	1	0.3	–	–	100	35
XVII	in.	in.	70	in.	150	2500	250	35
XIX	40	0.3	0.08	0.002	30	>50	200	0.004
XX	12	0.005	0.008	0.002	–	–	80	4
XXI	5000	0.05	0.02	0.03	0.001	0.5	100	1
XXII	in.	0.5	0.2	0.01	in.	–	2000	0.7
XXIII	in.	5	150	1	30	250	500	35
XXIV	<1000	200	12	50	150	500	in.	0.08

Expressed as ED-50 in µg/g; "in" – no effect was produced with 50 µg (Chrys., Trog., Car.) or 100 µg/spec. (the remaining species). Average weights of species used: Apis – 150 mg; Tenebrio, Leptinotarsa, Dermestes – 100 mg; Semiadalia – 40 mg; Caryedon – 20 mg; Crysopa – 8 mg; Ips – 5 mg; Trogoderma – 3 mg.

Apis (Table 8) and most flies (Table 7) were also sensitive to III and to aliphatic dienoates VII and VIII; a characteristic feature of flies was their greater sensitivity to the isopropyl

ester VIII than to the ethyl ester VII. All flies were also
highly responsive to IX and to oxa-analogs X and XI but only
slightly sensitive to compounds XVI, XVII, XXIII - XXVII (not
included in the Table). Species specific differences included
high activity of XV on Chironomus, of XIX on Ceratitis and of XX
on Calliphora; these compounds were much less active on other
flies. Apis resembled Ceratitis by its high sensitivity to X and
XIX but differed from the flies by being sensitive to XXIV. The
ED-50 doses of the most potent compounds were about 1 ng/g in
the case of Apis, Ceratitis and Musca, ten times higher in the
case of Calliphora and Sarcophaga, and $10^4 - 10^5$ times higher in
Dasyneura, Tabanus and Syrphus.

   Effects of Juvenoids on Diapause in Immature Stages. We have
confirmed that pupal diapause in Lepidoptera can be terminated
with juvenoids. For example, application of 200 µg of VIII a
week after pupal ecdysis in Mamestra produced pupal-adult intermediates within 2 months when the insects were kept at room temperature. Interestingly, a dose of 20 µg elicited formation of
pharate adults within 5 months although they died inside the pupal
exuvia. Pure acetone caused no effect and the animals remained
in diapause for at least seven months after the treatment. The
pupal diapause of Celerio was readily broken with the application
of 100 µg of III to pharate pupae, but only occasionally with
treatments applied after the pupal ecdysis. All activated insects
molted into pupal-adult intermediates. A precocious termination
of pupal diapause was also induced with juvenoids in the fly
Sarcophaga (Žd'árek and Denlinger, 1974).

   Diapause in other immature insect stages was insensitive to
juvenoids. The diapausing eggs of Malacosoma, diapausing first
instar larvae of Porthetria, and diapausing postfeeding larvae of
Carpocapsa and Cephus resisted the action of several selected
juvenoids.

   Effects of Juvenoids on Reproduction. Reproduction was
affected by juvenoids in different ways. An unequivocal response
was induction of reproduction in diapausing adults which normally
do not reproduce. A great majority of eggs laid by such activated
females were fertile and gave rise to a new generation. Oviposition
was induced with relatively low doses of many compounds in diapausing
Eurygaster, Chrysopa and Semiadalia but only with 100 µg of III in
Leptinotarsa (Table 9).

   Stimulation of reproduction was less obvious in non-diapausing
adults of Hylobius. Their rate of oviposition under optimal conditions was doubled after treatments with 2000 µg of XXVI but this
increase was accompanied by a decrease in egg fertility.

TABLE 9.   DOSES OF JUVENOIDS TERMINATING ADULT DIAPAUSE.

Compound	Eurygaster Activ.	Chrysopa Activ.	Ovip.	Semiadalia Activ.	Ovip.	Leptinotarsa Ovip.
I	70	–		in.		in.
II	0.5	–		100	?	>100
III	1	0.5	1	50	100	<100
IV	20	5	10	<100	100	>100
V	in.	3	10	100	?	in.
VI	100	–		50	100	in.
VIII	10	8	10	<100	<100	in.
X	1	10	?	50	100	in.
XII	–	3	10	50	100	–
XIII	–	0.8	1	1	1	–
XVIII	in.	–		100	?	>100
XIX	1	5	10	100	?	in.
XX	–	3	10	in.		–
XXI	in.	2	8	>100	?	>100
XXII	0.5	8	10	20	100	–
XXIII	2	10	100	–		>100
XXIV	0.05	–		in.		in.

    Doses in µg/spec. inducing egg development (Activ.) and actual
oviposition (Ovip.). Inactive compounds (in.) caused no effect at
100 µg/spec. and those active at >100 µg/spec. had only a slight
effect on egg development; ? indicates cases when 100 µg/spec.
clearly stimulated egg development and oviposition would certainly
be induced with higher doses (not tested). Compounds tested on
one species only are not included.

    A decrease in egg hatchability was caused by juvenoids in
many other insects. For example, treatment of reproducing adults
of Eurygaster with 200 µg of XXII decreased egg hatchability
without affecting fecundity whereas application of several juvenoids to the adults of Thermobia and Carpocapsa reduced both egg
hatchability and adult fecundity. The failure of eggs to hatch
could be due either to persistence of applied compounds in the

egg and the consequent block of embryogenesis or to defects in
the formation or fertilization of eggs within the treated female.
We did not distinguish between these two possibilities. On the
other hand, decrease in fecundity was traced to defects in the
ovaries of Thermobia and to the failure of egg deposition in
Carpocapsa.

Defects in ovaries also occurred after administration of
juvenoids to last instar larvae of Dysdercus and Dixippus and to
pupae of Galleria and Trogoderma. In some cases, juvenoids did
not inhibit differentiation of the ovaries but hindered formation
of oviducts and external genitalia. As a result, eggs were formed
but could not be laid. Such a failure to oviposit was seen in
many species and represented the only effect of juvenoids on
reproduction in Acyrthosiphon (understandably, the defective
adults did not contain eggs but embryos), Ceratitis and Syrphus.

DISCUSSION

Despite some exceptions we can generalize that juvenoids
inhibit insect differentiation at all developmental stages:
embryogenesis, phase formation, caste differentiation, diversification of seasonal forms in aphids, metamorphosis, and gametogen-
esis (including differentiation of trophocytes and follicular
cells). We are also justified in stating that juvenoids stimulate
ovarian development in diapausing adults. The inhibition of metamorphosis is commonly regarded as most important for the practical
use of juvenoids and will thus be the main topic of the following
discussion.

A complete block of metamorphosis resulting in development
of perfect superlarvae is exceptional and seems to occur predominantly in species with a large and variable number of larval
instars. In most exopterygotes, but probably also in some endopterygotes such as Chironomus and Plusia, the allometric changes
between consecutive larval instars seem to make survival and
metamorphosis of larvae of extra instars impossible. Therefore,
these species produce only non-viable, imperfect superlarvae.
Formation of superlarvae is apparently impossible in most endopterygotes because the larval-pupal transformation of some or of all
organs cannot be blocked with juvenoids. Juvenoids cause profound
morphological effects in Lepidoptera, in some families of Coleoptera and apparently also in some families of lower Diptera and
Hymenoptera but Diptera-Cyclorrhapha and Hymenoptera-Apocrita
seem insensitive to juvenoids. Limited sensitivity to juvenoids
is also typical for the pupae of most insects.

The responses of endopterygous larvae to juvenoids are often
a result of the interaction of applied compounds with the neuroendocrine system of the host. Through these interactions juvenoids significantly affect both the length of the last larval
instar and growth of the last instar larva and can elicit several
types of larval diapause. The character of these interactions
seems to be species specific. For example, in Dermestes juvenoids
applied to larvae inhibit molting, and larval-pupal intermediates
are thus never produced. In another dermestid, Trogoderma, the
inhibition of molting does not occur and both superlarvae and
intermediates are a common result of juvenoid application. In
most cases, however, species of the same family respond to juvenoids in a similar way.

The interactions with the neuro-endocrine system, the rates
of transport, metabolism and excretion, and the sensitivity of
target tissues appear to be critical for the activity of juvenoids
on each species. A few species respond to about 1 ng/g, most to
0.01 - 0.1 µg/g and some only to 1 - 100 µg/g. As mentioned above,
the larval-pupal transformation of many insects is totally insensitive to any dose of juvenoids. We believe that these differential
sensitivities reflect actual differences among target tissues:
although new compounds may be discovered, Syrphus will never be as
sensitive to juvenoids as Pyrrhocoris or Tenebrio.

Insect orders are also characterized by diversified sensitivity
to different compounds. For example, bugs appear to be more responsive than other insects to the dichloroderivative V, to aromatic
ethers XVII, XVIII, XXII - XXIV, and to the derivative of p-amino-
benzoic acid XXV. Lepidoptera respond relatively uniformly to
dienoates VII, VIII, XII - XIV and are also fairly sensitive to
III and IV. Beetles are most sensitive to VII (rarely to VIII)
and to aromatic compounds XX and XXI. The unifying feature of
Diptera is their high sensitivity to the dienoate VIII, to compound
IX and to oxa-analogs X and XI. In addition to these differences
between orders, there are probably no two families with identical
relative sensitivity to all juvenoids tested. The inter-familiar
differences are particularly obvious in bugs and it will be
interesting to learn if they reflect diversity in their JH. Inter-
generic differences within a family are usually small but by no
means negligible.

There is often only a limited correlation between activities
of juvenoids on different developmental stages of the same species.
For example, X is very active on the metamorphosis of Ceratitis
but is virtually inactive on embryogenesis. Similarly, XX and
XXI are very efficient in inhibiting the pupal-adult transformation
of Semiadalia but fail to terminate the imaginal diapause of this
species. Compound XXVI is an effective ovicide for Hylobius but
has no effect on its metamorphosis.

An important practical conclusion which follows from these observations is that development of new juvenoids should be closely linked with biological tests on several unrelated species. Development of juvenoids for practical use is unthinkable without tests on that stage of the respective pest which is being considered for treatment in field.

The decision as to which stage should be treated will obviously be based on the accessibility to sprays or other treatments and on the economic significance of different stages. Our study demonstrates, however, that speculations on the use of juvenoids against pests must take into account diversities in the responsiveness of different insects to juvenoids. For example, one cannot expect effective control of larvae of cockroaches with a short-term application of juvenoids. Only a fraction of the population (larvae in the second half of the last instar) would be affected irreversibly; younger larvae would be either unaffected or would produce superlarvae and metamorphose with a delay. Similarly, treatment of Leptinotarsa larvae would be partly successful only with high doses administered in the second half of the last larval instar. The treatment would never approach 100% efficiency and in most cases would be useless in regard to the damage caused both by the present and the next generation of the pest. Other similar deductions on the effectiveness of juvenoids should be possible on the basis of the data presented above.

REFERENCES

Daoud, S.D., and Sehnal, F., 1974, Bull. ent. Res. 64:643.

Gelbič, I., and Sehnal, F., 1973, Bull. Ent. Res. 63:7.

Hodek, I., Ružička, Z., and Sehnal, F., 1973, Experientia, 29:1146.

Jarolím, V., Hejno, K., Sehnal, F., and Šorm, F., 1968, Life Sci. (II), 8:831.

Kontev, C.A., Žďárek, J., Sláma, K., Sehnal, F., and Romaňuk, M., 1973, Acta ent. bohemoslov. 70:377.

Metwally, M.M., and Sehnal, F., 1973, Biol. Bull. 144:368.

Němec, V., Jarolím, V., Hejno, K., and Šorm, F., 1970, Life Sci. (II), 9:821.

Novák, V., Sehnal, F., Romaňuk, M., and Streinz, L., 1976, Z. ang. Ent. 78: in press.

Radwan, W., and Sehnal, F., 1974, Experientia 30:615.

Rohdendorf, E.B., and Sehnal, F., 1973, J. Insect Physiol. 19:37.

Ružička, Z., Sehnal, F., and Cairo, V.G., 1974, Z. ang. Ent. 76:430.

Sehnal, F., and Schneiderman, H.A., 1973, Acta ent. bohemoslov. 70:289.

Sehnal, F., and Žďárek, J., 1976, J. Insect Physiol. 22: in press.

Sehnal, F., Metwally, M.M., and Gelbič, I., 1976, <u>Z. ang. Ent</u>.
    78: in press.
Skuhravý, V., and Sehnal, F., 1973, <u>Z. ang. Ent</u>. 74:217.
Sláma, K., and Hodková, M., 1975, <u>Biol. Bull</u>.148:320.
Sláma, K., Romaňuk, M., and Šorm, F., 1974, "Insect Hormones and
    Bioanalogues", Springer Verlag, Wien New York.
Socha, R., 1974, <u>Acta ent. bohemoslov</u>. 71:1.
Staal, G.B., 1975, <u>Ann. Rev. Entomol</u>. 20:417.
Žďárek, J., and Denlinger, D., 1975, <u>J. Insect Physiol</u>; 21:1193.
Žďárek, J., and Haragsim, O., 1974, <u>J. Insect Physiol</u>. 20:209.
Yagi, S., and Fukaya, M., 1974, <u>Appl. Ent. Zool</u>. 9:247.

SUMMARY OF SESSION IV:

JUVENILE HORMONE EFFECTS AT THE MOLECULAR LEVEL (BINDING AND

TRANSPORT)

H. Emmerich

Zoologisches Institut der Technischen Hochschule
Darmstadt, W. Germany

Soon after the elucidation of the structure of the juvenile
hormones in the laboratories of Röller and Meyer, it was suggested
that the hemolymph of insects may contain carrier proteins for
these lipophilic molecules comparable to those found for many
hormones in vertebrate blood. We now know that such carrier
proteins are not necessary for solubilization of the juvenile
hormones (JH) because JH solubility in an aqueous medium is surprisingly high (Kramer et al.). However, the carrier proteins
appear to have several other important functions:

1) They protect JH from degradative attack by ubiquitous nonspecific esterases, e.g. in Manduca sexta (Kramer et al.; Nowock
et al.) and in Plodia interpunctella (Ferkovich et al.). In the
hemolymph of Manduca, however, a DFP-resistant, JH-specific esterase appears in the last larval instar, which is able to hydrolyze
both free and carrier-bound JH. This esterase probably arises
from the fat body (Nowock et al.) and plays an important role in
diminishing the JH titer prior to the pupal ecdysis. It remains
to be elucidated, however, whether the binding protein carrier has
a protective function within target cells as well. The data of
Ferkovich et al. demonstrate a reduced metabolism in homogenates
of the larval epidermis from Plodia if a binding protein fraction
from the hemolymph is added, but no data are available as to
whether such an effect also occurs in vivo. Indeed, it has not
been demonstrated that the carrier protein penetrates the target
cell membrane.

2) The carrier protein system represents a mechanism enabling
the storage of relatively large quantities of JH in the hemolymph.
In Manduca, the carrier may aid in controlling the actual titer of

323

these hormones since the titer of this protein changes during
larval development and drops during the pupal and pharate adult
stage (Nowock et al.).

The titer of JH in the hemolymph of insects is therefore
regulated in a very complicated manner by different systems:
synthesis and release of the JH by the corpora allata (see
Schooley et al., this volume; Tobe and Pratt, this volume),
occurrence of nonspecific and specific esterases in the hemolymph
(Nowock et al.); the amount of carrier protein in the hemolymph,
uptake and subsequent degradation in the tissues and then excretion (Nowock et al.); reflux from tissue compartments (Nowock
et al.) and by other factors.

3) The binding proteins may influence the biological activity
of the JH present in the hemolymph. In vertebrates, carrier bound
hormones are always in an inactive form and this may occur in
insects too (Ferkovich et al.). The uptake of the hormone into
Manduca fat body cells is reduced in the presence of the carrier
protein (Nowock et al.). Binding constants of cellular JH
receptors in the target cells are probably higher than those of
the carrier proteins as indicated by the dose response curves for
the biological activity of juvenoids. Therefore, in the vicinity
of the target cell, binding to cellular receptors may lead to a
local depletion of unbound hormone which elicits a rapid and
spontaneous dissociation of the carrier hormone complexes. The
presence of the carrier may even lead to the increased biological
activity of a given hormone titer by protecting the hormone from
hydrolytic degradation and by preventing excessive unspecific
binding to lipophilic compartments in the insect cells. Such a
function is indicated by the data of Ferkovich et al., although
these experiments were conducted by adding JH binding protein
fractions to epidermis homogenates of Plodia. This treatment
may give rise to many artifacts due to the interference of specific
and nonspecific binding as well as degradative metabolism.

The carrier protein system does not seem to be necessary for
the biological activity of JH however, since hormone administered
in vitro exerts its biological activity qualitatively in the same
manner in the absence of specific carrier proteins. Juvenoids
with high biological activity are not bound to the carrier of
Manduca (Goodman et al., this volume; Kramer et al.). Only the
properties of the carrier protein from Manduca hemolymph have been
elucidated so far. It is a small protein of about 2.8×10^4 daltons
(Kramer et al.; Nowock et al.) and is a single polypeptide without
lipoprotein characteristics. Its binding constant for the differ-
ent JHs ranges from $K_a = 8.6 \times 10^{-6}$ M^{-1} for JH 0 to $K_a = 1.0 \times$
10^{-6} M^{-1} for JH III (Kramer et al.; see also Goodman et al., this
volume). Interestingly, the binding constants are inversely
proportional to the solubility of these homologs in aqueous

solutions, i.e., the product of the association constants and the solubility is nearly constant. This may indicate that the increase in entropy occurring during the binding process contributes to the stability of the complex formed. Both the epoxy and ester groups of JH seem indispensable for binding to the carrier, since derivatives lacking these functional groups are not bound at all.

The presence of such a small carrier has been demonstrated in 10 species of Lepidoptera, two species of Coleoptera and one species of Hemiptera (Kramer et al.), but their physicochemical characterization has only just been initiated. There are, however, other binding proteins present in the hemolymph of the insects investigated so far. In the blood of many Lepidoptera there are lipoproteins of high molecular weight, which also bind JH with low affinity and exert a low specificity, but have a high binding capacity. In other insects (e.g. Tenebrio) proteins of similar size above 10^5 daltons are found, which have a high affinity to JH I, but do not seem to be hormone-specific. In the desert locust (Law, personal communication; Gellissen, personal communication) during all larval and adult stages, only a high molecular weight carrier has been found. We have to presume, therefore, that the hemolymph JH carriers in insects of various orders may be quite different, and that the JH titer in the hemolymph may be regulated and mediated in quite different ways.

The mechanism of action of JH at the morphogenetic level remains almost completely unknown at this time although more is known about JH action in eliciting vitellogenin synthesis during oogenesis (see Session V). Several more or less speculative models have been developed which include binding of the hormone to specific cellular "receptors" that possibly interact with the genome. The experimental evidence for the existence of such receptors, however, is relatively poor. Schmialek and his colleagues suggested the presence of a JH binding ribonucleoprotein in the pupal epidermis of Tenebrio, which was able to bind a JH analog, but this result may be due to an artifact caused by the use of Triton X-100 in the experiments (Ferkovich, personal communication).

The data presented by Ferkovich et al. raise the possibility of "specific" binding (i.e. stable in the presence of an excess of cold JH) in the nuclear and mitochondrial fractions. However, it is known that JH binds nonspecifically to all lipid-containing particles in homogenates and that degradative breakdown of the administered hormone takes place both during incubation and subsequent differential centrifugation. Therefore, these data on "specific" binding to subcellular components must be viewed as preliminary and a great deal more work must be conducted in this area.

The degradation of juvenoids resembles that of other xenobiotics (Hammock and Quistad). We have to anticipate, therefore,
that insect species treated with JH analogs of any structure may
develop resistant strains after a while as they do after treatment
with other insecticides. It therefore seems unrealistic to believe
that compounds can be developed against which insects could not
develop an effective degradative pathway.

The juvenoids possess, however, other properties which make
them very useful growth regulators: they have a relatively
pronounced specificity; often only a single order or family of
insects is attacked; low toxicity to mammals and excellent biological degradability, which prevents any accumulation of these compounds in the environment (see also Zurflueh, this volume; Sehnal,
this volume).

INTERACTION OF JUVENILE HORMONE WITH BINDING PROTEINS IN INSECT HEMOLYMPH

K.J. Kramer, P.E. Dunn, R.C. Peterson, and J.H. Law

Grain Marketing Research Center, 1515 College Avenue,
Manhattan, Kansas 66502 and Department of Biochemistry,
University of Chicago, Chicago, Illinois 60637

INTRODUCTION

Recently the interaction of juvenile hormone (JH) with hemo-lymph proteins has been the subject of intense investigation. There is an accumulating body of evidence indicating that specific carrier proteins interact with the hormone in the hemolymph to transport it from the site of synthesis, the corpora allata, to target cells. This transportation is necessary not because of the limited solubility of JH, but because of the presence of ubiquitous nonspecific esterases that degrade uncomplexed hormone. A specific JH-esterase has also been identified that hydrolyzes complexed hormone and lowers the JH titer at the end of larval growth, allowing metamorphosis to proceed.

For the past several years, we have studied the interactions of juvenile hormone with proteins from the tobacco hornworm, Manduca sexta. In the hemolymph of this insect, the hormone is present as a complex with a carrier protein (Kramer et al., 1974). This protein protects the hormone from attack by general carboxyl-esterases that hydrolyze the unbound hormone to form the inactive acid metabolite (Sanburg et al., 1975a; Hammock et al., 1975). The protein also exerts a synergistic action with JH at target tissues (Sanburg et al., 1975b).

When hemolymph of M. sexta larvae is fractionated, two forms of JH binding protein are found (Akamatsu et al., 1975). Both forms of the carrier protein have been purified to homo-geneity in an overall yield of approximately 30% using a combin-ation of gel filtration, ion-exchange chromatography, and iso-electric focusing procedures (Akamatsu et al., 1975). The

major hemolymph binding protein (CP-1) has been extensively studied
(Kramer et al., 1976). It consists of a single polypeptide chain
of molecular weight 2.8 x 10^4 as determined by SDS-gel electrophoresis, gel filtration, and meniscus depletion sedimentation experiments. The protein forms an equilibrium complex with JH having a
dissociation constant (K_d) of 3-4 x 10^{-7} M. The stoichiometry of
hormone binding has been studied using a filter paper adsorption
assay and one binding site for JH per 2.8 x 10^4 molecular weight
protein molecule has been measured.

In this paper we present evidence that the binding site on
the major M. sexta carrier protein is specific for the essential
structural features of JH. In addition, data are presented
describing the existence of specific carrier proteins for JH in
the hemolymph of other insect species including members of the
Lepidoptera, Coleoptera and Hemiptera.

MATERIALS AND METHODS

Animals. M. sexta eggs were the gift of Drs. R.A. Bell and
J.P. Reinecke, USDA, Fargo, N.D. and larvae were reared as described previously (Kramer et al., 1974). Hyloicus chersis were
collected locally on ash trees. Other insects were obtained from
the U.S. Grain Marketing Research Center Laboratory cultures.
Hemolymph was collected from M. sexta, and H. chersis according
to Kramer et al. (1974) and from other insects by cutting off one
of the thoracic legs and collecting the blood with a microcapillary
pipet. The sample was immediately transferred to a centrifuge
tube and spun at 10,000 x g. The resulting supernatant was frozen,
lyophilized and stored at -20°C.

Chemicals. Pure synthetic JH I (methyl-7,11-dihomojuvenate,
methyl-trans,trans,cis-3,11-dimethyl-7-ethyl-10,11-epoxy-2,6-
tridecadienoate, C$_{18}$JH) was purchased from Eco Control* and Regis
Chemical Co.   Labeled JH I (7-ethyl-1,2-[^{3}H],11.76 Ci/mM; 10-[^{3}H],
13.5 Ci/mM) was from New England Nuclear. Stock solutions of [^{3}H]-
JH mixed with unlabeled carrier JH were prepared in 5 mM Tris-HCl
buffer, pH 7.3 or 8.3. The final concentration was ascertained
by determination of the radioactivity. JH III (methyl juvenate,
methyl-trans,trans-3,7,11-trimethyl-10,11-epoxy-2,6-dodecadienoate),
JH III acid, and JH III diol (trans,trans-methyl farnesenate-10,11-
diol) were gifts of Drs. John A. Katzenellenbogen and Steven B.
Bowlus, Univ. of Illinois. JH II (methyl 11-homojuvenate,methyl-
trans,trans,cis-3,7,11-trimethyl-10,11-epoxy-2,6-tridecadienoate),

---

* Mention of a proprietary product does not constitute an endorse-
ment by the U.S. Department of Agriculture.

JH 0 (methyl-3,7,11-trihomojuvenate, methyl-trans,trans,cis-11-
methyl-3,7-diethyl-10,11-epoxy-2,6-tridecadienoate) and methoprene
(ZR-515) were gifts of Zoecon Corp., Palo Alto, Calif. 1-(4-
Ethylphenoxy)-3,7-dimethyl-6,7-epoxy-2-octene (R-20458) was kindly
provided by Stauffer Chemical. JH I acid was prepared according
to the procedure of Metzler et al. (1972) using a general esterase
preparation derived from gel filtered M. sexta hemolymph (Sanburg
et al., 1975a). Trans,trans farnesenic acid was prepared from
farnesol (Givaudan) and purified as the S-benzylthiopseudouronium
salt (Law et al., 1966). The free acid was generated by neutralization and extracted into organic solvent. After treatment with
diazomethane, the all trans methyl ester was further purified by
column chromatography on silica gel (SilicAR CC7, Mallinckrodt)
impregnated with silver nitrate (5%) using a linear gradient of
0-10% diethylether in hexane to resolve the geometrical isomers.
1-[^{14}C]-Tripalmitin (25 mCi/mM) was from Schwarz/Mann. Other
materials used were the same as described previously (Kramer et al.,
1974; Sanburg et al., 1975a) or of the highest purity commercially
available.

JH Binding Protein Assays. (a) Charcoal adsorption assay.
For the semi-quantitative determination of JH binding protein, we
adopted the estrogen binding assay of Korenman (1969). In this
assay, activated charcoal is used to adsorb JH and other lipophilic
compounds while leaving the ligand-binding protein complex in
solution. Norite A charcoal (Matheson, Coleman, and Bell) was
prepared by washing successively with 1 N HCl, water, 1% sodium
bicarbonate and water until neutral. In order to minimize protein
adsorption to charcoal, the Norite was coated with dextran C
(Schwarz/Mann) by dissolving 0.5 g of dextran in 100 ml of 0.01 M
Tris-HCl-0.0015 M EDTA-0.003 M sodium azide, pH 7.3, and adding 5 g
of washed Norite A. For the stock charcoal solution, the dextran-
Norite suspension was diluted eleven fold with 5 mM Tris-HCl,
pH 7.3. For routine analyses, the binding assay mixture consisted
of ten μl of an aqueous solution of labeled ligand (1-50 x 10^{-7} M)
and 10-50 μl of sample [10^{-3} M diisopropylphosphorofluoridate (DFP)
treated and gel filtered hemolymph] in 5 mM Tris-HCl buffer, 0.1 M
NaCl, pH 7.3. After a 30 min complexation period at 4°C with
occasional mixing, unbound ligand was removed by incubating the
solution with 10 μl of the stock charcoal dextran suspension for
15 min and centrifuging at 15,000 x g. To determine the amount
of bound ligand, an aliquot of the supernatant was removed and
counted in scintillation cocktail prepared from 1 L of toluene,
1 L of ethylene glycol monomethyl ether (Pierce) and 8 g of
Omnifluor (New England Nuclear Corp.). Incubation of the charcoal
suspension with radioactive ligand in the absence of binding protein
resulted in adsorption of 95% of the ligand in 2 min at 4°C.
Binding of tripalmitin (Applied Science Lab.) was measured in a
similar manner except that a chloroform solution of labeled tripalmitin was transferred to the assay vessel, evaporated to dryness

under nitrogen and then incubated with aliquots of gel-filtered
hemolymph for 2 hr.

(b) DEAE filter disc assay. For the quantitative determination
of the M. sexta JH binding protein, we have developed an ion-
exchange paper disc assay (Kramer et al., 1976). This assay differs
from the charcoal adsorption assay in that after incubating extracts
that contain binding protein with labeled hormone at pH 8.3, free
binding protein and hormone complex are adsorbed on ion-exchange
paper (Reeve Angel) and the free hormone is removed by washing.
The diethylaminoethyl (DEAE) filter disc with adsorbed radioactive
complex is then counted in a liquid scintillation counter using
scintillation cocktail prepared from 1 L of Triton X-100 (Research
Products Int.), 2 L of toluene, and 12 g of Omnifluor.

Determination of Aqueous Solubilities. For the determination
of aqueous solubilities, stock solutions of JH homologs, metabolites,
and mimics were prepared in benzene and aliquots transferred to
test tubes. After removing the solvent under nitrogen, 2 ml of
5 mM Tris-HCl, pH 8.3, was added to each tube. The tubes were then
tightly stoppered and shaken in a water bath at 25°C. After 48 hr,
the absorption spectrum (210-300 nm) of each solution was recorded
using one cm quartz curvettes and a Cary 15 spectrophotometer.
Absorbance maxima (λ max) were observed at 222 nm for all compounds
tested except trans,trans methyl farnesenate for which λ max occurred at 218 nm. To obtain the solubility limit, the absorbance
at λ max for each compound was plotted as function of the amount of
that compound added to the tube.

Competition Assay. The competition assay used was modified
from that described by Korenman (1970). In our procedure, a series
of incubation mixtures that contain a constant concentration of
binding protein and of excess tritiated JH I and varying concentrations of competitors were prepared as follows. A benzene solution
of the competitor was added to the incubation tube and the solvent
was removed under nitrogen. Buffer (10 μl) was added to the tube to
dissolve the competitor. After 2 hr, an aqueous solution of tritiated JH (10 μl) was added to the tube and incubated for an additional 2 hr. Binding protein solution (10 μl) was then added and,
after allowing 4 hr for equilibration, aliquots were removed for
the measurement of radioactive complex by the DEAE filter assay.
Using this procedure three parameters were measured: (1) the ratio
of free radioactive hormone to bound radioactive hormone in the
absence of any competitor (R); (2) the concentration of unlabeled
JH I required to inhibit 50% of the radioactive hormone binding
($I_{50,JH}$); and (3) the concentration of other unlabeled competitors
required to inhibit 50% of the radioactive hormone binding ($I_{50,C}$).
From these parameters, the ratio of association constants (RAC) for
the competitor and the radioactive hormone can be calculated using
equation (1).

$$RAC = \frac{K_C}{K_{JH}} = \frac{R \ [I_{50,JH}/I_{50,C}]}{R + 1 - [(I_{50,JH}/I_{50,C})]} \tag{1}$$

Gel Filtration. The lyophilized hemolymph (derived from 1 ml) was reconstituted in 0.2 ml of 0.02 M Tris-HCl, pH 7.3, 0.1 M NaCl, 10^{-4} M 1-phenyl-2-thiourea (PTU), 10^{-3} M DFP and incubated for 12 hr at 4°C. After centrifugation, the supernatant was placed on a 0.9 x 70 cm column of Sephadex G-100 (Pharmacia) equilibrated with the same buffer lacking PTU and DFP. One ml fractions were collected and analyzed for JH binding activity using the charcoal adsorption assay. The most active fractions were also tested for binding of the JH acid metabolite and tripalmitin.

Polyacrylamide Gel Electrophoresis. Electrophoresis of gel filtered hemolymph was carried out in 3.8% (w/v) acrylamide (Eastman) gels at pH 8.4 according to Ornstein (1964) and Davis (1964). For liquid scintillation counting, 2 mm gel slices were eluted with 0.4 ml of 2% sodium dodecyl sulfate (Pierce) at 37°C.

Determination of Molecular Weight. Calibrated gel filtration was carried out according to Whitaker (1963) and Fish et al.(1969) using the following standards: Blue dextran 2000, bovine serum albumin (6.8 x 10^4 , Sigma), ovalbumin (4.3 x 10^4 , Worthington), pepsin (3.5 x 10^4 , Worthington), chymotrypsinogen (2.5 x 10^4 , Worthington) and α-lactalbumin (1.5 x 10^4 , Worthington).

RESULTS

Binding Protein Assays

In order to monitor the purification and characterization of JH binding proteins, we have developed two binding assays. First, a semiquantitative charcoal binding assay carried out at pH 7.3 and second, a quantitative ion-exchange filter disc assay performed at pH 8.3. Both use radiolabeled JH I as ligand. The first exploited the fact that activated carbon will adsorb free JH from aqueous solution and leave protein-bound hormone in the supernatant. The relatively long exposure of the complex to charcoal, however, reduced the accuracy of this assay. The second method is more rapid and quantitative but was developed specifically for the M. sexta carrier protein and may not be suitable for all JH binding proteins. Proteins with higher isoelectric points than that of the Manduca carrier protein may not adsorb to DEAE filter discs under the conditions used in our procedure. Therefore, we have used the DEAE filter disc assay for experiments with the Manduca JH-

carrier protein and the pI independent, semiquantitative charcoal
assay for the initial characterization of binding proteins from
other insect species.

Specificity of the M. sexta JH Carrier Protein

The determinants of the specificity for hormone binding to
the M. sexta JH carrier protein have been studied using a competition assay. In this assay, hormones or analogs are allowed to
compete with a constant concentration of tritiated JH I for a
constant number of binding sites, and the amount of bound labeled
hormone plotted against the log of the competitor concentration.
Three typical competition curves are shown in Fig. 1.

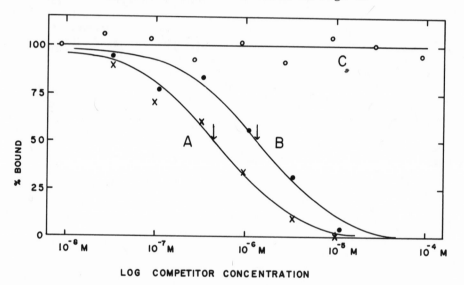

LOG COMPETITOR CONCENTRATION

Fig. 1.   Competition curves for JH homologs and a metabolite.
The curves drawn were calculated from theory: A.   JH I (x);
B.   JH III (o);   C.   JH III acid (o).

Curve A is a competition of unlabeled JH I with the same species
labeled with tritium. Curve B was obtained when JH III was used
as a competitor. The arrows indicate the concentrations of cold
competitor that gave 50% inhibition of labeled hormone binding.
An analysis of the shape of these competition curves using a
linear transformation demonstrated that the competitions had
reached equilibrium and that the test substances are in fact competitive inhibitors of labeled hormone binding. Curve C was
obtained when JH acid was used as a competitor and exemplifies the
data obtained when the compound tested has no affinity for the

hormone binding site. Using equation (1), the ratio of association constants for the cold competitor and the labeled hormone (K_C/K_{JH}) can be calculated. Table 1 summarizes the results obtained when homologs, metabolites, and analogs of JH I were used as competitors.

TABLE 1. SPECIFICITY OF THE M. SEXTA JH-CARRIER PROTEIN.

Compound	RAC[a]	Solubility M x 10^6
JH 0	2.6	17
JH I	1.0	34
JH II	0.36	100
JH III	0.30	>200
JH III diol	<.001	>300
JH III acid	<.001	>120
t,t-Methyl farnesenate	<.01	6
ZR-515	<.01	4.5[b]
R-20458	<.001	40

[a] In all incubations the ratio of free to complexed hormone (R) was greater than 8.3.

[b] J.B. Siddall, personal communication.

Each of the JH homologs tested competed for the hormone binding site. Increasing the length of the branches from the main carbon chain enhanced binding. W. Goodman (Gilbert, 1975) has demonstrated that JH 0, a compound that has not yet been found in any insect, has the highest affinity for the hormone binding site so far observed. We subsequently confirmed this finding and determined an affinity constant of more than 8.6×10^6 M^{-1} for this compound. Neither of the hormone metabolites, JH III acid nor JH diol, competed for the binding site when used as competitors ($K_a < 10^3$ M^{-1}).

Likewise, the hormone analogs, <u>trans,trans</u>-methyl farnesenate, ZR-515 and R-20458 did not compete with labeled hormone for the binding site.

Since a compound may show no competition because it is not soluble at the concentration used, the solubilities of the unlabeled competitors in 5 mM Tris-HCl (pH 8.3) have been measured by spectrophotometry. A plot of the data obtained with JH I is shown in Fig. 2. The measured solubilities of the unlabeled competitors are listed in the far right column of Table 1. The limiting solubilities of three of the competitors had not been reached at the highest concentration tested. The solubility determined for JH I is in good agreement with that reported earlier (Kramer et al., 1974) in a different aqueous buffer. It is clear that the juvenile hormones from true aqueous solutions at concentrations in excess of those found <u>in vivo</u> in insect hemolymph.

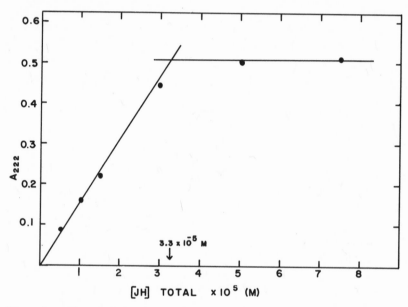

Fig. 2.   The solubility of JH I in 5 mM Tris-HCl, pH 8.3.

Using the solubility data obtained in this manner, upper limits have been established for the association constants of compounds that showed no competition by the method employed. Limits of 0.01 for the RAC's of <u>trans,trans</u>-methyl farnesenate and ZR-515, which have low solubilities, were calculated on the basis that saturated solutions of these compounds showed no competition. The limit of 0.001 for the RAC's of the diol and acid metabolites

and for R-20458 was calculated from the finding that no competition occurred with the highest concentration of competitor tested,
which was below the limiting solubility of the compound. As a
result of negative deviations inherent in our data treatment which
occur when the competitor has a higher affinity for the binding
site than the compound used as tracer (Rodbard and Lewald, 1970),
the RAC calculated for JH 0 may be lower than the true value.

Presence of JH Binding Proteins in Other Insects

A survey for JH binding proteins in the DFP treated larval
hemolymph of seven other insect species using the techniques of gel
filtration and the charcoal binding assay has revealed two general
profiles of hormone binding (Fig. 3). The top profile resembled
that obtained with the hemolymph of M. sexta (Kramer et al., 1974).
It consisted of two peaks with hormone binding activity, the first
with an elution volume corresponding to a polypeptide of molecular
weight $>10^5$ and the second eluting with a molecular weight of
3×10^4. This profile was obtained with hemolymph of the navel
orange worm, Paramyelois transitella (Fig. 3A). Other species
that gave similar profiles were the lepidopterans, Plodia interpunctella (Indian meal moth), Caudra cautella (almond moth),
Hyloicus chersis (chersis moth) and the coleopteran, Tenebroides
mauritanicus (cadelle beetle). The second profile consisted of
three fractions with hormone binding activity. In this case, the
three peaks eluted with apparent molecular weights of $>10^5$, 7×10^4
and 3×10^4, respectively. This profile was obtained with hemolymph of the coleopteran, Tenebrio obscurus (dark mealworm,
Fig. 3B) and the hemipteran, Oncopeltus fasciatus [large milkweed
bug, S. Bassi (personal communication)]. In all the Lepidoptera
examined, the low molecular weight fraction bound more hormone
under the conditions of the assay then did the high molecular
weight fraction. With the Coleoptera and Hemiptera, however,
the higher molecular weight components bound more hormone.

Specificity of the JH Binding Proteins of Other Species

The ability of the components in each of the most active
fractions from DFP-treated and gel-filtered hemolymph of some of
the above insects to bind lipophilic compounds is compared in
Table 2. All of the fractions bound JH to some extent at $1-5 \times$
10^{-7} M initial hormone concentration. Only the lowest molecular
weight fractions, however, displayed specificity for the hormone.
All higher molecular weight fractions also formed complexes with
the acid metabolite and a triglyceride found in the hemolymph,
tripalmitin (Chang and Friedman, 1971).

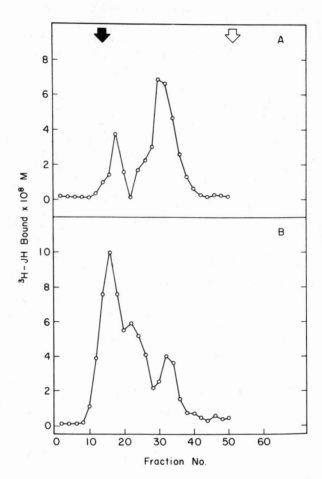

Fig. 3. Gel filtration patterns of insect hemolymph. 0.2 ml of
five times concentrated larval DFP-treated hemolymph was chromatographed on Sephadex G-100 (0.9 x 70 cm) in 0.02 M Tris buffer,
0.1 N NaCl, pH 7.3. One ml per fraction. JH binding was measured
by the charcoal adsorption assay using 5×10^{-7} M initial hormone
concentration. The solid arrow at the left of the figure indicates
the void volume of the column and the open arrow at the right, the
inclusion volume. A. Paramyelois transitella; B. Tenebrio
obscurus.

TABLE 2. BINDING OF JH, JH ACID AND TRIPALMITIN BY SEPHADEX G-100 FRACTIONS OF LARVAL HEMOLYMPH.

Radioactivity Bound in Supernatant (%)[a]

Compound	Manduca		Paramyelois		Plodia		Caudra		Hyloicus		Tenebroides		Tenebrio		
	F20	F31	F19	F30	F20	F31	F19	F31	F20	F32	F20	F32	F20	F25	F33
JH I	3	82	28	75	8	83	1	36	7	52	65	68	73	56	48
JH I acid	8	<1	17	<1	12	<1	3	<1	5	<1	62	<1	32	18	<1
Tripalmitin	15	<1	18	<1	24	<1	14	<1	26	<1	88	<1	78	53	<1

[a] Determined by charcoal assay on fractions preincubated with 10^{-3} M DFP (12 hr, 4°C). JH I = $1\text{-}5 \times 10^{-7}$ M, JH I acid and tripalmitin = $1\text{-}5 \times 10^{-6}$ M.

Electrophoretic Mobility of JH Binding Proteins

In order to assess the number of hormone binding components in the various molecular weight fractions, the peak tube from each fraction was mixed with labeled JH I and subjected to electrophoresis in 3.8% polyacrylamide gels at pH 8.4. In the case of the 3×10^4 molecular weight fraction, at least 80% of the hormone binding was due to a single electrophoretic component in all species. The mobility of this component varied with the species and ranged from 0.85 for H. cheris to 0.40 for P. transitella. Thus, these binding proteins, all having similar molecular weights, differ in overall charge at pH 8.4. The higher molecular weight fractions, on the other hand, were heterogeneous and displayed from three to five JH-binding components after electrophoresis depending on the species tested.

DISCUSSION

The existence of proteins in hemolymph of several insects species that bind JH both in vivo and in vitro is well established. Early reports described the binding of JH and JH analogs to high molecular weight lipoproteins in the hemolymph of pupae and adults of saturniid moths, Hyalophora gloveri, H. cecropia and Antheraea polyphemus (Whitmore and Gilbert, 1972), larvae of the yellow mealworm, Tenebrio molitor (Trautmann, 1972), and adult females of the locust, Locusta migratoria (Emmerich and Hartmann, 1973). However, a lower molecular weight binding protein found in larvae of M. sexta binds JH with a higher affinity than the lipoproteins also present in the hemolymph of this insect (Kramer et al., 1974). A similar protein has been described in the hemolymph of the larvae of the Indian meal moth, P. interpunctella (Ferkovich et al., 1975).

Since our first report of the existence of the high affinity binding protein in M. sexta, this protein has been purified and characterized and its role in the protection of JH from hydrolysis by nonspecific carboxylesterases was demonstrated (Sanburg et al., 1975a). A rapid and quantitative DEAE filter paper assay for the binding of JH has been developed for the M. sexta carrier protein. The availability of this assay and homogeneous protein has made possible a quantitative study of the determinants of specificity for hormone binding. The results demonstrate an absolute requirement for the presence of both the ester and epoxide functions and indicate that specific contacts occur between these functions and hydrophilic groups in the protein binding site. These interactions contribute significantly to the free energy of association of the complex since elimination of either abolishes binding. In addition, the enhancement in binding that occurs with an increase in the length of the branches from the main carbon chain points to a

participation of hydrophobic residues in the binding site. The
specificity, together with the affinity of the M. sexta carrier
protein for these essential structural elements of the natural
hormones, is in full agreement with the physiological role of the
protein in the protection and transport of the hormone.

The lack of binding with the insect growth regulators is not
surprising in view of the observed specificity of binding for the
epoxide ring and ester function. Since these biologically active
compounds are either resistant to rapid degradation by hemolymph
esterases (Weirich and Wren, 1973; Yu and Terriere, 1975) or are
devoid of an ester function entirely, they do not need a protective
carrier. In order to interpret the data for the specificity experiments, it was necessary to establish that all compounds used as
ligands were soluble in the assay buffer at the concentrations
used. The solubilities measured for the JH homologs increased
with decreasing length of aliphatic side chains. The smallest
member of the series, JH III, is at least ten times more soluble
than JH I.

Our experience with the Manduca carrier protein and esterases
enabled us to develop methods[*] for detecting similar binding proteins
in the hemolymph of other insects that may not have been observed
using previously reported techniques. We were able to identify
these binding proteins and examine their properties only after
separating them from other components in hemolymph also capable
of interacting with JH. As a further precaution, hemolymph was
treated with DFP before gel filtration to inhibit esterolytic
enzymes. If such treatments are not utilized, preincubation of JH
with a sample contaminated with degrading enzymes might erroneously
indicate the absence of binding proteins. Even in DFP treated
hemolymph, binding proteins may not be observed if they co-elute
with DFP-insensitive hydrolases or if their titer falls to very
low levels at particular times in the life cycle. Using methods
similar to ours, S. Mumby and B. Hammock (personal communication)
have detected JH binding proteins of 3×10^4 molecular weight in
the larval hemolymph of five additional species of Lepidoptera:
Spodoptera exigua (beet armyworm), Estigmene acraea (salt-marsh
caterpillar), Heliothis virescens (tobacco budworm), Trichoplusia
ni (cabbage looper) and Pectinophora gossypiella (pink bollworm).
Therefore, proteins of 3×10^4 molecular weight which bind JH have
been demonstrated in the larval hemolymph of thirteen insect species
from three different orders.

---

[*] In principle similar methods could be used to demonstrate proteins
that bind other insect hormones. Attempts to date using these
methods, however, have failed to detect the presence of an ecdysterone binding macromolecule in larval hemolymph of M. sexta.

In addition to demonstrating the presence of binding proteins of 3×10^4 molecular weight in several insect species, we have shown that these proteins display a specificity for JH. Other components of higher molecular weight were also observed which bind JH in vitro. In some species, the contribution of the high molecular weight components to hormone binding appeared to be more important. However, it remains a possibility that the lower molecular weight carrier protein had fallen to a low titer at the particular stage we chose to examine. The importance of the high molecular weight proteins as JH carriers is difficult to assess since they are heterogeneous and do not display true specificity for JH. They were also able to bind a hormone metabolite and a triglyceride. It is not clear whether these high molecular weight proteins or those reported by others that bind JH in vivo (Whitmore and Gilbert, 1972; Emmerich and Hartmann, 1973) have a physiological role in JH transport. They may simply bind JH nonspecifically when the low molecular weight carriers are saturated or absent.

SUMMARY

The determinants of the specificity for juvenile hormone (JH) binding to the carrier protein from Manduca sexta have been studied using a competition assay with [^{3}H]-JH I and a series of soluble homologs, analogs and related compounds. In this assay, hormone-protein complex formation is measured by adsorbing the complex to ion-exchange filter paper under conditions in which the free hormone is not adsorbed. The ester and epoxide functions, as well as the alkyl substituents on the main carbon chain, were found to be important for binding. Using a charcoal adsorption assay, binding proteins of similar molecular weight exhibiting specificity for JH have been detected in diisopropylphosphorofluoridate (DFP) - treated and gel filtered hemolymph from other species of Lepidoptera, Coleoptera and Hemiptera.

ACKNOWLEDGEMENTS

We thank H. Seballos, C. Childs, M. Cohn, D. Dill, and J. Ong for their assistance. This research was supported in part by National Science Foundation Grant BMS-74-21379 and National Institubes of General Medical Sciences Grant GM 13863.

REFERENCES

Akamatsu, Y., Dunn, P.E., Kézdy, F.J., Kramer, K.J., Law, J.H., Reibstein, D., and Sanburg, L.L., 1975, in "Control Mechanisms and Development" (R. Meints and E. Davies, eds.), pp. 123-149, Plenum Press, New York.

Chang, F., and Friedman, S., 1971, Insect Biochem. 1:63.

Davis, B.J., 1964, Ann. N.Y. Acad. Sci., 121:404.

Emmerich, H., and Hartmann, R., 1973, J. Insect Physiol. 19:1663.

Ferkovich, S.M., Silhacek, D.L., and Rutter, R.R., 1975, Insect Biochem. 5:141.

Fish, W.W., Mann, K.G., and Tanford, C.J., 1969, J. Biol. Chem. 244:4989.

Gilbert, L.I., 1975, Amer. Chem. Soc. 170th Meeting, Pesticide Div., Abst., No. 32.

Hammock, B., Nowock, J., Goodman, W., Stamoudis, V., and Gilbert, L.I., 1975, Molec. Cell. Endocrinol. 3:167.

Korenman, S.G., 1969, Steroids 13:163.

Korenman, S.G., 1970, Endocrinology 87:1119.

Kramer, K.J., Sanburg, L.L., Kézdy, F.J., and Law, J.H., 1974, Proc. Nat. Acad. Sci. USA 71:493.

Kramer, K.J., Dunn, P.E., Peterson, R.C., Seballos, H., Sanburg, L.L., and Law, J.H., 1976, submitted to J. Biol. Chem.

Law, J.H., Yuan, C., and Williams, C.M., 1966, Proc. Nat. Acad. Sci. USA 55:576.

Metzler, M., Meyer, D., Dahm, K.H., and Röller, H., 1972, Z. Naturforsch. 27B:321.

Ornstein, L., 1964, Ann. N.Y. Acad. Sci. 121:321.

Rodbard, D., and Lewald, J.E., 1970, Acta Endocrinol. Suppl. 147:79.

Sanburg, L.L., Kramer, K.J., Kézdy, F.J., and Law, J.H., 1975a, J. Insect Physiol. 21:873.

Sanburg, L.L., Kramer, K.J., Kézdy, F.J., Law, J.H., and Oberlander, H., 1975b, Nature 253:266.

Trautmann, K.H., 1972, Z. Naturforsch. 27B:263.

Weirich, G., and Wren, J., 1973, Life Sci. 13:213.

Whitaker, J.R., 1963, Anal. Chem. 35:1950.

Whitmore, E., and Gilbert, L.I., 1972, J. Insect Physiol. 18:1153.

Yu, S.J., and Terriere, L.C., 1975, Pest. Biochem. Physiol. 5:418.

THE BINDING OF JUVENILE HORMONE TO LARVAL EPIDERMIS: INFLUENCE OF

CARRIER PROTEIN FROM THE HEMOLYMPH OF PLODIA INTERPUNCTELLA

S.M. Ferkovich, D.L. Silhacek, and R.R. Rutter

Insect Attractants, Behavior, and Basic Biology Research
Laboratory, ARS, USDA, Gainesville, Florida 32604

INTRODUCTION

The role of juvenile hormone (JH) in growth and development
of insects has been discussed in several recent reviews (Menn and
Beroza, 1972; Burdette, 1974; Sláma et al., 1974). The hormone is
secreted by the corpora allata and is transported by the hemolymph
to target tissues. Lipoproteins and low molecular weight proteins
in the hemolymph have been implicated as carriers for JH (Trautmann,
1972; Whitmore and Gilbert, 1972; Emmerich and Hartmann, 1973;
Kramer et al., 1974a; Ferkovich et al., 1975). The low molecular
weight proteins which bind JH and have been isolated from both
Manduca sexta (Sanburg et al., 1975a,b) and Plodia interpunctella[*]
protect JH against nonspecific esterases in fourth- and early fifth-
instar larvae. JH specific esterases that degrade the protein-
bound JH also appear in the hemolymph during the late fifth instar
and may provide a mechanism for removing JH from the hemolymph to
permit metamorphosis.

In addition to its protective role in the hemolymph, JH carrier
protein may be involved in the interaction of JH with receptor sites
within the cells of target tissues. Although it has been postulated
that JH acts at the nuclear (genetic) (Sláma et al., 1974) or cytoplasmic level (Firstenberg and Silhacek, 1973), the primary site(s)
of action has not been identified.

---

[*] Kramer, K.J., USDA, ARS, Grain Market Research Center, 1515
College Ave., Manhattan, Kansas 66502, personal communication.

Sláma et al. (1974) have hypothesized that JH binds to receptors in target tissues that mediate characteristic developmental
responses in insects and "recognize" the hormone on the basis of
its size, shape, and physiochemical properties. Only two studies
directed toward the isolation of receptors for JH have been reported. Schmialek (1973) used a radiolabeled analogue of JH to isolate a Triton X-100 solubilized ribonucleoprotein receptor from
pupal epidermis of Tenebrio molitor. Also, Ferkovich et al.
(1974) examined the binding of JH to subcellular fractions of
whole larvae of P. interpunctella and showed that the hormone was
primarily bound by protein in fractions containing membrane fragments and RNA; however, binding by degradative enzymes was not
ruled out.

We have examined the effects of carrier protein fraction
(CPF) on JH binding to epidermal homogenates from larvae of the
Indian meal moth, P. interpunctella. We consider what role
carrier protein may play in protecting JH from degradation by
subcellular fractions of epidermis, and whether CPF is important
in recognition of JH by subcellular binding sites.

MATERIALS AND METHODS

The carrier protein fraction (CPF) was isolated from the
hemolymph of mid-fifth-instar larvae (12 mg/larva) by gel permeation chromatography as described by Ferkovich et al. (1975).
The JH binding in the fractions was measured by separating free
JH from bound JH by charcoal absorption (Sanburg et al., 1975a).
Esterases were monitored by measuring the 1-napthylacetate hydrolytic activity (Sanburg et al., 1975a) or by measuring metabolites
of [^{3}H]-JH by thin-layer chromatography (Slade and Zibitt, 1972).
A Sephadex G-100 column (1.5 cm X 90 cm long column at a 27 ml/hr
flow rate) separated the majority of the carrier protein from the
JH esterases.

Epidermal homogenates were prepared from mid-fifth instar
larvae (12 mg/larva) in a Teflon-pestle homogenizer (clearance
0.15-0.22 mm; 10 full strokes at 2500 rpm). The CPF or bovine
serum albumin (BSA) (less than 0.005% fatty acids, Sigma) was
incubated with [^{3}H]-JH (Hyalophora cecropia C_{18} JH [7-ethyl-1,2-
^{3}H(N)] 14.1 Ci/mM; New England Nuclear Corp.) for 15 min at room
temperature before incubating with the homogenized epidermis for
30 min. The homogenates were then fractionated according to the
scheme presented in Fig. 1.

The sucrose density gradient fractions were not homogeneous.
However, the predominant organelles contained in each fraction
were: F_1 and F_2, membrane vesicles; F_3, fragments of rough ER,

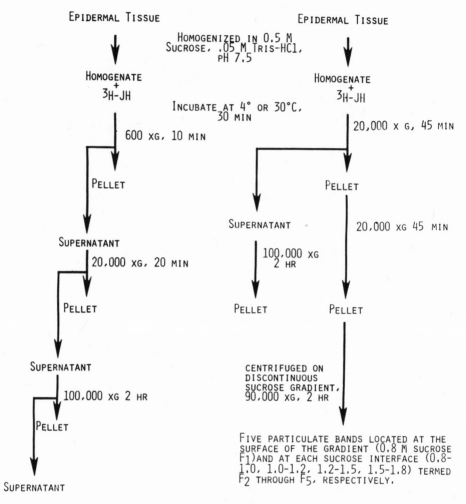

Fig. 1. Procedure for the centrifugation methods used for subcellular fractionation of epidermal tissue incubated with [^{3}H]-JH.

small mitochondria; F_4 rough ER, large mitochondria, lysosomes; F_5, nuclear material, membrane fragments (Ferkovich et al., 1975). To facilitate discussion of JH binding in the particulate fractions, the data obtained on each was combined and is reported as follows: MV, membrane vesicles (F_1 + F_2), M, mitochondrial (F_3 + F_4); and N, nuclear (F_5). The 20,000 X g supernatant is referred to as

M + S, the microsomal + supernatant fraction.

Specific binding in the subcellular fractions was determined according to King and Mainwaring (1974). Homogenized epidermis and [3H]-JH were incubated at 4°C for 15 min. An unlabeled mixture of H. cecropia JH isomers (Hoffmann-LaRoche) was then added to half of the homogenate and both portions were incubated at the appropriate temperature for an additional 15 min.

The protein content of the fractions was determined by a modification of the Lowry method described by Schacterle et al. (1973); and the radioactivity was determined by scintillation counting (Ferkovich et al., 1975).

In certain experiments the epidermal homogenates were incubated with RNase, 500 μg/ml, or DNase (Worthington Biochem. Corp.), 500 μg/ml + 0.06% MgSO$_4$ at 37°C for 30 min prior to incubation with radioactive JH.

RESULTS

Influence of CPF on Binding and Degradation of JH

The radioactivity/μg protein and the percentage of unmetabolized [3H]-JH in fractions obtained from homogenates of epidermis incubated with and without the CPF is shown in Fig. 2 (Ferkovich and Rutter, 1975b). The relative order of radioactivity/μg protein was nuclear (N) < mitochondrial (M) < membrane vesicles (MV) < microsomal + supernatant fractions (M + S). The majority of the label recovered (>94%) remained in the 20,000 X g supernatant that contained the microsomes and cytosol proteins. With the CPF [3H]-JH binding was diminished in all the particulate fractions, probably because of the solubilizing effect of CPF on JH in the supernatant.

In the nuclear and mitochondrial fractions there was more degradation of [3H]-JH in the presence of the CPF; whereas, in the membrane vesicle and microsomal + supernatant fractions the hormone was protected (Table 1). Analyses of the [3H]-JH bound in the nuclear and mitochondrial fractions indicated that the addition of CPF increased the percentage of all the metabolites (Table 2). The diol acid metabolite was highest in the membrane vesicle and microsomal + supernatant fractions and CPF had the most striking protective effect in these fractions.

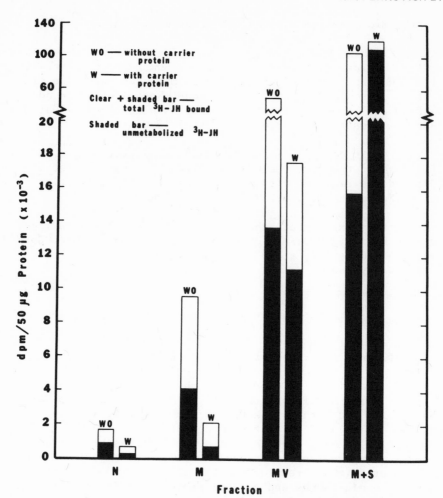

Fig. 2. Subcellular distribution of bound radioactivity after
incubation of homogenized epidermis with and without the CPF.
Epidermal homogenates (0.12 g wet wt/ml; 150 μg protein 0.1 ml)
were incubated with [^{3}H]-JH (8 x 10^{-8}M) with and without the CPF
400 μg/ml at 4°C for 30 min. The fractions were obtained by
sucrose gradient centrifugation of the 20,000 X g pellet (Ferkovich
and Rutter, 1975b).

Comparisons of CPF and BSA on Binding and Degradation of JH

    The binding of JH by BSA has been demonstrated by using UV
difference spectroscopy (Ferkovich et al., 1974) and more recently,

TABLE 1.   PERCENTAGE OF BOUND [^{3}H]-JH METABOLIZED IN THE SUBCELL-
ULAR FRACTIONS OF HOMOGENIZED EPIDERMIS INCUBATED WITH [^{3}H]-JH
IN THE PRESENCE OF CPF.[a]

Subcellular fraction	% Metabolized JH	
	Without CPF	With CPF
Nuclear	44	63
Mitochondrial	56	66
Membrane vesicles	75	35
Microsomal + supernatant	85	8

[a] Epidermal homogenates (0.12 g wet wt/ml; 150 µg protein/0.1 ml)
were incubated with [^{3}H]-JH (8 x 10^{-8} M) plus 400 µg/ml of CPF at
4°C for 30 min.

fluorescence spectroscopy.[*]   We also compared the relative effectiveness of BSA and CPF on binding and degradation of JH in subcellular fractions obtained by differential centrifugation of
epidermal homogenates.

In the presence of BSA or CPF the nuclear and microsomal +
supernatant fractions contained about 4 X 10^{10} molecules of JH/µg
of protein; the mitochondrial fraction contained about 10^{10} molecules of JH/µg of protein.   In the presence of BSA (nuclear and
mitochondrial) fractions of epidermal homogenates broke down 49%
of the JH compared with 90% by the microsomal + supernatant fractions during a 30 min incubation at 4°C.   Substitution of CPF for
BSA resulted in a 65% (nuclear and mitochondrial) to 10% (microsomal + supernatant) breakdown of JH during a similar incubation.
These data indicate that binding of JH to protein does not in
itself afford protection of the JH molecule in the microsomal +
supernatant fraction, but that other properties of the CPF confer
the protective effect.

[*]   Mayer, R., USDA, ARS, P.O. Drawer GE, College Station, Texas
77840, personal communication.

TABLE 2. PERCENTAGE OF JH DEGRADED IN SUBCELLULAR FRACTIONS OF
HOMOGENIZED EPIDERMIS WITH AND WITHOUT CPF.

	% JH or metabolite without CPF				% JH or metabolite with CPF			
Fraction	JH	Diol ester	Epoxy acid	Diol acid	JH	Diol ester	Epoxy acid	Diol acid
Nuclear	61.2	16.1	10.8	11.8	42.0	19.4	15.9	22.7
Mitochondrial	50.0	11.9	15.3	22.7	41.2	18.2	15.8	24.9
Membrane vesicles	29.7	12.9	15.7	41.7	69.1	5.5	5.2	20.0
Microsomal + supernatant	15.2	5.1	4.0	75.8	92.9	–	–	7.1

Determination of High-Affinity Binding

To further characterize the nature of the binding in the
particulate fractions, homogenates labeled with [^{3}H]-JH were
treated with a 1000-fold excess of unlabeled JH. In essence, JH
bound to low-affinity sites should be displaced by the unlabeled
JH whereas JH bound to high-affinity sites should not be displaced
(King and Mainwaring, 1974). Interpretation of the data summarized in Table 3 was complicated by the breakdown of JH simultaneous with its binding to the subcellular components in the homogenate. The excess unlabeled JH decreased the radioactivity in
the membrane vesicles fraction indicating low-affinity binding.
However, the expected increase in radiolabel in the supernatant
did not occur; rather, radiolabel increased in the nuclear, mitochondrial, and microsomal fractions. Our interpretation of these
data is that rapid binding of labeled JH to both low- and high-
affinity sites occurred initially, but that the amounts of JH
bound decreased during the remainder of the 30 min incubation due
to breakdown of the hormone. The addition of excess unlabeled JH
at 15 min displaced labeled hormone from low-affinity sites and
saturated the degradative enzymes. The net result of adding
unlabeled JH would be to decrease the radiolabel in those fractions

TABLE 3. SUBCELLULAR DISTRIBUTION OF [³H]-JH AFTER INCUBATION OF
HOMOGENIZED LARVAL EPIDERMIS FOR 30 MIN AT 4°C WITH AND WITHOUT
THE ADDITION OF EXCESS UNLABELED JH AT 15 MIN.

| Fraction[a] | % [³H]-JH bound/50 µg protein per 30 min | |
	No addition[b]	Unlabeled JH added[c]
Nuclear	0.6	2.3
Mitochondrial	3.2	9.9
Membrane vesicles	71.0	56.0
Microsomes	4.1	10.9
Supernatant[d]	21.1	20.7

[a] Particulate fractions (F_1-F_6) were obtained after discontinuous
sucrose gradient centrifugation of the 20,000 X g pellet.

[b] Homogenized epidermis was incubated with [³H]-JH (3.36×10^{-8} M)
for 30 min at 4°C.

[c] Homogenized epidermis was incubated with [³H]-JH (3.35×10^{-8} M)
for 15 min at 4°C then with unlabeled JH (3.36×10^{-4} M) for an
additional 15 min at the same temperature.

[d] Free JH was not separated from bound hormone.

having low-affinity sites and increase the radiolabel in those
fractions having high-affinity sites. On this basis, we have
tentatively concluded that the nuclear, mitochondrial, and microsomal fractions contain high-affinity binding sites. However,
these experiments are certainly not conclusive and we are currently
investigating this aspect further.

Influence of RNase and DNase on [³H]-JH Binding

Epidermal homogenates were treated with RNase and DNase prior
to incubation with [³H]-JH to determine if intact nucleic acids
were involved in the binding of hormone. RNase and DNase catalyze
the fragmentation of RNA and DNA (to nucleotides, nucleosides, and

and possibly nucleoprotein particles in the case of DNase) (Mahler and Cordes, 1966; Oosterhof et al., 1975). The effect of these treatments on the distribution of protein and isotope in subcellular fractions of epidermis are summarized in Fig. 3. Neither enzyme was effective in converting nuclear material into small soluble particles as evidenced by the lack of increase in protein and radioactivity in the microsomal + supernatant fraction. However, both enzymes fragmented nuclear material (as shown by the decreased protein in the nuclear fraction) into particles that sedimented primarily with the mitochondrial fraction and to a lesser extent with the membrane vesicle fraction. Fragmentation of material in the mitochondrial fraction by RNase is possible but could not be discerned because of the particles formed by the fragmentation of material in the nuclear fraction.

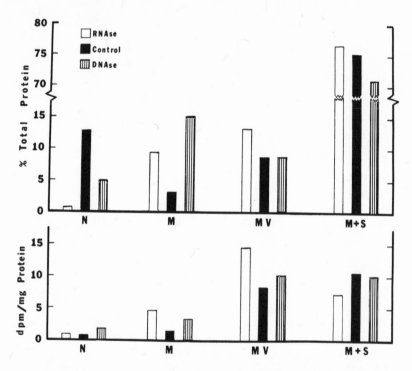

Fig. 3. Effect of incubation with RNase and DNase on [³H]-JH binding in epidermal homogenates. Homogenized epidermis (0.09 g wet wt/ml) treated with RNase and DNase (see Materials and Methods) was incubated with 2.8 X 10⁻⁸ M [³H]-JH for 30 min at 4°C. N, nuclear; M, mitochondrial; MV, membrane vesicles, and M+S, microsomal + supernatant. Values for dpm/mg protein were calculated on a relative basis.

The nuclear fragments which resulted from RNase incubation and which sedimented with the mitochondrial and membrane vesicle fractions were extremely effective in binding [^{3}H]-JH as shown by the increase in radioactivity/mg protein (calculated on a relative basis) (Fig. 3). On the other hand, DNase did not have such a pronounced effect and, in fact, the loss of "DNase fragments" increased the [^{3}H]-JH/mg protein in the remaining nuclear fraction. We conclude that DNase has some effect on the JH binding sites in the epidermal tissue, but that RNase has a more profound effect on both the size of the particle containing the binding sites and on the availability of the sites to bind JH.

DISCUSSION

Evidence has been presented that the carrier protein in Manduca sexta (Sanburg et al., 1975a,b) and P. interpunctella[*] functions in the transport and protection of JH from degradation by enzymes in the hemolymph. The protection of JH by the carrier protein has been extended to other tissues in Manduca sexta (Hammock et al., 1975). Our data reported here indicate that epidermal target tissue contains similar enzymes which degrade JH and that the carrier protein influences such catabolism in vitro. However, we do not know whether such protection also occurs in vivo because it has not been demonstrated that the carrier protein-JH complex penetrates the target cell membrane.

JH was degraded most rapidly in the microsomal + supernatant, and membrane vesicle fractions and slowest in the mitochondrial and nuclear fractions. Addition of CPF largely protected JH from degradation in the cytosol, microsomal, and membrane vesicle fractions but not in the mitochondrial and nuclear fractions. In fact, more hormone was catabolized in the presence of the CPF in the latter two fractions. Our interpretation of these observations is that the fractions contained high-affinity binding sites for JH (possibly receptors) which selectively removed JH from the protective influence of carrier protein. This permitted the hormone access to degradative enzymes in the presence of the carrier protein. On the other hand, the microsomal plus cytosol fractions may not have contained high-affinity binding sites which allowed the carrier protein to exert its protective influence. The BSA binding experiment which demonstrated that general protein binding did not protect JH supports the speculation that high-affinity sites could remove JH from carrier protein but not protect the hormone. However, the possibility that the nuclear and mitochondrial fractions contained enzymes comparable to the carrier-bound-JH esterases in the hemolymph of Manduca sexta (Sanburg et al., 1975) has not been ruled out.

---

[*] Kramer, K.J., USDA, ARS, Grain Market Research Center, 1515 College Avenue, Manhattan Kansas 66502, personal communication.

The final step in the processing of JH is the transfer of the hormone from carrier protein to a receptor in the target tissue. This could be achieved by receptor recognition of the carrier protein-JH complex or of JH alone. However, the similar binding patterns in epidermal homogenates with JH alone or with the carrier protein-JH complex indicate that the carrier protein does not participate in subcellular recognition of JH.

Although the carrier protein-JH complex may not be required for subcellular recognition of JH, it may function by protecting the hormone until the high affinity receptors remove JH from the complex. Our binding data disclosed that nuclear, mitochondrial, and perhaps the microsomal fractions may contain such high-affinity binding sites. Furthermore, both isolated nuclei (Sekeris, 1972) and mitochondria (Firstenberg and Silhacek, 1973) have been shown to respond directly to JH by alterations in enzymatic activity. We therefore conclude that at least these two subcellular fractions should contain receptors for JH.

Although the chemical nature of JH receptors is not known, evidence is now accumulating that ribonucleoproteins are either receptors themselves or are closely associated with receptors for JH. This is supported by the work of Schmialek (1973) and by our observation that in the nuclear fractions RNase digestion and, to a lesser degree, DNase digestion caused formation of smaller nucleic acid fragments having a higher JH binding capacity. The specificity of these sites and their possible physiological significance remains to be determined. It is tempting to speculate that JH may bind in much the same way as steroids which bind nuclear and cytoplasmic ribonucleoprotein particles in target cells (Liang and Liao, 1974).

SUMMARY

A juvenile hormone (JH) carrier protein fraction (CPF) isolated from the hemolymph of Plodia interpunctella reduced binding but did not alter the relative binding patterns of JH in fractions of homogenized epidermis separated by differential and sucrose density gradient centrifugation. Bovine serum albumin (BSA) also reduced JH binding to the epidermal fractions. The CPF but not BSA protected the hormone from degradation by epidermal enzymes. Evidence is presented that suggests that high affinity binding sites for JH exist in the nuclear and mitochondrial fractions and that the JH carrier protein does not participate in recognition of the binding sites.

ACKNOWLEDGEMENTS

We thank Hoffmann-LaRoche for providing the unlabeled juvenile hormone and Dorothy Blackwell for typing the manuscript.

REFERENCES

Anderson, J.N., Peck, E.J., and Clark, J.H., 1974, J. Steroid
   Biochem. 5:103.
Burdette, W.J., 1974, "Invertebrate Endocrinology and Hormonal
   Heterophylly," Springer-Verlag, New York.
Emmerich, H., and Hartmann, R., 1973, J. Insect Physiol. 19:1663.
Ferkovich, S.M., Rutter, R.R., and Anthony, D.W., 1974, J. Insect
   Physiol. 20:1943.
Ferkovich, S.M., Silhacek, D.L., and Rutter, R.R., 1975a, Insect
   Biochem. 5:141.
Ferkovich, S.M., and Rutter, R.R., 1975b, Wilhelm Roux' Arch.,
   in press.
Firstenburg, D.E., and Silhacek, D.L., 1973, Experientia 29:1420.
Hammock, B., Nowock, J., Goodman, W., Stamoudis, V., and Gilbert,
   L.I., 1975, Molec. Cell. Endocrinol. 3:167.
King, R.J.B., and Mainwaring, W.I.P., 1974, "Steroid-Cell Inter-
   actions," University Park Press, Baltimore.
Kramer, K.J., Sanburg, L.L., Kézdy, F.J., and Law, J.H., 1974a,
   Proc. Nat. Acad. Sci. USA, 71(2):493.
Liang, T., and Liao, S., 1974, J. Biol. Chem., 249:4671.
Mahler, H.R., and Cordes, E.H., 1966, "Biological Chemistry,"
   Harper and Row, New York.
Menn, J.J., and Beroza, M., 1972, "Insect Juvenils Hormones:
   Chemistry and Action," Academic Press, New York.
Oosterhof, D., Hozier, J.C., and Rill, R.L., 1975, Proc. Nat.
   Acad. Sci. USA, 72(2):633.
Sanburg, L.L., Kramer, K.J., Kézdy, F.J., and Law, J.H., 1975a,
   J. Insect Physiol. 21:873.
Sanburg, L.L., Kramer, K.J., Kézdy, F.J., Law, J.H., and Oberlander,
   H., 1975b, Nature, 253:266.
Schacterle, G., 1973, Anal. Biochem. 51:654.
Schmialek, P., 1973, Nature, 245:267.
Sekeris, C.E., 1972, Gen. Comp. Endocrinol. Suppl. 3:149.
Slade, M., and Zibitt, C.H., 1972, in "Insect Juvenile Hormones"
   (J.J. Menn and M. Beroza, eds.), pp. 155-176, Academic Press,
   New York.
Sláma, K., Romañuk, M., and Šorm, F., 1974, "Insect Hormones and
   Bioanalogues," Springer-Verlag, New York.
Trautmann, K.H., 1972, Z. Naturforsch, 27b:263.
Whitmore, E., and Gilbert, L.I., 1972, J. Insect Physiol. 18:1153.

THE BINDING PROTEIN AS A MODULATOR OF JUVENILE HORMONE STABILITY

AND UPTAKE

J. Nowock, B. Hammock, and L.I. Gilbert

Department of Biological Sciences, Northwestern
University, Evanston, Illinois 60201

INTRODUCTION

Past data on the corpora allata and juvenile hormone (see
Gilbert and King, 1973) suggest a highly intermeshed regulatory
system which must allow for rapid and dramatic changes in hormone
titer as well as fine tuning. An important compartment in this
control system is the hemolymph since it serves as the shuttle
system between the corpora allata and peripheral tissues. This
compartment represents the central mixing point for those factors
determining the juvenile hormone (JH) titer, the latter determining
the hormone concentration at the target tissues. The processes
responsible for the hemolymph pool of JH are: synthesis and release
of the hormone by the corpora allata (e.g. Röller and Dahm, 1970;
Judy et al., 1973; Pratt and Tobe, 1974); under certain conditions
there may be a reflux from tissues; metabolic inactivation; and
tissue uptake with subsequent degradation and/or excretion.

Recent studies have demonstrated that JH binds specifically
to certain hemolymph proteins (Whitmore and Gilbert, 1972; Emmerich
and Hartmann, 1973; Goodman and Gilbert, 1974; Kramer et al., 1974).
By interaction with these proteins, JH can exist in both a bound
and free state, the ratio of which is a function of the total amount
of binding proteins present and their binding constants (see
Goodman et al., Kramer et al., this volume). As a consequence,
this must affect properties of the pool such as metabolic rate,
rate of equilibration with other pools, and pool size. In the
present paper we report on the roles of the high affinity binding
protein in Manduca sexta in the regulation of JH content in hemolymph and other tissues. Since the fat body plays a crucial role
in JH metabolism (e.g. Hammock et al., 1975), we have utilized the
fat body in organ culture as our primary test system.

MATERIALS AND METHODS

Experimental Animals. Manduca sexta larvae were reared on an artificial diet as described previously (Sroka and Gilbert, 1971) under a 16L:8D photoperiod.

Juvenile Hormone. [7-Ethyl-1,2-^{3}H(N)]-JH I (11.17 Ci/mmol) was purchased from New England Nuclear while [^{3}H]methoxy-JH I (isomeric mixture, 4.34 Ci/mmol) was a generous gift from K.H. Trautmann (Dr. Maag AG). Radiolabeled JH was used directly or diluted with cold JH I (95% t,t,c, courtesy of K. Dahm). Stock solutions were prepared in benzene:hexane (4:1) or methanol and concentrations were determined by liquid scintillation spectrometry of [^{3}H]-JH or from absorbence values of methanolic solutions (cold JH) using a molar extinction coefficient ε_{220} = 13,830 l· mol^{-1}·cm^{-1} (Trautmann et al., 1974). Highly concentrated stock solutions of JH (10^{-3} M) were obtained by weighing. [^{3}H]-JH acid was prepared from chain labeled JH I by esterase hydrolysis in fifth instar larval hemolymph.

In Vitro Procedures. Fat body was removed from larvae as two coherent organs, rinsed twice in Grace's medium and finally washed in the medium to be used for incubation. Under short-term conditions ($\leq$2 hr), fat body was cultured in Grace's medium at pH 6.6 (Grand Island Biological Co.) or in a Ringer's solution containing 1.5 mM Na$_2$HPO$_4$, 54.4 mM KCl, 10.5 mM CaCl$_2$ and 60.4 mM MgCl$_2$·6 H$_2$O, adjusted with HCl to pH 6.6. For incubations over longer periods of time a specially devised Manduca culture medium (Nowock et al., 1975) was used and all procedures carried out under sterile conditions.

All incubations involving JH were carried out in glass vials coated with carbowax 20 M (polyethylene glycol; M$_r$ 20,000) or silicone to reduce adsorption of the hormone. JH was transferred from stock solutions into the culture vials and after evaporation of the organic solvent under N$_2$, medium was added. The hormone was resolubilized by sonication, or by preincubation for at least 30 min when the medium contained proteins. Incubations were carried out in a shaking water bath at 26°C unless otherwise stated.

Binding Protein Preparation. Hemolymph from day 3 and day 4 fifth instar larvae was used to prepare a partially purified binding protein fraction (BPF). Two fractionation procedures were employed, the first already described in detail by Hammock et al. (1975) and involving treatment of cell-free hemolymph with diisopropyl fluorophosphate (DFP) to inhibit esterases, followed by filtration through Sephadex G-150 (0.1 M phosphate, pH 7.4). Fractions with binding activity were concentrated and passed through Sephadex G-75 (0.1 M phosphate, pH 7.4). Binding activity in the eluted fractions was determined by charcoal assay using [^{3}H]-JH I as ligand (Hammock

et al., 1975). The active fractions from the G-75 column were
lyophilized and desalted by passage through a Sephadex G-25
column equilibrated with Grace's medium. For in vitro studies,
the binding activity was adjusted by charcoal assay to the activity
of DFP-treated L5D3 hemolymph. (L denotes instar while D indicates
day of instar in the previous notation, i.e. L5D3.)

An improved two-step procedure was used to prepare BPF for
the studies on JH-uptake by fat body (see Goodman et al., this
volume). Cell-free hemolymph was saturated to 70% with $(NH_4)_2SO_4$.
The precipitate was redissolved in 0.05 M phosphate buffer, pH 6.6
containing 0.05 M KCl, dialyzed and passed through Sephadex G-150
(same buffer). The active fractions were concentrated by $(NH_4)_2SO_4$
precipitation, the proteins redissolved and dialyzed against 0.05 M
acetate buffer, pH 5.0 with 0.01 M KCl, and applied to a Whatman
CM 52 column equilibrated with the acetate buffer. BP is uncharged
at pH 5.0 and was therefore not retained by the resin. The active
fractions were concentrated and dialyzed against Ringer's solution.
In this preparation, BP was contaminated by three or four other
proteins as determined by gel electrophoresis and no esterase
activity was detected.

The concentration of binding sites in this preparation was
determined by a saturation procedure (Westphal, 1967) using charcoal as adsorbent to separate bound from free hormone. Optimal
conditions for the charcoal adsorption were found to be the same as
described by Goodman et al. (this volume).

Extraction, Separation and Characterization of Metabolites.
Extraction procedures, separation and identification of JH and
metabolites were described previously (Hammock et al., 1975).
Recovery was >90% of the total radioactivity applied.

Determination of Esterase Activity. The hydrolytic activity
of hemolymph samples and fat body culture media was assayed with
α-naphthyl acetate (α-NA) and JH as substrates to distinguish in
a crude way between "general" and "JH specific" esterases (Sanburg
et al., 1975). α-NA hydrolytic activity was determined spectro-
photometrically (van Asperen, 1962). JH-hydrolytic activity was
assayed by one of the following methods. Using chain-labeled JH
as substrate, quantitation was achieved by extraction of the reaction mixture with ethyl acetate followed by thin layer chromatographic (TLC) separation and subsequent radioassay of the appropriate
zones of the silica gel plates (Weirich et al., 1973). A less
laborious method was developed with $[^3H]$-methoxy-JH I. This assay
is based on the partition of the generated methanol in a two-phase
system (see also Sanburg et al., 1975). Under standard assay
conditions 170 µl of 10^{-5} M JH in 0.05 M phosphate buffer (pH 6.6)
were mixed with 50 µl of appropriately diluted enzyme solution in
carbowaxed test tubes (7 x 20 mm) and incubated in a shaking water

bath at 25°C for 30 min. The reaction was terminated by adding
200 μl methanol and 500 μl chloroform. The mixture was vortexed
and the phases separated by centrifugation at 2,000 g for 5 min.
The phases separate in the ratio 19:27 (v/v; epi-/hypophase).
The methanol (80 ± 2%) appeared in the epiphase. Partition of
unreacted JH into this phase was negligible. The assay showed
proportionality between hydrolysis rate and enzyme concentration
at up to 10% substrate conversion.

Polyacrylamide gel electrophoresis was performed with an
Ortec 4200 system as described (Nowock et al., 1975). Protein was
determined by a modified Lowry procedure (Low et al., 1969).

Determination of JH Uptake. Fat body from L4D2 larvae (0.90-
0.95 g) was used for the uptake studies. Tissues were preincubated
for 30 min at room temperature with two changes of medium to allow
for degradation of endogenous JH. Since endogenous titers were
not measured, all calculations assume that residual hormone levels
after preincubation are negligible. Fat body was finally rinsed
for 5 min in Ringer's solution at incubation temperature. One
fat body strand was placed in a culture vial containing 2 ml
incubation mixture from which 20 μl aliquots were taken at various
time intervals. Uptake into tissue was calculated from the depletion of radiolabel from the medium. Incubation of fat body with
inulin-[carboxyl-^{14}C] demonstrated the absence of extracellular
spaces in which label could distribute and be retained. Initial
uptake rates were estimated by the 1 min uptake data.

Reversibility of JH-uptake was determined by pre-incubating
fat body for 30 min in medium containing JH and reincubating the
tissue in JH-free medium after an intermittant rinse in ice cold
Ringer's for 20 sec. In several cases media and tissues were
extracted with ethyl acetate at the end of the incubation period
and the extent of JH degradation was determined (Hammock et al.,
1975).

RESULTS

Effect of BP on JH Metabolism

Distribution of JH and Metabolites in Fat Body and Medium.
To glean preliminary information on the uptake and metabolism of
JH and the possible role of binding protein in these processes,
fat body from fourth and fifth instar larvae were incubated with
[^{3}H]-JH in the presence or absence of BPF. BPF was replaced by an
equal quantity of bovine serum albumin (crystalline BSA; Miles,
Inc.) in the control samples since BSA binds JH only weakly and

unspecifically. The distribution of total label as well as JH and
metabolites was monitored as a function of time. The data reveal
that in the absence of BPF, label is rapidly taken up by the tissues
(Fig. 1). Differences exhibited between the two developmental
stages presumably reflect differences in the amount of tissue. The
cultured fat body metabolized JH very efficiently with fifth instar
tissue exhibiting the higher rate of degradation. However, at both
developmental stages a fairly constant level of JH was not degraded
during the incubation period indicating that some JH is retained in
compartments not accessible to degradative enzymes. In the presence
of BPF both the uptake of JH by the tissues and its subsequent metabolism were remarkably reduced.

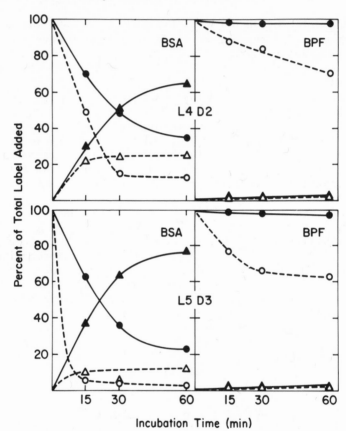

Fig. 1. Time course analysis of the distribution of JH and metabolites between fat body and culture medium. Fat body from L4D2 or
L5D3 larvae was incubated in Grace's medium, pH 6.6 containing 10^{-8} M
[^{3}H]-JH and BPF or BSA. Media and tissue homogenates were extracted
and analyzed by TLC for JH and metabolites. Curves represent content
of total [^{3}H] in medium (●—●) or fat body (▲—▲) and [^{3}H] as JH in
medium (o--o) or fat body (Δ--Δ) expressed as percent of total label
added.

To determine the nature of the JH metabolites, extracts of
media and tissue homogenates were analyzed by TLC both before and
after derivatization. The major metabolites noted were the acid
and diol acid while diol was only found in small amounts. Trace
amounts of polar metabolites remained at the origin under the TLC
conditions employed (Fig. 2) and no oxidative products were detected.
To generally localize the catabolic enzymes of the tissue, fat body
homogenates were subjected to differential centrifugation. JH
degradative activity resided primarily in the soluble fraction and
was a consequence of esterase activity, whereas epoxide hydratase
activity was associated with the microsomal fraction (Hammock
et al., 1975).

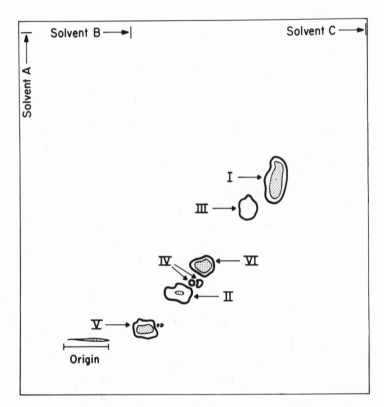

Fig. 2. Scheme of two-dimensional thin-layer chromatography of JH
metabolites (I) JH, (II) diol, (III) diepoxide, (IV) tetrahydrofuran diols, (V) acid diol and (VI) acid. Solvent A = hexane:
ethyl acetate (2:1); solvent B = benzene:n-propanol (4:1); solvent
C = benzene:n-propanol (10:1). Open spots designate metabolite
standards while stippled areas represent the result of autoradiographic determination of metabolites from [^{3}H]-JH (from Hammock
et al., 1975).

Although the metabolic pattern was qualitatively identical in all cases, quantitative differences were noted. After 1 hr of incubation the fat body yielded approximately equal amounts of acid and diol acid as did the media devoid of exogenous BPF. With increasing time of incubation a shift towards the diol acid was observed and the diol acid appears to be the final, primary metabolite in this system. However, when BPF is present in the medium, JH acid was the predominant metabolite. The reduction in the quantity of diol and diol acid in the medium in the presence of BPF suggests that BPF inhibits the uptake of JH by the fat body since epoxide hydratases are microsomal enzymes. The occurrence of JH acid in the medium can be a result of either selective transport out of the tissue or the hydrolysis of JH in the medium by esterases released from the fat body. The former alternative can be excluded because a substantial amount of acid was present in the fat body after incubation in a BPF-free medium.

Release of Esterases by the Fat Body. To test the possibility that the fat body actively secretes esterases into the medium, tissues were preincubated for 1 hr, removed, and the preconditioned media were divided into aliquots and combined with an equal amount of fresh medium containing JH and BPF or BSA. When the metabolic activity of these media was assayed, it was found that they degraded the JH to the acid with essentially no other identifiable metabolites being present (Fig. 3). Enzyme activity was greatest in preconditioned medium derived from L5D3 fat body and the presence of BPF dramatically reduced the rate of ester hydrolysis. In media derived from L4D2 fat body, acid formation was nearly completely suppressed whereas some residual activity remained in the L5D3 cultures. JH was not protected by BP from this residual esterase activity. Analysis of the preconditioned media with α-naphthyl acetate as substrate revealed a higher activity in the L4D2 medium. Since BP appears to protect JH from the attack of "general" esterases but offers only minimal protection from "JH specific" esterases (Sanburg et al., 1975; Hammock et al., 1975) we conclude that L4D2 fat body secreted a predominence of "general" esterases into the medium whereas L5D3 fat body also released a substantial quantity of "JH specific" esterases. We are using the term "general" esterase as defined by Sanburg et al. (1975) as those esterases which are capable of degrading free JH, but show little preference for JH over α-NA as a substrate. It should be kept in mind that the juvenile hormones are enoate esters and thus are quite stable to base hydrolysis (Sanburg et al., 1975), and esterases from many sources are not capable of hydrolyzing JH (unpublished information). Esterases from several insects show very different elution profiles upon Sephadex chromatography when p-nitrophenyl acetate, α-NA and JH are used as substrates and their relative inhibition by many compounds is quite different. Thus, these "general" esterases show specificity for JH but are unable to attack JH when it is complexed to BP.

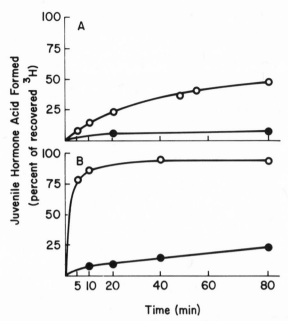

Fig. 3. Hydrolytic activity of esterases released from the fat body into the culture medium. Tissue of approximately the same weight from L4D2 (A) and L5D3 larvae (B) was incubated for 1 hr, after which the tissue was removed. Aliquots of the preconditioned media were combined with equal volumes of medium containing [^{3}H]-JH in the presence (●—●) or absence (o—o) of BPF, and JH acid formation was monitored (from Hammock et al., 1975).

The activity patterns in the preconditioned media were in accord with those found in hemolymph from the same developmental stage (Weirich et al., 1973; Sanburg et al., 1975). This suggests that hemolymph esterases originate in the fat body. To test this possibility a developmental approach was utilized in which we compared the hydrolytic activity of hemolymph and media preconditioned with fat body, both derived from the same stage of the last larval instar. The developmental pattern of hemolymph esterases exhibited a maximum at days 3 and 4, irrespective of whether α-naphthyl acetate or JH served as substrate (Fig. 4). In the preconditioned media, total and DFP-resistant (i.e. "JH specific" according to Sanburg et al., 1975) hydrolytic activity increased during the first 3 days of the instar and decreased thereafter. α-NA hydrolytic activity peaked at day 4 and seems to be separately controlled. Slab gel electrophoresis and subsequent staining with α-NA and fast blue revealed the same pattern of "general" esterases in hemolymph and incubation medium from the same stage. Indeed, the R_f values for particular esterase bands were identical. In

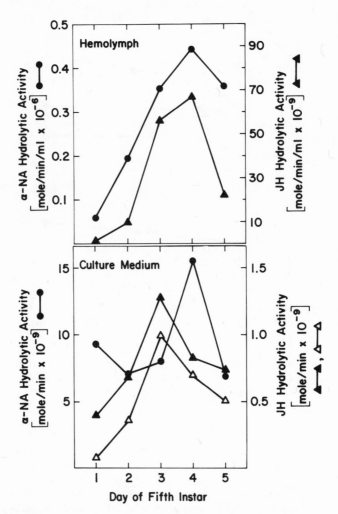

Fig. 4. Developmental analysis of esterase activity in hemolymph and culture medium preconditioned with fat body. Hydrolytic activity was determined with α-NA and chain-labeled [^{3}H]-JH (hemolymph) or [^{3}H]-methoxy-JH (culture media) as substrates as described in Materials and Methods. Fat body of various stages was cultured in vitro for 24 hr and the media analyzed for esterase activity. Closed triangles indicate total JH hydrolytic activity. A 15 min treatment with DFP (10^{-4} M) revealed DFP resistent (i.e. "JH specific") esterase activity. Hydrolytic activity in culture media is expressed per mg fat body protein.

contrast, analysis of fat body homogenates yielded a more complex
esterase pattern indicating that the esterases found in the medium
are specifically released from the fat body and are not a result
of leakage or cell breakdown.

Both the developmental pattern of hydrolytic activity and the
electrophoretic mobility of the "general" esterases support the
assumption that hemolymph esterases originate in the fat body.
However, additional information is required to definitively prove
that the esterases released in vitro are identical to those occurring in the hemolymph.

Effect of BP on JH Uptake Into Tissues

Effect of JH Concentration and Temperature on Uptake. The
ideal experimental situation for the study of JH transport between
hemolymph and fat body precludes the utilization of heterogeneous
tissue, the existence of extracellular spaces and degradation of
the hormone. Unfortunately, this ideal model was not found in
Manduca, especially in regard to JH metabolism. Theoretically,
one might circumvent the latter problem with the use of juvenoids
possessing high metabolic stability (see Hammock and Quistad,
this volume). However, these juvenoids do not bind to the hemolymph
BP or show such weak affinity (see Goodman et al., this volume) that
they would be useless for studying BP involvement on hormone uptake.
As a compromise, we therefore selected fourth instar fat body
despite its metabolic activity and the fact that it releases both
esterases and BP into the medium (Nowock et al., 1975). It does
however have the appropriate morphological properties for such
studies and can be obtained easily in sufficient quantities.

Fat body was incubated in Ringer's solution containing either
8×10^{-9}, 10^{-7}, 10^{-6} or 10^{-5} M [^{3}H]-JH, and changes in the radiolabel of the medium were monitored over time at both 4° and 26°C.
When the fractional uptake of radiolabel was plotted against time,
no differences between the various JH concentrations were noted
within a particular temperature series. Therefore, the data of
the 4° and 26°C regimens respectively, were pooled (Fig. 5). At
26°C label was taken up maximally after 10 min, but subsequently
there tended to be a net back flow into the medium. At the lower
incubation temperature radiolabel was taken up at a slightly decreased
rate, although the difference is not statistically significant, and
equilibrium was achieved after about 30 min. By extending the incubation time, a net reflux at a low rate was noted here as well.
Analysis of extracts of the media revealed that up to 70% of the
initial JH was metabolized at 26°C, mainly to the acid and acid
diol after 30 min. Incubation at 4°C did not inhibit degradation
completely as 25% of the JH was still metabolized, predominantly
to the acid. The net reflux of radiolabel from the fat body into

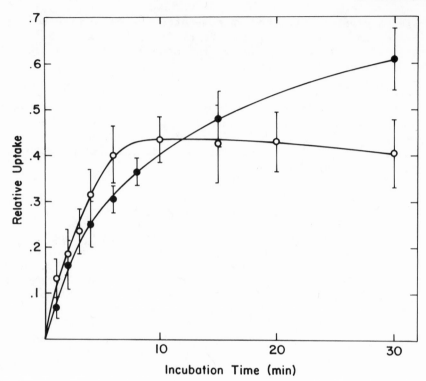

Incubation Time (min)

Fig. 5.   Time course of radiolabel uptake by fat body incubated in
[^{3}H]-JH.   Tissue of comparable weight was incubated in Ringer's
solution containing various concentrations of JH (8 x 10^{-9}, 10^{-7},
10^{-6}, and 10^{-5} M) and uptake was assayed as described in Materials
and Methods.   Each concentration series was run at 4°C (●—●) and
26°C (o—o).   As data for fractional uptake showed no significant
differences between the JH concentrations tested at a certain temperature, they were pooled and expressed as means ± S.D.   n = 8
(4°C) and 12 (26°C), respectively.

the medium seems to be principally a function of JH degradation
since it was not observed with isopropyl-11-methoxy-3,7,-11-
trimethyl-trans-2,trans-4 dodecadienoate (Altosid, Zeocon Corp.)
which was extremely stable under our in vitro conditions.   Interaction with BP which might arise from the fat body during incubation cannot be excluded as a possibility, but it should not have a
significant effect, especially at higher JH concentrations.

The similarity of fractional uptake within a concentration
series indicated that uptake was not saturable over the tested
range, but rather that the rate of uptake was directly proportional
to the external concentration of JH.   This corrects an earlier

preliminary observation (Hammock et al., 1975). When the initial
uptake rate, for which the 1 min data were used as approximations,
was plotted against the initial JH concentration, a linear relationship was obtained having the first-order rate constants of
0.77×10^{-4} min^{-1} (4°C) and 1.20×10^{-4} min^{-1} (26°C) (Fig. 6).

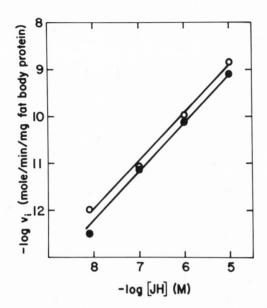

Fig. 6. Initial uptake rate as a function of external JH concentration. Data are from the same experiment as presented in Fig. 5.
Determinations were performed at 4°C (●——●) or 26°C (o——o).

However, the difference in rate constants at the two temperatures
is not statistically significant. Since uptake was not greatly
influenced by temperature, further experiments were performed at
4°C to minimize interference due to JH degradation and the possible
release of BP into the medium.

When fat body was incubated in 10^{-8} M [^{3}H]-JH acid, both the
initial uptake rate (0.38×10^{-12} mol/min/mg) and total uptake (20%)
were significantly reduced when compared to JH. After 30 min of
incubation, about 26% of the acid was further metabolized to the
acid diol.

Effect of BP. When free JH rather than JH-BP complex is taken
up by the fat body, or when the rate of uptake of the complex is
low when compared to free JH, the uptake rate in the presence of BP
should be proportional to the concentration of unbound hormone. To

test this assumption, 10^{-8} M JH was incubated with increasing
amounts of BP. The controls received equal amounts of BSA. After
equilibration, fat body was added and the initial uptake rates
determined. From the dissociation constant of the JH–BP complex,
K_D = 3 x 10^{-7} M (Kramer et al., 1974), the concentration of free
hormone can be calculated for a particular [JH]/[BP] ratio. Using
these concentrations and the estimate of the first order rate
constant of the uptake process (0.77 x 10^{-4} min^{-1}), the initial
uptake rates can be obtained and compared with the experimental
data. Figure 7 shows that a decrease in the [JH]/[binding sites]
ratio, i.e. a decrease in the concentration of free hormone,
resulted in a lower uptake rate although somewhat greater than
expected from theoretical considerations. This may be a consequence
of experimental variability, deviation from the linear relationship
between JH concentration and uptake rate at lower [JH]/[BP] ratios,
or to the interference of the uptake process by non-binding proteins present in the BPF-preparation. Relatively high concentrations of BSA also slightly reduced the rate of uptake.

Reversibility of Uptake. Fat body was preincubated in 10^{-8} M
[^{3}H]-JH for 30 min, briefly rinsed and transferred to fresh
Ringer's solution or medium containing BPF or BSA. Similar amounts
of radiolabel were taken up by the tissues (47-56%) during preincubation and the wash fluid contained $\leq$1.5% of the total label
initially applied. Under all conditions tested a reflux of radiolabel occurred (Fig. 8) and was most prominent in the media containing BPF. After 90 min, 75% and 84% of the label taken up was
released again depending on the number of binding sites present
(10^{-8}; 10^{-7} M, respectively). This suggests that only a minute
amount of JH is tightly bound to the cells. Testing media with
two concentrations of BSA yielded a similar rate and amount of
reflux. (It should be noted that up to 5% of the total label
released into Ringer's solution was adsorbed to the vial wall
compared to $\leq$1% when proteins are present in the medium.) As JH
metabolism still occurred during incubation at 4°C, it was
necessary to ascertain whether BPF favored the preferential reflux
of JH or its metabolites. Fat body was incubated in the presence
of 10^{-8} M [^{3}H]-JH for 30 min and reincubated as noted above for
another 30 min. Tissue homogenates and media from both incubation
periods were extracted with ethyl acetate and analyzed by TLC.
In all cases the dominant metabolite was the acid. The data per-
taining to the distribution between JH and total metabolites in
both compartments (Table 1) are difficult to interpret since degradation occurs both within the tissue and in the medium, creating
a complex pattern of fluxes of the various compounds. However, it
is evident that under all experimental conditions there was con-
sistently more JH than metabolites in the fat body. This differs
from the situation in the incubation media and confirms our earlier
observation that JH tends to partition into compartments within
the fat body where it is not easily accessible to degradative

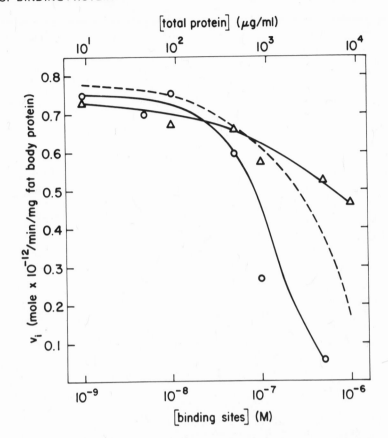

Fig. 7. Effect of BP on the rate of uptake of JH by the fat body. [³H]-JH (10^{-8} M) was equilibrated with increasing concentrations of BPF, fat body was added and the initial uptake rate determined (o—o). In the controls, BPF was replaced by BSA (△—△). Assuming that the uptake rate is proportional to the concentration of unbound JH, initial uptake rates were calculated for various [JH]/[binding sites] ratios using a dissociation constant of $K_D = 3 \times 10^{-7}$ M and a rate constant for the uptake of 0.77×10^{-4} min^{-1} (dashed line).

enzymes. When protein-free Ringer's solution was utilized as the second incubation medium, a substantial amount of metabolites was retained in the tissue. Reincubation in either protein-containing medium resulted again in a low level of metabolites being present in the fat body. This observation was also true of the BPF medium, whereas the BSA medium contained slightly more metabolites (mainly acid) than JH. The accumulation of acid could be due to: (1) generation from JH in the medium mediated by esterases originating

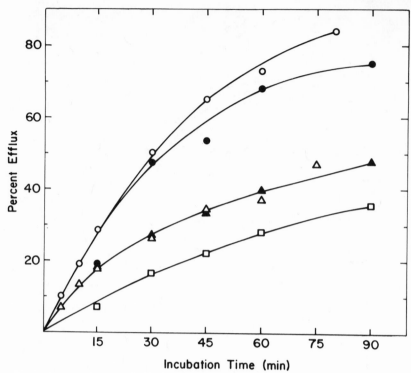

Fig. 8. Reversibility of uptake. Fat body was incubated for
30 min in 10^{-8} M [^{3}H]-JH, briefly rinsed and re-incubated in
either protein-free Ringer's solution (squares) or in Ringer's
containing BPF or BSA. Concentration of binding sites: 10^{-8} M
($\bullet$—$\bullet$) and 10^{-7} M (o—o); concentration of BSA: 85.5 µg/ml
($\blacktriangle$—$\blacktriangle$) and 855 µg/ml ($\triangle$—$\triangle$).

from the fat body; (2) release from the tissue by an enhanced
efflux in the presence of BSA in the medium; (3) a combination of
both processes.

Upon analyzing the reversibility of uptake after pre-incubating
fat body in the presence of [^{3}H]-JH acid, we found that of the
radiolabel initially taken up, 60% was released into Ringer's
solution, 79% into BSA-medium and 86% into BPF medium after 30 min.
Thus, as with the case of JH, the presence of proteins in the
medium enhances the reflux of the acid and acid diol formed during
incubation. Under all conditions utilized here the rate of radiolabel release was greater when fat body was pre-incubated with
[^{3}H]-JH acid as compared to [^{3}H]-JH. In this context, it is of
interest that after pre-incubation with JH only a small amount of
metabolites was released when the fat body was transferred to fresh
Ringer's solution devoid of proteins (Table 1). However, a rapid
reflux was noted after pre-incubation with JH acid. This suggests

TABLE 1. DISTRIBUTION OF JH AND METABOLITES AFTER PRE-INCUBATION OF FAT BODY WITH JH AND RE-INCUBATION IN DIFFERENT MEDIA.

Percent of extractable label

Incubation medium	1st Incubation Medium JH	Met.	1st Incubation Fat body JH	Met.	Incubation medium	2nd Incubation Medium JH	Met.	2nd Incubation Fat body JH	Met.
	33	12	47	8					
[³H]-JH in Ringer's	39	16	(45)		Ringer's	6	15	45	34
	30	5	(65)		Ringer's + BSA	21	26	41	13
	37	15	(48)		Ringer's + BPF	47	11	29	13

Tissues were incubated in 10^{-8} M [³H]-JH in Ringer's solution for 30 min and transferred to protein-free medium or medium containing BPF (10^{-8} M binding sites) or the same amount of BSA for an additional 30 min. Media and fat body homogenates were extracted and analyzed by TLC. The distribution of JH and metabolites (Met.) between medium and tissue is expressed as percent of total label extractable from each incubation (medium + fat body). Data are the means of two determinations.

that metabolites generated internally from JH and those taken up from an external pool are contained in different cellular compartments within the fat body.

DISCUSSION

Fat body inactivates JH in vitro by ester hydrolysis and epoxide hydration, which agrees with the pathways established in a variety of insects (Slade and Zibitt, 1972; White, 1972; Ajami and Riddiford, 1973; Slade and Wilkinson, 1974). Oxidative pathways accounted for negligible metabolism in the fat body, although they may be more important in other tissues such as midgut. Conjugate formation was of no significance, at least under the short-term in vitro conditions employed here. This is not surprising since conjugation occurs with great efficiency in gut and Malpighian tubules and is probably related to excretion (Slade and Wilkinson, 1974; Erley et al., 1975). Since epoxide hydratases are membrane bound (Hammock et al., 1975), diol formation requires the uptake of JH into the tissues. The esterase situation was more complex since ester hydrolysis of JH occurred in the medium due to the release of carboxyl esterases from the fat body as well as in the fat body itself. The data strongly suggest that the esterases normally present in the hemolymph are identical to those detected in the incubation medium and that the fat body is the source of these enzymes.

In our in vitro system the presence of BP in the medium affected JH metabolism in two ways. First, BP decreased the uptake of JH into the fat body and thereby reduced intracellular degradation. Second, it protected JH against extracellular esterases with the exception of the so-called "JH specific" esterases (Sanburg et al., 1975) which could mediate JH hydrolysis even when the hormone was complexed to BP. Since JH is transported from the corpora allata to peripheral tissues via the hemolymph, the concentration of BP and activity of esterases in the hemolymph determine to a great extent the JH titer at, or within, target cells. The presence of BP in fourth and fifth instar larvae is well established (Kramer et al., 1974; Goodman and Gilbert, 1974; Gilbert et al., 1976) as are "general" esterases of moderate or high activity (Weirich et al., 1973; Sanburg et al., 1975). "JH specific" esterases are barely detectable in the hemolymph of fourth instar larvae, but exhibit a remarkable increase in activity at the middle of the fifth instar which is followed by a rapid decline. These data are supportive of accepted views of JH requirements for larval development and metamorphosis (Gilbert and King, 1973). The very low activity of "JH specific" esterases and the protection afforded by BP against general esterases in fourth instar hemolymph favor the maintenance of a relatively high JH titer which is requisite for a larval-larval molt. The dramatic

increase in "JH specific" esterase activity during the fifth instar
clears JH from the hemolymph as would be required for a larval-
pupal molt. However, the situation is more complex and the above
explanation is rather simplistic. When assaying the hemolymph of
ligated fourth instar larvae by JH bioassay, Fain and Riddiford
(1975) determined the half life of endogenous JH to be the surprisingly low value of 90 min. Although the bioassay data alone
do not allow a decision between JH decline due to inactivation or
tissue uptake, they do demonstrate that the presence of BP is
not in itself sufficient for the extended persistence of JH in
the hemolymph. The observation that the half life of JH in fifth
instar larvae (when esterase activity is high) is even shorter
(<30 min), nevertheless indicates a protective role for BP.
Nijhout and Williams (1975) found that the JH titer of fifth
instar larvae begins to decline when the animals attain a weight
of approximately 5 g. The decline requires slightly greater than
24 hr and coincides exactly with the stage at which we found "JH
esterase" activity to be maximal. However, since endogenous JH
at this stage does not persist longer than 2 hr (Nijhout, 1975),
JH release by the corpora allata must still continue during the
time that the hemolymph titer is gradually decreasing. This
indicates that the control of JH production and release may be as
important a component of the JH titer regulating system as the
esterase, BP components. It is obvious that the roles of the "JH
specific" esterases must be investigated further.

The uptake of JH by the fat body followed first-order
kinetics, i.e. the initial uptake rate was proportional to the
hormone concentration in the medium over the range of 8×10^{-9} to
10^{-5} M. Due to both the nature of the experimental regimes and
the relatively low specific activity of the $[^3H]$-JH, lower JH
concentrations could not be tested. When JH was applied together
with BP, the uptake rate was dependent on the concentration of
unbound hormone. It is not yet known whether BP or the JH-BP
complex enters the fat body cells. If they do however, uptake
should occur at a much lower rate than with free JH. JH titers
in fourth instar larvae can be in the range of 5×10^{-8} M (Fain
and Riddiford, 1975; Goodman et al., this volume). Assuming a
BP concentration of 3×10^{-6} M (Gilbert et al., 1976) and a
dissociation constant of 3×10^{-7} M (Kramer et al., 1974), about
10% of the hemolymph JH may be unbound. (It should be noted that
the K_D was determined at 4°C and that it may be different at the
rearing temperature.) Since we utilized slightly higher hormone
concentrations, it is still possible that under physiological
conditions a saturable uptake mechanism exists for free JH.

JH acid exhibited a lower rate of uptake into the fat body
than JH. Whether this is due to a difference in charge or polarity between the two molecules is conjectural at this time although
with molecules of similar size the more non-polar (i.e. JH)

molecule can transit the cell membrane more readily. Most of the
JH taken up by the fat body was only weakly bound since uptake was
to a great extent reversible when fat body was transferred to
fresh medium containing BPF. Reflux was even more rapid in the
case of metabolites and did not appear to require proteins with
high binding affinity and specificity. We therefore assume that
hemolymph lipoproteins possessing low affinity and high capacity
(see Gilbert et al., 1976) may bind JH-metabolites and thus enhance
or elicit their exit from the tissues.

From the above, one can readily see that the maintenance of
JH titers requires a most complex regulatory system and that BP,
lipoproteins, degradative enzymes and activity of the corpora
allata may all be critically involved. Information on most of
the above is still preliminary and it is therefore obvious that
a great deal of research on the topic remains to be conducted.
However, it is likely that BP probably does not play a role in
target cell recognition and that unbound JH is probably the form
that enters peripheral tissues. This supposition is based on the
observation that juvenoids with potent JH morphogenetic activity
bind very weakly or not at all to the BP (see Goodman et al.,
this symposium).

ACKNOWLEDGEMENTS

Joachim Nowock was supported by a post-doctoral fellowship
from the Deutsche Forschungsgemeinschaft and Bruce Hammock by a
post-doctoral fellowship from the Rockefeller Foundation. This
work was supported by grants AM-02818 from the National Institutes
of Health and BG-27574 from the National Science Foundation.

REFERENCES

Ajami, A.M., and Riddiford, L.M., 1973, J. Insect Physiol. 19:635.
Emmerich, H., and Hartmann, R., 1973, J. Insect Physiol. 19:1663.
Erley, D., Southard, S., and Emmerich, H., 1975, J. Insect Physiol.
    21:61.
Fain, M.J., and Riddiford, L.M., 1975, Biol. Bull.149:506.
Gilbert, L.I., and King, D.S., 1973, in "The Physiology of Insecta"
    (M. Rockstein, ed.) vol. I, 2nd ed. pp. 249-370, Academic Press,
    New York and London.
Gilbert, L.I., Goodman, W., and Nowock, J., 1976, in "Biosynthesis,
    Metabolism and Cellular Action of Invertebrate Hormones,"
    Colloque Intl. du C.N.R.S., in press.
Goodman, W., and Gilbert, L.I., 1974, Amer. Zool. 14:1289.
Hammock, B., Nowock, J., Goodman, W., Stamoudis, V., and Gilbert,
    L.I., 1975, Molec. Cell. Endocrinol. 3:167.
Judy, K.J., Schooley, D.A., Dunham, L.L., Hall, M.S., Bergot, B.J.,
    and Siddall, J.B., 1973, Proc. Nat. Acad. Sci. USA 70:1509.

Kramer, K., Sanburg, L., Kézdy, F., and Law, J., 1974, Proc. Nat. Acad. Sci. USA 71:493.

Low, R.B., Wool, I.G., and Martin, T.E., 1969, Biochim. Biophys. Acta 194:190.

Nijhout, H.F., 1975, Biol. Bull.49:568.

Nijhout, H.F., and Williams, C.M., 1974, J. Exp. Biol. 61:481.

Nowock, J., Goodman, W., Bollenbacher, W., and Gilbert, L.I., 1975, Gen. Comp. Endocrinol. 27:230.

Pratt, G.E., and Tobe, S.S., 1974, Life. Sci. 14:575.

Röller, H., and Dahm, K.H., 1970, Naturwiss. 57:545.

Sanburg, L.L., Kramer, K.J., Kézdy, F.J., and Law, J.H., 1975, J. Insect Physiol. 21:873.

Slade, M., and Wilkinson, C.F., 1974, Comp. Biochem. Physiol. 49B:99.

Slade, M., and Zibitt, C.H., 1972, in "Insect Juvenile Hormones: Chemistry and Action" (J.J. Menn and M. Beroza, eds.) pp. 155-176, Academic Press, New York and London.

Sroka, P., and Gilbert, L.I., 1971, J. Insect Physiol. 17:2409.

Trautmann, K.H., Schuler, A., Suchý, M., and Wipf, K.-H., 1974, Z. Naturforsch. 29c:161.

Weirich, G., Wren, J., and Siddall, J.B., 1973, Insect Biochem. 3:397.

Westphal, U., 1967, Arch. Biochem. 118:556.

White, A.F., 1972, Life Sci. 11:201.

Whitmore, E., and Gilbert, L.I., 1972, J. Insect Physiol. 18:1153.

THE DEGRADATIVE METABOLISM OF JUVENOIDS BY INSECTS

B.D. Hammock and G.B. Quistad

Division of Toxicology and Physiology, Department of
Entomology, University of California, Riverside 92502
and Zoecon Corp., 975 California Avenue, Palo Alto,
California 94304

INTRODUCTION

Juvenoids (juvenile hormone mimics) comprise the major class
of insect growth regulators and hold promise as insect control
agents. Thus, the fate of juvenoids in target and nontarget
insects is of interest. Knowledge of the degradative metabolism
of juvenoids is important for the rational development of more
effective compounds of varying selectivity, the use of possible
synergists, the prediction of application schedules, and the cir-
cumvention of resistance. The effectiveness of juvenoids as pest
control agents is dependent upon a myriad of factors, but degradative metabolism by the insect is an important consideration.
Hormone mimics are also of value in more basic studies of hormone
action, and for these uses knowledge of their metabolism is necessary.

Juvenoid metabolism will be discussed in this chapter with
some reference to mammalian comparative metabolism. Some juvenoids
seem to be metabolized by pathways important in the degradation of
the juvenile hormones (JH) while the high activity of many juvenoids is due, in part, to their not entering normal pathways of JH
degradation. Reference to the metabolism of the natural hormones
is thus requisite.

MATERIALS AND METHODS

Preliminary degradative work with juvenoids demonstrated the
inadequacy of thin-layer chromatography (TLC) for the resolution
and quantitation of metabolites from many systems because a very

large number of organosoluble metabolites were produced. Several
approaches have been used to attack the resolution problem including the use of 2-dimensional TLC or multiple solvent systems which
allow both the resolution of very similar compounds and compounds
with vastly differing polarities on the same plate. Other procedures for structural verification of metabolites involve the classical techniques of microscale derivatization indicative of certain
functional groups. Radio-gas-liquid chromatography (GLC) has been
used infrequently in JH or juvenoid work in spite of its resolving
power (Slade and Zibitt, 1972; White, 1972; Kamimura et al., 1972).
Terriere and Yu (1973) found that GLC with an electron capture
detector is quite sensitive for detecting juvenoids with conjugated
systems as methoprene, but unfortunately this procedure is not
useful for examining metabolites of variable polarity or metabolites from systems with contaminating materials. Hoffman et al.
(1973) made extensive use of electron impact mass spectroscopy
(MS) for metabolite identification, and chemical ionization MS is
very useful for some polyhydroxylated juvenoid metabolites
(Hammock, unpublished). One can predict more extensive use of MS
and especially GLC/MS as such instruments become more sensitive
and accessible.

Initiation of the investigation of juvenoid metabolism coincided with the first intensive use of high resolution liquid
chromatography (HRLC), and recent research has applied this power-
ful tool to the JHs and juvenoids (e.g. Dunham et al., 1975;
Schooley et al., this volume). Gel permeation, adsorption, and
reversed-phase HRLC are useful in the purification of juvenoid
metabolites and their tentative identification by co-chromatography.
The detection limit at the present time for UV absorbing juvenoids
is in the nanogram range which is useful for monitoring some
relatively clean enzyme systems as epoxide hydratases (Hammock,
unpublished), but is generally not sensitive enough for most metabolism work.

The procedures involved in the elucidation of pathways involved in the metabolism of JHs and juvenoids are common to most
studies of xenobiotic metabolism with a heavy emphasis on radio-
tracers. The recurrent problem through most JH and juvenoid studies is that the specific activities of the radiolabels used are in
many cases not high enough to permit study at low dose levels consistent with the physiological action of these compounds.

RESULTS AND DISCUSSION

Metabolism of Functional Groups of Juvenoids

Metabolism of the Ester Functionality. A major route of
juvenile hormone (JH) degradation in all insects examined except
possibly the higher Diptera is ester cleavage (Slade and Zibitt,
1971, 1972; White, 1972; Ajami and Riddiford, 1973). As an example,
JH metabolism in Manduca sexta rapidly converts JH I in vivo to
its JH acid (2, Fig. 1) and JH acid diol (3, Fig. 1). The hemolymph of M. sexta contains high levels of soluble esterases capable
of hydrolyzing JH (Weirich et al., 1973; Sanburg et al., 1975)
while the fat body contains both esterases and epoxide hydratases.
The 12,000 g pellet ("mitochondria"), 100,000 g pellet ("microsomes") and 100,000 g supernatant (soluble) fractions of M. sexta
fat body are all capable of hydrolyzing JH to JH acid (Hammock et
al., 1975b) indicating the presence of both membrane-bound and
soluble esterases.

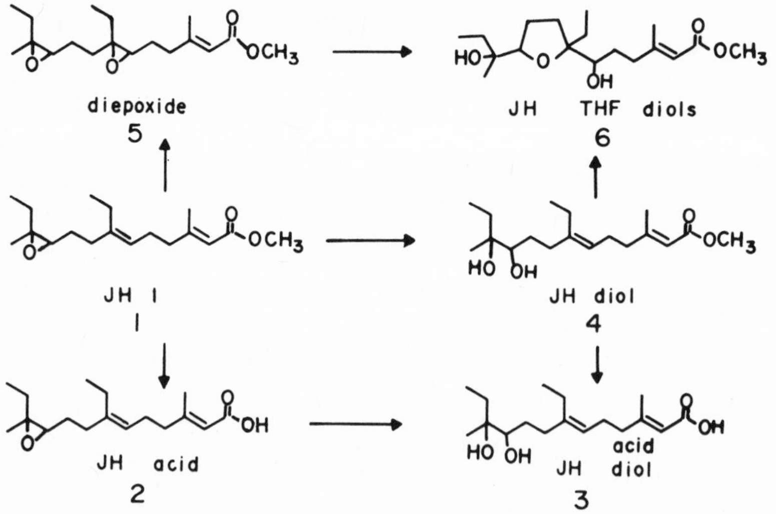

Fig. 1. Established pathways of JH metabolism in insects. Conjugation of the diol and acid diol also may occur.

The first report of a correlation between activity and metabolic stability of juvenoids was made by Weirich and Wren (1973).
They reported that the esterases in M. sexta hemolymph rapidly
hydrolyzed JH I and that the ethyl ester of JH I (3, Fig. 2)
appeared to be hydrolyzed more slowly than the methyl ester while

the isopropyl ester of the juvenoid methoprene (1, Fig. 2) was not
hydrolyzed. This observation points to a general trend in esterases that hydrolysis rate is often methyl > ethyl > propyl >>
isopropyl although hydrolysis of the isopropyl ester, methoprene,
cannot be compared directly with the methyl and ethyl esters of JH.
Weirich and Wren also found that the natural 2E isomer of JH I was
very rapidly hydrolyzed while the 2Z isomer was not. This finding
has been confirmed in several other Lepidoptera for methyl farnesoate and indicates a rather marked substrate specificity for JH
esterases.

    The ester functionalities of JHs are distinct in that they
are conjugated with the 2,3-double bond providing more chemical
stability than corresponding unconjugated esters, and this conjugation may explain the marked stability of JH to many general
esterases as reported for M. sexta hemolymph by Sanburg et al.
(1975). The stability is also exemplified by the lack of degradation of JH by mammalian hepatic soluble esterases (Hammock et al.,
1974c), Drosophila pseudoobscura locus E5 esterases (Harshman,
unpublished observation), and partially purified esterases from
Musca domestica, Blaberus giganteus, and Heliothis zea which
rapidly hydrolyze α-naphthol acetate and p-nitrophenyl acetate
(Mumby and Hammock, unpublished observations). With regard to
D. pseudoobscura and M. domestica it should be remembered that
Slade and Zibitt (1972) found that ester cleavage is a very minor
route of JH I metabolism in another Dipteran Sarcophaga bullata.

    In most insects the hydrolysis of the methyl ester is a major
route of JH metabolism, especially in those stages when the insect
appears most sensitive to JH or juvenoid treatment. Therefore,
other alcohols have been used in synthetic juvenoids partially to
circumvent the facile hydrolysis of methyl esters. Several compounds such as hydroprene and Ro-8-4314 (4, 5 respectively Fig. 2)
are ethyl esters, while reminiscent of estrogen work, other compounds (6, 7, Fig. 2) are alkynyl esters. One of the most active
juvenoids of many insects is the isopropyl ester methoprene (1,
Fig. 2). Unsaturated bonds at the 2 and 4 positions in methoprene
and several other juvenoids, result in a conjugated dienoate system
which is even more stable to chemical hydrolysis than the conjugated enoate system of the natural JHs. By monitoring the disappearance of methoprene by electron capture GLC from housefly enzyme
systems without NADPH, Yu and Terriere (1975) have tentatively
indicated the importance of esterases in the metabolism of ethyl,
propynyl and thioesters (4, 8, 6, 7, and 9, Fig. 2). The isopropyl ester methoprene is not only refractory to degradation by
M. sexta hemolymph esterases (Weirich and Wren, 1973) but also
esterases from other M. sexta tissues (Nowock et al., 1975).
Solomon and Metcalf (1974) found that a compound co-chromatograph-
ing with the methoxy acid was the major in vivo metabolite of
methoprene in the milkweed bug Oncopeltus fasciatus and a minor

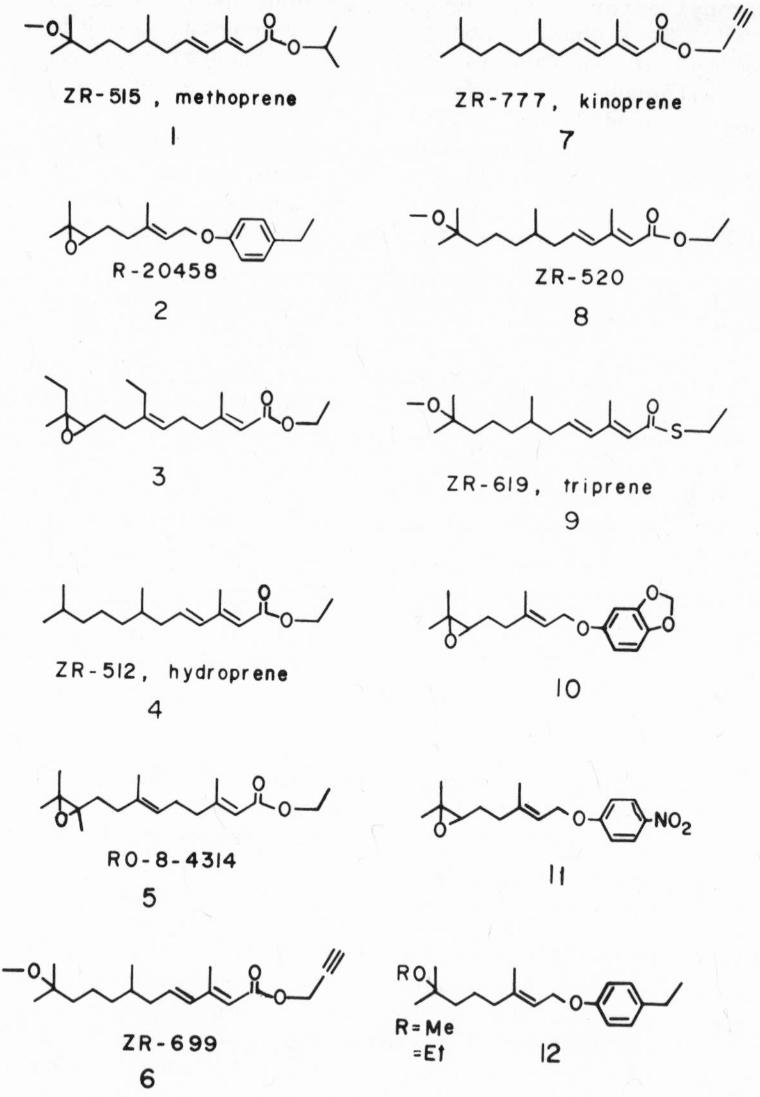

ZR-515 , methoprene

1

R-20458

2

3

ZR-512 , hydroprene

4

RO-8-4314

5

ZR-699

6

ZR-777 , kinoprene

7

ZR-520

8

ZR-619 , triprene

9

10

11

R=Me
=Et   12

Fig. 2. Structures of juvenoids involved in metabolism studies.
compounds 1, 4, 6, 7, 8, 9 are from Zoecon Corporation; 2 from
Stauffer Chemical Company; and 5 from Hoffman La Roche.

metabolite in the yellow meal worm <u>Tenebrio molitor</u>. These authors
also showed that treatment with tri<u>orthocresylphosphate</u> (TOCP),
which is a known inhibitor of many insect esterases after metabolic
activation, decreased the proportion of acidic metabolites of
methoprene in the test insects. Quistad <u>et al</u>. (1975a) found that

in <u>Aedes</u> <u>aegypti</u> and <u>Culex</u> <u>pipiens</u> <u>quinquefasiatus</u> larvae the methoprene ester is cleaved to give both the methoxy and the hydroxy acids (2, 3 respectively, Fig. 3). Additional evidence for the importance of metabolism of methoprene by esterases in mosquitoes is provided by the slight increase in larval mortality in methoprene-treated larvae when co-treated with TOCP (Quistad et al., 1975a).

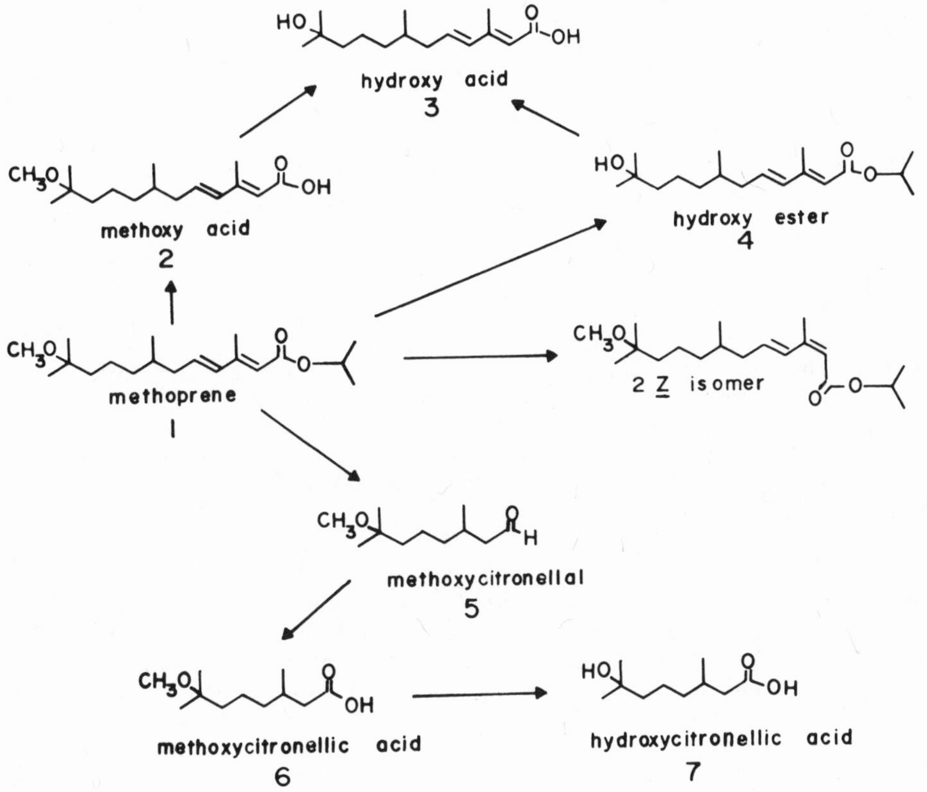

Fig. 3. Major nonconjugative pathways of methoprene metabolism in insects.

Quistad <u>et al</u>. (1975a) found that the major nonpolar metabolite of methoprene from <u>M</u>. <u>domestica</u> post-feeding third-instar larvae is the hydroxy acid. Hammock <u>et al</u>. (1975c) have found that the hydroxy-ester is the major early <u>in vivo</u> metabolite of methoprene in post-feeding larvae of <u>M</u>. <u>domestica</u>, but that more polar products become more important with time. The <u>in vitro</u> systems fail to predict completely the <u>in vivo</u> products as Yu and Terriere (1975) and Hammock <u>et al</u>. (1975c) have failed to find significant ester cleavage of methoprene in homogenates or

subcellular fractions of larvae or adult M. domestica. Gel filtration of the soluble fraction from larval M. domestica separated
two major esterase fractions, neither showing significant metabolism of methoprene (Hammock et al., 1975c). It is possible that
oxidative ester cleavage of methoprene may occur in M. domestica,
but this has not been demonstrated in vitro. Perhaps only a brief
burst of esterase activity capable of degrading JH or methoprene
exists in M. domestica.

Inhibitors of metabolism are commonly used in toxicology as
synergists and to indicate the importance of metabolic pathways
affecting biologically active compounds. Interpretation of such
data requires cognizance of the possible multiplicity of action of
such inhibitors. The probable high cost of manufacturing juvenoids also makes the development of an effective low cost synergist
attractive. A number of workers, most recently Ajami (1975), have
suggested esterase inhibitors as possible synergists, and the specificity of JH esterases suggests that highly specific JH esterase
inhibitors can be found.

Of the juvenoids with an ester functionality so far investigated the isopropyl dienoate structure of methoprene appears to be
the most stable; however, many active juvenoids no longer contain
the ester functional group. Of these juvenoids, extensive metabolism work has only been reported for the geranyl phenyl ester
series.

Metabolism of the Ethylphenyl Moiety. The discovery by
Bowers et al. (1969) of the JH activity of the geranyl phenyl
ethers has led to the synthesis of a number of very active juvenoids. Although cursory studies were made on unlabeled methylenedioxyphenyl and labeled p-nitrophenyl compounds (Singh and Hammock,
unpublished information, 10, 11, Fig. 2), essentially all of the
insect metabolism work with this series has been with the p-ethylphenyl compounds (Hammock et al., 1974a, 1975a). Based on earlier
work (Eto and Abe, 1971), it was not difficult to predict the
major routes of metabolism of the p-ethylphenyl moiety in insects.
M. domestica, Stomoxys calcitrans (stable fly), and Schistocerca
americana (American locust) hydroxylate the p-ethylphenyl moiety in
the α-position in vivo. The metabolism of the p-ethylphenyl juvenoid R-20458 (2, Fig. 2) has been most extensively studied in the
housefly microsomal enzyme system (Fig. 4). Alpha-hydroxylation
of the ethyl group is the predominate oxidative pathway while
β-hydroxylation is less significant. β-Hydroxylation is more
pronounced in housefly microsomes than mammalian liver microsomes.
Further oxidation of the β-hydroxyethyl compounds to phenylacetic
acids is minor if it occurs at all in housefly microsomes, although
this is a pathway in mammals. Addition of NAD and the soluble
fraction to the microsomes increases dehydrogenation of α-hydroxy-
ethylphenyl to acetophenyl derivatives, although this pathway is

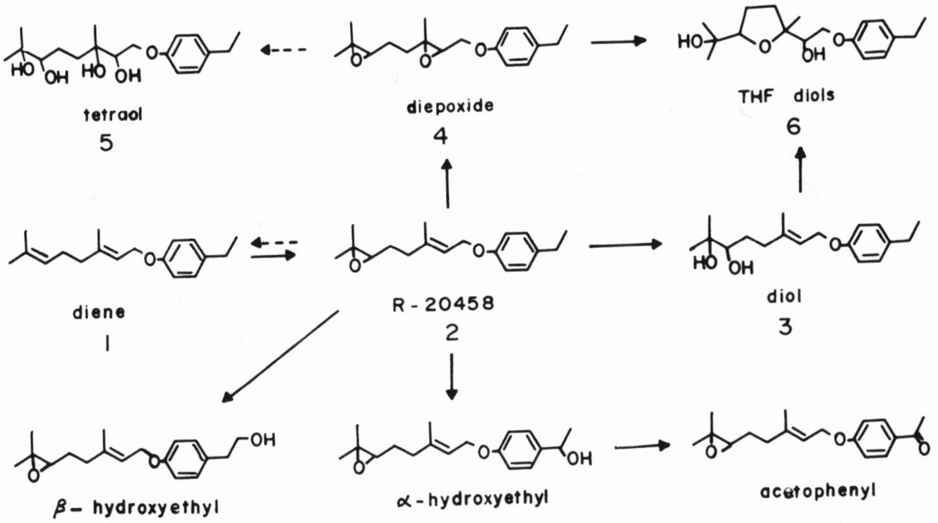

Fig. 4. Major routes of R-20458 metabolism in insects. Metabolism of both the geranyl and phenyl moieties can occur on the same molecule. Dotted lines indicated metabolic routes not established in insects.

not as important as with mammalian liver enzymes. S. bullata abdomen microsomes produce the acetophenyl derivative as their major metabolite of R-20458. No cleavage of the geranyl phenyl ether was detected in housefly microsomes and such cleavage appears to be minor in most insects with the possible exception of P. americana, while in mammals in vivo ether cleavage and α-oxidation are major degradative pathways (Gill et al., 1972; Hoffmann et al., 1973; Gill et al., 1974; Ivie et al., 1975). Ring hydroxylations have not been described for aromatic juvenoids in any insects or mammals examined.

The p-ethylphenyl moiety of some juvenoids is rapidly metabolized in every mammalian species examined, while its stability varies greatly in insects. It is quite refractory to degradation in T. molitor, but it is very susceptible to α-hydroxylation and other metabolic pathways in M. domestica.

Metabolism of the Terminal Functionality. The epoxide or oxirane moiety of the juvenile hormones is chemically labile, but it was found quite early to be necessary for high activity in many insects (Bowers et al., 1965). Although several functional groups have been used to replace the epoxide moiety, only the alkoxides have progressed very far in development (Staal, 1975).

Metabolism of Epoxides. Some of the major work on juvenoid and JH epoxide metabolism in insects has been to establish the absence of certain pathways. Reductive opening of epoxides to tertiary alcohols, rearrangement to allylic alcohols, or cyclization analogous to conversion of squalene 2,3-oxide to lanosterol have not been reported in insects or mammals, although some of the pathways are involved in environmental degradation (Ajami and Riddiford, 1973; Hoffmann et al., 1973; Gill et al., 1974; Hammock et al., 1974a,b). Glutathione conjugation is a well-known pathway of epoxide metabolism in mammals, but it is apparently insignificant in R-20458 metabolism in mammals and insects, possibly due to the trisubstituted nature of the epoxide (Gill et al., 1974). Apparent conversion of an epoxide to an olefin was reported in rats by Hoffmann et al. (1973) and this pathway has been recently confirmed in steers (Ivie et al., 1975) apparently mediated by microorganisms in the rumen (Ivie, 1975). Epoxide reduction has been found also in the gut contents of mice, rats, and man but not in those of Gromphadorhina portentosa (Madagscar cockroach). Such a conversion of an epoxide to an olefin has not been reported in insects although in T. molitor unidentified metabolites of a higher Rf than the starting epoxide have been noted on TLC (Hammock, unpublished).

Epoxides are hydrated in insects by hydratases which are membrane-bound for the cyclodienes, JHs and juvenoids so far examined as substrates. Most activity is in the microsomal fraction, but some activity is associated with cellular debris and the mitochondrial fraction. It is not known if the activity in these two fractions is intrinsic or an artifact due to contamination (Brooks et al., 1970; Hammock et al., 1974a, 1975b; Slade et al., 1975). With most substrates, mammalian epoxide hydratase activity is also membrane-bound with the exception of a highly active soluble fraction in liver and kidney with JH or juvenoid substrates (Gill et al., 1972, 1974; Hammock et al., 1974c). Possibly the soluble epoxide hydratase in mammals is involved in degrading metabolic by-products from steroid biosynthesis, and insects, which do not synthesize steroids, have no need for such enzymes.

Using the cyclodiene epoxide, HEOM, Slade et al. (1975) report a pH optimum of 9.0 for Calliphora erythrocephala and Prodenia eridania, and 8.1 for G. portentosa microsomal epoxide hydratases. Slade et al. (1975) also noticed a pH optimum 7.9 for JH hydration in P. eridania suggesting that different epoxide hydratases are responsible for JH and HEOM metabolism although there certainly may be overlap in activity. Hammock et al. (1974a) found distinct peaks of epoxide hydratase activity in housefly microsomes at pHs 6.8, 8.1 and 8.9 using R-20458 and its diepoxide as substrates suggesting involvement of multiple enzymes. As with mammalian microsomal hydratases, insect hydratases are notable in that they are active at relatively high pHs.

Lack of significant inhibition of R-20458 hydration by 1,1,1-
trichloropropene oxide and cyclohexene oxide suggests that the
housefly microsomal epoxide hydratase system is quite distinct
from the mammalian epoxide hydratases responsible for arene oxide
metabolism (Hammock et al., 1974a). Slade et al. (1975) also
presented evidence for multiple epoxide hydratases in insects since
1,1,1-trichloropropene oxide inhibits HEOM hydration without
stabilizing JH in several insect systems.

The metabolism of a juvenoid by epoxide hydration in insects
was first demonstrated when the diol of R-20458 (3, Fig. 4) was
detected in the feces of S. americana (Gill et al., 1972). Brooks
(1973) reported circumstantial evidence for the degradation of a
juvenoid (7 ethyl homolog of 11, Fig. 2) in homogenates of Calliphora erythrocephala and T. molitor to a diol while the related
radiolabeled compound (11, Fig. 2) is metabolized by epoxide hydration (Hammock, Gill, and Casida, unpublished results) in housefly
microsomes. Epoxide hydration of R-20458 has been shown in vivo
in nymph and adult P. americana and S. americana and in larvae,
pupae, and adults of T. molitor, M. domestica and Stomoxys calcitrans (Hammock et al., 1974a; 1975a,c), while tissues in vitro or
tissue homogenates of M. domestica larvae and adults, S. bullata
adults, S. calcitrans adults, Trichoplusia ni larvae, M. sexta
larvae and T. molitor larvae and pupae convert R-20458 to its
corresponding diol in varying amounts (Hammock et al., 1974a;
1975a,c; Nowock et al., 1975). Epoxide hydratase activity varies
from quite low in adult S. bullata to very high in larval T. ni
homogenates, although undoubtedly much of this variation exists
due to the developmental stage of the insects examined and methods
of homogenate preparation. An intensive study of the nature and
properties of the epoxide hydratases of several insects is needed
to better understand their role in JH and juvenoid metabolism.
Such a study has probably not been completed due to the difficulties encountered in working with membrane-bound enzymes and the
lack of rapid assays for epoxide hydratase activity.

The epoxide hydratases are thus important in the degradation
of JH and epoxide containing juvenoids. Extending the alkyl
branches of the epoxide moiety may serve to stabilize the epoxide
by interfering with its fit at the hypothetical active sites of
hydratases. The tetrasubstituted epoxide (Ro-8-4314) (5, Fig. 2)
is more stable to housefly epoxide hydratase than R-20458 or JH I
indicating that the addition of the methyl group at C-10 may
sterically prevent binding at the enzyme active site or interfere
with the hydration mechanism. Instead of being stabilized, the
epoxide has been either eliminated or replaced with other functional groups such as the alkoxy moiety in some juvenoids.

Metabolism of Alkoxides. The use of the alkoxide moiety to
replace the epoxide was suggested by Šorm (1971) and has led to

several highly potent juvenoids in which a conformation alkoxide
probably mimics the epoxide. Since the alkoxides are not highly
strained ethers they are not subject to acid- or base-catalyzed
hydrolysis or enzymatic hydration, yet they are subject to enzymatic O-demethylation. Solomon and Metcalf (1974) found that methoprene is O-demethylated by O. fasciatus and T. molitor (Fig. 3) and
that this reaction is inhibited by piperonyl butoxide. Since the
tertiary alcohol produced upon O-demethylation of methoprene is
more active in a few insects than the alkoxide, O-dealkylation can
sometimes be considered an activation reaction. Yu and Terriere
(1975) provided circumstantial evidence for their hypothesis of
in vitro O-demethylation of methoprene in M. domestica. Quistad
et al. (1975a) established that the tertiary alcohol is a primary
metabolite of the mosquito larvae Aedes aegypti and Culex pipiens
quinquefasciatus and the larvae of M. domestica. Hammock et al.
(1975a) have shown that both the methoxide and ethoxide of a
geranyl phenyl ether (12, Fig. 2) are metabolized by O-dealkylation in T. molitor larvae and adults of S. calcitrans and M. domestica. As expected Hammock et al. (1975a) found that the O-dealkylation is predominantly a NADPH dependent microsomal reaction in
S. calcitrans and M. domestica and that ethoxides are dealkylated
more slowly than methoxides both in vitro and in vivo.

Conjugation of the Terminal Functionality. Polar conjugates
of JH and juvenoids have not been thoroughly investigated and in
most metabolism studies a mixture of enzymes is used to cleave the
conjugates before analysis. Slade and Wilkinson (1974) present
data that in P. eridania in vivo and in vitro reaction of the diol
functionality of JH diol or JH acid diol with sulfate is the predominate conjugation pathway. This suggests that the diol of
epoxides or the tertiary alcohol of alkoxides may be similarly
conjugated, but sulfate conjugation will usually occur preferentially at secondary rather than tertiary alcohols. Only the
aglycone portions of other conjugates have been described.

Metabolism of Olefinic Isopropylidenes. A possibility for
avoiding the epoxide moiety in juvenoids is to treat insects with
the precursor olefin assuming that the insect will epoxidize the
olefin to a more active compound (Jacobson et al., 1972). This
conversion is difficult to show in vivo in insects since very
minor conversion could account for a large increase in activity
and the epoxide is labile to metabolism. The conversion of the
R-20458 diene to R-20458 (1 to 2, Fig. 4) has been demonstrated
as a NADPH-dependent reaction in M. domestica microsomes; however,
the major metabolic pathways are totally different than those
involved in R-20458 metabolism. The diene is probably oxidized
on a methyl alpha to the 6,7-double bond to give an alcohol,
aldehyde and acid at a very high rate; however, this rapid degradation of the diene was not observed in enzymes from S. bullata,
P. americana, or M. sexta. The diene is converted cleanly to its

corresponding epoxide by corpora allata homogenates from M. sexta
and Blaberus giganteus (cockroach) when NADPH is present (Hammock
and Gilbert, 1974), and in this conversion the juvenoid is probably
mimicking a JH precursor.

Metabolism of the Terpenoid Chain. Although the functional
groups most susceptible to metabolic modification are usually at
the extremities of the juvenoid molecule, several metabolic pathways involve the terpenoid chain itself.

Olefinic Isomerization. Just as in JH I (Schwieter-Peyer,
1973), a 2E double bond is essential for high insecticidal activity
in the dienoate methoprene (Henrick et al., 1973, 1975). Quistad
et al. (1975a) have shown that the 2,3 double bond of methoprene
is susceptible to biological isomerization by mosquitoes and house-
flies (Fig. 3). Primary metabolites and recovered methoprene had
been substantially isomerized at C-2 to the 2Z isomer which is
less active by 100-fold on A. aegypti. Hence, economic biological
activity was essentially eliminated by olefinic isomerization, and
the mechanism for this provocative transformation remains unknown.
The same authors showed that although methoprene was isomerized
by M. domestica larvae to the 2Z isomer, the converse reaction for
formation of methoprene from the 2Z isomer did not occur. The
findings of Quistad et al. (1975a) have been confirmed in three
strains of M. domestica by Hammock et al. (1975c), and one wonders
if this olefinic isomerization is involved in JH metabolism in the
Diptera. Such isomerization has not been reported in the metabolism of other juvenoids or JH, probably because the investigators
lacked the techniques to easily quantitate the isomers.

Olefinic Oxidation. Oxidative scission of the 4,5 double
bond of methoprene by M. domestica results in 7-methoxycitronellal
(5, Fig. 3) which is evolved as a volatile metabolite (Quistad
et al., 1975a). Further oxidation and O-dealkylation of this
aldehyde results in 7-methoxycitronellic acid and 7-hydroxycitron-
ellic acid (6, 7 respectively, Fig. 3) which have been detected
as methoprene metabolites in M. domestica, Culex pipiens, and
Aedes aegypti. This oxidative scission is apparently less important in insects than in other degradative systems such as plants
(Quistad et al., 1974a).

An important mammalian degradation pathway for methoprene
involves exhaustive catabolism of the terpenoid chain to [14C]
acetate which is then reincorporated into a plethora of natural
products (Quistad et al., 1974b, 1975b,c). Primary metabolites
from methoprene in mammals are found only in excrement while the
radiolabeled residues in tissues consist mostly of biochemicals
such as [14C] fatty acids. The metabolic capability of insects
appears to be quite different for methoprene since only primary
metabolites could be characterized and degradation of methoprene

to [^{14}C] acetate by insects apparently did not occur as evidenced by lack of radiolabeled natural products and the small amounts of ^{14}CO$_2$ detected. It is likely that the geranyl moiety of geranyl phenyl ether juvenoids undergoes similar metabolism following cleavage, but such findings have not been reported because the available radiolabels are lost from the geranyl moiety following ether cleavage.

Oxidation of the isopropylidene olefin to the corresponding epoxide has already been discussed with regard to activating some juvenoids and as a potential step in the biosynthesis of juvenile hormone. The 2,3-olefinic bond in the geraniol derived juvenoids and the 6,7-olefinic bond in farnesol derived materials are much less susceptible to chemical and biological oxidation than is the isopropylidene group. The diepoxide of JH has been reported as a possible metabolite of D. melanogaster although the investigators admitted their lack of definitive evidence (Ajami and Riddiford, 1973) (5, Fig. 1). The diepoxide of R-20458 (4, Fig. 4) is a minor metabolite in microsomal systems from M. domestica abdomens and P. americana fat body (Hammock et al., 1974a), while in mammalian microsomes the diepoxide is a more important NADPH-dependent metabolite than in insects (Gill et al., 1974). Housefly microsomes with NADPH are also able to convert JH I to small amounts of the corresponding 6,7-10,11-diepoxide (Hammock, unpublished results). Although the 4,5-epoxide of methoprene is a known photoproduct (Quistad et al., 1975d), neither the epoxide nor its corresponding diol could be found as an in vivo metabolite in A. aegypti, C. pipiens or M. domestica.

Since a monoepoxide is readily converted to its corresponding diol by chemical or enzymatic hydration, it was assumed that a diepoxide would be hydrated to its corresponding tetraol (5, Fig. 4). This assumption led to several mistakes (Gill et al., 1972; Ajami and Riddiford, 1973), because the diol epoxides resulting from the hydration of one epoxide of a diepoxide rapidly cyclize to form a variety of products (Hammock et al., 1974a,b). The most abundant products are the corresponding cis and trans tetrahydrofuran diols shown to be in vitro metabolites of R-20458 in M. domestica (6, Fig. 4). The corresponding tetrahydrofuran diols of JH I were not detected in M. sexta tissues in vitro (Hammock et al., 1975b) but are minor metabolites of JH I in housefly microsomes. An olefinic diol when epoxidized either chemically or enzymatically also undergoes cyclization to yield tetrahydrofuran diols and such reactions have been found in vitro in several insects (Hammock et al., 1974a). The question then arises if tetraols are ever formed from compounds of the general structure 1,2-epoxy-5-hexene. Although very minor metabolites of R-20458 in M. domestica microsomes do co-chromatograph with the tetraol of R-20458, 2,3,6,7-tetraols were disallowed by derivatization and co-crystallization studies (Hammock et al., 1974a). In mammals

such tetraols occur from juvenoid metabolism in vivo in addition
to tetrahydrofuran diols possibly arising from a coupled oxygenase-
hydratase system (Hoffman et al., 1973; Gill et al., 1974).

Other Metabolism of the Terpenoid Chain. Metabolic reduction
of the 2,3-double bond in geraniol has been shown in mammals (Kuhn
et al., 1936). Thus, reduced metabolites of geranyl phenyl ethers
such as R-20458 might be expected, but none have been reported.
Methoprene is reduced to totally saturated hydroxy and methoxy
acids by chickens (but not insects) which are then largely stored
by glycerides (Quistad et al., 1975e). Other metabolic pathways
for the terpenoid chain have also not been established in insects.
The phenols from mammalian metabolism of geranyl phenyl ethers
probably result from rearrangement of an unstable intermediate
which has been hydroxylated on carbon 1 of the terpenoid side chain.
Hydroxylations on carbons alpha to olefins are common degradative
reactions and such hydroxylated metabolites of R-20458 have been
identified on the basis of MS in the rat (Hoffmann et al., 1973).
Such hydroxylated compounds have not been identified as metabolic
products from any JH or juvenoid in insects, but they probably
account, in part, for minor metabolites often found during degradation in juvenoids by some insects.

Juvenoid Modification of JH Metabolism

It has been suggested (Slade and Wilkinson, 1973) that juvenoids possess minimal inherent juvenile hormone activity but rather
are morphogenically disruptive because they inhibit the JH degradation system. Hence JH is synergistically protected from degradation and remains at a high titer in the relevant tissues. The
generalization that juvenoid activity is synergistic rather than
intrinsically hormonal has been refuted by several papers (e.g.
Solomon and Walker, 1974), yet Slade and Wilkinson's paper did
acknowledge some interesting problems. For instance, Slade and
Wilkinson (1973) refer to propyl 2-propynylphenylphosphonate
(NIA 16388) as a juvenile hormone analog (JHA) due to its activity
in a morphogenetic assay (Bowers, 1968) while a more restrictive
definition of juvenile hormone mimic or juvenoid requires that the
compound restores a normal JH mediated function in an allatectomized insect. However, Solomon et al. (1973) have shown that at
least piperonyl butoxide has low inherent JH activity. Data
indicating that juvenoids can stabilize JH at physiological levels
is lacking. One can, however, easily predict that some juvenoids
such as the methylenedioxyphenyl ethers will act as microsomal
oxidase inhibitors and possibly retard their own metabolism
(Brooks, 1973), and the high doses of some known synergists
required to produce a morphogenetic response may actually stabilize
intrinsic JH (among other actions). Also, interesting is the low
but detectable inhibition of insect epoxide hydratase by some

juvenoids (Brooks, 1973; Slade et al., 1975).

The structural similarity of most juvenoids to JH suggests that juvenoids may mimic JH in several systems and that competitive metabolism is quite probable. Undoubtedly there are many sites of action for JH in an insect and each site might be slightly different; a given compound may be a JH mimic, antagonist, or have no effect at the various active sites. Part of the remarkably disruptive activity of some juvenoids may result from their mimicking the natural hormone at some, but not all, of its sites of action.

Terriere and Yu (1973) and Yu and Terriere (1974a) reported that high doses of some juvenoids and synergists modify xenobiotic metabolism in M. domestica adults. Of more interest were the findings by these workers (Yu and Terriere, 1974b) of substantial changes in the microsomal oxidase levels of M. domestica during development and that these changes may be, in part, under the control of JH and thus influenced by juvenoids.

Metabolism Activity Correlations

Both activation and inactivation of virtually every class of pesticides determines their selective toxicity to different organisms or various stages of the same organism. The great differences in juvenoid activity suggest that JH receptors vary greatly in their requirements for juvenoid fit at the active site(s). Since receptors have not been characterized, we cannot sharply distinguish between selectivity caused by metabolism and action at the active site. Hammock et al. (1974b, 1975a) suggested that a refractory pool (possibly specific carrier molecules) may be important in explaining juvenoid action, and this subject is examined in several papers presented in this book. However, assuming equal response at the active sites some correlations can be drawn between metabolism and activity.

Using parabiosis experiments Reddy and Krishnakumaran (1972) presented evidence that several juvenoids were more stable and more active in T. molitor than JH I. Quistad et al. (1975a) explored the age-dependent response by A. aegypti mosquitoes to methoprene. These mosquito larvae are known to be most sensitive to methoprene as late (24-48 hr) fourth instars. Contrary to expectations the total amount of methoprene metabolized by A. aegypti increased with age in third, early fourth (0-24 hr), and late fourth instars and activated metabolites were contraindicated by the decreased activity of known metabolites. However, other factors such as selective cuticle penetration may have been operative since the actual amount of unmetabolized methoprene inside the most sensitive late fourth-instar larvae was 2-3 times greater than in less susceptible younger larvae. The same workers also found that Culex larvae are apparently

less sensitive to methoprene than A. aegypti because they inactivate it faster.

In a comparative study using alkoxides of geranyl phenyl
ethers (10, Fig. 2) Hammock et al. (1975a) suggested that the high
activity of ethyl and propyl alkoxides on T. molitor is due in part
to the enhanced stability of the alkoxides over the epoxides. The
same authors also warned that in insects with high levels of O-
dealkylating enzymes but low levels of epoxide hydratases as M.
domestica or S. calcitrans, the opposite relationship between structure and activity might be found.

In most cases metabolism of juvenoids results in products with
considerably reduced morphogenic activity. A possible exception to
this rule may be O-dealkylation of methoprene to the hydroxy ester
which appears to be an activated metabolite in O. fasciatus (Solomon
and Metcalf, 1974) and is inherently more active against some
insects under bioassay conditions (Henrick C.A. and Staal, G.B.,
unpublished). Another exception is the diene of R-20458 which has
already been discussed as a probable metabolic precursor of the
more potent R-20458 (Hammock et al., 1974a).

Juvenoid Metabolism as a Resistance Mechanism. The development
of resistance to a pest control agent by the target species is a
major problem in insect control. Some workers predicted that juvenoids would be immune to the problem of resistance although most
entomologists feel that resistance can develop to any agent provided
there is intense selection pressure and an adequate genetic pool.
Thus, it was not surprising when cross-resistance between juvenoids
and insecticides was found (Dyte, 1972; Cerf and Georghiou, 1972,
1974; Benskin and Vinson, 1973; Plapp and Vinson, 1973; Vinson and
Plapp, 1974; Kadri, 1975). Laboratory selection pressure with
juvenoids has resulted in the development of resistance in several
insect species (Brown and Brown, 1974; Brown et al., 1974;
Georghiou et al., 1974; Georghiou, 1975). Although there are numerous mechanisms of resistance to insecticides, it has been noted
that strains of M. domestica with high cross-resistance to juvenoids have high microsomal oxidase levels (Plapp and Vinson, 1973;
Cerf and Georghiou, 1974; Vinson and Plapp, 1974). Terriere and
Yu (1973) indicated the involvement of the microsomal oxidase
system to the degradation of methoprene by M. domestica enzymes
and further showed that the degradation was more rapid in enzymes
from resistant strains. Hammock et al. (1974a) found that R-20458
is more rapidly metabolized (a susceptible SCR strain) in enzymes
from a R-Bagon (propoxur) strain than a susceptible SCR strain.
The R-Bagon strain shows cross-resistance to several juvenoids and
has a high microsomal oxidase level (Vinson and Plapp, 1974). The
R-Dimethoate strain of M. domestica shows a high degree of cross-
resistance to juvenoids (40X), and it and other strains have been
further selected with methoprene to obtain highly resistant strains

(R-methoprene, 1190X; Georghiou, 1975). Both the R-Dimethoate and R-methoprene strains show a high level of microsomal oxidase. Larvae, pupae, and adults of these strains are capable of rapidly metabolizing methoprene and R-20458 both in vitro and in vivo with O-demethylation and α-hydroxylation being the predominant respective pathways. Although the in vitro metabolism of methoprene by R-Dimethoate and R-methoprene microsomes is similar, a slight decrease in penetration coupled with a slight increase in in vivo metabolism in white larvae results in a lower peak level of methoprene in larvae of the R-methoprene strain when low topical dose levels are used. No differences in epoxide hydratase or esterase activity were detected in any of the strains. It is unlikely that a complete explanation has been found for the resistance present in the R-methoprene strain (Hammock et al., 1975b), but possibly a narrow window of susceptibility accounts in part for the shallow slope of the dose-response regression lines noted by Georghiou (1975) or the resistance may be due in part to the R-methoprene strain no longer recognizing methoprene as a JH mimic.

It is obvious that insects can become resistant to juvenoids and indeed that cross-resistance already exists. Whether resistance will become a problem depends upon the intensity of selection pressure caused by juvenoid use. Laboratory work with mosquitoes is encouraging in that only minimal resistance has been induced under intense selection pressure (Georghiou et al., 1974); however, studies with M. domestica may be indicative of future resistance problems. It is likely that enhanced metabolism of juvenoids will be an important factor in resistance.

SUMMARY

In general, insects metabolize juvenoids in the same manner as other xenobiotics. Ester cleavage, epoxide hydration, olefinic isomerization, and various oxidative pathways have been established for juvenoid metabolism with metabolic pathways largely determined by the functionality available for biochemical modification in each juvenoid. It cannot be generalized that a given metabolic pathway is most important for all juvenoid degradation, since biologically active juvenoids exist in the absence of specific functional groups (e.g. R-20458 lacks ester and methoprene lacks epoxide). Thus, the basis for JH activity in each juvenoid must be considered separately and the importance of various metabolic routes can only be measured empirically until more is known about the molecular structure-activity relationships for natural juvenile hormones.

REFERENCES

Ajami, A.M., 1975, J. Insect Physiol. 21: 1017.

Ajami, A.M., and Riddiford, L.M., 1973, J. Insect Physiol. 19:635.

Benskin, J., and Vinson, S.B., 1973, J. Econ. Entomol. 66:15.

Bowers, W.S., 1969, Science 164:323.

Bowers, W.S., Thompson, M.J., and Uebel, E.C., 1965, Life Sci. 4:2323.

Brooks, G.T., 1973, Nature 245:382.

Brooks, G.T., Harrison, A., and Lewis, S.E., 1970, Biochem.Pharmacol. 19:255.

Brown, A.W.A., Brown, T.M. and Hooper, G.S.H., 1974, in "Report of the Pesticide Research Center," p. 29, Michigan Agricultural Expt. Station, East Lansing, Michigan.

Brown, T.M., and Brown, A.W.A., 1974, J. Econ. Entomol. 67:799.

Cerf, D.C., and Georghiou, G.P., 1972, Nature 239:401.

Cerf, D.C., and Georghiou, G.P., 1974, Pestic. Sci. 5:759.

Dunham, L.L., Schooley, D.A., and Siddall, J.B., 1975, J. Chromat. Sci. 13:334.

Dyte, C.E., 1972, Nature 238:48.

Eto, M., and Abe, M., 1971, Biochem. Pharmacol. 20:967.

Georghiou, G.P., 1975, World Health Organization Committee on Resistance of Vectors and Reservoirs to Pesticides, Working Paper (September, 1975) p. 1.

Georghiou, G.P., Lin, C.S., Apperson, C.S., and Pasternak, M.E., 1974, Pro. Calif. Mosq. Control Assoc. 42:117.

Gill, S.S., Hammock, B.D., Yamamoto, I., and Casida, J.E., 1972, in "Insect Juvenile Hormones: Chemistry and Action" (J.J. Menn and M. Beroza, eds.), pp. 177-189, Academic Press, New York.

Gill, S.S., Hammock, B.D., and Casida, J.E., 1974, J. Agric. Food Chem. 22:386.

Hammock, B.D., Gill, S.S., and Casida, J.E., 1974a, Pestic. Biochem. Physiol. 4:393.

Hammock, B.D., Gill, S.S., and Casida, J.E., 1974b, J. Agr. Food Chem. 22:379.

Hammock, B.D., Gill, S.S., Stamoudis, V., and Gilbert, L.I., accepted for publication, 1974c, Comp. Biochem. Physiol.

Hammock, B.D., Gill, S.S., Hammock, L., and Casida, J.E., 1975a, Pestic. Biochem. Physiol. 5:12.

Hammock, B., Nowock, J., Goodman, W., Stamoudis, V., and Gilbert, L.I., 1975b, Molec. Cell. Endocrinol. 3:167.

Hammock, B.D., Mumby, S.M., Lee, P.W., Fukuto, T.R., and Georghiou, G.P., 1975c, in preparation.

Henrick, C.A., Staal, G.B., and Siddall, J.B., 1973, J. Agric. Food Chem. 21:354.

Henrick, C.A., Willy, W.E., Garcia, B.A., and Staal, G.B., 1975, J. Agric. Food Chem. 23:396.

Hoffman, L.J., Ross, J.H., and Menn, J.J., 1973, J. Agric. Food Chem. 21:156.

Ivie, G.W., submitted for publication, 1975, Science.

Ivie, G.W., Wright, J.E., and Smalley, H.E., submitted for
    publication, 1975, J. Agric. Food Chem.

Jacobson, M., Beroza, M., Bull, D.L., Bullock, H.R., Chamberlain,
    W.F., McGovern, T.P., Redfern, R.E., Sarmiento, R., Schwarz, M.,
    Sonnet, P.E., Wakabayashi, N., Waters, R.M., and Wright, J.E.,
    1972, in "Insect Juvenile Hormones: Chemistry and Action" (J.J.
    Menn and M. Beroza, eds.), pp. 43-68, Academic Press, New York.

Kadri, A.B.H., 1975, J. Med. Entomol. 12:10.

Kamimura, H., Hammock, B.D., Yamamoto, I., and Casida, J.E., 1972,
    J. Agric. Food Chem. 20:439.

Kuhn, R., Kohler, F., and Kohler, L., 1936, Z. Physiol. Chem. 242:
    171.

Nowock, J., Hammock, B.D., and Gilbert, L.I., 1975, in preparation.

Plapp, F.W., and Vinson, S.B., 1973, Pestic. Biochem. Physiol.
    3:131.

Quistad, G.B., Staiger, L.E., and Schooley, D.A., 1974a, J. Agric.
    Food Chem. 22:582.

Quistad, G.B., Staiger, L.E., and Schooley, D.A., 1974b, Life Sci.
    15:1797.

Quistad, G.B., Staiger, L.E., and Schooley, D.A., 1975a, Pestic.
    Biochem. Physiol. 5:233.

Quistad, G.B., Staiger, L.E., Bergot, B.J., and Schooley, D.A.,
    1975b, J. Agric. Food Chem. 23:743.

Quistad, G.B., Staiger, L.E., and Schooley, D.A., 1975c, J. Agric.
    Food Chem. 23:750.

Quistad, G.B., Staiger, L.E., and Schooley, D.A., 1975d, J. Agric.
    Food Chem. 23:299.

Quistad, G.B., Staiger, L.E., and Schooley, D.A., 1975e, J. Agric.
    Food Chem., accepted for publication.

Reddy, G., and Krishnakumaran, A., 1972, J. Insect Physiol. 18:2019.

Sanburg, L.L., Kramer, K.J., Kézdy, F.J., and Law, J.H., 1975,
    J. Insect Physiol. 21:873.

Schwieter-Peyer, B., 1973, Insect Biochem. 3:275.

Slade, M., and Wilkinson, C.F., 1973, Science 181:672.

Slade, M., and Wilkinson, C.F., 1974, Comp. Biochem. Physiol. 49B:
    99.

Slade, M., and Zibitt, C.H., 1971, Proc. II Int. IUPAC Congr. Pest.
    Chem. 3:45.

Slade, M., and Zibitt, C.H., 1972, in "Insect Juvenile Hormones:
    Chemistry and Action" (J.J. Menn and M. Beroza, eds.), pp. 155-
    176, Academic Press, New York.

Slade, M., Brooks, G.T., Hetnarski, H.K., and Wilkinson, C.F.,
    1975, Pestic. Biochem. Physiol. 5:35.

Solomon, K.R., and Metcalf, R.L., 1974, Pestic. Biochem. Physiol.
    3:127.

Solomon, K.R., and Walker, W.F., 1974, Science 185:461.

Solomon, K.R., Bowlus, S.B., Metcalf, R.L., and Katzenellenbogen,
    J.A., 1973, Life Sci. 13:733.

Šorm, F., 1971, Mitt. Schweiz. Entomol. Ges. 44:1.
Staal, G.B., 1975, Ann. Rev. Entomol. 20:417.
Terriere, L.C., and Yu, S.J., 1973, Pestic. Biochem. Physiol.
   3:96.
Vinson, S.B., and Plapp, F.W., 1974, J. Agr. Food Chem. 22:356.
Weirich, G., and Wren, J., 1973, Life Sci. 13:213.
Weirich, G., Wren, J., and Siddall, J.B., 1973, Insect Biochem.
   3:397.
White, A.F., 1972, Life Sci. 11:201.
Yu, S.J., and Terriere, L.C., 1974a, Pestic. Biochem. Physiol. 4:
   160.
Yu, S.J., and Terriere, L.C., 1974b, J. Insect Physiol. 20:1901.
Yu, S.J., and Terriere, L.C., 1975, Pestic. Biochem. Physiol. 5:
   418.

DISCOVERY OF INSECT ANTIALLATOTROPINS

W.S. Bowers

New York State Agricultural Experiment Station, Cornell
University, Geneva, New York 14456

INTRODUCTION

The juvenile hormones (JH) were quickly focused upon as potential insecticide substitutes because of their selective morphogenetic effects on insects and their simple chemical structures
(Fig. 1). The insect molting hormones (ecdysones) have received
less attention from those scientists alert to insecticidal potentials due to their complex steroidal chemistry (Fig. 1). However,
some investigators have developed non-steroidal inhibitors of
molting hormone synthesis which may also lead to methods of insect
control (Robbins et al., 1975). The interest of agricultural
scientists in hormonally based insecticides becomes clear when one
considers that the juvenile hormones and analogs show a wide range
of effectiveness against many insects and in addition they are
nonpersistent and reasonably selective to insects. Problems of
resistance in field populations have not occurred although resistance of a non-selective sort has been demonstrated in insecticide
resistant laboratory cultures (Vinson and Plapp, 1974). Although
JH analogs do show some ovicidal activities against certain insects
(Slama and Williams, 1966) the principle utility is their ability
to interrupt metamorphosis. The juvenile hormones are present in
insects throughout all life stages with the exception of those
periods during which cellular differentiation is directed towards
maturation to the adult stage. During the final stages of maturation JH must be absent to permit adult differentiation. Therefore
if exogenous JH is supplied to an insect during this critical
period, differentiation is deranged resulting in a creature consisting of a mosaic of adult and juvenile morphology which is
incapable of further development and soon perishes. The natural

JUVENILE HORMONES

J.H. I   Roller et al 1967
                (♂ Adult Cecropia)

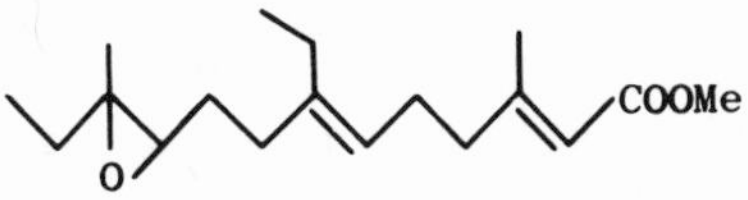

J.H. II   Meyer et al 1968
                (♂ Adult Cecropia)

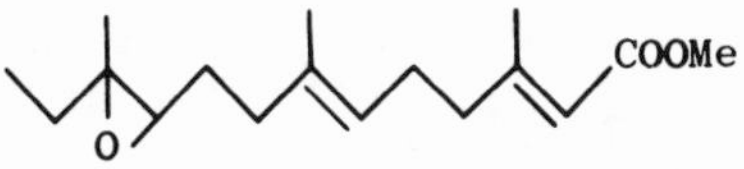

J.H. III   Bowers et al 1965
                Judy et al 1973
                (Manduca)

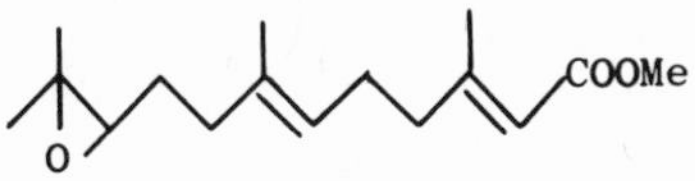

MOLTING HORMONES

α-Ecdysone   Huber & Hoppe 1965
                (Bombyx)

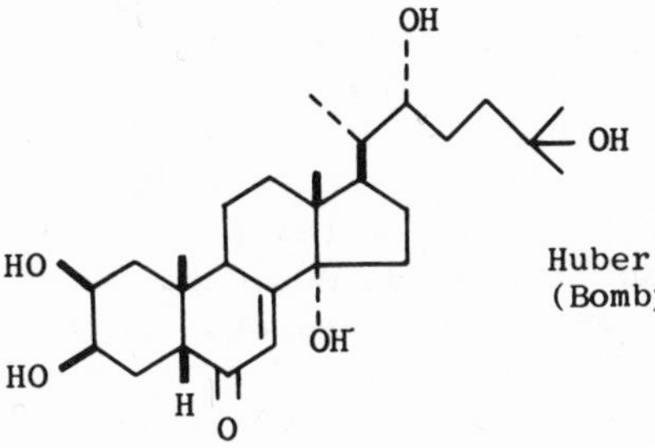

Ecdysterone   Hocks & Wiechert 1966
                Hoffmeister 1966
                (Bombyx)

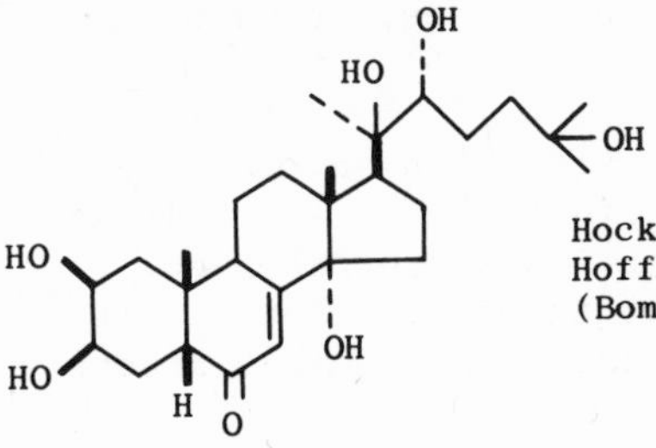

Fig. 1. Natural insect hormones.

juvenile hormones were quickly found to be too labile in practical
field situations for insect control, although hundreds of analogs
were prepared which were somewhat more stable and much more active
than the natural hormones. However, the narrow developmental period
or "gate" during which excess JH could effect a lethal derangement
has proven a drawback to the development of these hormones as widely
successful pesticides. Perhaps the most serious deficiency in this
method of insect control, is the fact that the immature and adult
insects are unaffected by an excess of JH and are able to produce

their damage (i.e., plant destruction, disease transmission).

"Methoprene" is a highly active JH analog developed by Zeocon Corporation (Hendrick et al., 1973) which is registered for control of floodwater mosquitoes, and as a cattle food additive for the control of manure breeding flies. It has also been registered for use in the silk industry to increase the yield of silk. Other JH analogs have been demonstrated to be effective for the control of certain greenhouse pests (Staal et al., 1973; Nassar et al., 1973). Doubtless, additional specialized applications for JH based insecticides will develop. However, most insect control situations contain mixed populations of insect pests in all stages of development which are not susceptible to the application of exogenous JH analogs.

From the classical background of insect endocrinology we know that surgical ablation of the corpora allata which produce the juvenile hormones causes: precocious metamorphosis of immature insects (Pflugfelder, 1937), while adult female insects are rendered sterile (Wigglesworth, 1936; Chen et al., 1962) and some insects are unable to produce their characteristic sex pheromones (Barth, 1962). Without JH many insects which normally enter diapause in the larval state due to a high titer of JH (Fukaya and Mitsuhashi, 1958; Chippendale and Yin, 1973) would be forced to complete their development irrespective of the season. Certain adult insects enter diapause when JH ceases to be produced. Removal of the corpora allata puts them in diapause (deWilde and deBoer, 1961), and treatment of diapausing beetles with JH III terminates diapause (Bowers and Blickenstaff, 1966).

Visualizing this situation several years ago I felt that the narrow gate for JH activity would severely limit the effectiveness of any JH insecticide. On the other hand, since JH is necessary for insect development throughout the immature stages, for development of the ovaries, for sex pheromone production and for larval diapause, a means of stopping JH secretion or activity seemed to me to hold more promise as an endocrinologically effective insecticide than the juvenile hormones themselves. In view of the foregoing I began searching for an antiallatotropin many years ago and urged others to do so frequently (Bowers, 1968, 1971a, 1971b). Although a great deal of study has been done on the control of JH secretion in the insect, the actual regulatory mechanism(s) remain obscure (Gilbert and King, 1973). Since plants have been shown to contain compounds with both juvenile(Bowers et al., 1966; Cerny et al., 1967) and molting hormone activity (Nakanishi et al., 1966; Galbraith and Horn, 1966; Takemoto et al., 1967) I wondered if plants might have gone a step further and developed compounds with antihormonal activities. The literature (Heal et al., 1950; Jacobson, 1958, 1975) reveals that considerable effort has been devoted to the search for conventional insecticides in plants. The extraction and biological testing methods however seemed very

limited and stereotyped. Most of the extraction methods employed
polar solvents such as water or alcohol since the investigators
were looking for polar alkaloidal insecticides such as rotenone
and ryania. The biological assays were devoted to testing almost
exclusively on adult insects for rapid toxicity within 24 to 48 hr.
Clearly, such approaches would not reveal an antihormone.

RESULTS

My approach has been to extract plants with less polar solvents
such as ether-acetone and to test upon very immature insects by
contact throughout all of the developmental stages and through one
reproductive cycle. Such an approach is difficult and laborious
but will reveal the presence of an antihormone. By this approach
I discovered two antiallatotropic compounds in the common bedding
plant Ageratum houstontonianum. My initial results were obtained
by exposing 2nd instar milkweed bug nymphs, Oncopeltus fasciatus
to a residue of the oily extract of this plant at a concentration
of 7 μg/cm^2 in a petri dish. The nymphs molted to apparently
normal third and fourth instar nymphs and then molted precociously
to tiny adults, omitting the fifth nymphal stage.

Isolation and Synthesis of Antiallatotropins

Figure 2 details the identification of two antiallatotropins
from Ageratum: 7-methoxy-2,2-dimethyl chromene and 6,7-dimethoxy-
2,2-dimethyl chromene. In view of their induction of precocious
metamorphosis I named them precocene I and II, respectively. Two
synthetic methods of general utility are shown in Fig. 3.

Subsequently, I found that precocene I and II have previously
been identified and synthesized (Alertson, 1955; Huls, 1958;
Livingstone and Watson, 1957; Kasturi and Manithomas, 1967).

Biological Activities of the Precocenes

Following purification and synthesis, the precocenes were
examined in a variety of biological systems for antiallatotropic
activity.

Precocious Metamorphosis. Milkweed bug nymphs treated by
contact with a residue of the precocenes in a petri dish consist-
ently molted into precocious adults (Table 1). Precocene I is
about one magnitude less active than precocene II. With the pure
precocenes I found that precocious maturation in the milkweed bug
occurred from the second, third, and fourth instars following
treatment of the first, second, and third instars, respectively.

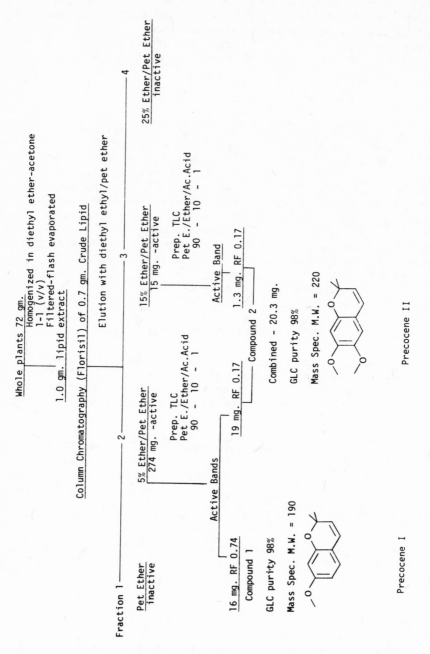

Fig. 2.   Isolation and identification of two antiallatotropins from Ageratum.

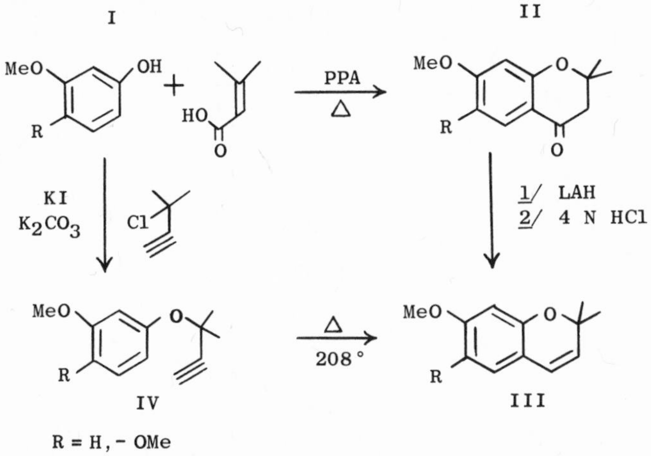

Fig. 3.* Synthesis of Antiallatotropins.
* Reaction of the appropriate phenol I with a slight excess of
3,3-dimethyl acrylic acid in polyphosphoric acid (PPA) on the steam
bath gives on work-up a nearly quantitative yield of the crystalline
chromanone II. The chromanone in ether is reduced with excess lithium aluminum hydride (LAH) and then stirred with 4 N HCl to promote
dehydration to the chromene. Overall yield is about 80%. An al-
ternate synthesis (Hug et al., 1971) via the propargyl ether IV,
gives a lower yield.

Thus in order to induce precocious maturation it is necessary to
treat the previous nymphal stage. Rarely, treatment of a very
early fourth stage nymph is followed by precocious maturation from
that instar. Figure 4 shows the precocious adults obtained from
the second, third, and fourth instar following treatment of first,
second, and third stage nymphs. Perhaps an intercallary instar is
necessary following the induced cessation of JH to allow the expression of some intermediate genetic information prior to the expression of the genetic information requisite to adult development.
The intercallary stage may assume the status of the normal fifth
instar nymph which by analogy must functionally resemble the pupal
stage of the holometabola. Alternatively, the precocenes may simply
require an extended period of time to reduce the JH titer sufficient
to permit adult differentiation. Examination of the precocious
adults reveals that the morphology and coloration are identical to
the normal adult insect. The lateral lobes on the fifth abdominal
segment of the ventral aspect of the abdomen which are useful in
separating the adult sexes are present in the precocious adults.
The wings are not as large and proportional to the body in the precocious adults as in the normal adults. Apparently, under the en-
forced telescoped developmental sequence there is insufficient time
for normal wing cell elongation to occur. Under magnification,

TABLE 1.   INDUCTION OF PRECOCIOUS METAMORPHOSIS IN THE MILKWEED BUG.

Concentration[1]/ $\mu g/cm^2$	Mortality %	Precocious adults % [2]/
Precocene I		
3.9	55.0	100
1.9	10.0	0.0
Precocene II		
0.7	10.0	90.0
0.4	5.0	15.0
Control	0.0	0.0

[1]/ Twenty 2nd instar nymphs were confined in a 9 cm petri dish
   containing the specified residue of antiallatotropin.

[2]/ Surviving nymphs molted to 3rd and 4th instars and then molted
   to precocious adults.

however, the coloration, texture and veination are distinctly adult-
oid. Dissection reveals adult testis and ovaries which appear
grossly to be functional. The ovaries of the female precocious
adults do not undergo development or yolk deposition and the females
are sexually unreceptive despite the very positive attentions of the
precocious males.

    An interesting developmental characteristic of Hemiptera is the
possession of 2 segmented tarsi throughout the nymphal stages which
give rise to 3 segmented tarsi in the adult stage. The precocious
adult arising from a 4th stage nymph develops normal adult 3 segmented tarsi, however precocious adults arising from 2nd and 3rd
stage nymphs retain 2 segmented tarsi. Thus, certain morphological
features appear to require some fixed number of developmental stages.

    It is also possible to induce precocious maturation by fumiga-
tion of milkweed bug eggs with precocene II. In order to minimize
contamination the eggs are removed from the fumigation chamber just

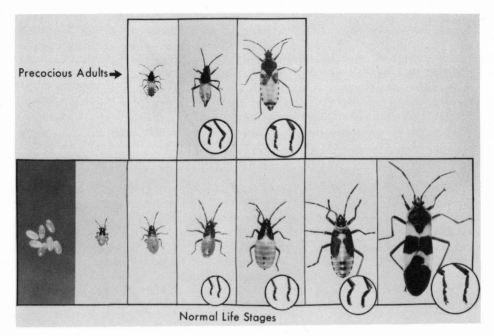

Fig. 4.   Induction of precocious metamorphosis in milkweed bugs
with precocene II.

prior to hatching and the empty chorions removed immediately
following hatching. Subsequently, normal nymphal development
proceeds to the third instar and then precocious adults are formed.
Probably very little of the precocene penetrates into the embryo
and therefore a lengthy period is required to effectively lower
the JH titer. Other Hemiptera are sensitive to the precocenes and
precocious metamorphosis has been induced in Lygaeus kalmii Stal.
and Dysdercus cingulatus (F.).

   Antigonadotropic Activity. Precociously developed females
have well formed ovaries which remain essentially undeveloped
throughout their short life. Topical treatment of these females
with JH III induces ovarian development to the stage of tiny yolked
eggs, however oviposition does not take place. The micropyles of
these eggs were formed toward the middle of the eggs rather than
on the end, indicating that the eggs were less than perfectly
formed. These results establish that the ovaries represent a
poised target capable of responding to the gonadotropic activity of
JH III (Bowers et al., 1965), and also indicate that the inhibitory
actions of the precocenes are not at the target tissue receptor site.

Treatment of a variety of adult insects with the precocenes soon after emergence prevents ovarian development. Table 2 summarizes the antigonadotropic activity of precocene II. If inseminated gravid milkweed bugs are treated with the precocenes, the eggs are resorbed and the ovaries regress to the undeveloped condition. If completely mature eggs are present at the time of treatment they are deposited and hatch normally, but undergo precocious metamorphosis from the third instar. This result parallels the effect of fumigation of eggs described earlier. Precocene I is also antigonadotropic on the milkweed bug, but is only about one-half as active as precocene II.

Sex Pheromone Inhibition. Barth (1962) has shown that the gonadotropic hormone is necessary for sex pheromone production in cockroaches. Since development of the ovaries and sexual attraction are coincidental, the use of the gonadotropic hormone as a pheromonotropic hormone seems a reasonable biological economy for the insect. Topical treatment of pheromone producing virgin female cockroaches, Periplaneta americana L., with precocene II resulted in termination of sex attractant secretion within 5 days whereas untreated controls continued to produce pheromone for at least 15 days.

Diapause Induction. deWilde and deBoer (1961) showed that diapause could be induced in the adult potato beetle, Leptinotarsa decemlineata, by allatectomy. Bowers and Blickenstaff (1966) demonstrated that adult diapause in the alfalfa weevil, Hypera postica (Gyllenhal), could be terminated by topical application of JH III, proving that diapause in adult beetles is the result of an insufficiency of JH.

Topical treatment of potato beetles (held under non-diapause inducing conditions) with precocene II induced a substantial percentage of the beetles to leave their food plants, burrow into the soil and enter diapause. In Table 3, the most effective concentration was the lowest dosage used (i.e., 100 µg topical). At the higher concentrations activity and feeding were severely depressed, although the beetles were alive three weeks after treatment. Perhaps the insects remaining on the plants also went into diapause but so rapidly that they were unable to leave the plant and burrow into the soil.

Ovicidal Activity. Sláma and Williams (1966) and Riddiford and Williams (1967) have shown that JH and analogs block insect embryogenesis at different stages depending upon the time of hormone application. The antiallatotropins are also ovicidal by fumigation of the eggs of milkweed bugs and the Mexican bean beetle, Epilachna varivestis Mulsant (Table 4). Topical treatment of Mexican bean beetle eggs on bean leaves was also successful (Table 5). The ovicidal action occurs late in embryonic development since fully

TABLE 2. STERILIZATION OF INSECTS WITH ANTIALLATOTROPIN.

Concentrations (Precocene II) necessary to prevent ovary development	Insect
7.0 μg/cm^2 [1]	Adult Milkweed Bug (Oncopeltus fasciatus)
7.0 μg/cm^2 [2]	Cotton Stainer (Dsydercus cingulatus)
1.5 μg/cm^2 [3]	Apple Maggot (Rhagoletis pomonella)
1000 ppm spray [4]	Mexican Bean Beetle (Epilachna varivestis)

[1] Eight newly-emerged females confined to treated 9 cm petri dish for 48 hr. Ovaries examined for development after 6 days.

[2] Ten newly-emerged females confined to treated 9 cm petri dish for 72 hr. Ovaries examined for development after 13 days.

[3] Forty-five newly-emerged apple maggot females were confined to a 9 cm petri dish containing a residue of the test compound for 30 hr. After treatment flies were held in oviposition cages and examined for ovarian development when control insects began oviposition.

[4] Ten newly-emerged Mexican Bean Beetle females were sprayed while feeding on a bean plant with emulsified Precocene II. Ovaries were dissected out and examined for development when controls began oviposition.

developed embryos could be seen through the chorion. A few nymphs did emerge but died within a few hours. Since the endocrinological events during insect embryogenesis are quite obscure it is untimely to speculate upon the mode of ovicidal action.

Reversal of Antiallatotropic Activity. If the precocenes bring about their antihormonal activity by directly or indirectly inter-fering with allatal activity, their biological actions should be reversible by treatment of blocked insects with juvenile hormones. Although I have not yet attempted to reverse all of the antiallato-tropic activities I have been able to prevent precocious metamor-phosis of milkweed bug nymphs by treatment with precocene II and JH III. Similarly, adult milkweed bugs treated with precocene II

TABLE 3.   INDUCTION OF DIAPAUSE IN ADULT COLORADO POTATO BEETLES.

Synthetic Precocene II Topical (µg)[1]	Insects Entering Diapause % [2]
500	40
250	38
100	75
Control	0

[1] Insects were treated topically on the abdomen with the anti-allatotropin in 1 µl of acetone and placed upon potato plants growing in soil.

[2] After 21 days insects remaining on the potato plants were judged non-diapausing. Insects which had entered the soil and become quiescent were screened from the soil and judged to be diapausing.

TABLE 4.   OVICIDAL ACTIVITY OF ANTIALLATOTROPIN BY FUMIGATION.

Concentration of Precocene II (mg) necessary to prevent egg hatch or survival of 1st instar insects[1]	Insect
0.5 mg	Milkweed Bug
0.2 mg	Mexican Bean Beetle

[1] Newly-laid eggs were confined in watch glasses exposed to vapors of compound placed on lid of watch glass. A small percentage of nymphs or larvae sometimes emerged from the eggs but died within a few hours.

TABLE 5. OVICIDAL ACTIVITY OF ANTIALLATOTROPIN BY CONTACT SPRAY
ON MEXICAN BEAN BEETLE EGGS.

Concentration of Precocene II in spray [1]	% Hatch	% Dead Larvae
100 ppm	2.7	100
10 ppm	18.4	76.7
Control	80.0	0.0

[1] Four-day-old eggs on bean leaves were sprayed with Precocene II
in emulsion formulation.

and JH III undergo normal mating, ovarian development and oviposition.

Mode of Action. Since the precocenes terminate allatal function in both immature and adult stages, affecting a variety of
physiological functions dependent upon JH, a common mode of action
is suggested, despite the conflicting evidence over the natural
mechanism(s) of allatal regulation (Gilbert and King, 1973).
Although the mode of action of the precocenes is unknown, several
possibilities may be considered, such as: 1. Interference with
biosynthesis within the corpora allata; 2. Disruption of brain
regulation of allatal activity; 3. Induction of enzymes responsible
for the metabolism of JH or a precursor; 4. Disruption of, or
competition for hormone-protein bindings (Whitmore and Gilbert,
1972); 5. Mimicry of some hormone metabolite responsible for feedback control of allatal secretion; and 6. Competition with JH at
the sites of activity. Since precocious metamorphosis and antigonadotropic activities can be reversed with exogenous JH, the
last hypothesis seems the least worthy. Each of these possibilities would be subject to experimental determination and, in any
event, represent worthwhile areas of investigation for the discovery
and development of additional methods of hormone antagonism.

An additional action of the precocenes which I have occasionally observed is that a few treated nymphs fail to molt into precocious adults and remain as nymphs without further molting until
death. If the precocenes are acting via the brain to inhibit
allatal activity perhaps they also interfere with the release of

prothoracotropin from the brain. Without prothoracotropin the
insect would be unable to produce the molting hormone and further
development would be impossible. An alternative explanation would
be that the precocenes act directly upon the prothoracic gland to
prevent ecdysone synthesis or secretion.

DISCUSSION

The discovery of antiallatotropins raises the question of
whether the precocenes are natural plant protectants or curious
products of plant metabolism of indefinable benefit to the plant.
In our limited experience, Ageratum as a cultivated ornamental,
is not highly susceptible to insect attack although occasional
insect feeding damage has been observed. In the greenhouse
Ageratum is readily attacked by white flies and requires protection. However, it should be noted that white flies feed upon the
plant sap and the precocenes may not be present in the plant fluids.
On the other hand, leaf feeding insects would probably consume
sufficient of the antiallatotropins to be affected if absorption
through the gut occurs. More depthful study will be necessary to
determine whether the precocenes are genuine plant protectants.

Since both the developmental and adult stages are sensitive
to the antiallatotropins the possibility of utilizing such compounds for insect control are intriguing. Indeed all of our biological evaluations with the precocenes indicate disruption of
normal embryogenesis, feeding, development, reproduction, sexual
attraction, diapause and behavior. It does not seem unreasonable
to speculate that if the use of juvenile hormones and analogs,
with their limited actions against insects are considered 3rd
generation insecticides (Williams, 1967), then the antiallatotropins may be entertained as a 4th generation concept. Even if the
present compounds prove to be of insufficient activity against a
broad spectrum of insect pests the scope of a new mode of insect
control has been revealed. The principle of antagonizing allatal
function and thus interrupting multiple vital functions encompassing
the whole of insect life is demonstrated. Since chemical structure
optimization studies resulted in JH analogs or derivatives greatly
exceeding the activity of the natural juvenile hormones, it might
be anticipated that similar effort with the antiallatotropins will
be successful.

Although the precocenes represent the first insect endocrine
antagonists to be discovered, and appear to hold promise for future
development of insect control agents, we should not neglect the
possibility that plants might have evolved other endocrinologically
specific defense mechanisms, such as antiecdysones. Numerous other
aspects of insect endocrine functions are recognized which may be
exploited through deliberate study of insect-plant interactions.

ACKNOWLEDGEMENTS

I am deeply indebted to The Rockefeller Foundation and to Hoffmann LaRoche Inc. for their generous support of these investigations.

REFERENCES

Alertson, A.R., 1955, Acta Chem. Scand. 9:1725.

Barth, R.H., 1962, Gen. Comp. Endocrinol.2:53.

Bowers, W.S., 1971a, Bull. Soc. Ent. Suisse. 44:115.

Bowers, W.S., 1971b, in "Naturally Occurring Insecticides" (M. Jacobson and D. Crosby, eds.), pp 307-332, Marcel Dekker Inc. New York.

Bowers, W.S., and Blickenstaff, C.C., 1966, Science 154:1673.

Bowers, W.S., Thompson, M.J., and Uebel, E.C., 1965, Life Sci. 4:2323.

Bowers, W.S., Fales, H.M., Thompson, M.J., and Uebel, E.C., 1966, Science 154:1020.

Cerny, V., Dolejs, L., Labler, L., and Šorm, F., 1967, Coll. Czech. Chem. Commun. 32:3926.

Chen, D.H., Robbins, W.E., and Monroe, R.E., 1962, Experientia 18:577.

Chippendale, G.M., and Yin, C.M., 1973, Nature 246:511.

Fukaya, M., and Mitsuhashi, J., 1958, Jap. J. appl. Ent. Zool. 2:223.

Galbraith, M.N., and Horn, D.H.S., 1966, Chem. Commun. 905.

Gilbert, L.I., and King, D.S., 1973, in "The Physiology of Insecta" Vol. 1, 2nd Ed., Academic Press, Inc., New York.

Heal, R.E., Rodgers, E.F., Wallace, R.T., and Starnes, O., 1950, Lloydia 13:89.

Henrick, C.A., Staal, G.B., and Siddall, J.B., 1973, J. Agr. Food Chem. 21:354.

Huber, R., and Hoppe, W., 1965, Chem. Ber. 98:2403.

Hochs, P., and Wiechert, R., 1966, Tetrahedron Letters, 2989.

Hoffmeister, H., 1966, Z. Naturforsch. 21b:335.

Hug, R., Frater, G., Hansen, H.J., Schmid, H., 1971, Helv. 54:306.

Huls, R., 1958, Bull. Soc. Chim. Bolg. 67:22.

Jacobson, M., 1958, Agri. Handbook No. 154, USDA.

Jacobson, M., 1975, Agri. Handbook No. 461, USDA.

Kasturi, T.R., and Manithomas, T., 1967, Tetrahedron Letters 27: 2573.

Livingston, R., and Watson, R.B., 1957, J. Chem. Soc., 1509.

Meyer, A.S., Schneiderman, H.A., Hanzman, E., and Ko, J.H., 1968, Proc. Nat. Acad. Sci. USA 60:853.

Nakanishi, K., Koreeda, M., Sasaki, S., Chung, M.L., and Hsu, H.Y., 1966, Chem. Commun., 915.

Nassar, S., Staal, G.B., and Ormanious, N.I., 1973, J. Econ. Entomol., 66:847.

Pflugfelder, V.O., 1937, Zeitschr. Wiss. Zool., 149:477.

Riddiford, L.M., and Williams, C.M., 1967, Proc. Nat. Acad. Sci. USA 57:595.

Robbins, W.E., Thompson, M.J., Svoboda, J.A., Shortino, T.J., Cohen, C.F., Duthy, S.R., and Duncan, O.J., 1975, Lipids 10:353.

Röller, H.K., Dahm, H., Sweeley, C.C., and Trost, B., 1967, Angew. Chem. 79:190.

Sláma, K., and Williams, C.M., 1966, Nature 210:329.

Staal, G.B., Nassar, S., and Martin, J.W., 1973, J. Econ. Entomol., 66:851.

Takemoto, T., Hikino, Y., Arai, T., Kawahara, M., Konno, C., Arihara, S., and Hikino, H., 1967, Chem. Pharm. Bull. (Tokyo) 15:1816.

Vinson, S.B., and Plapp, F.W., 1974, J. Agr. Food Chem. 22:356.

Whitmore, E., and Gilbert, L.I., 1972, J. Insect Physiol. 18:1153.

Wigglesworth, V.B., 1936, Quart. J. Microscop. Sci. 79:91.

deWilde, J., and deBoer, J.A., 1961, J. Insect Physiol. 6:152.

Williams, C.M., 1967, Sci. American 217:13.

SUMMARY OF SESSION V:

EFFECTS OF JUVENILE HORMONE AT THE MOLECULAR LEVEL (PROTEIN

SYNTHESIS)

J. Ofengand

Roche Institute of Molecular Biology, Nutley,
New Jersey 07110

As the title of the session indicates, this portion of the
conference was devoted to studies which attempted to elucidate
the mechanism of action of juvenile hormone (JH) at the molecular
level. The common thread connecting all of the papers was the
belief that JH acts by regulating macromolecular synthesis, in
particular, the synthesis of specific proteins. In its morphogenetic role, it has usually been considered that JH acts nega-
tively to block the synthesis of proteins characteristic of later
developmental stages, and an example of this effect was given in
the paper by Fristrom et al.

The system used by these workers was the leg imaginal disc
of Drosophila melanogaster which undergoes evagination in culture
when treated with ecdysone. This morphogenetic effect can be
blocked by JH (at 100 µg/ml) and this dose of JH also reversibly
inhibits protein synthesis. Other inhibitors of overall protein
synthesis such as puromycin or cycloheximide also block evagination. However, when evagination is induced by trypsin treatment
instead of by ecdysone, puromycin and cycloheximide fail to inhibit,
and JH is not inhibitory either. Since JH dramatically inhibits
the basal level of protein synthesis in the incubated discs as
well as the increased synthesis due to ecdysone, it appears that
the JH effect on evagination reflects only the general inhibition
of protein synthesis and not a specific blocking of protein synthesis required for evagination. This view is supported by the
dose-response curves which were stated to be the same for the
effects of JH on evagination and on overall protein synthesis.
The mechanism by which JH reversibly inhibits general protein synthesis in these discs is not known, but would be an interesting
area for study in view of the dramatic effects obtained. However,

409

since general protein synthesis is not affected during morphogenesis, this system may not be the most suitable for understanding the mechanism of action of JH during development.

The action of JH during the reproductive phase of adult life, on the other hand, has been shown in several cases to proceed by direct stimulation of specific protein synthesis. Consequently, since stimulatory effects are more readily analyzed than inhibitory ones and the proteins induced in the reproductive cycle systems are more readily characterizable, JH induction of reproductive phase proteins has been used by several groups as a model system to study JH action. There is the additional advantage that cause-effect relationships are generally easier to analyze in these systems than in morphogenetic situations although the advances reported by Riddiford (this volume) indicate that here also analysis at the molecular level may soon be possible.

In general, JH control of protein synthesis could occur either at the transcriptional (mRNA synthesis) level, or at the translational (protein synthesis) level. The paper by White et al. dealt with translational control by JH; namely, whether or not the availability of the appropriate aminoacyl-tRNA could regulate the synthesis of specific proteins. This hypothesis has existed for over a dozen years, but despite many attempts to show such a regulatory role for aminoacyl-tRNA in both prokaryotes and eukaryotes, the only evidence so far has been a series of experiments by Ilan and co-workers in Tenebrio molitor. These experiments consisted of three parts; first, evidence showing that mRNA for adult cuticular proteins already exists in early pupae; second, in vitro synthesis of cuticle protein dependent on a tRNA fraction from late pupae which was inactive if the pupae serving as the source of tRNA had been first treated with JH; and third, the demonstration that a new leucine tRNA and synthetase appeared late in normal pupation. Presumably, this new tRNA was the regulatory tRNA whose appearance was blocked by JH. This last aspect was reinvestigated in the paper by White et al. In summary, no changes were found either in the amount of leucine tRNA, the isoacceptor distribution, or in leucyl-tRNA synthetase level or specificity for tRNA isoacceptors when larval, early or late pupal, or adult stages of Tenebrio were examined. These results, obtained in vitro, were confirmed by in vivo labeling of leucine tRNA. The contradictory findings previously reported probably arose from failure to use optimal aminoacylation conditions, as results similar to the earlier work were obtained by White et al. when the previously reported conditions were used.

A critique of the other experimental evidence in support of the translational control theory was also given in this paper. Considering all the data, it appears that translational control via tRNA as a mode of action of JH is still no more than hypothetical,

and future efforts would be more profitably directed toward examining the role of JH in the control of transcription. Such systems
were described in the papers by Engelmann, Koeppe and Ofengand,
and Chen et al.

These papers described studies on the JH induced synthesis of
yolk protein in Leucophaea maderae and Locusta migratoria. These
systems appear to be very suitable for analysis of the mechanism
of action of JH for several reasons. First, the site of synthesis
of yolk protein is known to be the fat body where it is synthesized in a precursor form termed vitellogenin (Vg). (Chen et al.
have proposed to reserve the term vitellogenin for the precursor
form of the mature oocyte yolk protein, vitellin.) Second, fat
body can be maintained in organ culture and their vitellogenin
products, both intracellular and secreted, can be labeled with
radioactive amino acids. Third, not only does fat body from mated
females actively synthesize Vg in culture while tissue from virgin
females, males, or pregnant females does not, but virgin females
can be induced to synthesize Vg by treatment with JH and derivatives. Both insect systems are very similar in many respects so
that the choice of one over the other depends on personal preference and on whether or not the laboratory is located in the United
States where raising Locusta is illegal.

The vitellin product has been more carefully characterized in
Leucophaea, but the nature of the immediate translation product
seems more clear in Locusta. In Leucophaea, the secreted Vg
exists already as a complex of 550,000 daltons which is taken up
by the oocyte and then trimerized to a 1,600,000 dalton complex,
while in Locusta, it is 520,000 (470,000 according to Emmerich and
co-workers) and does not further aggregate upon uptake into the
oocyte. In both cases, however, oocyte vitellin consists of several
discrete smaller glycopeptides. Locusta vitellin contains 3 major
and 5 minor polypeptides by SDS-gel electrophoresis varying in size
from 55,000 to 130,000 daltons but no unique stoichiometry could
be deduced. In Leucophaea, on the other hand, the picture appears
much simpler. The trimer form of vitellin contains just 3 polypeptides in the ratio of $A_1B_3C_2$ which together account for the
molecular weight of a complete monomer. However, the native monomer
form, found only in young oocytes, contains an additional polypeptide D, slightly larger than B, and has the stoichiometry
$A_1B_1D_2C_2$. Peptides A-D range in size from 58,000 to 118,000 daltons.
Koeppe and Ofengand have proposed that D is converted to B^D by a
proteolytic process whose function is to allow trimerization to
occur in order to increase the protein content per volume of the
oocyte, but no experimental evidence was presented to support this
proposal. In both Locusta and Leucophaea, analysis of the Vg
produced by fat body showed that the complex of medium-sized polypeptides found in the oocyte was originally synthesized as one or
more large precursor proteins. In Locusta, short-term pulse-chase

experiments showed a single 260,000 dalton polypeptide to be the earliest detectable intracellular Vg product. This polypeptide began to undergo both proteolysis to smaller polypeptides and dimerization to a 520,000 dalton complex even before secretion. By the time the protein was extracellular, it consisted completely of smaller pieces in a 520,000 dalton complex. More cleavage must occur in the hemolymph since oocyte vitellin showed evidence of further proteolysis. In Leucophaea, there is clear evidence for a 260,000 dalton precursor as well as one of 180,000 daltons both intracellularly in the fat body and as part of the 550,000 dalton secretion product. There is circumstantial evidence for proteolytic processing intracellularly and direct evidence for processing in the hemolymph. Although it should have been possible to deduce precursor-product relationships in this system in view of the clear stoichiometry of Leucophaea vitellin subunits, this has not been possible so far. A hypothetical scheme for synthesis and processing of Vg was presented by Koeppe and Ofengand, whose principal feature involved the existence of two primary gene products. The best that can be said for this scheme is that it is at least consistent with all the known facts. By contrast, Locusta Vg would appear to be made as a single gene product, although even this conclusion must remain tentative until at least the stoichiometry of Locusta vitellin is determined.

In both systems, induction by JH results in a massive increase in Vg synthesis. In Leucophaea, the stimulation was over 50-fold at the peak and accounted for 80% of the total protein synthesized and secreted. Equivalent induction in ovariectomized isolated abdomens showed that ovarian hormones were not involved. The extent of stimulation of Vg synthesis in Locusta was not given, although at the peak of the response, 40-60% of the total protein secreted was Vg compared to 1-2% before induction. The kinetics of induction in both systems were similar, the peak of response occurring approximately 72 hr after induction. Chen et al. suggested that the JH analog they used, ZR515, is rather stable in Locusta since the relative capacity for Vg synthesis expressed as percent of total synthesis did not decrease from 72 to 120 hr after induction. However, as they did not show the total amount of either Vg or total protein synthesis, their results could be similar to those of Koeppe and Ofengand who found that while the relative percent of protein synthesis which was Vg did not decrease, the total amount of synthesis had decreased to one half by 120 hr. The dose of JH needed in the locust system for half-maximal response (75 μg/animal) was even higher than that needed for Leucophaea (20 μg/animal gave 65% of the maximum response).

Chen et al. presented a series of cytological observations on fat body which was or was not synthesizing Vg by virtue of either allatectomy or induction with externally applied JH. By the use of ferritin labeled antibody to Vg, they were able to localize

those intracellular Vg molecules sufficiently large enough to
possess antigenic sites. Large amounts of Vg were found in the
rough endoplasmic reticulum, in the Golgi apparatus where presumably the carbohydrate is attached, and in channels of the plasma
membrane. In other words, the distribution found was that of a
typical export protein as seen, for example, in a liver or
pancreas cell.

The cell free synthesis of Vg was attempted by both groups
and also by Engelmann, who used Leucophaea. Engelmann showed that
a 12-hr treatment with α-amanitin blocked the appearance of Vg
polypeptides on microsomes from fat body of vitellogenic females,
and concluded from this that the mRNA for Vg must turn over by
12 hr in vivo. He also presented some evidence that a 5-12S RNA
species may be the mRNA for Vg, but in view of the above discussion
concerning the size of Vg precursors, the mRNA is in all probability
closer to 30-45S in size. Perhaps the 5-12S material arose from
nucleolytic cleavage of larger molecular weight mRNA. Engelmann
was able to obtain some cell-free synthesis of protein immunologically identifiable as Vg by addition of total RNA from vitellogenic females to a system containing ribosomes from vitellogenic
females and a membrane fraction. Vg synthesis was stimulated
5-fold by the RNA addition but the percent of the total protein
synthesized which was Vg could not be determined from the data
presented. A membrane fraction was necessary to get good cell-
free synthesis.

Koeppe and Ofengand were less successful in their attempt at
cell-free synthesis of Leucophaea Vg. Although poly A containing
fat body RNA was most active in a wheat germ or ascites cell-free
systems, no protein immunologically identifiable as Vg could be
found. On the other hand, Chen et al. were able to synthesize
Locusta Vg in a wheat germ cell free system supplemented with RNA
from fat body. At best, a 2-fold stimulation over the blank level
was obtained, corresponding to 6% of the total protein synthesized,
but the RNA stimulation observed was strictly dependent on the
use of vitellogenic female fat body for the source of RNA. Allatectomized females, males, or Drosophila fat body RNA failed to
stimulate. Further characterization of the in vitro immunologically
reactive product and purification of the mRNA species involved
remain to be done.

The final paper of the session described the development of
a JH-induced protein synthesis system in Drosophila. Postlethwait
et al. demonstrated that JH was required for the synthesis and/or
uptake of vitellogenin into oocytes by measuring oocyte development cytologically. They went on to show that of 7 sterile mutants,
2 were defective in the synthesis and/or secretion of JH while the
remaining 5 were blocked at some stage in vitellogenin uptake,
since these ovaries, transplanted into normal hosts could not

develop. So far, no mutants defective in JH induction of Vg synthesis have been detected. The phenotype of such a mutant should
be one whose ovaries are capable of development when transplanted,
but which do not respond to externally added JH. Such mutants
would be extremely useful for dissection of the mechanism of JH
induction.

Another system described by Postlethwait et al. should also
prove very interesting. An ovarian acid phosphatase was identified and shown to be inducible by JH. Transplantation experiments
showed that the enzyme is synthesized in the ovary. Most significantly, one of the sterile mutants whose ovaries do not develop
even when transplanted into a normal host, did not make this
enzyme with or without JH treatment. In contrast, a mutant defective in JH production and which also lacks the enzyme activity
could be restored to normal levels by JH. The latter observation
confirms the regulatory role of JH in acid phosphatase activity.
The former experiment suggests the possibility of a mutant defective in some step of the JH-induced protein synthetic process,
since acid phosphatase is apparently made in the ovary. This
situation is to be distinguished from the failure of oocytes to
develop which reflects a failure to take up Vg rather than a block
in its synthesis. Since JH is also required for Vg uptake by
oocytes, the blockage of both effects by a single mutation raises
the exciting possibility of a mutation in the JH receptor protein
or uptake process.

It appears evident that at least three systems now exist that
are capable of analysis of the mode of action of JH at the molecular level. The cockroach and locust systems are very much alike
and are being analyzed in rather similar ways. Their main advantage lies in the abundance of material available for biochemical
studies. On the other hand, the Drosophila system has the advantage of mutants. If the lessons of bacterial genetics have any
relevance here, one would be tempted to predict that the availability of mutants in this system will eventually turn out to be the
key to understanding the mechanism of JH action.

In any case, several areas appear to be of central interest.
It is hypothesized that there is a cellular receptor for JH which
is not the hemolymph binding protein (see Kramer et al., this
volume; Goodman et al., this volume) and which plays a role similar to that for steroid hormone action (see O'Malley, this volume).
The identification of this receptor will be a major advance. Probably, the receptor-JH complex activates the chromatin of fat body
cells to allow synthesis of Vg mRNA which in turn is responsible
for the massive synthesis of Vg. There is no necessity to devise
a mechanism to turn off the synthesis of other proteins, since there
is little overall protein synthesis going on at this time in fat
body. A similar mechanism should exist for acid phosphatase

synthesis in the ovary. Consequently, isolation and characterization of these mRNAs and the demonstration that their level in the cell goes up upon JH induction will be another key area of research. However, it may be difficult to completely analyze these systems so long as JH needs to be added in vivo. Clearly, it is important to be able to induce fat body cells in culture with JH, and this aspect is now under active study in several laboratories.

A final question, however, has still to be considered. Adult females are able to respond to JH induction, but males are not. This situation is in contrast to that in chickens or frogs, where males given estradiol, will begin to synthesize vitellogenin. At what point in their life cycle is the chromatin of these male insects permanently inactivated? Can we manipulate their chromatin in vitro so that the putative JH-receptor complex can now turn on Vg mRNA synthesis? Will the knowledge of how to inactivate and reactivate chromatin tell us anything about morphogenesis? Perhaps some of the answers to these questions will be available at the next Juvenile Hormone Conference.

AN ANALYSIS OF LEUCINE tRNAs DURING THE DEVELOPMENT OF TENEBRIO MOLITOR

B.N. White, N.J. Lassam, and H. Lerer

Department of Biology, Queen's University, Kingston
Ontario Canada

INTRODUCTION

The biosynthesis of a protein can be regulated at the level of transcription of the genetic information into messenger RNA or at the level of translation of that information into the amino acid sequence of the protein. When studying the mode of regulation of protein synthesis by juvenile hormone at the molecular level it is important first to define whether this regulation is being exerted at the translational level, at the transcriptional level or possibly both.

A model has been proposed postulating a role for juvenile hormone in the regulation of the biosynthesis of pupal and adult cuticular proteins at the translational level in Tenebrio molitor (Ilan et al., 1972). This model and the evidence that gave rise to it are not only important in the context of juvenile hormone action but also, more generally, in relation to tRNA mediated translational control, as they stand as the clearest support for such a mechanism in a eukaryotic system. The model relies on several lines of evidence. Firstly, the effects of actinomycin D on RNA synthesis in pupae indicated that the messenger RNAs for adult cuticular proteins were synthesized early in pupal development (Ilan et al., 1966). The messenger RNAs for these proteins appeared to be very stable and were not translated until late in pupal[*] life when the level of juvenile hormone decreased. Further

---

[*] The terms pupae or pupal life are used to include pharate adults up to the time of the pupal-adult ecdysis.

416

studies, utilizing an in vitro protein synthesizing system from
Tenebrio pupae, demonstrated differences between the tRNAs and
aminoacyl-tRNA synthetase extracted from early and late pupae
(Ilan et al., 1970). The demonstration of these differences
relied heavily on the supposition that adult cuticular proteins
have a "high" tyrosine content, as the criterion used for the
in vitro synthesis of these proteins was a high ratio of radioactive tyrosine:leucine incorporated into trichloroacetic acid
precipitable material. These studies further indicated that there
was a substantial increase in total leucine tRNA in late pupae and
that this new leucine tRNA fraction was aminoacylated only by an
enzyme preparation extracted from late pupae (Ilan et al., 1970).

A more detailed analysis of the leucyl-tRNA synthetase from
early and late pupae indicated that the only difference appeared
to be the inability of the enzyme from early pupae to aminoacylate some leucine tRNAs that appeared late in pupation (Ilan
and Ilan, 1975a). This study also examined the leucine tRNA
species present in early and late pupae and reported that the new
leucine tRNA found in late pupae represented the major species at
this developmental stage. These studies support a classical
translational control model in which the translation of a preformed
messenger RNA is controlled by the availability of a leucine tRNA,
the synthesis and aminoacylation of which appear to be influenced
by juvenile hormone.

In Drosophila, as well, a marked change has been observed
during development in the pattern of tRNAs that recognize codons
of the type XA_U^C (White et al., 1973). The onset of this change
is also during pupal life and so there was reason to believe that
there might be correlation with the observations made in Tenebrio.
Although the molecular basis of the change in Drosophila was determined to be a modification of a nucleoside in the first position
of the anticodon, no regulatory role has yet been assigned to this
change.

The Tenebrio system was thus unique in that an observed tRNA
change appeared to have a definite regulatory role. We re-examined
the system with the intention of purifying and analyzing the leucine
tRNA that apparently was synthesized in late pupae and which appeared
to play such an important function in the control of development.

METHODS

Growth of Tenebrio. Tenebrio molitor larvae were obtained
from Carolina Biological Supply Co., Burlington, North Carolina
and reared at 29°C according to the method of Patterson (1957).
Under these conditions the pupal period lasted for 6 days with
adults emerging on the 7th.

Preparation of tRNA and Aminoacyl-tRNA Synthetases. Transfer
RNA was prepared by the phenol method of Kirby (1956) and DEAE-
cellulose chromatography procedures of Kelmers et al. (1965).
Aminoacyl-tRNA synthetases were prepared by a modification of the
method of Twardzik et al. (1971) as described elsewhere (Lassam
et al., 1975b).

Aminoacylation of tRNA in vitro. The conditions for the
esterification of amino acids to Tenebrio tRNA are fully described
elsewhere (Lassam et al., 1975a,b). Aminoacyl-tRNA for reversed
phase chromatography was isolated from the components of the
reaction mixture by DEAE-cellulose chromatography as described
by Yang and Novelli (1968).

Aminoacylation of tRNA in vivo. Approximately 25 pupae were
injected with 5 μCi (in approximately 5 μl) each of [^{3}H]-leucine
(50 Ci/mmol) and left 3-5 min. The pupae were then homogenized in
equal volumes of 88% phenol and a buffer containing 0.25 M NaCl,
0.01 M MgCl$_2$, 0.001 M 2-mercaptoethanol, 0.01 M sodium acetate,
pH 4.5. The aqueous phase was removed and the phenol re-extracted
with an equal volume of buffer; the pooled aqueous phases were
applied to a DEAE-cellulose column equilibrated with the same
buffer. The column was washed with this buffer and [^{3}H]-leucyl-
tRNA eluted with the buffer containing 0.75 M NaCl. This material
was diluted to 0.375 M NaCl and applied to the RPC-5 columns.

Reversed-phase Chromatography of Aminoacyl-tRNA. The RPC-5
system of Pearson et al. (1971) was used.

RESULTS

Properties and Assay Conditions for the Leucyl-tRNA Synthetase

Before any comparison can be made of tRNA species from different developmental stages it is important that an assay system be
established which allows rapid and complete aminoacylation of the
tRNA. We found that the conditions previously reported for the
esterification of leucine to tRNA^{Leu} species were not optimal
(Ilan et al., 1970; Ilan and Ilan, 1975a) and therefore we re-
examined the properties of the leucyl-tRNA synthetase. This was
important not only in the context of the assay but because of the
reported functional difference between the enzyme extracted from
early and late pupae (Ilan and Ilan, 1975a).

When attempting to aminoacylate tRNAs fully and rapidly the
concentration of the amino acid in the assay system should be at
least several fold higher than the aminoacyl-tRNA synthetase's
apparent K_m for that amino acid. We therefore determined the K_m

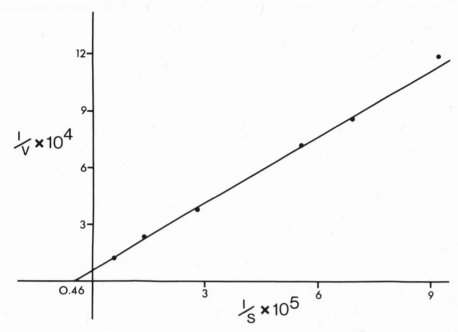

Fig. 1. Variation of [^{14}C]-leucine concentration in the esterification reaction. Reactions were terminated after 2 min with 10%
TCA and the precipitate collected on Millipore filters. The
reaction velocity, $\underline{V}$, is expressed as cpm incorporated/2-min
reaction time; substrate concentration ($\underline{S}$) is expressed as µM of
[^{14}C]-leucine.

of the leucyl-tRNA synthetase (Fig. 1) and found it to be rather
high, 22 x 10^{-6} M, although the value is 10 fold lower than the
previously reported K_m of 2 x 10^{-4} M (Ilan and Ilan, 1975a).
Therefore the minimum concentration of leucine used to aminoacylate
tRNA^{Leu} species was 50 µM. This compares with the 4-10 µM concentration used in previous studies (Ilan et al., 1970; Ilan and Ilan,
1975a).

    The rate of aminoacylation is also markedly affected by the
ratio of magnesium to ATP in the incubation mixture. When this
was examined (Fig. 2) it was found that a ratio of 0.63 Mg^{2+}:ATP
was optimal for the leucyl-tRNA synthetase. This compares closely
with other eukaryotic leucyl-tRNA synthetases (Smith and McNamara,
1971; White and Tener, 1973) but not with the previously used ratio
of 2.5 for this enzyme (Ilan et al., 1970; Ilan and Ilan, 1975a).

    The monovalent cation and pH requirements of the enzyme were
also studied. The observed pH optimum of 8.5 in Tris-HCl buffer
is in good agreement with that found by Ilan and Ilan (1975a).

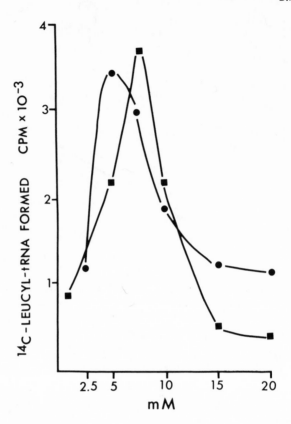

Fig. 2. Effect of ATP and MgCl$_2$ concentration on the initial rate of [^{14}C]-leucyl-tRNA formation. Reactions were terminated after 2 min with 10% TCA and the precipitate collected on Millipore filters. Solid squares, ATP; o———o MgCl$_2$.

The effect of varying the monovalent cation concentration on the initial rates of [^{14}C]-leucyl-tRNA formation is shown in Fig. 3. Potassium chloride was stimulatory in the range 8-20 mM while NaCl was generally less effective. Ammonium chloride produced a marked stimulation of the initial rate when present at relatively low concentrations but at higher concentrations became inhibitory.

This study led to an assay system for the leucyl-tRNA synthetase which contained per ml: Tris-HCl, 50 μmol; (pH 7.5 or 8.5); 2-mercaptoethanol, 5 μmol; MgCl , 5 μmol; ATP, 8 μmol; [^{14}C]-amino acid, 50 μmol; KCl, 10 μmol; aminoacyl-tRNA synthetase preparation, 1 mg of protein; and tRNA, 0.3-0.8 mg.

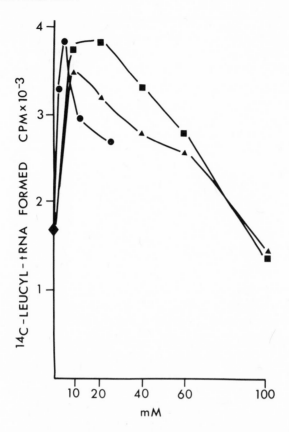

Fig. 3. Effect of monovalent cation concentration on the initial
rate of [^{14}C]-leucyl-tRNA formation. Reactions were terminated
after 2 min with 10% TCA and the precipitate collected on
Millipore filters. o——o NH$_4$Cl; Δ——Δ NaCl; solid squares, KCl.

Quantitative Acceptance of [^{14}C] Amino Acids by tRNA from Different
Developmental Stages

     Using improved assay conditions for the esterification of
leucine to tRNA^{Leu} species and similar conditions established for
phenylalanine and tyrosine esterifications (Lassam et al., 1975b)
a comparison was made of the levels of tRNAs for these amino acids
during the development of Tenebrio.

     One of the previously reported indications that the proposed
translational control mechanism involved tRNA^{Leu} species was an
increase in the acceptance of leucine by tRNA preparations from
late pupae compared with those from early pupae (Ilan et al., 1970;
Ilan and Ilan, 1975a) which was apparent only when the aminoacyl-

tRNA synthetase was prepared from late pupae. In this study
however we found the level of leucine tRNAs when compared with
either tyrosine or phenylalanine tRNAs remained essentially the
same during development (Table 1). The same result was found
using aminoacyl-tRNA synthetase preparations from early or late
pupae.

TABLE 1.   ACCEPTANCE OF AMINO ACIDS BY tRNA FROM DIFFERENT
DEVELOPMENTAL STAGES.

Amino Acid	Larval	pmol of Amino Acid accepted per A_{260} unit of total tRNA			Ratio of 1st-day pupal: last day pupal
		1st day pupal	Last day pupal	adult	
Leucine	44.3	55.3	55.1	38.1	1.07
Phenylalanine	17.7	26.3	25.4	15.2	1.02
Tyrosine	19.3	24.7	22.7	15.8	1.09

The aminoacyl-tRNA synthetase preparation used was from last-
day pupae.

RPC-5 Chromatography of [^{14}C]-leucyl-tRNAs from Different
Developmental Stages

Although no increase in the acceptance of [^{14}C]-leucine by
tRNA from late pupae was observed, there conceivably could have
been a new leucine tRNA formed with a concomitant decrease in the
other isoaccepting species. To examine this possibility [^{14}C]-
leucyl-tRNAs from different developmental stages were chromatographed on RPC-5 columns (Fig. 4). The same profile was also
obtained if the aminoacyl-tRNA synthetase used was prepared from
early pupae or from late pupae as a "microsomal wash" by the
method described by Ilan et al. (1970).

Even if a new leucine tRNA co-chromatographed with a pre-
existing one a quantitative change might be expected in the peaks
of the profile. No such change was observed.

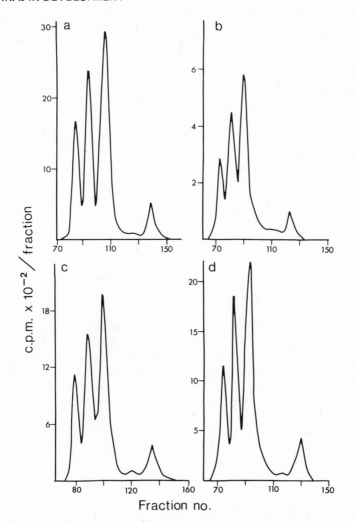

Fig. 4.   Chromatography of [^{14}C]-leucyl-tRNAs from the developmental stages on RPC-5 columns. Elution was by 100 ml linear
gradients from 0.55 to 0.675 M NaCl on jacketed columns (13 x
0.9 cm).   The flow rate was 15 ml per hr with 0.5 ml fractions
being collected. a) larval; b) first-day pupal; c) last-day
pupal (pharate adults); d) adult.

Rates and Extent of [^{14}C]-leucyl-tRNA Formation Under Limiting Conditions

A major difference between the analysis of the leucine tRNAs
in this study and those reported previously (Ilan et al., 1970;
Ilan and Ilan, 1975a) are the conditions for aminoacylation. We
therefore analyzed various parameters to see how they might affect

both the initial rates of aminoacylation as well as the final
extents. As the concentration of [^{14}C]-leucine in the reaction
mixture was reduced below 25 μM there was not only a decrease in
the initial rate but also in the final extent of aminoacylation
(Fig. 5a). Reduction in the amount of leucyl-tRNA synthetase
resulted in a similar decrease (Fig. 5b). Sub-optimal cation and
ATP concentrations produced a similar result (Fig. 5c). When
conditions similar to those used by Ilan et al. (1970) were
analyzed in this way a very slow rate of aminoacylation was found
with a significantly reduced final extent of [^{14}C]-leucyl-tRNA
formation (Fig. 5c). The plateau levels of aminoacylation reached
under these sub-optimal conditions obviously cannot be used to
estimate the amount of leucine tRNA present in the assay system
as a large amount of the tRNA^{Leu} remains uncharged.

Chromatography [^{14}C]-Leucyl-tRNAs Aminoacylated Under Sub-optimal Conditions

The reduction in the final extent of aminoacylation under
sub-optimal conditions could result either from a proportional
decrease for all tRNA^{Leu} species or from a selective decrease in
some over others. To examine this, the chromatographic profiles
of [^{14}C]-leucyl-tRNAs aminoacylated under various sub-optimal
conditions were analyzed (Fig. 6). Under all sub-optimal conditions tested peaks 1 and 2 were markedly reduced when compared
with peak 3. This differential rate of aminoacylation of the
various tRNA^{Leu} species presumably reflects some differences in
the affinity of the leucyl-tRNA synthetase for the various species.

RPC-5 Chromatography of [^{3}H]-Leucyl-tRNAs Aminoacylated In Vivo

The analysis of the leucyl-tRNAs in vitro relies on good
extraction procedures for the tRNA and aminoacyl-tRNA synthetases
as well as assay conditions that closely reflect the in vivo
situation. To verify the in vitro results an analysis was performed on [^{3}H]-leucyl-tRNAs aminoacylated in vivo. Larvae, first
and last-day pupae and adults were injected with [^{3}H]-leucine,
left 3-5 min and [^{3}H]-leucyl-tRNA extracted as described in
"Methods". The chromatographic profiles obtained (Fig. 7) are
virtually identical and are very similar to the profiles obtained
in the analysis in vitro. The only difference in the results is
the very small peak found between peaks 3 and 4 which is not
apparent in the in vivo profile. This could reflect the unavailability of the injected [^{3}H]-leucine to tissues where this tRNA
species is found. It could represent mitochondrial tRNA^{Leu} that
is not detected by the in vivo analysis or perhaps an in vitro
peak artifact. The most important point about this analysis,

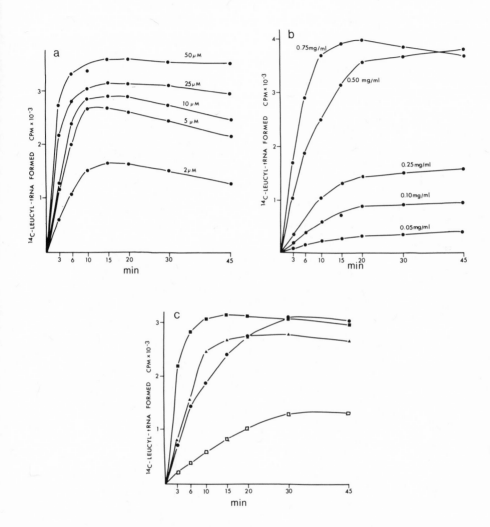

Fig. 5.   Effect of various conditions on the rates and extent
of [^{14}C]-leucyl-tRNA formation.   a) effect of varying the concentration of [^{14}C]-leucine; b) effect of varying the concentration of the aminoacyl-tRNA synthetases; c) solid squares, complete;
▲————▲ minus KCl; ●————● 10 mM MgCl$_2$ and 4 mM ATP; open squares,
10 mM MgCl$_2$, 4 mM ATP and 50 mM imidazole pH 7.0.

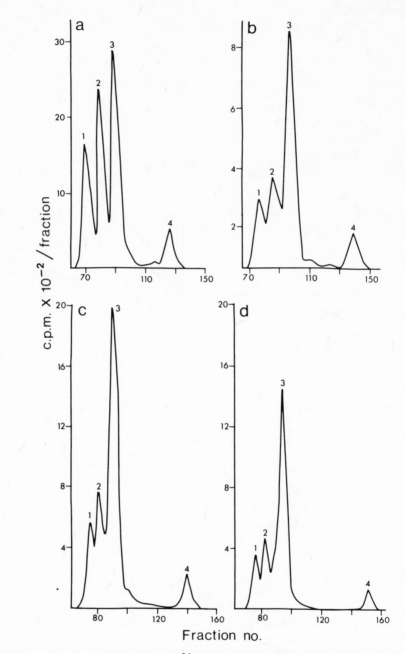

Fig. 6. Chromatography of [^{14}C]-leucyl-tRNAs aminoacylated under sub-optimal conditions. a) complete; b) 10 mM MgCl$_2$, 4 mM ATP and 50 mM imidazole pH 7.0; c) 2 μM [^{14}C]-leucine; d) 0.25 mg/ml aminoacyl-tRNA synthetase preparation.

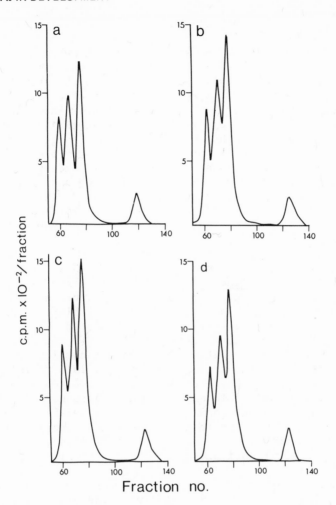

Fig. 7.   Chromatography of [³H]-leucyl-tRNAs aminoacylated in vivo.   Pupae were injected with [³H]-leucine and the [³H]-leucyl-tRNA extracted as described in the "Methods".   Elution was by 500 ml linear gradients from 0.55-0.675 M NaCl on jacketed columns (50 x 1.25 cm).   The flow rate was 75 ml per hr with 2.5 ml fractions being collected.   a) larval; b) first-day pupal; c) last-day pupal (pharate adult); d) adult.

however, is that no difference was observed between early and late pupae, and this confirms the conclusions from the in vitro work.

DISCUSSION

The potential of a translational control mechanism mediated by a transfer RNA has been recognized for a long time (Ames and Hartman, 1963; Stent, 1964). In these models the absence or low level of a particular tRNA species was postulated to prevent or regulate the rate of translation of a messenger RNA containing codons which could be recognized only by that tRNA species. It is now well established that there are often changes in tRNAs when a cell passes from one physiological state to another. Such changes have been observed at different developmental stages of various eukaryotes (Lee and Ingram, 1967; Yang and Comb, 1968; Molinaro and Mozzi, 1969; Bagshaw et al., 1970; DeWitt, 1971; White et al., 1973), but only in Tenebrio molitor has an actual control mechanism been suggested for such a tRNA change (Ilan et al., 1970).

The evidence for tRNA mediated translational control of adult cuticular proteins in Tenebrio molitor is of three kinds. The first requirement of course for any translational control model is the demonstration of the existence of a stable messenger RNA that is synthesized well before it is translated. The evidence for this in Tenebrio rests partly with the observation that injection of actinomycin D into pupae does not completely prevent the synthesis of adult cuticle (Ilan et al., 1966). This observation was taken to indicate that the mRNAs for adult cuticular proteins are synthesized in early pupae but are not translated until later in pupal development. However injection of actinomycin D inhibited only 51-67% of labeled uridine incorporation into RNA. Therefore it cannot be assumed that the synthesis of the mRNAs coding for the adult cuticular proteins was completely blocked.

The second series of experiments suggesting that the synthesis of the mRNAs for adult cuticular proteins occurs in early pupae involves a cell-free protein synthesizing system (Ilan et al., 1970). The definition of the in vitro synthesis of adult cuticular proteins in this system relies on the claim that adult Tenebrio cuticular proteins are high in tyrosine while pupal cuticular proteins contain little tyrosine (Ilan et al., 1970). This claim is apparently based on published analyses on the beetle Agrianome spinicollis (Hackman and Goldberg, 1958). However it appears from the more recent work of Patel (1972) and Anderson et al. (1973) that Tenebrio adult cuticular proteins have in fact less tyrosine than pupal cuticular proteins. This makes interpretation of the results obtained in the Tenebrio cell-free protein synthesizing system very difficult (Ilan et al., 1970). If the criterion for in vitro adult cuticular protein synthesis is incorrect then the apparent change in the ratio of tyrosine: leucine incorporation with certain combinations of components in the assay system, cannot be used to support the suggestion that the mRNAs for these proteins are present in early pupae. It is this same cell-free system and the same criterion that provides the strongest support for the involvement of juvenile hormone in the proposed translational control model (Ilan et al., 1972). Again the interpretation of the data clearly changes if the criterion for adult cuticular protein synthesis is not valid.

The final requirement for a tRNA mediated translational control system is the demonstration of the absence or low level of the controlling aminoacyl-tRNA at a stage when the particular mRNA is not being translated and its presence when the mRNA is being translated. The first indication that there was a difference in the tRNA and aminoacyl-tRNA synthetases came from data derived from the cell-free protein synthesizing system and the problems with this assay have already been discussed. The evidence that leucine tRNA and leucyl-tRNA synthetase change during development comes from aminoacylation assays (Ilan et al., 1970). In these assays tRNAs and aminoacyl-tRNA synthetases from first day and seventh day pupae were mixed in certain combinations. A marked increase in the acceptance of [^{14}C]-leucine by seventh day pupal tRNA was found when the source of the enzyme was also seventh day pupae. This type of assay rests heavily on optimal aminoacylation conditions because the validity of the interpretation relies on complete reaction and meaningful plateau levels. In the study reported here we found that the assay conditions used previously were far from optimal (Ilan et al., 1970; Ilan and Ilan, 1975a) and that they did not produce plateau levels representing complete aminoacylation.

When optimal conditions were used for the aminoacylations we observed no change in the level of leucine tRNAs during development when compared with phenylalanine and tyrosine tRNAs. Further we found no evidence for a functional difference in the leucyl-tRNA synthetase isolated from different developmental stages.

A previous indirect analysis of the leucine tRNA species revealed five peaks in first day pupae and six in seventh day pupae (Ilan and Ilan, 1975a). The new peak found in seventh day pupae represented the major chromatographic form at this stage. This analysis was not performed on the tRNA molecules themselves but on T$_1$ RNase digestion products of labeled leucyl-tRNAs. As with the quantitative determinations it is essential that complete aminoacylation be achieved for this type of analysis. When such a dual labeled comparison of different tRNA preparations is made, the concentration of the differently labeled amino acids in the respective charging mixtures must be identical and the comparison made reciprocally to avoid isotope artifacts (Waters and Novelli, 1961). When we used the improved charging conditions and direct examination of [^{14}C]-leucyl-tRNAs by reversed-phase 5 chromato-

graphy, no change was observed during development. It is possible
that more than one tRNA^{Leu} species is present under the peaks
observed but even so a relative quantitative change should occur
in the profile.

Chromatography of [^{14}C]-leucyl-tRNAs aminoacylated under
sub-optimal conditions produced profiles in which peaks 1 and 2
were very much reduced compared to peak 3. This apparent differential affinity of the leucyl-tRNA synthetase for the tRNA^{Leu}
species may explain much of the data that led Ilan et al. (1970)
to propose the translational control model. For example, it is
conceivable that the unavailability of certain leucyl-tRNA species
might produce some kind of translational control in a cell-free
protein synthesizing system. This apparent regulation would be
due to lack of the aminoacyl-tRNA rather than the tRNA itself.
When aminoacylation assays are carried out under sub-optimal conditions a stimulation resulting from addition of another enzyme
preparation may be due to the presence of monovalent cations in
this preparation (Ilan et al., 1970). The increase in aminoacylation in this case would have nothing to do with a functional
difference in the enzymes.

The earlier evidence for a change in leucine tRNAs and leucyl-
tRNA synthetase (Ilan et al., 1970; Ilan and Ilan, 1975a) and most
of the evidence against it reported here and elsewhere (Lassam
et al., 1975a,b) depend on in vitro assays. The fact that leucyl-
tRNAs aminoacylated in vivo in this study show no change during
development and are virtually identical to the in vitro charged
tRNAs, provides strong evidence against any new leucine tRNA
appearing late in Tenebrio pupae.

In conclusion the translational control mechanism proposed
for Tenebrio (Ilan, 1969; Ilan et al., 1970, 1972; Ilan and Ilan,
1974, 1975a,b) adult cuticular proteins rests on three major points.
    1) The demonstration of the presence of mRNAs for adult cutic-
ular proteins in early pupae;
    2) A high tyrosine content of the adult cuticular proteins
which has been used as an index of the in vitro synthesis of these
proteins;
    3) The appearance of a new leucine tRNA and a new or modified
leucyl-tRNA synthetase late in pupation.

The evidence and discussion presented here cast such doubt on
each of these points that we feel that there is no solid evidence
that supports the previously proposed translational control model.
Therefore in relation to the mode of action of juvenile hormone at
the molecular level we believe future effort should be directed
towards its role in the control of transcription.

REFERENCES

Ames, B., and Hartman, P., 1963, Cold Spring Harb. Symp. Quant. Biol. 28:349.

Anderson, S.O., Chase, A.M., and Willis, J.H., 1973, Insect Biochem. 3:171.

Bagshaw, J.C., Finamore, F.J., and Novelli, G.D., 1970, Devel. Biol. 23:23.

Dewitt, W., 1971, Biochem. Biophys. Res. Commun. 43:266.

Hackman, R.H., and Goldberg, M., 1958, J. Insect Physiol. 2:221.

Ilan, J., 1969, Cold Spring Harb. Symp. Quant. Biol. 34:787

Ilan, J., and Ilan, J., 1973, Ann. Rev. Ent. 18:167.

Ilan, J., and Ilan, J., 1974, in "Physiology of Insecta" (M. Rockstein, ed.), Vol. 4, pp. 355-422, Academic Press, New York.

Ilan, J., and Ilan, J., 1975a, Devel. Biol. 42:64.

Ilan, J., and Ilan, J., 1975b, Curr. Top. Develop. Biol. 9: (in press).

Ilan, J., Ilan, J., and Quastel, J.H., 1966, Biochem. J. 100:441.

Ilan, J., Ilan, J., and Patel, N.G., 1970, J. Biol. Chem. 245:1275.

Ilan, J., Ilan, J., and Patel, N.G., 1972, in "Insect Juvenile Hormones Chemistry and Action" (J.J. Menn and M. Beroza, eds.), pp. 43-68, Academic Press, New York.

Kelmers, A.D., Novelli, G.D., and Stulberg, M.P., 1965, J. Biol. Chem. 240:3979.

Kirby, K.S., 1956, Biochem. J. 64:405.

Lassam, N.J., Lerer, H., and White, B.N., 1975a, Nature 256:734.

Lassam, N.J., Lerer, H., and White, B.N., 1975b, Devel. Biol. (in press).

Lee, J.C., and Ingram, V.L., 1967, Science 158:1330.

Patel, N.G., 1972, in "Molecular Genetic Mechanism in Development and Aging" (M. Rockstein and G.T. Baber, eds.) pp. 145-198, Academic Press, New York.

Patterson, P., 1957, J. Mol. Biol. 65:729.

Pearson, R.L., Weiss, J.F., and Kelmers, A.D., 1971, Biochem. Biophys. Acta 228:770.

Smith, D.W.E., and McNamara, A.L., 1971, Science 171:577.

Stent, G., 1964, Science 144:816.

Twardzik, D.R., Grell, E.H., and Jacobson, K.B., 1971, J. Mol. Biol. 57:231.

Waters, L.D., and Novelli, G.D., 1971, in "Methods in Enzymology" (L. Grossman and K. Moldave, eds.) 20:39.

White, B.N., and Tener, G.M., 1973, Can. J. Biochem. 51:896.

White, B.N., Tener, G.M., Holden, J., and Suzuki, D.T., 1973, J. Mol. Biol. 74:635.

Yang, S.S., and Comb, D.G., 1968, J. Mol. Biol. 31:139.

Yang, W., and Novelli, G.D., 1968, Biochem. Biophys. Res. Commun. 31:534.

THE EFFECTS OF JUVENILE HORMONE ON IMAGINAL DISCS OF <u>DROSOPHILA</u>

<u>IN</u> <u>VITRO</u>:   THE ROLE OF THE INHIBITION OF PROTEIN SYNTHESIS[1]

J.W. Fristrom, C.J. Chihara[2], L. Kelly and
J.T. Nishiura[3]
Department of Genetics, University of California
Berkeley, California 94720

INTRODUCTION

We have been studying the genetic, molecular and developmental processes involved in the evagination of imaginal discs of
<u>Drosophila</u> <u>melanogaster</u>. These studies have been facilitated, on
the one hand, by the fact that evagination occurs <u>in</u> <u>vitro</u> in a
variety of culture media in response to ecdysones (Mandaron, 1973;
Chihara <u>et al</u>., 1972; Fristrom <u>et al</u>., 1973; Milner and Sang, 1974),
and, on the other hand, by the availability of large numbers of
imaginal discs (600,000–800,000 per day) isolated by a preparative
procedure (Fristrom, 1972). Thus we have the advantage of both
defined <u>in</u> <u>vitro</u> conditions and sufficient material for biochemical
analysis. The morphology of an unevaginated and fully evaginated
leg disc is compared in Fig. 1. Ecdysone-induced evagination is
inhibited by all three of the known, naturally occurring juvenile
hormones (JHs) and some, but not all, juvenile hormone analogs
(Siegel and Fristrom, in press). In addition to inhibiting evagination, juvenile hormone also inhibits, under some conditions,
DNA synthesis (Logan <u>et al</u>., 1975), RNA synthesis (Chihara and
Fristrom, 1973) and protein synthesis, and affects some membrane
properties (e.g., uridine transport, Chihara and Fristrom, 1973).

[1]Previously unpublished work reported in this paper was supported
in part by a grant from the National Institute of General Medical
Sciences to J.W.F. (GM-19937).
[2]Current Address: Department of Molecular Biology, University of
California, Berkeley.
[3]Current Address: Biology Department, Brooklyn College, Brooklyn,
New York.

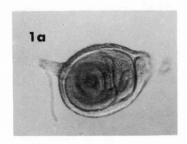 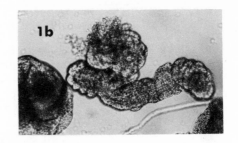

Fig. 1. (a) An unevaginated and (b) evaginated leg disc after
culture for 20 hr at 25°C in Robb's medium respectively without
and with 0.1 µg of β-ecdysone/ml. To express the degree of
evagination quantitatively usually 10 discs are scored on a scale
from 0 (unevaginated) to 10 (fully evaginated) (The Evagination
Index).

There are two generalizations that can be made about the different
effects of juvenile hormones on Drosophila discs. First, the dose
response curves for all of the effects are similar if not identical, i.e., for the C-16 hormone there is no effect below a concentration of 10 µg/ml when the hormone is added in 95% ethanol to
Robb's (1969) tissue culture medium. Second, analogs which do not
inhibit evagination have not been found to produce any of the other
JH effects. Our interest in the inhibition of evagination by JH
stems from the possibility that the effects of the hormone that we
find in vitro with the Drosophila disc system may, at least in
part, involve genuine physiologic activities of the hormone. In
addition an understanding of the mechanisms by which JH acts to
inhibit evagination may provide insight into the mechanism of
evagination.

RESULTS

Inhibition of Protein Synthesis

Juvenile hormone has a variety of effects on macromolecular
synthesis in imaginal discs, the most pronounced of which is the
inhibition of protein synthesis. This is documented in Fig. 2
depicting an experiment in which discs were preincubated using
different hormonal regimens for 3 hr after which the incorporation
of [^{3}H]-leucine into protein was monitored. Incorporation is lowest
in the presence of JH and highest in the presence of β-ecdysone.
We conclude that the inhibition of incorporation by JH represents
true inhibition of protein synthesis and does not result from a

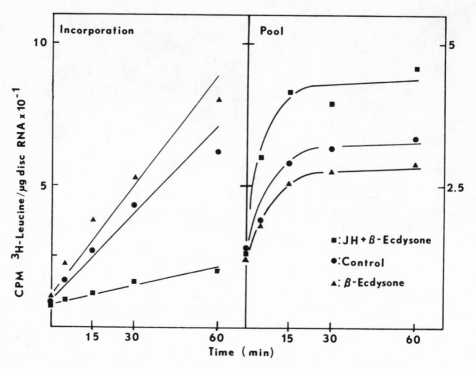

Fig. 2. The effect of JH on protein synthesis. Discs were incubated with β-ecdysone (1 μg/ml); JH (100 μg/ml) and β-ecdysone
(1 μg/ml) or without any hormone for 3 hr and then with [³H]-
leucine (15.6 mCi/mmol; 0.64 mM) for the times indicated. Incorporation into protein and uptake into the "pool" (acid-soluble
material) was then determined as previously described (Fristrom
et al., 1974). Points are the average of 2 determinations.

spurious effect such as inhibition of transport of leucine into
cells. This conclusion is based, in part, on previous studies
(Fristrom et al., 1974) demonstrating that approximately 95% of
the leucine used for protein synthesis in imaginal discs is
derived, under standard culture conditions, from exogenous leucine.
Hence, when the rate of protein synthesis increases it is necessary
for the rate of transport of leucine into disc cells to increase.
Further, the level of "free" leucine in disc cells is regulated by
the rate of removal of leucine from the pool by protein synthesis
and the entry of the amino acid into the cells by diffusion along
a concentration gradient. Therefore, to increase leucine transport into discs it is necessary to reduce the level of intracellular "free" leucine and create a steeper concentration gradient
relative to the culture medium. Such a reduction in the leucine
pool is seen in Fig. 2 where the level of leucine is lowest in

discs incubated with β-ecdysone. From such results one can predict
that if protein synthesis were inhibited the level of intracellular
"free" leucine should increase. This indeed is the case for discs
incubated with JH (Fig. 2). Further, the levels in the pool are
inversely proportional to the rates of protein synthesis. Because
the results with JH agree with expectations for inhibition of protein synthesis, we feel that the conclusion that protein synthesis
is inhibited by JH is justified.

The inhibition of protein synthesis in insect cells by JH
has been reported elsewhere (Cohen and Gilbert, 1973). However,
in contrast to the previous report we find that the inhibition of
protein synthesis in Drosophila discs is readily reversible. This
is documented in Fig. 3 in which results from incubating discs
under different hormonal regimens for 4 hr followed by removal of
the hormones are presented. Prior to removal of the hormones it
can be seen that protein synthesis is inhibited by JH both in the
presence and absence of β-ecdysone. Following removal of JH by
serial dilution, protein synthesis resumes rapidly. This is seen
in the case where discs were first incubated with JH and then with
β-ecdysone where the rate of synthesis is "immediately" (i.e. as
soon as we can measure it) equal to that in control discs. It
should also be noted that a transitory increase is found when
discs are reincubated with JH and β-ecdysone indicating a temporary
return to control level protein synthesis. Both of these observations indicate that the JH is rapidly removed from its receptor
sites, either by dissociation, as a result of metabolism (Chihara
et al., 1972) or both. Further, it can be seen that discs which
have been preincubated with JH are completely capable of responding
to β-ecydone with increased levels of protein synthesis with kinetics similar to those of control discs. Finally, it can be noted
that discs which have been preincubated with both β-ecdysone and
JH attain the β-ecdysone stimulated level of protein synthesis
much more rapidly than discs which have been incubated with JH
alone. This observation indicates that JH has little effect on the
binding of β-ecdysone to its receptors in disc cells, and hence the
ecdysone-stimulated effects occur rapidly after the removal of JH
from the culture medium. This conclusion is supported by direct
observations made in our laboratory by Yund (personal communication). It was previously demonstrated (Yund and Fristrom, 1975a,b)
that there are specific, high affinity binding sites for [^{3}H]-β-
ecdysone in disc cells that are ultimately found in the nuclei.
These observations, coupled with work from other laboratories on
cytoplasmic receptors for ecdysones (e.g., Emmerich, 1970) suggest
to us that the mechanism of ecdysone action in discs parallels
that found for vertebrate steroid hormones (for review, Edelman,
1975) where the steroid enters the cell cytoplasm, is bound to a
receptor protein, and the steroid-receptor complex then enters
the nuclei and is bound to chromatin. In studies with JH, Yund

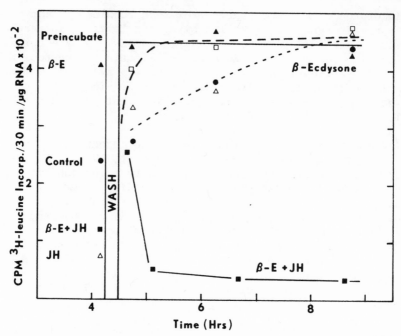

Fig. 3. The reversibility of JH inhibition of protein synthesis. Discs were incubated with different hormones at the concentrations used in Fig. 2 as indicated. Hormones were removed by centrifuging the discs six times through 15 ml volumes of cold <u>Drosophila</u> Ringers. Incorporation into protein was determined after a 30 min incubation with [³H]-leucine as previously described (Fristrom <u>et al</u>., 1974). The symbols used in the post-wash incubation are the same as in the preincubation except that open squares represent preincubation with JH and β-ecdysone followed by incubation with β-ecdysone.

has found that there is no reduction in [³H]-β-ecdysone in disc cells, but there is a 20-40% reduction in binding to the nuclei (Yund, personal communication). This observation indicates that JH does not prevent binding of [³H]-β-ecdysone to its receptors, but it may affect movement of the ecdysone-receptor complex to the nuclei or its interaction with the chromatin. At this point there is no reason to suspect that this effect on β-ecdysone distribution is related to the inhibition of protein synthesis by JH.

Inhibition of Evagination

It is evident that the inhibition of protein synthesis inhibits evagination. Results from previous studies (Fristrom <u>et al</u>.,

1973) presented in Table 1, demonstrate that inhibitors of protein synthesis inhibit evagination. Hence, inhibition of evagination can be explained by inhibition of protein synthesis by JH. However, the possibility exists that JH might act directly on the mechanism of evagination per se. This possibility can be tested under conditions in which evagination is accelerated by trypsin treatment (Poodry and Schneiderman, 1971). Partially evaginated discs are exposed to 0.1% trypsin for 10 min at the end of which time substantial evagination has occurred. Such a time period is too short for macromolecular synthesis to be important and inhibitors of protein synthesis do not block evagination (Table 1). Neither does JH block evagination under these conditions. Hence, there is no indication that JH blocks the mechanics of evagination per se.

TABLE 1.   EFFECTS OF INHIBITORS OF PROTEIN SYNTHESIS ON ACCELERATED AND UNACCELERATED EVAGINATION.

Incubation Conditions	Evagination Index	
	Unaccelerated Evagination	Accelerated Evagination
Control	1-2	2-4
Standard	10	7.7
Puromycin	1-2	7.2
Cycloheximide	1-2	7.5
JH(C-16)	1-2	7.2

Control: No hormone; no trypsin. Standard: 0.1 µg/ml β-ecdysone or 0.1% trypsin. Puromycin: 10 µg/ml. Cycloheximide: 1 µg/ml. JH: 100 µg/ml. The evagination index is described in the legend to Fig. 1. The degree of evagination was determined after a 20 hr incubation in Robb's at 25°C for unaccelerated evagination and after 10 min in trypsin for accelerated evagination. Data from Fristrom et al., 1973 and Fekete et al., 1975.

Effects of JH on RNA Synthesis

In addition to inhibiting evagination, the inhibition of
protein synthesis by JH has an important effect on the synthesis
of RNA in discs. In a previous study Chihara and Fristrom (1973)
reported that concentrations of JH that inhibit evagination inhibit
RNA synthesis in discs incubated with β-ecdysone, but have little
effect on the rate of RNA synthesis in discs not incubated with
β-ecdysone. This is seen in Fig. 4 where JH reduces the rate of
RNA synthesis from the β-ecdysone-stimulated level to the control
level, but does not effect the rate of synthesis in control discs.
We concluded at the time that JH preferentially inhibits β-ecdysone-
stimulated RNA synthesis. However, in the light of recent work by
Nishiura and Fristrom (1975) we now put forward a different explanation.

Nishiura and Fristrom (1975) examined the activity of RNA
polymerases in discs incubated in the presence and absence of
β-ecdysone. Since the stimulation in RNA synthesis caused by
β-ecdysone mainly involves ribosomal RNA synthesis (Petri et al.,
1971 and unpublished results) it was not surprising to find that
there is about a 1.8-1.9 fold stimulation in the activity of RNA
polymerase I (the polymerase that synthesizes rRNA) in discs in
the presence of β-ecdysone. What was surprising was the discovery
that a comparable increase in activity occurred in the presence
of cycloheximide or JH even in the absence of β-ecdysone. These
observations have led us to formulate a simple model for the regulation of polymerase I activity in discs (Nishiura and Fristrom,
1975). The main characteristics of this model are as follows:

1) That there is an inhibitory protein present which inhibits
the activity of RNA polymerase I. That such an inhibitor exists
is suggested by mixing experiments in which discs separately incubated in the presence or absence of β-ecdysone are extracted
together and RNA polymerase I activity is assayed and found to be
the same as in discs incubated without β-ecdysone and not at an
intermediate value (as would be expected if β-ecdysone produced
a stable activation of polymerase I).

2) The inhibitory protein turns over rapidly. This is suggested by the fact that inhibition of protein synthesis causes an
increase in polymerase I activity. It is further supported by the
observation that when JH is removed from the culture medium and
protein synthesis resumes, that the level of polymerase I activity
returns to the control, no hormone, level.

3) β-Ecdysone irreversibly inhibits synthesis of the mRNA for
polymerase I. This conclusion is supported by evidence indicating
that β-ecdysone acts at the level of the genome (Yund and Fristrom,

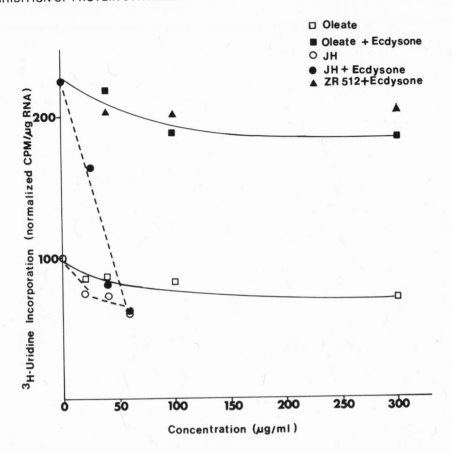

Fig. 4.   The effect of different concentrations of JH on incorporation into RNA. Discs were incubated as indicated for 1 hr in
Robb's culture medium with and without β-ecdysone (1 μg/ml) and
then for 30 min with [³H]-uridine. Incorporation into RNA was
determined and normalized with control values set at 100 as previously described (Chihara and Fristrom, 1973). From Chihara
and Fristrom, 1973.

1975b) and by observations on puffing in polytene chromosomes
showing that some puffs are turned off by incubation with β-ecdysone
(Ashburner et al., 1973). Also, removal of β-ecdysone from the
culture medium does not result in a reduction in polymerase I
activity for at least 7 hr.

    Assuming that this model is true, it follows that the level
of polymerase I activity will be the same in discs incubated with
β-ecdysone, JH or β-ecdysone and JH, since the inhibitory protein

will be absent under all three conditions. If JH acts in the same manner in the presence and absence of β-ecdysone then it follows that the rate of RNA synthesis should be the same in discs incubated with inhibitory concentrations of JH alone and with JH plus β-ecdysone. This, indeed, is the case as is shown in Fig. 4. Also the fact that JH can inhibit RNA synthesis in the presence of β-ecdysone or that RNA synthesis does not increase as a result of increasing polymerase I activity indicates that JH is inhibiting RNA synthesis independently of the inhibition of protein synthesis. This conclusion is supported by data on incorporation of [3H]-uridine into RNA in the presence and absence of β-ecdysone and cycloheximide (Table 2). As can be seen the levels of RNA synthesis in discs in the presence of cycloheximide, β-ecdysone, or β-ecdysone and cycloheximide together are 1.3-1.5 times that of the level found in untreated discs. An increase in RNA synthesis is expected assuming an increase in polymerase I activity using the above model, and shows that inhibition of RNA synthesis is not causally tied to inhibition of protein synthesis (in fact, the opposite is suggested - i.e., that stimulation of RNA synthesis is causally tied to the inhibition of protein synthesis). Thus, the effects of JH on RNA synthesis are separable into two categories, an increase in polymerase I activity resulting from inhibition of protein synthesis, and inhibition of RNA synthesis independent of polymerase I activity and independent of the inhibition of protein synthesis.

TABLE 2.   EFFECTS OF CYCLOHEXIMIDE ON RNA SYNTHESIS OF DROSOPHILA DISCS.

Incubation Condition	CPM [3H]-Uridine Incorporated into RNA/30 min/μg RNA
Control	278
Cycloheximide (1 μg/ml)	370
β-Ecdysone (1 μg/ml)	417
β-Ecdysone plus Cycloheximide	393

Discs were incubated in Robb's for 2 hr at 25°C as indicated and incorporation of [3H]-uridine into RNA was then measured following the procedure described in Chihara and Fristrom, 1973. The results are the average of two experiments.

Membrane Effects

Juvenile hormone affects membrane properties in discs in a manner which also appears to be independent of the inhibition of protein synthesis. One such effect shown in Fig. 5 involves the transport of uridine into disc cells. Here the data of Chihara and Fristrom (1973) are plotted using a Lineweaver-Burke plot which demonstrates that there are two transport systems operative in the uptake of uridine by discs; a low K_m system ($K_m = 2 \times 10^{-4}$ M) and a high K_m system ($K_m = 3 \times 10^{-3}$ M). In the presence of β-ecdysone, JH or the two hormones together the low K_m system is affected so that there is increased transport of uridine into disc cells, but there is no detected effect on the high K_m system. It should be emphasized that both hormones have the same effect, there being no significant difference between the points in the presence of any hormonal combination, but all of the hormonal points are significantly different from the control values. Since the change in the rate of uridine transport occurs within 30 min after addition of the hormones to the culture medium we conclude this effect is independent of the inhibition of protein synthesis.

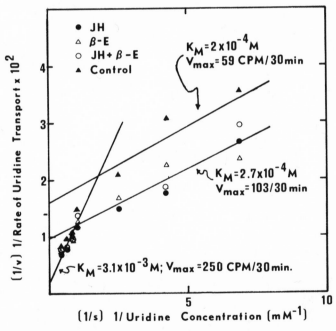

Fig. 5. The effect of JH and β-ecdysone on the transport of [3H]-uridine into disc cells. The data of Chihara and Fristrom (1973) are expressed as a Lineweaver-Burke plot. Each point represents the average of from 4-12 determinations.

A second membrane component which has been investigated is the membrane bound Na^+/K^+ dependent ATPase, the protein involved in the transport of Na^+ across cell membranes. This enzyme is of particular interest in insect tissues because it has been implicated in the activity of both β-ecdysone (Kroeger, 1963; Kroeger et al., 1973) and JH (Lezzi and Gilbert, 1972). We have examined the effects of the hormones on this enzyme in two ways; first, by examining [^{3}H]-ouabain binding (ouabain is a steroid glucoside that is a specific inhibitor of Na^+/K^+ dependent ATPase) and the activity of the ATPase after incubation of discs with β-ecdysone and JH. The results from the study of [^{3}H]-ouabain binding are presented as a Scatchard plot in Fig. 6. β-Ecdysone has no effect on the binding of [^{3}H]-ouabain either in the presence or in the absence of JH. In contrast, JH acts as a non-competitive inhibitor of ouabain binding. Since binding of ouabain to the Na^+/K^+ dependent ATPase results in inhibition of the enzyme, reduction in binding of the inhibitor might indicate that there is an increased amount of enzyme in an active configuration in the presence of JH. This possibility is confirmed by assays on Na^+/K^+ dependent ATPase in discs performed after a 1 hr incubation with different hormonal regimens (Table 3). When the difference in ATPase activity in the presence and absence of Na^+ and K^+ is determined we find that there is more activity in discs incubated with JH or JH with β-ecdysone than in control discs or discs incubated with β-ecdysone alone. Incubation of discs with cyclo-heximide has no effect on binding of [^{3}H]-ouabain, and thus the effect of JH on the ATPase is independent of the inhibition of protein synthesis caused by this hormone.

However, it seemed reasonable to believe that a stimulation in Na^+/K^+ dependent ATPase by JH, with a resultant increase in efflux of Na^+ from disc cells, might inhibit protein synthesis and subsequently inhibit evagination. Therefore, we investigated whether conditions which affected the ATPase activity also affected the inhibitory activity of JH. In one set of experiments we investigated whether high concentrations of ouabain altered the inhibitory effects of JH. The results, presented in Table 4, show that JH inhibits evagination to the same degree in the presence or absence of 0.5 mM ouabain. A second approach involved the utilization of a temperature sensitive mutant, shibire. This mutant was recovered as a paralytic mutant at high temperatures and current evidence indicates (Kelly, unpublished) that it has a defective Na^+/K^+ dependent ATPase at high temperatures (29°C and above). Therefore we have investigated the ability of JH to inhibit evagination of shibire and wild type discs cultured at 33.5°C (the highest temperature we have found that allows evagination of wild type discs). We find no difference between wild type and shibire discs at this temperature (Table 4), and hence again no indication that the effect of JH on the Na^+/K^+ dependent ATPase is causally related to the inhibition of protein synthesis. It must be

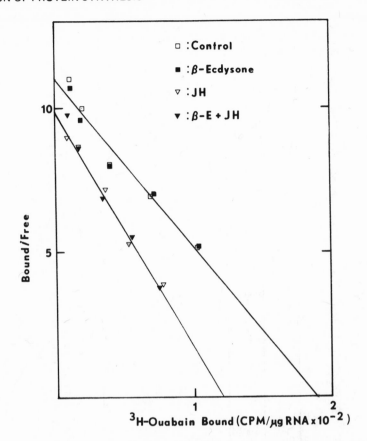

Fig. 6. The effect of JH and β-ecdysone on binding of [³H]-ouabain by discs. Discs were incubated at a concentration of 10,000 discs/ml in Robb's culture medium in Nalgene flasks at 25°C. Discs were incubated for 30 min with the hormones using the concentrations described for Fig. 2, and then with varying concentrations of [³H]-ouabain for an additional 3 hr. Unbound ouabain was removed by 6 centrifugations through 15 ml volumes of cold Drosophila Ringer's solution. Each value is the average of 4-16 determinations.

emphasized that we have not demonstrated in either of these experimental conditions that the effect of JH on Na⁺ efflux has been altered, and hence the possibility that the inhibition of protein synthesis by JH results from an effect on the Na⁺/K⁺ dependent ATPase has not been eliminated.

TABLE 3. THE EFFECTS OF HORMONES ON K^+/Na^+ DEPENDENT ATPase ACTIVITY OF DROSOPHILA IMAGINAL DISCS.

Incubation	ATPase Activity[1]		
Condition	I. with K^+/Na^+	II. without K^+/Na^+	K^+/Na^+ Dependent (I-II)
Control	1.68	1.52	0.16
β-Ecdysone (1 μg/ml)	1.66	1.51	0.15
JH (100 μg/ml)	1.75	1.43	0.32
JH + β-Ecdysone	1.80	1.44	0.36

[1] μmol phosphate/hour/mg protein. Data are the average of 2 determinations. Discs were incubated for 1 hr at 25°C in Robb's tissue culture medium with and without hormones prior to homogenization. Incubations were done in Tris (30 mM), pH 7.5; with $MgCl_2$ (12 mM) and ATP (1.25 mM) and when present, KCl (20 mM) and NaCl (100 mM) for 60 min at 25°C.

TABLE 4. INHIBITION OF EVAGINATION OF WILD TYPE DISCS BY JH IN THE PRESENCE OF OUABAIN OR IN SHIBIRE DISCS AT A RESTRICTIVE TEMPERATURE.

Concentration of JH (C-16)	Evagination Index					
	Wild Type			Shibire		
	Plus Ouabain	Minus Ouabain				
	22°C	22°C	33.5°C	22°C	29°C	33.5°C
None	9.5	10.0	9.1	8.5	9.0	6.0
50 µg/ml	2.3	2.3	1.0	0.5	--	0.4
100 µg/ml	0	0	0	0	0	0

Discs were incubated at elevated temperature or with ouabain (0.5 mM) in Robb's with 0.1 µg of β-ecdysone/ml for 5 hr prior to the addition of JH to provide ample opportunity for inhibition or denaturation of the Na^+/K^+ dependent ATPase. Wild type discs were preparatively isolated; shibire discs were obtained by dissection from larvae grown at 22°C. Incubations were for 20 hr prior to determination of the Evagination Index. The Evagination Index is described in Fig. 1. Determinations on wild type involved 10 discs, on shibire involved 4-10 discs.

DISCUSSION

In conclusion we find that some of the effects of JH on discs result from the inhibition of protein synthesis; included are inhibition of evagination and an increase in RNA polymerase I activity. In contrast the effects of JH on membrane properties, e.g., uridine transport and the Na^+/K^+ dependent ATPase, and the inhibition of RNA synthesis do not result from the inhibition of protein synthesis. The data offer little help in identifying a primary site of JH action. The possibility that the effect of JH on the Na^+/K^+ dependent ATPase is responsible for the inhibition of protein synthesis and evagination is not supported by the data. The putative primary site for the action of JH in Drosophila discs remains to be determined. However, the observation that there are similar dose response relationships for all of the effects of JH on discs suggests that such a site exists.

It can be noted that there is no single relationship which emerges when one compares the effects of JH and β-ecdysone on specific processes. Juvenile hormone inhibits protein synthesis while β-ecdysone produces a stable increase in protein synthesis (Fristrom et al., 1974). Both hormones increase RNA polymerase I activity, but by different mechanisms. Both hormones increase uridine transport, but only JH affects the Na^+/K^+ dependent ATPase. There is no indication to date that the two hormones ever act directly on the same processes. Rather, it is better to view the effect of β-ecdysone and juvenile hormone as involving a set of independent responses resulting from activity via independent mechanisms. As such one expects, and finds, that the two hormones do not have competitive effects on any cellular function or synthetic activity.

It is necessary to consider whether the effects of juvenile hormone on Drosophila discs in vitro represent physiological activities of the hormone, or simply effects peculiar to the in vitro system. Two major objections to this in vitro system are the high concentrations of JH required and the fact that mature imaginal discs are probably never exposed to JH in vivo. The high concentrations of JH are necessitated by the low solubility of the hormone (as we use it) and its rapid metabolism by discs (Chihara et al., 1972). The type of responses found, the reversible inhibition of protein synthesis and evagination, seem reasonable since these are exactly the responses that would maintain the discs in a juvenile state. Also, although mature larval discs are presumably not exposed to JH in vivo, there is no reason to conclude that a mechanism for response is not still present as a holdover from early larval development. Hence, we find no reason to conclude that the site of action of JH in the disc system in vitro is necessarily different than the site in vivo. It should be borne in mind that although the type of response to JH will vary from tissue to tissue,

that the mechanisms for responding to the hormone may be very
similar. Such a situation is clearly documented in studies in
vertebrates in which different target tissues respond to the same
steroid hormone with different synthetic responses, but by a similar response mechanism, i.e., via a cytoplasmic receptor protein.
Hence, if we are ever able to identify the hypothetical receptor
sites for JH in Drosophila discs in vitro it would not surprise us
if similar receptor sites were also found in other organisms in
traditional target tissues.

ACKNOWLEDGEMENTS

The authors are indebted to Mrs. Odessa Eugene and Mrs. Susie
Kuniyuki for expert technical assistance. We also acknowledge
gratefully the comments made by Dr. Dianne Fristrom during the
preparation of this manuscript.

REFERENCES

Ashburner, M., Chihara, C., Meltzer, P., and Richards, G., 1973,
    Cold Spring Harbor Symp. Quant. Biol. 38:655.
Chihara, C., and Fristrom, J., 1973, Develop. Biol. 35:36.
Chihara, C., Petri, W., Fristrom, J., and King, D., 1972, J.
    Insect Physiol. 18:1115.
Cohen, E., and Gilbert, L., 1973, J. Insect Physiol. 19:1857.
Edelman, I.S., 1975, J. Steroid Biochem. 6:147.
Emmerich, H., 1970, Z. vergl. Physiol. 68:385.
Fekete, E., Fristrom, D., Kiss, I., and Fristrom, J., 1975,
    Wilhelm Roux' Arch. Entwicklungsmech. Organ. 178:123.
Fristrom, J., 1972, in "The Biology of Imaginal Discs" (H. Ursprung
    and R. Nöthiger, eds.) pp. 109-154, Springer-Verlag, Berlin.
Fristrom, J., Logan, W., and Murphy, C., 1973, Develop. Biol. 33:
    441.
Fristrom, J., Gregg, T., and Siegel, J., 1974, Develop. Biol. 41:
    301.
Kroeger, H., 1963, Nature 200:1234.
Kroeger, H., Trosch, W., and Muller, G., 1973, Exp. Cell Res. 80:
    329.
Lezzi, M., and Gilbert, L., 1972, Gen. Comp. Endocrinol. Suppl.
    3:159.
Logan, W., Fristrom, D., and Fristrom, J., 1975, J. Insect Physiol.
    21:1343.
Mandaron, P., 1973, Develop. Biol. 31:101.
Milner, M., and Sang, J., 1974, Cell 3:141.
Nishiura, J., and Fristrom, J., 1975, Proc. Nat. Acad. Sci., USA
    72:2984.
Poodry, C., and Schneiderman, H., 1971, Wilhelm Roux' Arch.
    Entwicklungsmech. Organ. 168:1.

Robb, J., 1969, J. Cell Biol. 41:876.

Siegel, J., and Fristrom, J., 1976, in "Biology of Drosophila,"
    (M. Ashburner and T. Wright, eds.) v. II, Academic Press,
    New York, in press.

Yund, M., and Fristrom, J., 1975a, Develop. Biol. 43:287.

Yund, M., and Fristrom, J., 1975b, Proc. ICN, UCLA Squaw Valley
    Conf. Develop. Biol., in press.

A GENETIC APPROACH TO THE STUDY OF JUVENILE HORMONE CONTROL OF

VITELLOGENESIS IN DROSOPHILA MELANOGASTER

J.H. Postlethwait, A.M. Handler, and P.W. Gray

Department of Biology, University of Oregon
Eugene, Oregon 97403

INTRODUCTION

Juvenile hormone (JH) is a key agent in the regulation of insect development, and thus it must effect proper gene management during ontogeny. How this occurs is as yet a mystery, but JH can alter chromosome function as assayed by puffing in polytene chromosomes of Diptera (Lezzi and Gilbert, 1969) and models based on gene regulation have been presented for its action (Williams and Kafatos, 1971). Therefore it seems appropriate to approach the study of JH action from a genetic standpoint. Mutant genes causing defects in JH action in Drosophila might be expected to develop lethal morphogenetic irregularities during embryonic, larval, or pupal stages (Dearden, 1964; Bryant and Sang, 1968; Ashburner, 1970; Madhavan, 1973; Postlethwait, 1974). Studying the biochemistry of these moribund animals might reveal only frustrating artifacts. A more convenient system might involve a tissue which responds to JH but which is unnecessary for the individual's survival. For example, in many insects JH is required for some step in oogenesis (see Doan, 1973; Engelmann, 1970). A genetic approach to the hormonal control of oogenesis in Drosophila seems appropriate due to several factors. Drosophila females can lay up to 60% of their weight per day in eggs (King et al., 1955; David et al., 1968), which indicates the prodigious biosynthetic activity directed toward oogenesis. Furthermore, the ultrastructure of normal oogenesis has been very carefully described by King (1970). Finally, a great many female sterile mutants are known (King and Mohler, 1975). Our investigation exploits these properties to approach the following three points regarding JH control of oogenesis: (1) What is the endocrinology of oogenesis in Drosophila melanogaster? (2) How are female sterile mutants altered in JH action? (3) How

449

does JH act to control the appearance of specific proteins in
developing oocytes?

MATERIALS AND METHODS

Drosophila melanogaster were cultured on standard yeast-
agar-molasses medium at 25°C unless otherwise noted. Female
sterile mutants were carried in stocks balanced over SM1, SM5,
or TM3 balancing chromosomes and were obtained from the Cal. Tech.
Stock Center, Drs. A. Bakken and D. Mohler (Bakken, 1973), or
M. Landers (personal communication). For some experiments these
heterozygous siblings provided controls. Specific female sterile
mutants will be described in the results section. For experiments
which required experimental treatments at precise times near
eclosion, animals were cultured at 25.0 ± 0.5°C in a 12 hr light:
12 hr dark photoperiod (see Skopik and Pittendrigh, 1968). Under
these conditions 95% of the females collected simultaneously as
white prepupae just before lights-off eclosed within 2 hr of
lights-on 108 hr later. Thus, experimental treatments could be
administered prior to eclosion to animals of known age. Ages are
given with respect to the time the adult ecloses. Standard ligation and transplantation procedures were employed (Ursprung, 1967;
Schneiderman, 1967) using Chan and Gehring (1971) Ringers.
Protein and enzyme assays were performed as described earlier
(Postlethwait and Gray, 1975). Ovarian stages are given in King
(1970). Animals were treated with 0.15 µg of JH analog ZR-515
(isopropyl-11-methoxy-3,7,11-trimethyl-dodeca-2,4-dienoate)
dissolved in acetone.

RESULTS

The Endocrinology of Oogenesis

Cephalic Factor. In order to ascertain the phenotype one
might expect of a female sterile mutant possessing defective JH
metabolism, we first needed to learn the phenotype of an ovary not
exposed to juvenile hormone during the adult stage. We also felt
it would be useful to understand the role of any other endocrine
or nutritional factors involved in ovarian maturation during the
adult stage. Figures 1 and 2 show the morphological and temporal
course of vitellogenesis for control flies. To test the effect of
any cephalic endocrine, neural, or neuroendocrine factors required
for vitellogenesis, we decapitated flies at defined ages either
before or after eclosion and counted the number of ovarian oocytes
which were at early (stages 8 and 9), intermediate (stages 10-12),
or late (stages 13 and 14) phases of vitellogenesis (Fig. 1) by
72 hr after eclosion. Decapitated flies can live for several days,

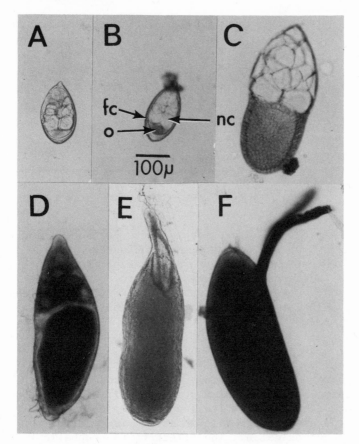

Fig. 1. Eggs at various stages of vitellogenesis from Drosophila
melanogaster. (fc) follicle cells, (nc) nurse cells, (o) oocyte.
(A) stage 7 (non-vitellogenic), (B) stage 8 (early stage) (note
the opacity of the oocyte caused by yolk deposition), (C) stage 10
(intermediate stage), (D) stage 11 (intermediate stage), (E) stage
14 (late stage), (F) oviposited egg. Stages are taken from King
(1970).

spend most of their time standing upright, and display a remarkable
repertoire of normal behavior (Spieth, 1966). To control for the
fact that headless flies don't eat, we removed the labial palps
from flies at different ages before and after eclosion and scored
the flies at 72 hr after eclosion to test the effect of adult
nutrition on oocyte maturation.

Figure 2 shows that removal of mouth parts to prevent eating
does not inhibit the schedule of vitellogenesis over the time
period examined. On the other hand, removal of the head from fully

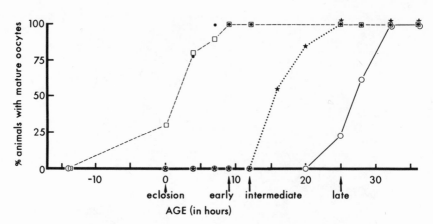

Fig. 2. Appearance of mature oocytes in experimental and control preparations. Unoperated controls were dissected at the times indicated. Operations were performed on experimental preparations at the times indicated and the ovaries were dissected at 72 hr after eclosion. Open circles, unoperated controls; squares, decapitated animals; stars, isolated abdomens; filled circles, depalped animals.

developed pharate adults several hours before eclosion prohibits vitellogenesis in all the animals and decapitation less than 10 min after eclosion permits vitellogenesis in only 30% of the animals. Apparently a cephalic influence is required for vitellogenesis at about eclosion. The effect of the cephalic factor is a persistent one, as is demonstrated by the data in Table 1, which shows the distribution of vitellogenic oocytes in control and experimental preparations. At 7 hr, control animals contain no vitellogenic follicles. Yet removing the labial palps or the head at this time does not prevent mature oocytes from forming by 72 hr. Oogenesis is continuing in ovaries in these lesioned flies since they are still producing early and intermediate vitellogenic follicles at 72 hr. These results show that the effects of the cephalic influence can persist to permit the continued formation of early vitellogenic stages and their subsequent maturation.

   Thoracic Factor. To test for the effect of a thoracic factor required for vitellogenesis we ligated abdomens at various times after eclosion and dissected them at 72 hr after eclosion to check for vitellogenesis. Figure 2 shows that abdomens ligated before 16 hr fail to develop mature oocytes. However, isolating abdomens at 16 hr permits late vitellogenic stages to form even though control flies have not yet formed any mature oocytes at the time of ligation. Apparently a thoracic factor is required prior to 16 hr to permit completion of vitellogenesis.

TABLE 1. THE AVERAGE NUMBER OF FOLLICLES PER OVARY IN VARIOUS
STAGES OF VITELLOGENESIS IN CONTROL AND EXPERIMENTAL PREPARATIONS.

| | | | Vitellogenic | | |
| | Age at Operation | | | Stage[a] | |
Preparation	(hr)	N	8-9	10-12	13-14
Unoperated control	7	24	0	0	0
Depalped	7	12	1.2	0.0	1.6
Decapitated	7	20	0.6	0.05	2.8
Isolated abdomens	7	20	0	0	0
Unoperated control	24	18	4.3	3.2	0.4
Depalped	24	7	1.0	0	3.0
Decapitated	24	11	1.0	0.7	5.4
Isolated abdomens	24	19	0	0	2.0

[a] Experimental preparations were examined at 72 hr after eclosion.

The thoracic factor differs from the cephalic not only by
being required later in development but also in that it seems to
be continuously necessary for normal vitellogenesis instead of
merely providing a trigger. Table 1 shows that at 24 hr, unoperated
control animals have several early and intermediate vitellogenic
stages and an average of 0.4 late vitellogenic stages per ovary.
Abdomens isolated from the thorax and head at 24 hr have about two
late vitellogenic stages per ovary but no early or intermediate
stages. In none of the 160 isolated abdomens prepared at 12 time
points which were examined in this study were any early or intermediate stages found by 72 hr. It can be concluded that in isolated
abdomens some follicles which have already initiated vitellogenesis

can complete the process, but that removing the thorax along with
the head inhibits the appearance of new vitellogenic follicles.

Corpus Allatum Transplants. In order to find whether the
relevant vitellogenic factor missing from isolated abdomens could
be supplied by the corpus allatum, the source of JH, we transplanted
corpora allata (along with corpora cardiaca) from 60-72 hr old
adult females into abdomens isolated within 15 min after eclosion,
and at 72 hr we dissected and staged the follicles. The results,
shown in Table 2, indicate that implanted corpora allata cause the
appearance of early, intermediate, and even late vitellogenic
follicles in isolated abdomens. These results show that the corpus
allatum can provide a factor necessary for vitellogenesis in isolated abdomens.

Juvenile Hormone Analog Treatments. The next question then
is whether a JH analog can induce vitellogenesis in an isolated
abdomen. Table 2 shows that ovaries in isolated abdomens treated
with ZR-515 can develop mature oocytes while isolated abdomens
alone, or those injected with Ringer's fail to undergo vitellogenesis. It can thus be concluded that isolated abdomens require
only the addition of JH to promote vitellogenesis. These results
also serve as a control for the health of the preparation since
they demonstrate that given the right conditions, isolated abdomens can support vitellogenesis. The results with isolated abdomens allow us to define the phenotype expected of a mutant possessing low titers of available JH, or a mutant whose ovaries are not
able to respond to JH. This phenotype consists of ovaries containing follicles up to stage 7, but no vitellogenic follicles (see
Fig. 1).

Female Sterile Mutants

Ovarian Phenotypes. Having identified the ovarian phenotype
one would expect of a mutant lacking JH in the adult, we can now
designate non-vitellogenic female sterile mutants as candidates
for being involved in some aspect of juvenile hormone regulation
of oogenesis. Table 3 lists seven mutants, each of which has the
non-vitellogenic phenotype expected to accompany a JH defect.
In fs(3)L8^{ts} cultures at 29°C and fs(2)A17, follicles develop to
stage 7 and then begin to degenerate. In fs(3)L1 and fs(2)A18
some vitellogenesis begins, but then ceases. In fs(3)L3 follicles
do not develop to stage 7. In apterous-4 (ap^4, see Butterworth
and King, 1965, for description of this locus) and fs(3)A1, stage
7 follicles accumulate and occasionally a vitellogenic follicle is
observed.

TABLE 2. INDUCTION OF VITELLOGENESIS IN ISOLATED ABDOMENS.

Preparation	Treatment	Number of preparations	Percentage with vitellogenesis	Vitellogenic Stage		
				early vfpo[a]	intermediate vfpo	late vfpo
Unoperated control	none	10	100	3.3	1.0	8.3
Isolated abdomen	none	14	0	0	0	0
Isolated abdomen	Ringer injection	12	0	0	0	0
Isolated abdomen	Corpus allatum—corpus cardiacum—implantation	22	46	0.11	0.02	0.38
Isolated abdomen	0.15 µg ZR-515	9	100	1.1	0.22	4.5
Isolated abdomen	Ringer injection plus 0.15 µg ZR-515	11	82	0.90	0.32	2.8

[a] Average number of vitellogenic follicles per ovary.

<u>Ovarian Transplants</u>. The first problem to be approached is
to decide whether a given non-vitellogenic mutant is defective in
the internal environment required for proper ovarian maturation,
or if the genetic lesion is intrinsic to the ovary. The question
of ovary autonomy can readily be answered by ovary transplants
(King and Bodenstein, 1956). We transplanted ovaries from freshly
eclosed female sterile mutants into newly eclosed normal females
(either Oregon R wild type, or heterozygotes for the particular
mutant) and dissected the hosts three days later to see if mutant
ovaries are cured by a wild type host. We also transplanted wild
type ovaries into mutant hosts to find whether the mutant environment
can support vitellogenesis. Since an implanted ovary fails to
attach to the host's oviduct, a donor ovary is readily distinguished from the host's ovaries.

The results are shown in Table 3, and they allow us to place
the mutants into one of two categories: either ovary autonomous
or ovary non-autonomous. Ovaries of four mutants <u>(fs(2)A17,</u>
<u>fs(3)L3</u>, <u>fs(3)L1</u>, <u>fs(2)A18</u>) failed to undergo enhanced vitellogen-
esis in normal hosts (Table 3, column 3), and so they are classi-
fied as ovary autonomous. In these cases the genetic lesion
resides in the ovary since normal hemolymph does not cure the
defect. Furthermore, the body cavities of these mutants are normal
since they permit implanted wild type ovaries to develop mature
oocytes (Table 3, column 4). This result shows that the internal
environment, including hormonal conditions, is normal in these
mutants.

A fifth ovary autonomous mutant is <u>fs(3)L8^{ts}</u>. As Table 3
shows, <u>fs(3)L8^{ts}</u> flies are non-vitellogenic when cultured at 29°C,
but their ovaries develop normally and they are fertile at 22°C.
This conditional mutant is ovary autonomous since <u>fs(3)L8^{ts}</u> ovaries
will not mature in a wild type host cultured at 29°C, but <u>fs(3)L8^{ts}</u>
females at 29°C will permit implanted wild type ovaries to complete
vitellogenesis. Ovaries of <u>fs(3)L8^{ts}</u> will undergo vitellogenesis
when implanted into a wild type male cultured at 22°C (data not
shown). A second phenotype of <u>fs(3)L8^{ts}</u> is a recessive temperature
sensitive larval lethality. Although the biochemical defect in
this mutant is as yet unknown, the two phenotypes of larval lethality and female sterility would be expected of a mutant interfering in a JH involved process common to both larval and adult JH
functions.

One of the ovary autonomous mutants seems of special interest.
<u>fs(2)A17</u> Females provide an especially favorable environment for
vitellogenesis. As Table 3, column 4 shows, normal ovaries develop
about twice as many vitellogenic follicles when implanted into
<u>fs(2)A17</u> hosts as when transplanted into any other host, including
wild type. We can therefore use these flies as hosts for a more
sensitive test of non-autonomy.

TABLE 3a. OVARIAN TRANSPLANTATIONS TO TEST FOR OVARY AUTONOMY IN FEMALE STERILE MUTANTS.

1	2 Unoperated control		3 Indicated ovary implanted into normal host				4 Normal ovary implanted into indicated host			
			Donor		Host		Donor		Host	
Genotype	# ov.[a]	vfpo[b]	# ov.	vfpo	# ov.	vfpo	# ov.	vfpo	# ov.	vfpo
Oregon R	24	17.6	18	10.2	36	14.2	18	10.2	36	14.2
fs(2)A17	30	0	17	0	34	15.8	20	20.0	40	0
fs(3)L3	20	0	16	0	32	13.4	15	5.5	30	0
fs(3)L8^{ts},29°	28	0	8	0	16	22.8	15	8.94	30	0
fs(3)L8^{ts},22°	16	18.6	17	0.47[c]	24	13.3	14	10.8	28	4.9
fs(3)L1	10	2.4[c]	6	0.5[c]	12	12.2	3	7.0	6	1.3[c]
fs(2)A18	36	0.4	22	0	44	14.7	6	8.2	12	0.2[c]
ap^4	34	0.06	12	7.0	24	17.7				
fs(3)A1	33	0.09	15	0.07	29	19.2	10	0.9	20	0.1

[a] Number of ovaries. [b] Vitellogenic follicles per ovary. [c] All vitellogenic follicles in early stages.

TABLE 3b. OVARIAN TRANSPLANTATIONS TO TEST FOR OVARY AUTONOMY IN FEMALE STERILE MUTANTS.

	5 — Indicated ovary implanted into fs(2)A17 host				6 — ap^4 ovary implanted into indicated host				7
	Donor		Host		Donor		Host		
Genotype	# ov.[a]	vfpo[b]	# ov.	vfpo	# ov.	vfpo	# ov.	vfpo	Classification
Oregon R	20	20.0	40	0	12	7.0	24	17.7	Normal
fs(2)A17	7	0	14	0	16	21.1	32	0	Autonomous
fs(3)L3	2	0	4	0	7	4.7	14	0	Autonomous
$fs(3)L8^{ts}$,29°	3	3.7	6	0					Autonomous
$fs(3)L8^{ts}$,22°	6	2.0	12	0					Autonomous
fs(3)L1	4	3.0	8	0	1	7.0	4	2.8[c]	Autonomous modifiable
fs(2)A18	16	21.1	32	0	7	7.4	14	0.07	Autonomous modifiable
ap^4									Non autonomous
fs(3)A1	8	7.75	16	0	9	1.2	18	0.11	Non autonomous

[a] Number of ovaries. [b] Vitellogenic follicles per ovary. [c] All vitellogenic follicles in early stages.

Table 3 (column 4) shows that fs(3)L1 and fs(2)A18 can support the development of normal ovaries and so they are classified as ovary autonomous mutants. But when their ovaries are implanted into fs(2)A17 females, the donor ovaries contain more vitellogenic follicles than normal, and even a few intermediate stages are formed. Thus, these mutants are ovary autonomous but environmentally modifiable.

The second class of female sterile mutant is the ovary non-autonomous class, including ap^4 and fs(3)A1. Ovaries from ap^4 females implanted into either wild type or fs(2)A17 hosts mature normally, confirming the earlier results of King and Bodenstein (1956), and indicating that the normal host provides ap^4 ovaries with all they need to mature. As we have shown previously (Postle-thwait and Gray, 1975), ovaries from heterozygous ap^4/SM5 flies, which usually develop normally, fail to develop when implanted into ap^4 homozygous females. This indicates that the ap^4 internal environment fails to provide a requirement for normal ovarian maturation.

While ovaries from fs(3)A1 fail to develop in wild type hosts, this mutant can also be classified as non-autonomous since its ovaries each develop about 7 vitellogenic follicles, including some mature oocytes, when implanted into the autonomous female sterile fs(2)A17. Furthermore, Table 3 (columns 4 and 6) shows that fs(3)A1 is a very poor host for both wild type and ap^4 ovaries. These results suggest that fs(3)A1 is sterile due to a defect in its internal environment.

Juvenile Hormone Analog Treatments. The transplantation tests identify at least two of the mutants examined as possessing ovaries capable of vitellogenesis but with a non-permissive internal environment. One wonders whether the defect in this mutant environment can be cured by an application of a JH analog. Table 4 shows the results of topically applying 0.15 µg ZR-515 to 2 hr old flies and scoring for vitellogensis at 72 hr. The mutants classed as ovary autonomous by the transplantation procedure failed to respond to the JH analog by maturing. The ovary autonomous modifiable female steriles possessed a few more vitellogenic follicles after ZR-515 treatment. The mutants which fell in the ovary non-autonomous class responded to the hormone by forming mature stage 14 oocytes. This result suggests that these mutants may be sterile due to the unavailability of juvenile hormone in the adult.

Corpus Allatum Transplants. Unavailability of juvenile hormone to the mutant ovaries might involve failure of the mutant corpus allatum to synthesize or secrete the hormone, lack of adequate hormone transport, or enhanced juvenile hormone degrada-tion. To test whether the corpora allata from the mutants were normal, we implanted three mutant or wild type corpus allatum-

TABLE 4. THE RESPONSE OF NON-VITELLOGENIC FEMALE STERILE MUTANTS
TO TOPICAL APPLICATIONS OF ZR-515.

Genotype	Untreated control		ZR-515	
	Number of ovaries	Vit. foll. per ovary	Number of ovaries	Vit. foll. per ovary
fs(2)A17	30	0	12	0
fs(3)L3	20	0	10	0
fs(3)L1	10	2.4^a	10	7.0^a
fs(2)A18	36	0.4^b	22	1.5^c
ap^4	34	0.06^a	8	9.5^c
fs(3)A1	33	0.09^a	26	1.5^c

a All early vitellogenic stages.

b Includes one late vitellogenic stage

c Includes late vitellogenic stages.

corpus cardiacum complexes from 3 day old flies into each Oregon R
wild type abdomen isolated from the thorax less than 2 hr after
eclosion. Table 5 shows that 46% of the surviving abdomens implanted with wild type corpora allata formed vitellogenic oocytes.
Glands from four mutants classified as ovary autonomous and auton-
omous-modifiable supported vitellogenesis in 20 to 29% of the
abdomens, indicating that their corpora allata can function to
support vitellogenesis. This result is consistent with the finding
(Table 3) that these mutants can support development of wild type
ovaries. On the other hand, the corpora allata from the two non-
autonomous mutants stimulated very little vitellogenesis. They
were each effective in only one abdomen out of 24 and 25 opportun-
ities. This result indicates that the corpus allatum of ap^4 and
fs(3)A1 fails to function normally. This conclusion is supported
by the transplants in Table 3 and in Postlethwait and Gray (1975)
which show that ap^4 and fs (3)A1 are not adequate hosts for implanted wild type ovaries.

TABLE 5. A TEST FOR FUNCTION OF THE CORPORA ALLATA FROM FEMALE
STERILE MUTANTS.

Genotype of corpus allatum	Number of surviving abdomens	Number of abdomens with vitellogenesis	Percentage of abdomens with vitellogenesis
Oregon R	22	10	46
fs(2)A17	29	6	21
fs(3)L3	25	5	20
fs(3)L1	14	4	29
fs(2)A18	32	7	22
ap^4	24	1	4
fs(3)A1	25	1	4

Ovarian Proteins

Ovarian Enzyme Levels. After having identified JH as an
important controlling factor in Drosophila vitellogenesis, and
having categorized two mutants defective in the supply of JH and
several others as defective in the ability to respond, we next need
to turn to how JH exerts its effect on ovarian function. One way
to do this is to identify specific ovarian proteins associated
with vitellogenesis and then to study their genetics and developmental physiology. One set of proteins appropriate for study is
the yolk proteins (Gelti-Douka et al., 1974; H. Teitelbaum, personal communication), and a second group is ovarian enzymes.

It has long been known that Drosophila ovaries contain high
levels of acid phosphatase (Yao, 1950; Giorgi, 1974). We have
dissected ovaries from developing normal flies of two genotypes
(Acph-1^C and ap^4/SM5) at various stages after eclosion and assayed
them for acid phosphatase activity and protein levels. Table 6
shows that acid phosphatase activity per ovary increases about 70-
fold over the first two days of adult life. Protein per ovary also
increases during this period, and so specific activity is given in
Fig. 3. The specific activity of acid phosphatase in normal ovaries

TABLE 6.   OVARIAN ACID PHOSPHATASE IN CONTROL ANIMALS AND IN
ISOLATED ABDOMENS.

Preparation	Age(h)	Number of determinations	Ave. number of ovaries per determination	Activity of acid phosphatase[a]
Unligated control	1	6	9.8	1.05
Unligated control	48	8	2.1	77.2
Ligated control	48	19	5.6	0.86
Ligated + ZR-515	48	19	1.9	28.9

[a] μmol p-nitrophenylphosphate/min/ovary (X 10^4).

is roughly constant for the first 12-15 hr of adult life. At about
the time that intermediate vitellogenic follicles start appearing,
the acid phosphatase specific activity begins to increase, and by
the third day after eclosion it has increased by a factor of about
three. These results indicate that the acid phosphatase activity
increases more rapidly than general protein amount, and that it is
a function of vitellogenic stage.

Identifying the Gene Responsible for Ovarian Acid Phosphatase
Activity. In salivary band 99DE (Morrison, 1973) is a gene which
controls the electrophoretic mobility of acid phosphatase and to
which null activity mutants map (MacIntyre, 1966; Bell et al.,
1973). Does this locus control ovarian acid phosphatase? Figure 3
shows that the specific activity of acid phosphatase in the null
allele Acph-1^{n4} is very low and remains constant during vitellogenesis. This indicates that the gene MacIntyre (1966) has identified is responsible for the increase in acid phosphatase activity
in the ovary.

Tissue of Synthesis. Since proteins can be taken into Drosophila oocytes by pinocytosis (Mahowald, 1972), we cannot be certain

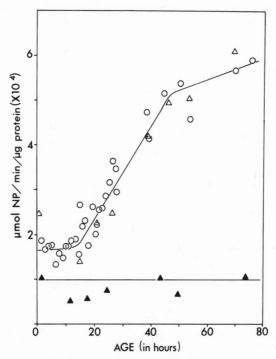

Fig. 3. Specific activity of acid phosphatase. Circles, normally developing ovaries of $ap^4/SM5$ genotype; open triangles, normally developing ovaries of the wild type $Acph-1^C$; filled triangles, normally developing ovaries of the null activity allele, $Acph-1^{n4}$.

that the ovary synthesizes its acid phosphatase. To answer this question we have implanted ovaries from the null activity allele $Acph-1^{n4}$ into the wild type $Acph-1^C$ and vice versa, and then two days later we dissected out the mature stage 14 oocytes and assayed their acid phosphatase activity. Table 7 shows that the enzyme titer of an oocyte depends solely on its genotype and not on the genotype of its host. We can thus conclude that the ovary is responsible for the acid phosphatase found within it.

Extra Ovarian Factors Controlling Acid Phosphatase Activity. In a preceding section we argued that a thoracic factor, probably JH, was necessary for the formation of stage 14 oocytes. Here we ask whether JH is required for development of acid phosphatase activity. Isolated abdomens were prepared from 1 hr old wild type $Acph-1^C$ flies to deprive them of their corpora allata, and either treated with ZR-515 or untreated. $Acph-1^C$ and $ap^4/SM5$ served as unligated controls. Table 6 shows that ovaries in isolated abdomens two days old had enzyme levels characteristic of ovaries from

TABLE 7. OVARIAN TRANSPLANTS TO ASCERTAIN THE SITE OF ACID
PHOSPHATASE SYNTHESIS.

	Donor		Host	
Genotype	Number of transplants	Average Acph activity[a]	Genotype	Average Acph activity[a]
--	8^b	--	C^c	6.3
--	7^b	--	n^{4d}	0.64
C	11	5.2	C	5.6
n^4	7	0.83	n^4	0.48
C	20	5.7	n^4	0.41
n^4	18	0.42	C	5.0

[a] μmol p-nitrophenol/min/ovarian oocyte (x 10^4).

[b] Number of non-transplanted control animals.

[c] Acph-1^C, a wild type stock.

[d] Acph-1^{n4}, a null activity stock.

freshly eclosed animals. However, when isolated abdomens were
treated with ZR-515 the ovarian acid phosphatase activity increased
about thirty-fold, and approached the control level. These
experiments show that JH has profound effects on the acid phosphatase activity found in ovaries.

Acid Phosphatase Activity in Female Sterile Mutants. We
selected one ovary autonomous (fs(3)L3) and one ovary non-autonomous
(ap^4) female sterile mutant to test the response of their acid
phosphatase to ZR-515. Figure 4 shows that ap^4 ovaries respond to
ZR-515 by the production of nearly normal levels of acid phosphatase activity. On the other hand, fs(3)L3 ovaries have the same
low amount of activity either with or without ZR-515. These
findings emphasize the conclusion that JH plays a regulatory role
in the appearance of ovarian acid phosphatase activity, and they
show that although ap^4 and fs(3)L3 have similar gross phenotypes,

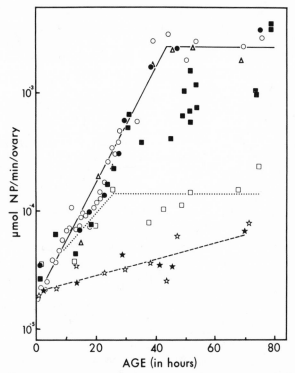

Fig. 4. The effects of treatment with the JH analog ZR-515. Open circles, ap[4]/SM5 normally developing untreated control; filled circles, ap[4]/SM5 treated; open squares, ap[4]/ap[4] untreated; filled squares, ap[4]/ap[4] treated; open star, fs(3)L3/fs(3)L3 untreated; filled stars, fs(3)L3/fs(3)L3 treated. Hormone was applied at eclosion and ovaries were dissected out and assayed at the indicated times.

that they alter different steps in the regulation of acid phosphatase activity and vitellogenesis.

DISCUSSION

A cephalic factor necessary for vitellogenesis at about eclosion was revealed by decapitation, and a thoracic factor required to permit complete vitellogenesis until 16 hr after eclosion was identified by preparation of isolated abdomens. The corpus allatum-corpus cardiacum complex, or a JH analog can substitute for the thorax, indicating that JH is probably the vitellogenic factor derived from the thorax. This result is consistent with the finding that corpora allata are important vitellogenic factors

in other species of Drosophila (Vogt, 1943; Bodenstein, 1947); that
JH can cure a nonvitellogenic mutant (Postlethwait and Weiser,
1973); that it can break diapause in a cave dwelling Drosophila
species (Kambysellis and Heed, 1974); that corpora allata implanted
into pharate adults causes premature increase in ovary size
(Manning, 1967); and that application of JH can cause precocious
appearance of mature oocytes in D. melanogaster (Bouletreau-Merle,
1973). Since abdomens isolated after intermediate stages have
appeared can form a few mature oocytes, but stop forming new early
vitellogenic stages, we can conclude that JH seems necessary to
permit stage 7 oocytes to become vitellogenic. The identity of
the cephalic factor is unknown, but in some other flies the medial
neurosecretory cells function to promote oogenesis (deWilde and
deLoof, 1973). More precise ablation experiments will permit the
identification of the vitellogenic cells in the head. Our results
are consistent with a factor produced by the head shortly before
eclosion which signals the corpus allatum to make JH available.
If this were so, one would expect corpus allatum-corpus cardiacum
complexes from pharate adults to be unable to provoke vitellogenesis in isolated abdomens.

Animals deprived of a JH source are non-vitellogenic, and so
mutants affecting JH availability, or those eliminating a response
to JH would also be expected to be non-vitellogenic. Five such
mutants were categorized as ovary autonomous non-vitellogenic
mutants by ovary transplants. Since these mutants have a normal
internal environment, they must be defective in an ovarian function such as JH reception, interpretation, or excution of the developmental program leading to vitellogenesis.

Two female steriles behaved as ovary non-autonomous non-
vitellogenic mutants in ovarian transplants. Genetic alterations
of the internal milieu might include aberrant JH production,
secretion, transport, or degradation, as well as abnormal nutrition, protein synthesis, fat body function and other defects. Two
of these mutants were shown to have defective corpora allata.
This might be due to a failure of the corpus allatum to be activated in the adult stage (perhaps by the brain), or a defect in JH
synthesis or secretion. A mutant involving aberrant control of
the corpus allatum has already been described in Manduca sexta
(Safranek and Riddiford, 1975). Since the ovary non-autonomous
mutants metamorphose normally, we can conclude that either the
adult and larval JH are different chemical entities, or that they
are governed by separate genetic control mechanisms. However,
some JH related processes may be under similar controls in adults
and larvae as evidenced by our temperature sensitive mutant.
Temperature shifts at appropriate developmental stages show that
something required for larval survival is also necessary for
vitellogenesis, but not for adult survival.

In order to approach the molecular and genetic basis of JH action in vitellogenesis, we have begun to study a specific ovarian protein. Ovarian acid phosphatase activity increases dramatically in normal ovaries over the first three days of adult life. Immuno-assay of material cross-reacting to antibody against acid phospha-tase shows that the increase in activity is due to synthesis of the enzyme (Sawicki and MacIntyre, 1975). Abdomens deprived of JH by ligation or mutation do not develop acid phosphatase activity. Supplying these preparations with a source of JH activity by applying the JH analog ZR-515, causes acid phosphatase activity to approach normal levels. JH also causes acid phosphatase activi-ty to increase in degenerating muscle of the Douglas Fir beetle (Sahota, 1975). Whether the hormone acts directly on the ovary or the response is mediated by another tissue is not known, but in vitro culture will provide an answer to this question. The signi-ficance of acid phosphatase to the economy of the fly is not known. Animals homozygous for the null activity allele are viable and fertile. Yet the developmental controls governing acid phosphatase demonstrated in this report argue that it has developmental signi-ficance, and that it can be used as a tool to investigate gene-hormone interactions. Isolation and investigation of mutants which are either non-vitellogenic (i.e. don't deposit yolk protein) but still increase in acid phosphatase activity, or mutants which are vitellogenic but fail to increase in acid phosphatase should further the understanding of the way JH regulates ovarian matura-tion, as well as how hormones and genes interact to effect develop-ment.

SUMMARY

This report addresses three problems concerning genetic and hormonal regulation of vitellogenesis in Drosophila. (1) The endocrinology of normal vitellogenesis; (2) the effect of mutants altering this endocrinology or the ovary's response to it; and (3) the control of ovarian acid phosphatase activity. Decapitation and preparation of isolated abdomens indicated that a cephalic factor triggers vitellogenesis at about the time of adult eclosion, and that a thoracic factor was continuously necessary to permit the formation of early and intermediate vitellogenic follicles. Trans-plantations of the corpus allatum-corpus cardiacum complex and treatments with the juvenile hormone analog ZR-515 strongly suggest that the juvenile hormone is the thoracic factor.

Seven female sterile mutants were examined which had an ovar-ian phenotype similar to that of ovaries not exposed to juvenile hormone. Reciprocal ovarian transplants between mutant and wild type flies showed that in five of these mutants the defect resides in the ovary (ovary autonomous female sterile mutants). In the

other two, development of the mutant ovary is enhanced when transplanted to permissive environments (ovary non-autonomous female
sterile mutants). The two non-autonomous mutants possess corpus
allatum-corpus cardiacum complexes which are faulty at inducing
vitellogenesis in a wild type isolated abdomen.

Ovarian acid phosphatase activity was investigated to find an
enzyme whose activity was a function of oocyte maturation. Ovarian
acid phosphatase specific activity increased by a factor of about
three in the first three days of adult life. The gene Acph-1 was
identified as the structural gene for the acid phosphatase found
in the ovaries. Transplantation of ovaries between wild type and
a null activity allele showed that the tissue responsible for
ovarian acid phosphatase is the ovary itself. Preparation of
isolated abdomens and treatments with ZR-515 indicated that acid
phosphatase activity was dependent on juvenile hormone. Investigation of two female sterile mutants showed that enzyme levels
increased in response to ZR-515 in a non-autonomous mutant, but
were unchanged in an autonomous mutant. The results suggest that
this might be a useful system in which to study gene-gene and
gene-hormone interactions.

ACKNOWLEDGEMENTS

We thank Jennifer White and Douglas Sears for technical assist-
ance, Christine Kirby and Harry Teitelbaum for helpful discussions,
Zoecon Corporation for ZR-515, and NIH for grant GM 21548 and a
Career Development Award, GM-00119-01.

REFERENCES

Ashburner, M., 1970, Nature 227:187.
Bakken, A.H., 1973, Dev. Biol. 33:100.
Bell, J., and MacIntyre, R., 1973, Biochem. Genet. 10:39.
Bodenstein, D., 1947, J. Exptl. Zool. 104:101.
Bouletreau-Merle, J., 1973, C.R. Acad. Sci. Paris 277:2045.
Bryant, P., and Sang, J., 1968, Nature 220:393.
Butterworth, F., and King, R., 1965, Genetics 52:1153.
Chan, L., and Gehring, W., 1971, Proc. Nat. Acad. Sci. USA 68:2217.
David, J., and Merle, J., 1968, Dros. Inf. Serv. 43:122.
Dearden, M., 1964, J. Insect Physiol. 10:195.
Doane, W., 1973, in "Developmental Systems: Insects) (S.J. Counce
    and C.H. Waddington, eds.), Vol. 2, pp. 291-497, Academic Press,
    London.
Englemann, F., 1970, "The Physiology of Insect Reproduction",
    Pergamon Press, Oxford.
Gelti-Douka, H., Gingeras, T., and Kambysellis, M., 1974, J. Exptl.
    Zool. 187:167.

Giorgi, F., 1974, Histochem. J. 6:71.
Kambysellis, M., and Heed, W., 1974, J. Insect Physiol. 20:1779.
King, R.C., 1970, "Ovarian Development in Drosophila melanogaster",
    Academic Press, New York.
King, R., and Bodenstein, D., 1965, Z. Naturforsch. 20B:292.
King, R., and Mohler, J., 1975, in "Handbook of Genetics" (R.C.
    King, ed.), Vol. 3, Plenum Press.
King, R., and Wilson, L., 1955, J. Exp. Zool. 130:71.
Lezzi, M., and Gilbert, L.I., 1969, Proc. Nat. Acad. Sci. USA
    64:498.
MacIntyre, R., 1966, Genetics 53:461.
Madhavan, K., 1973, J. Insect Physiol. 19:441.
Mahowald, A., 1972, J. Morphol. 137:29.
Manning, A., 1967, Anim. Behav. 15:239.
Morrison, W.J., 1973, The Isozyme Bulletin 6:13.
Postlethwait, J.H., 1974, Biol. Bull. 147:119.
Postlethwait, J., and Gray, P., 1975, Devel. Biol. 47:196.
Postlethwait, J., and Weiser, K., 1973, Nature New Biol. 244:284.
Safranek, L., and Riddiford, L., 1975, J. Insect Physiol. 21:1931.
Sahota, T.S., 1975, J. Insect Physiol. 21:471.
Sawicki, J., and MacIntyre, R., 1975, The Isozyme Bulletin 8:35c.
Schneiderman, H., 1967, in "Methods in Developmental Biology",
    (F. Wilt and N. Wessells, eds.), pp. 753-766, Crowell, New York.
Skopik, S., and Pittendrigh, C., 1967, Proc. Nat. Acad. Sci. USA
    58:1862.
Spieth, H.T., 1966, Anim. Behav. 14:226.
Ursprung, H., 1967, in "Methods in Developmental Biology" (F. Wilt
    and N. Wessels, eds.), pp. 485-492, Crowell, New York.
Vogt, M., 1943, Biol. Zentr. 63:467.
deWilde, J., and deLoof, A., 1973, in "The Physiology of Insecta",
    (M. Rockstein, ed.), Vol. 1 (2nd ed.), pp. 97-158, Academic
    Press, New York.
Williams, C.M., and Kafatos, F., 1971, Mit. Schweiz. Entomol. Ges.
    44:151.
Yao, T., 1950, Quart, J. Microsc. Sci. 91:79.

INDUCTION OF THE INSECT VITELLOGENIN IN VIVO AND IN VITRO

F. Engelmann

Department of Biology, University of California
Los Angeles, California 90024

INTRODUCTION

It has been documented for several species of insects belonging to different orders that the predominant yolk protein precursor, or vitellogenin, is synthesized by an extraovarian tissue, namely the fat body, of the female. The fat body produces this female specific protein and exports it into the hemolymph, from which it is taken up by the growing oocytes via pinocytosis. Vitellogenins from the fat body are immunologically identical to those of the fully grown eggs (see Engelmann and Ladduwahetty, 1974). The juvenile hormones (JH) are found to direct vitellogenin synthesis in all species studied in detail with the exception of the Saturniid moths (Telfer, 1965) and presumably Aedes aegypti (Hagedorn, 1974). In Leucophaea, the three known naturally occurring JHs can direct the de novo synthesis of vitellogenin in vivo (Engelmann, 1971). Analysis of the dose-response relationship showed that JH III, presumably the species' own JH, was least effective. In this species, as well as in others, JH appears to be an essential rate-limiting component for vitellogenin synthesis.

In the production of fully grown eggs, JH probably acts at several additional sites, all eventually contributing to the production of vitellogenin and its incorporation into the growing oocytes (Engelmann, 1976). While it is well established that the JH directs the de novo synthesis of vitellogenin, the details of how this is accomplished are less well understood. This report is thus intended to further elucidate the action of JH in directing vitellogenin synthesis in the cockroach Leucophaea, both in vivo and in vitro.

MATERIALS AND METHODS

Females of Leucophaea maderae during their first reproductive cycle were used exclusively. Methods of rearing the animals, bleeding, allatectomy, or preparation of antibodies, non-specific and specific for vitellogenin, were outlined in Engelmann (1971) or earlier papers.

Details of the procedure used for tissue preparation and microsome identification are described in Engelmann (1974a,b), Engelmann and Barajas (1975).

For in vitro protein synthesis the 1.0 ml of incubation medium contained the following: 10 μmol MgAc, 150 μmol KCl, 5 μmol ATP, 0.2 μmol GTP, 2 μmol creatine phosphate, 50 μg creatine phosphokinase, about 1 mg of rRNA in the ribosomes collected from females by Na deoxycholate dissociation of the endoplasmic reticulum (ER), about 1 mg of protein in the pH 5 fraction obtained from rat liver, and 0.2 μmol of each of 18 amino acids. The medium contained 100 μmol tris-HCl adjusted to pH 7.6. 0.2 μCi of [^{14}C]-leucine was added to each of the incubation mixtures. After a 30 min incubation at 37°C, Triton X-100 was added to make a final 1% solution for dissociation of the membranes. Proteins in aliquots of the medium were then precipitated by trichloroacetic acid (TCA), non-sex-specific antisera, and anti-vitellogenin. Antibody precipitates were washed in saline once, followed by two TCA washes. The final precipitates were then redissolved in 1 N NaOH and radioassayed.

RESULTS

Vitellogenin Synthesis on Fat Body Endoplasmic Reticulum

In Leucophaea vitellogenin appears to be synthesized exclusively on rough surfaced endoplasmic reticulum (rER) of the fat body cells. This ergastoplasm of a vitellogenic female is heavily studded with ribosomes which appears to result in greater density of the microsomes than those of non-vitellogenic females during homogenization of the tissues (Engelmann, 1974b, 1976; Engelmann and Barajas, 1975). Vitellogenin is secreted through the membranes into the cisternae of the rER from which it is later released into the hemolymph. During this synthesis of vitellogenin, nascent polypeptide chains anchor the ribosomes to the membranes: treatment with concentrated KCl is not sufficient to release these ribosomes (Fig. 1). This mode of polypeptide synthesis is fundamentally different from that for proteins used intracellularly and is generally found in systems that produce proteins used outside the cells which synthesize them (see Tata, 1973; Adelman et al., 1973).

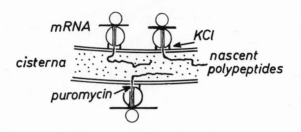

Fig. 1. Model for synthesis of vitellogenin on endoplasmic reticulum. All evidence shows that ribosomes are bound to the mem-
brains by the nascent polypeptide during translation. In addition
all ribosomes are bound by a mechanism that is sensitive to high
concentrations of KCl (Engelmann and Barajas, 1975).

In other words, vitellogenin is an exportable protein discharged
vectorially into the cisternae.

Treatment of the microsomes with increasing concentrations
of K^+ results in a release of primarily monosomes and small amounts
of polyribosomes (Engelmann, 1974a). These released monosomes and
polysomes are apparently not engaged in vitellogenin synthesis.
Additional monosomes and small polysomes (particularly dimers and
trimers) are freed after dissociation of the membranes with 0.7
to 1.0 % Na deoxycholate (Fig. 2). The amount of monosomes freed
is more than twice that of all the small polysomes together, as
measured by the area below the peaks. Similar results were
obtained when other detergents, such as Triton X-100, were used.
One can interpret this finding to mean that the majority of ribosomes are bound to membranes as monosomes, assuming that the mild
detergent treatment had not disrupted polysome integrity.

Effects of α-Amanitin on Vitellogenin Synthesis

Normal vitellogenic females were injected with 20 μg of α-
amanitin per female. Three hours later both treated and control
females were labeled with [^{32}P] and then bled 1 hr thereafter.
Vitellogenin is a phosphoprotein, and even though the phosphorus
content is low (Engelmann and Friedel, 1974), it was thought that
any effect of the drug would be readily revealed since vitellogenin
is the major phosphoprotein. Incorporation of [^{32}P] into hemolymph
vitellogenin was indeed reduced to less than 50% during the 4 hr
of α-amanitin exposure compared to values obtained from untreated
females (Table 1). One to several of the non-sex-specific proteins
also incorporate [^{32}P] but their rate of synthesis is considerably
lower than that of the vitellogenin (Engelmann et al., 1976).

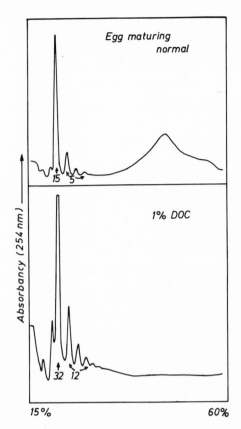

Fig. 2. Polyribosomal and microsomal profiles of 200 mg of fat
body from vitellogenic females on linear sucrose gradients of 15
to 60%. The 100,000 g microsome pellet was digested with 1% Na
deoxycholate and the digest layered onto the sucrose gradient.
The figures below the peaks are arbitrary planimeter units which
indicate that twice as many monosomes have been freed from the
membranes than the total of all polysomes.

Consequently the effect of the drug on these proteins was relatively small.

The inhibition of vitellogenin synthesis by α-amanitin was
further shown when incorporation of [^{14}C]-leucine into microsomal
vitellogenin was monitored (Fig. 3). Following a 12 hr α-amanitin
and a 1 hr [^{14}C]-leucine pulse, polysomes and microsomes were prepared for sucrose density gradient centrifugation as shown earlier
(Engelmann, 1974b). The label contained in the vitellogenin of
each of the fractions was determined by antibody precipitation.
As in untreated tissues, newly synthesized vitellogenin in

TABLE 1.   INCORPORATION OF [^{32}P] INTO HEMOLYMPH PROTEINS OF NORMAL
AND α–AMANITIN TREATED VITELLOGENIC FEMALES.

	Label in trichloroacetic acid precipitable proteins			
	Spec. activity			
	cpm/mg protein	% of normal	cpm	% of normal
Normal	203,400		137,200	
α–Amanitin	140,000	68.9	83,300	60.7

	Label in antibody precipitable proteins			
	Non-sex-specific		Anti-vitellogenin	
	cpm	% of normal	cpm	% of normal
Normal	18,500		122,500	
α–Amanitin	15,500	83.9	59,100	48.3

0.5 mCi of [^{32}P] (as orthophosphate) was injected into vitellogenic females 1 hr prior to bleeding. 20 µg α–amanitin per
female was injected 4 hr prior to bleeding in seven of the females.
12.5 µl of the pooled sera were worked up in either normal or
α–amanitin treated females. The protein concentration of the
pooled sera in either group was practically identical.

α–amanitin treated females was nearly exclusively associated with
the microsomes (Fig. 3). However, the incorporation of leucine was
now reduced to about 10 percent of the normal values. Also, a
small amount of vitellogenin was located on top of the sucrose
gradient. This is the portion of the vitellogenin which presumably
was freed from the microsome vesicles by mechanical damage during
preparation of the microsomal pellet. The same is seen in normal
tissues.

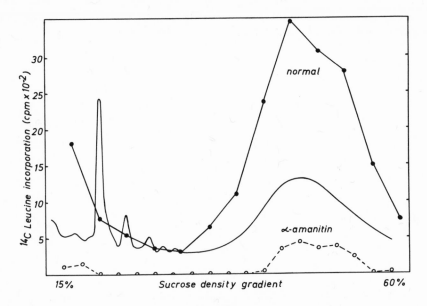

Fig. 3. Polyribosomal and microsomal profile of 500 mg of fat body from normal and α-amanitin treated vitellogenic females. Females were injected with 20 μg of α-amanitin 12 hr prior to sacrifice and had received 1 μCi of [^{14}C]-leucine 1 hr before the fat body was collected. The incorporation of [^{14}C]-leucine into anti-vitellogenin precipitable label was assayed for each fraction after the microsomes had been digested with 1% Triton X-100.

Incorporation of Nucleic Acid Precursors into Microsomal RNAs

Vitellogenin as well as non-sex-specific serum proteins are synthesized on rER in a presumably identical manner. If this is so, one should be able to identify on the microsomes the mRNAs that code for both vitellogenin and non-sex-specific proteins. With this in mind, vitellogenic or allatectomized females were pulse labeled with uridine for various times, and the polysomes and microsomes prepared as before (Engelmann, 1974a). Within 1 hr the total amount of uridine label in RNA of microsomes from vitellogenic females had reached a level which appeared to stay relatively stable for a period of 8 hr, which was as long as the animals were observed. Under identical conditions microsomes of allatectomized females incorporated only about one third of the label compared to vitellogenic females (Table 2). The amount of label above that of allatectomized females probably reflects the presence of newly synthesized messenger that codes for vitellogenin.

TABLE 2.   INCORPORATION OF [^{14}C]-URIDINE INTO RNA OF MICROSOMES
FROM FAT BODIES OF VITELLOGENIC AND ALLATECTOMIZED FEMALES.

Pulse duration hr	Vitellogenic cpm	Allatectomized cpm
1	1416	474
1 1/2	1225	–
2	2050	750
4	1640	–
8	1840	–

2 μCi of [^{14}C]-uridine (247 mCi/mmol) was injected into the
females at the indicated times prior to sacrifice. In every case
400 mg of the fat body was worked up and fractionated as published
earlier (Engelmann, 1974a). Fractions containing the microsomes
were digested with SDS, the RNA precipitated in the cold with ethyl
alcohol, the pellet redissolved and radioassayed.

More detailed information on the type of labeled RNA following
various pulse durations was obtained through sucrose density gradient centrifugation of the RNA from dissociated microsomes. RNAs
from these SDS digested microsomes were centrifuged through sucrose
density gradients of 5 to 20% or 10 to 40%. As shown (Fig. 4), a
2 hr in vivo pulse with orotic acid resulted primarily in labeling
of heterogenous RNA with sedimentation coefficients between 5 to
12 S. At this time practically no label was seen in ribosomal or
larger species of RNA. However, distinct peaks of rRNAs were seen
after a 4 hr pulse and the amount of label in these RNA species
progressively increased during the observation period of 18 hr.
Regardless of the duration of in vivo labeling a prominent incorporation of orotic acid occurred into RNAs heavier than 4 S but
lighter than 18 S (Fig. 4). Among these species of RNAs may be
the mRNA that codes for the vitellogenin.

In order to gain information on the flow of RNA between the
cellular compartments, fat body cells of vitellogenic females were
fractionated various time intervals after [^{14}C]-uridine labeling.

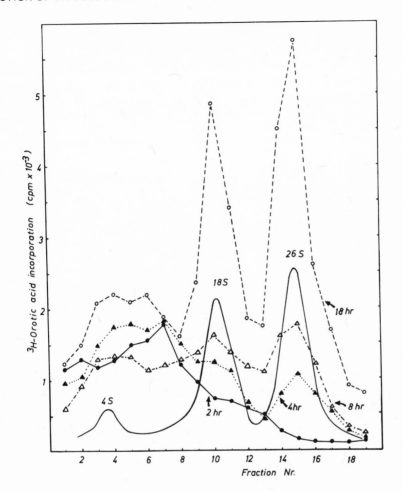

Fig. 4. Profiles of incorporation of 5-[³H]-orotic acid (spec.
act. 11.5 Ci/mmol) into microsomal RNA of fat body from vitellogenic females after various pulse durations. 20 μCi had been
injected into each female. The 149,000 g pelleted microsomes were
digested with 1% SDS and 1 mM EDTA, the digest layered onto a 5 to
20% linear sucrose gradient, and spun for 18 hr at 4°C in a SW 27.1
rotor at average 76,000 g. The sucrose gradients contained 0.05%
SDS and 1 mM EDTA in acetate buffer pH 5.0. Two ml fractions were
collected, the RNA precipitated with ethyl alcohol, redissolved in
tris buffer, and radioassayed.

The RNAs of nuclei, the 100,000 g supernate, and the microsomes
were extracted with SDS-phenol at 60°C, and the specific activity
determined. As is seen (Fig. 5), the specific activity in the

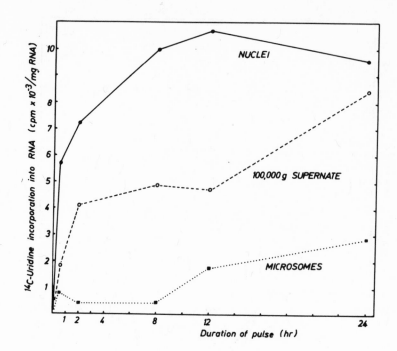

Fig. 5. Specific activity of [^{14}C]-uridine labeled RNA from
nuclei, 100,000 g supernate, and microsomes obtained from fat body
of vitellogenic females after various pulse durations.

nuclear fraction rose dramatically within one half hour and
continued to rise for the next 8 to 12 hr. Microsomes were
labeled only to a modest degree during the first 8 hr, but thereafter considerably more labeled uridine was incorporated. This
increase in incorporation of the RNA precursor following an initial
lag period was also seen earlier (Fig. 4). It represents an accumulation of labeled ribosomes on the endoplasmic reticulum.

Application of 20 μg of α-amanitin per female 3 hr prior to a
2 hr orotic acid pulse reduced the label in the heterogenous small
size class of RNA from microsomes (Fig. 6). An effect of the drug
on rRNA synthesis cannot be seen at this time because very little
labeled rRNA is extractable even from the untreated tissues. From
these data on short term precursor incorporation one may conclude
that α-amanitin probably inhibited transcription of the genetic
information for vitellogenin. Vitellogenin messenger appears to
be synthesized at a rate of at least three times faster (Table 2)
than those mRNAs of other exportable serum proteins. It remains
to be shown, however, that RNA of the 5 to 12 S region can be

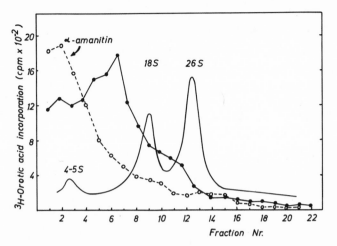

Fig. 6. Profiles of incorporation of 5-[³H]-orotic acid (spec.
act. 11.5 Ci/mmol) into microsomal RNA of fat body from α-amanitin
(20 μg/female) treated and untreated vitellogenic females.
α-Amanitin was given 3 hr prior to a 20 μCi orotic acid pulse of
2 hr. The fat body preparations and procedures were the same as
outlined in Fig. 4, except that a 10 to 40% sucrose gradient was
used here.

translated into vitellogenin. Following a long duration pulse
(15 hr) of orotic acid simultaneously with an application of
α-amanitin, the label in both small heterogenous RNA species and
in rRNA was reduced or nearly absent (not shown here). However,
a very prominent incorporation of label into 4 to 5 S RNA is seen
in all cases.

Vitellogenin Synthesis in a Cell Free System

It was shown earlier (Engelmann, 1974a,b; Engelmann and
Barajas, 1975) that in Leucophaea, vitellogenin is synthesized on
endoplasmic reticular membranes in vivo. This finding suggested
that for a successful in vitro synthesis of the molecule the
membrane component may be essential. In earlier attempts using a
traditional in vitro system, synthesis of vitellogenin and its
immunological identification had given spurious results. In most
cases only a small amount of immuno-precipitable label was produced by the complete system that contained only free ribosomes
harvested by Na deoxycholate digestion of the microsomes (Table 3).
However, if membranes from female fat body cells, degranulated by
incubation with 500 mM KCl and 1 mM puromycin, had been added to
the system, significantly more antibody precipitable label was

TABLE 3.   INCORPORATION OF [^{14}C]-LEUCINE IN VITRO INTO ANTIBODY
PRECIPITABLE PROTEINS.

	Non-sex-specific proteins	Vitellogenin
	cpm	cpm
Medium	420	140
Medium plus 180 μg RNA	390	310
Medium plus degranu- lated ER	1190	490
Medium plus degranu- lated ER and 180 μg RNA	2210	2450

The complete medium included ribosomes from fat body of
vitellogenic females collected by homogenization in NaOH-glycine
buffer containing 1% Na deoxycholate. The added RNA was obtained
from vitellogenic females by SDS-phenol extraction at 60°C.

obtained. The added bulk RNA had been extracted from fat body of
vitellogenic females by SDS-phenol at 60°C and presumably contained
messenger for both non-sex-specific proteins and the vitellogenin.

Incubation of rER that was untreated, i.e. still contained
the ribosomes, yielded a rather modest amount of vitellogenin.
After the addition of SDS-phenol extracted RNA from fat body of
vitellogenic females vitellogenin was produced as efficiently as
in the reconstituted system shown in Table 3 (data not shown here).
Probably the isolated rER contained only a limited amount of
messenger and only after the addition of exogenous mRNA was the
system working at full capacity. This finding needs further ex-
ploration.

DISCUSSION

Vitellogenin, an exportable protein synthesized exclusively
by the fat body of vitellogenic females, can be identified on rER
of this tissue (Engelmann, 1974a,b; Engelmann and Barajas, 1975).
The mode of synthesis involves the secretion of the nascent

polypeptide into the cisternae of the rER which anchors the ribo-
some to the membrane. The insect fat body which produces vitellogenin on rER is, in other words, basically similar to several
known vertebrate tissues which synthesize exportable proteins
(see Tata, 1973). Thus, it appears that this mode of exportable
protein synthesis is quite universal in the animal kingdom. Ribo-
some-membrane association may convey a unique specificity to the
system and may be advantageous for the rapid synthesis of large
quantities of a protein such as the major yolk protein precursor.
What this advantage really is still needs to be elucidated.

We have as yet been unable to identify polysomes that are
engaged in vitellogenin synthesis in Leucophaea by antibody precipitation. The two reasons for this failure are, (1) those polysomes which are freed from the membranes by high K^+ concentrations
would not carry any vitellogenin since they had not been anchored
to the membranes by the nascent polypeptide; (2) polysomes engaged
in vitellogenin synthesis would still be anchored to the membranes
and anti-vitellogenin precipitated microsomes would also carry
down the non-sex-specific antigens because the latter proteins are
synthesized on membranes side by side with vitellogenin (Engelmann,
1974b). The amount of precipitated label is therefore not representative of vitellogenin synthesis alone. Dissociation of microsomal membranes with detergents freed mostly monosomes (Fig. 2),
and virtually no vitellogenin could be identified with either the
freed monosomes or polysomes (unpublished). Either the ribosomes
had lost the antigen attached to them during digestion of the
membranes or the nascent polypeptide (the incomplete vitellogenin),
still attached to the ribosomes, is not recognized by the antibody.
The result that primarily monosomes were obtained after dissociation of the endoplasmic reticulum may be noteworthy. In the same
context, in many electron micrographs of microsomes extremely few
polysomes were seen. This could be due to a fleeting presence of
polysomes engaged in the synthesis of vitellogenin on membranes,
but the finding that most ribosomes remain anchored to the membranes
after treatment with KCl argues for monosome involvement in vitellogenin synthesis.

Following a short term uridine pulse one finds label associated with the microsomes, and since vitellogenin synthesis takes
place on rER one may assume that the label was primarily contained
in mRNA coding for vitellogenin (Table 2) (Engelmann, 1976).
Longer uridine pulse durations may result, as expected, in heavy
labeling of rRNA as well. Further identification of the RNAs was
made by sucrose density gradient centrifugation of SDS dissociated
microsomes, a method which had also been used for mouse myeloma
cells (Zauderer et al., 1973). The early appearance of labeled
RNA in a region between 5 and 12 S (Fig. 4) may be tentatively
taken as evidence that this is indeed the size class of messengers
coding for serum proteins, primarily vitellogenin. At this early

time no larger RNA species, such as rRNA, contained appreciable label (Fig. 6). Also, after long duration radiopulses virtually no RNA larger than 26 S was labeled (data not shown).

Vitellogenin synthesis can be inhibited in Leucophaea by injection of α-amanitin into active vitellogenic females (Table 1, Fig. 3). Within four hours synthesis was reduced to 50 percent of that of untreated animals and at 12 hr practically no more vitellogenin synthesis took place on rER. Furthermore, within 5 hr of treatment very little nucleic acid precursor was incorporated into what is presumed to be mRNA extractable from microsomes. These data argue for the hypothesis that one of the first steps of JH action is on transcription of the genetic information for vitellogenin. In addition one can deduce from the same data (Table 1) that the half-life of vitellogenin messenger is about 4 hr in a highly active tissue. In tissues that are less active the half-life may be longer. This finding is somewhat at variance with the earlier report that actinomycin D at a 2 μg dose allowed vitellogenin synthesis to continue unabated for a least three more days, suggesting a rather long half-life of the messenger (Engelmann, 1971). The latter observation is of course open to several interpretations due to the pleiotropic nature of actinomycin D action. Results following α-amanitin treatment can be more realistically evaluated, i.e. the observed 50 percent decrease in vitellogenin synthesis at 4 hr probably represents the true half-life of the messenger.

α-Amanitin blocked DOPA-decarboxylase synthesis in vivo in Calliphora vicina (Shaaya and Sekeris, 1971) probably via inhibition of transcription. In this species the drug inhibited incorporation of precursors into small heterodisperse RNA species (Shaaya and Sekeris, 1971) as well as into 28 S and 18 S RNA (Shaaya and Clever, 1972). In Leucophaea, incorporation of precursors into rRNAs was likewise inhibited by α-amanitin, which suggests that the drug is not only affecting synthesis of mRNA but also of rRNAs. Its action here appears to be similar to that reported for rat liver tissues (Niessing et al., 1970). On the other hand, these findings from in vivo observations may need to be reinterpreted in view of results obtained from in vitro cultured chromosomes of Chironomus thummi (Serfling et al., 1972). In the latter case α-amanitin blocked transcription of presumed mRNA but did not inhibit rRNA synthesis; it only slowed the processing of rRNAs. From the results in this species it appears that α-amanitin may have a post-transcriptional effect as well. At present no additional data are available for Leucophaea that could allow a decision on whether analogous mechanisms exist in this species.

One of the final experiments needed to give the experimental evidence of hormonal involvement in transcriptional events is

certainly the translation of newly synthesized mRNA for vitellogenin in a cell free system. As pointed out earlier (Engelmann,
1972), RNA extracted from fat body or microsomes of vitellogenic
females could indeed be translated in a cell free system. That
from inactive females or males, on the other hand, could not.
The product obtained can be reliably identified immunologically
(Engelmann, 1974b). In this study, microsomes rather than ribosomes had been used for the successful translation of the mRNA.
As is shown in the present report, an in vitro system that contains
free ribosomes only was rather inefficient in producing antibody
precipitable label (Table 3), but a reconstituted system including
free ribosomes and degranulated ergastoplasmic membranes plus RNA
extracted from vitellogenic females was most efficient. This
illustrates that the membranes from fat body are an essential
component for the translation of the vitellogenin messenger.

Rat liver cells produce serum albumin preferentially on rER
(Redman, 1967, 1969; Andrews and Tata, 1971; Tanaka and Ogata,
1972), and even after the bound polysomes had been freed they were
5 to 6 times more efficient in synthesizing this exportable protein
in vitro than were the "free" polysomes (Takagi and Ogata, 1968).
Protein synthesis in a cell free system reconstituted from plasma-
cytoma cells of mice was 3 to 8 times greater when membranes had
been added than by polysomes alone (Kubota et al., 1973). In the
latter case the data point to the importance of the membrane component in the in vitro system and may reflect the situation found
in vivo. Efficiency was greatly enhanced but no information is
available on specificity of the product made in this system. The
data supplied on vitellogenin synthesis in Leucophaea may additionally suggest that the ER conveys specificity.

It has now also been shown conclusively that ecdysone, the
insect molting hormone, induces DOPA-decarboxylase synthesis in
Calliphora (Fragoulis and Sekeris, 1975a,b). In a series of
reported experiments it is made clear that this hormone causes
the rise in the level of this key enzyme in the integument at
pupariation; the enzyme is identified both by its enzymatic characteristics and by antibody recognition. In this case it is not
known whether the rER of the hypodermal cells is involved in a
similar manner as shown for vitellogenin synthesis in the fat body
of Leucophaea.

SUMMARY

In Leucophaea maderae the juvenile hormone induced vitellogenin is synthesized in fat body cells on rough surfaced endoplasmic reticulum and secreted into the cisternae. Ribosomes are
primarily bound to the membranes as monosomes. After 4 hr α-
amanitin treatment vitellogenin synthesis is reduced to 50% of its

normal rate and practically no more synthesis occurs on microsomes after a 12 hr exposure to the drug. A class of 5 to 12 S rapidly labeled RNA is found on microsomes of normal vitellogenic females. α-Amanitin inhibits the incorporation of orotic acid into these RNA species; it also inhibits rRNA synthesis. Microsomes of alla-tectomized females incorporated labeled uridine at about one third the rate of normal vitellogenic females. The cell free synthesis of immunologically identifiable vitellogenin was greatly stimulated by the addition of degranulated endoplasmic reticulum. These membranes appear to be an essential component for efficient vitello-genin synthesis.

ACKNOWLEDGEMENTS

The research results reported here were greatly facilitated by the support through NSF Grant GB 14965 and a Biomedical Science Support Grant. This support is gratefully acknowledged. I also thank Mr. Sam Oda for technical assistance. I thank Prof. Th. Wieland for a gift of α-amanitin.

REFERENCES

Adelman, M.R., Sabatini, D.D., and Blobel, G.J., 1973, J. Cell Biol. 56:206.
Andrews, T.M., and Tata, J.R., 1971, Biochem. J. 121:683.
Engelmann, F., 1971, Archs. Biochem. Biophys. 145:439.
Engelmann, F., 1972, Gen. Comp. Endocrinol. Suppl. 3:168.
Engelmann, F., 1974a, Insect Biochem. 4:345.
Engelmann, F., 1974b, Amer. Zool. 14:1195.
Engelmann, F., 1976, Coll. Intern. Centre Nat. Rech. Sci. Paris (in press).
Engelmann, F., and Barajas, L., 1975, Exp. Cell Res. 92:102.
Engelmann, F., and Friedel, T., 1974, Life Sci. 14:587.
Engelmann, F., and Ladduwahetty, M., 1974, Zool. Jahrb. Physiol. 78:289.
Engelmann, F., Friedel, T., and Ladduwahetty, M., 1976, Insect Biochem. (in press).
Fragoulis, E.G., and Sekeris, C.E., 1975a, Biochem. J. 146:121.
Fragoulis, E.G., and Sekeris, C.E., 1975b, Eur. J. Biochem. 51:305.
Hagedorn, H.H., 1974, Amer. Zool. 14:1207.
Kubota, K., Yamaki, H., and Nishimura, T., 1973, Biochem. Biophys. Res. Commun. 52:489.
Niessing, J., Schnieders, B., Kunz, W., Seifert, K.H., and Sekeris, C.E., 1970, Z. Naturforsch. 25b:1119.
Redman, C.M., 1967, J. Biol. Chem. 242:761.
Redman, C.M., 1969, J. Biol. Chem. 244:4308.
Serfling, E., Wobus, U., and Panitz, R., 1972, FEBS Lett. 20:148.

Shaaya, E., and Clever, U., 1972, Biochim. Biophys. Acta 272:373.
Shaaya, E., and Sekeris, C.E., 1971, FEBS Lett. 16:333.
Takagi, M., and Ogata, K., 1968, Biochem. Biophys. Res. Commun.
    33:55.
Tanaka, T., and Ogata, K., 1972, Biochem. Biophys. Res. Commun.
    49:1069.
Tata, J.R., 1973, Karolinska Symp. Res. Meth. Reprod. Endocrinol.
    6:192.
Telfer, W.H., 1965, Ann. Rev. Entomol. 10:161.
Zauderer, M., Liberti, P., and Baglioni, C., 1973, J. Mol. Biol.
    79:577.

JUVENILE HORMONE INDUCED BIOSYNTHESIS OF VITELLOGENIN IN

LEUCOPHAEA MADERAE FROM LARGE PRECURSOR POLYPEPTIDES

J.K. Koeppe and J. Ofengand

Department of Biochemistry, Roche Institute of
Molecular Biology, Nutley, New Jersey 07110

INTRODUCTION

Despite the extensive research in recent years on insect
juvenile hormone (JH), its mode of action at the molecular level
is still not understood. One reason for this gap in our knowledge
is that the usual morphological assays of JH action do not lend
themselves readily to analysis by molecular biological tools. For
this reason, we have chosen to study the mode of action of JH in a
system which, although lacking in morphogenetic changes, is nevertheless amenable to biochemical analysis. In this approach, we
assume that the mechanisms of action we may discover in this system
will be equally applicable to JH-regulated morphogenetic processes,
but of course this remains to be proven.

The system we have studied is the JH-induced biosynthesis of
vitellogenin, the major yolk protein of oocytes in the organism
Leucophaea maderae. The life cycle of this cockroach is outlined
in Fig. 1. For our purposes, the most important part of the cycle
is the adult reproductive phase. Upon mating, the corpora allata
are activated to synthesize JH which then induces a very large
increase in the synthesis of vitellogenin which is needed for the
maturing oocytes. Subsequent to fertilization, egg case synthesis,
and retraction of the egg case back into the abdomen, the level of
JH drops and vitellogenin synthesis then ceases.

Previous work in a number of insects has localized the site
of hormone-induced vitellogenin synthesis to the fat body tissue
(Doane, 1973) and in Leucophaea induction of vitellogenin synthesis
in vivo by externally added JH and synthesis of vitellogenin by
fat bodies in organ culture had been shown (Brookes, 1969;

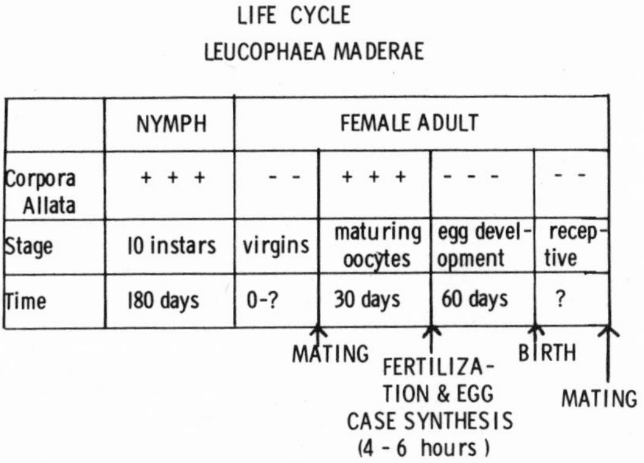

LIFE CYCLE
LEUCOPHAEA MADERAE

	NYMPH		FEMALE ADULT		
Corpora Allata	+ + +	- -	+ + +	- - -	- -
Stage	l0 instars	virgins	maturing oocytes	egg devel- opment	recep- tive
Time	180 days	0-?	30 days	60 days	?

MATING FERTILIZA-   BIRTH
        TION & EGG   MATING
        CASE SYNTHESIS
        (4 - 6 hours)

Fig. 1.   The life cycle of Leucophaea maderae.

Engelmann, 1971; reviewed by Engelmann, 1974). Moreover, a number
of properties of the vitellogenin protein product had already been
determined.

    Leucophaea vitellogenin, synthesized by the fat body and
secreted into the hemolymph for transport to the ovaries, is a
lipophosphoprotein (Engelmann and Friedel, 1974) which sediments
at 14 S in sucrose gradients (Brookes, 1969). It is incorporated
into the ovary as a 14 S (560,000 daltons) complex which then
apparently trimerizes to a 28 S complex (1,590,000 daltons). A
common amino acid composition, identical antigenic activities,
and pulse-chase experiments all support the precursor-product
relationship between hemolymph 14 S, oocyte 14 S, and oocyte 28 S
vitellogenin (Brookes and Dejmal, 1968; Brookes, 1969, Engelmann,
1971; Dejmal and Brookes, 1972; Engelmann and Friedel, 1974).
Mature oocyte vitellogenin contains 6.9% lipid, 8.2% carbohydrate,
and was said to consist of at least three polypeptide chains
(Dejmal and Brookes, 1972). However, the exact number of peptides,
their molecular weights, their stoichiometry, or the degree of
coordination of their synthesis was unknown when we began our
studies.

    Since Leucophaea fat body in organ culture synthesizes and
secretes at least 50% of its protein as vitellogenin (Brookes,
1969; Koeppe and Ofengand, 1976), it is well suited for analysis,
at the molecular level, of the mode of action of JH in inducing the
synthesis of specific proteins.

    In this report, we describe: (1) The in vivo induction by JH
of vitellogenin synthesis in fat bodies from ovariectomized females;
(2) The characterization of oocyte vitellogenin in terms of number
of peptides, molecular weights, and stoichiometry; (3) The variation

in polypeptide subunits of vitellogenin synthesized in the fat body
and secreted in organ culture and possible precursor-product relationships; and (4) The development of a system for searching for
the specific mRNA for vitellogenin.

MATERIALS AND METHODS

mRNA Isolation and In Vitro Translation. All techniques
used for RNA isolation, sucrose gradient purification, oligo-dT-
cellulose chromatography, and in vitro translation of RNA in the
wheat germ and ascites S-30 system are after the methods described
in Green et al. (1976).

All other procedures and preparations can be found in Koeppe
and Ofengand (1976).

RESULTS

Juvenile Hormone Induction of Vitellogenin Synthesis
in Ovariectomized Females

The necessity to clearly demonstrate the lack of participation
of an ovarian hormone in vitellogenin synthesis in Leucophaea
maderae was essential in view of the discovery that ecdysone produced by ovarian tissue, and not JH, stimulated vitellogenin
synthesis in Aedes aegypti (Fallon et al., 1974). Utilizing organ
culture techniques, tissue from juvenile hormone-induced and uninduced adult virgin females (with and without ovaries) were
analyzed to determine the comparative rates of vitellogenin synthesis.

The plan of the experiment is illustrated in Fig. 2. Isolated
abdomens of virgin adult females were prepared, and in one set, the
ovaries were also removed. Both groups were then given a single
dose of the Hoffmann-La Roche mixed isomer preparation of JH I in
olive oil, and at various times thereafter, fat body was removed
and assayed for the ability to synthesize vitellogenin by incubation with radioactive leucine in organ culture. An incubation
period of 5 hr was chosen since kinetic experiments had shown that
after this time, 45-50% of the total protein synthesized was vitellogenin. Only the vitellogenin secreted into the medium was taken
since the medium contained 80% of the total vitellogenin synthesized
in this time period and was substantially less contaminated by other
newly-synthesized proteins.

The results (Fig. 3) clearly show massive induction of vitellogenin synthesis in both groups of females (with and without
ovaries) after a lag of 18 hr. The subsequent rise to a maximum

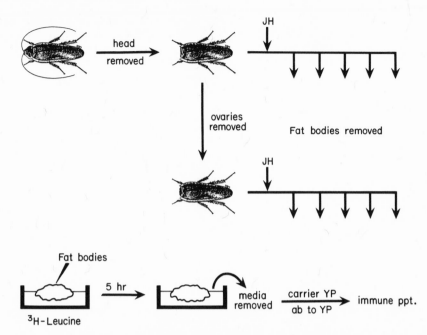

Fig. 2. Diagram of the experimental plan for determining the
kinetics of juvenile hormone induction of vitellogenin synthesis.
JH, juvenile hormone; YP, yolk protein (vitellogenin). Note that
only protein secreted into the medium was examined.

after 72-96 hr with a slower decline has also been observed in
other similar systems (Fallon et al., 1974; Clemens, 1974). Since
in vivo vitellogenin synthesis and oocyte maturation continues for
30 days, the decline observed here is probably due to catabolism
of the active form of JH in this single dose experiment. Ovariectomy had no effect when 100 μg of JH were given and only a small
effect at the lower dose. It appears that neither brain, corpora
allata and associated organs, or ovaries are directly required for
the response to JH.

The induction is somewhat dose dependent, as 20 μg JH yielded
a lesser response than 100 μg. However, the overall kinetics were
similar in both sets of hormone stimulated tissue. Thus, the
percent of the total newly-synthesized protein which was vitellogenin varied with the time of induction in the same way (Insert to
Fig. 3). Note that at the peak of stimulation (72 hr), vitellogenin
accounts for 75-80% of the total newly-synthesized protein which is
secreted. Non-vitellogenin protein synthesis is also stimulated by
JH with a sharp ten-fold rise between 18 and 48 hr which then stays
approximately constant to 72 hr followed by a slow decline. The

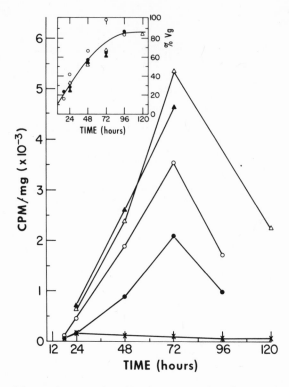

Fig. 3. Vitellogenin synthesis <u>in vitro</u> by fat body from headless
virgin females with and without ovaries after an <u>in vivo</u> injection
of juvenile hormone. Virgin females were decapitated and, as
indicated, ovaries removed from the terminal abdominal segment by
dissection. After a two-day resting period, the animals were given
a single injection of juvenile hormone dissolved in olive oil, and
at the indicated times thereafter, were sacrificed and the fat
body removed. The capacity of the fat body to synthesize vitellogenin was assayed by immunoprecipitation after a 5 hr incubation
with [^{3}H] leucine in organ culture. The results are expressed as
radioactive vitellogenin per mg of fat body in the culture dish
versus time of exposure to juvenile hormone. The insert shows the
fraction of total protein made after juvenile hormone stimulation
which is vitellogenin. Total protein was determined as hot TCA
precipitable material. Δ, females with ovaries (100 μg hormone);
▲, ovariectomized females (100 μg hormone); 0, females (20 μg
hormone); ●, ovariectomized females (20 μg hormone); X, sham
(olive oil) females with ovaries.

insert shows that the <u>fraction</u> of protein synthesized which was
vitellogenin was independent of the dose of JH, although the <u>amount</u>
of vitellogenin produced by a given amount of tissue was dose-

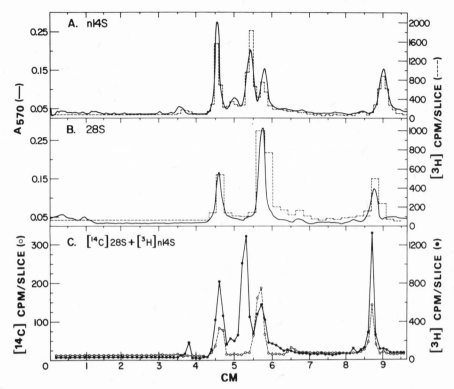

Fig. 4. SDS gel analysis of [³H] 14 S and 28 S vitellogenin. 14 S
and 28 S vitellogenin were labeled, purified, dissociated, electro-
phoresed for 4 hr, and analyzed (Koeppe and Ofengand, 1976). The
running gel begins at 0 cm. No isotope was found in the stacking
gel. Top: SDS gel of 14 S vitellogenin from both 2 mm and 5 mm
oocytes which were identical. Middle: SDS gel of 28 S vitellogenin
obtained from 2 mm and 5 mm oocytes. Bottom: [¹⁴C] 28 S and [³H]
14 S were dissociated, mixed, and co-electrophoresed. The peptides,
from left to right, are denoted A, D, B, C.

dependent. The fractional synthesis of vitellogenin increased with
time following JH induction but did not decrease once the maximal
level was reached even when the total synthetic capacity had
decreased by half. Induction of isolated abdomens of Leucophaea
by injection of JH has also been reported by Brookes (1969) and
Engelmann (1971), but the induction kinetics have not been previously described nor has the lack of effect of ovariectomy been quantitated.

Oocyte Vitellogenin Subunit Composition

One goal of this program is to isolate and identify the vitellogenin proteins which are synthesized and secreted from fat body upon induction by JH. In order to compare this induction product with oocyte vitellogenin it was first necessary to characterize the oocyte protein itself.

As stated in the Introduction, oocyte vitellogenin can exist in two forms, a 14 S (559,000 dalton) monomer or a 28 S (1,590,000 dalton) trimer. Both 14 S and 28 S are found in about the same amount in young (1-2 mm) oocytes but older (5-6 mm) oocytes contain almost exclusively the 28 S form (Dejmal and Brookes, 1972; Koeppe and Ofengand, 1976). The subunit composition of both of these forms was examined by the technique of SDS gel electrophoresis.

[^{3}H] or [^{14}C] Labeled oocyte vitellogenin was prepared by injecting either [^{3}H] or [^{14}C] leucine for several days before the maturing oocytes were removed and the 28 S and 14 S vitellogenin purified (Koeppe and Ofengand, 1976). The peptides were dissociated by heating to 100°C for 10 min in the presence of SDS-mercaptoethanol and analyzed by SDS gel electrophoresis (Fig. 4). Both radioactivity and protein staining were used to locate the polypeptide bands. It is evident that the 28 S complex consists of only three dissimilar peptides while the 14 S has one additional peptide (D). The double labeled experiment (Bottom Panel) confirmed that three peptides (A, B, and C) from each complex had identical mobilities while the fourth (peptide D) was unique to the 14 S complex. The same results were obtained when 8 M urea was present in addition to the SDS and mercaptoethanol both during dissociation and electrophoresis. Thus, the conversion of 14 S to 28 S does not involve simple trimerization of 14 S as originally proposed by Dejmal and Brookes (1972). Moreover, the occurrence of an additional band in 14 S was surprising since 14 S has been shown to be the precursor to the 28 S component (Brookes and Dejmal, 1968). A possible explanation of this apparent paradox derived from a consideration of the stoichiometry and molecular weights of the peptides involved will be discussed below.

The molecular weights were determined from a series of gels of varying pore size to eliminate the influence of carbohydrate on the mobility of the protein through the gel (Segrest and Jackson, 1972). Figure 5 indicates the results obtained from three representative experiments and Table 1 is a summary of the estimated molecular weights for each peptide. Although the pore size and running times were different for each gel, the estimated molecular weights of the unknown proteins remained within 1-2% of each other, thus confirming the size of each peptide. The effect of carbohydrate on the mobility of these SDS-polypeptides is negligible under these conditions.

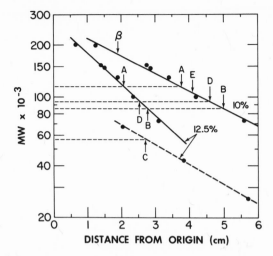

Fig. 5. Molecular weight determination of vitellogenin subunits by
SDS gel electrophoresis. 28 S and 14 S were prepared, electro-
phoresed, and analyzed as described in Koeppe and Ofengand (1976).
Bands were located by staining. Bands E and β were only present
in [^{3}H] vitellogenin isolated as an organ culture product (see
Fig. 6) and were detected by radioactivity. Radioactive band A also
present in the [^{3}H] vitellogenin preparation co-migrated with band
A of 14 S. The reference proteins used were myosin (200,000),
reo virus proteins (153,000; 148,000; 73,000), β-galactosidase
(130,000), phosphorylase a (100,000), bovine serum albumin (68,000),
ovalbumin (43,000), and α-chymotrypsinogen (27,500). Molecular
weights are from Weber et al. (1972) except for the reo virus
protein values (Both et al., 1975), (——) 9 hr electrophoresis,
(-----) 4 hr electrophoresis.

     Using these calculated molecular weights, the stoichiometry
of each complex was determined (Table 2). As shown in the Table,
treatment with iodoacetamide after dissociation did not affect the
results, and was omitted in some analyses. This was to be expected
since the amino acid analysis of Dejmal and Brookes (1972) showed
that cysteine was virtually absent. In fact, mercaptoethanol can
be omitted from the dissociating buffer with no effect on the gel
pattern (Gordon, E. and Ofengand, J. unpublished experiments).
Also, the stage of oocyte development between 2-5 mm did not
affect the size or distribution of peptides. Based on these molar
ratios, a stoichiometry of $A_1B_1D_2C_2$ was found for 14 S, and $A_1B_3C_2$
for 28 S. These stoichiometries correspond to a minimum molecular
weight of 549,000 for 14 S and 530,000 for 28 S after correction
for the 6.8% lipid content found in vitellogenin from mature oocytes
(Dejmal and Brookes, 1972). It is assumed that the same percent

TABLE 1.   MOLECULAR WEIGHT OF POLYPEPTIDE SUBUNITS OF 14 S and 28 S
VITELLOGENIN.

| Peptide | Apparent Molecular Weight | | | |
	7.5% Gel	10% Gel	12.5% Gel	Average
A	118,000	117,000	118,000	118,000
B	88,000	87,000	87,000	87,000
C	59,000	57,300	57,800	57,500
D	96,000	96,000	97,000	96,000
E	109,000	107,000	–	108,000
β	178,000	180,000	–	179,000

Polyacrylamide gel electrophoresis with reference proteins,
detection by staining or radioactivity, and calculation of molecular
weights was done as described by Koeppe and Ofengand (1976).   Peptides A, B, and C from either 14 S or 28 S gave the same molecular
weight values.   Bands β and E were only present in vitellogenin
isolated from the medium (see Fig. 5) as radioactive material.

lipid is also associated with the 14 S component.   These results
are in excellent agreement with the reported values of 559,000 for
14 S and 1,590,000 for 28 S (Dejmal and Brookes, 1972), assuming
the true 28 S stoichiometry to be $A_3B_9C_6$ (1,590,000 daltons).

The Precursor-Product Relationship Between Fat Body
and Oocyte Vitellogenin

Once we were able to establish the peptide composition of
oocyte 14 S and 28 S vitellogenin, it was possible to make comparative studies on the polypeptide subunits of vitellogenin secreted
from fat body in vitro.   SDS gel electrophoresis was used for this
purpose.   To ensure that the material isolated was only fat body
vitellogenin, two different procedures were used.   The first
sample (panel A, Fig. 6) was isolated by sucrose gradient fractionation of organ culture medium after a 5-hour incubation with

TABLE 2. MOLAR RATIO OF POLYPEPTIDE SUBUNITS OF 14 S and 28 S
VITELLOGENIN.

Peptide	Oocyte Size (mm)	Percent Total[a]		Molar Ratio[b]	
		14 S	28 S	14 S	28 S
A	2	23(22)	21	1.1	1.0
	5	28(30)	23(23)	1.4	1.0
D	2	41(37)		2.5	
	5	35(34)		2.1	
B	2	15(16)	57	1.0	3.7
	5	15(14)	54(55)	1.0	3.2
C	2	21(25)	21	2.1	2.1
	5	22(22)	24(24)	2.2	2.1

[a] Expressed as 100 x cpm or absorbance units in each peptide divided
by the sum of the cpm or absorbance units in peptides A + B + C + D.
The values in parentheses were obtained from iodoacetamide treated
samples. Only the cpm values were used for 14 S, while cpm and
absorbance values were averaged for 28 S.

[b] Calculated as the average percent total divided by molecular
weight and normalized to 1.0 for the smallest value. Data were
obtained by averaging from 2 to 4 electrophoretic runs for each
size oocyte and each size vitellogenin.

fat body. All of the radioactive vitellogenin secreted into the
medium sediments at 14 S in such an experiment and comprises 80%
of the total protein secreted. The remaining radioactivity can be
found between the 2-10 S region of the gradient. Most importantly,
no 28 S material is found (Brookes, 1969; Koeppe and Ofengand, 1976).
An aliquot from the 14 S peak was dissociated and analyzed without
further processing by SDS gel electrophoresis. The second sample
of organ culture medium was immunoprecipitated with vitellogenin
antiserum before dissociation and analysis by SDS gel electrophoresis (panel B, Figure 4). To ascertain if some of the products

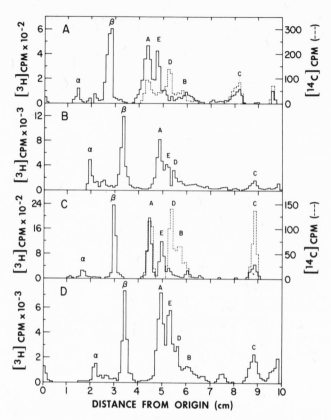

Fig. 6. SDS gel analysis of vitellogenin synthesized in organ
culture. Organ culture synthesis of [³H] leucine labeled vitellogenin was carried out with fat body from active females (4 mm
oocytes) for a period of 5 hr. A. The entire medium from an incubation which yielded 1980 cpm in vitellogenin/mg of fat body by
the anti-serum assay was layered on a sucrose gradient and fractionated. The 14 S peak was collected and used directly. All
solutions except the organ culture medium contained 10⁻⁵ M phenylmethyl sulfonyl fluoride (PMSF). The recovery of radioactivity
from the gel was 94%. B. Antibody precipitation was used to
purify vitellogenin from a second incubation as in part A which
yielded 2031 cpm in vitellogenin/mg of fat body by the antibody
assay. PMSF was absent from the solutions in this experiment.
The recovery of radioactivity was 95%. C. Vitellogenin was prepared from the medium of a third incubation as in part B, but in
this case 10⁻⁵ M PMSF was present in all solutions, including the
incubation medium. This decreased synthesis of vitellogenin to 962
cpm/mg fat body. Radioactivity recovery from the gel was 119%.
D. Vitellogenin was prepared from the fat body used in the incubation of part B by the antibody method. The recovery of radioactivity
from the gel was 76%. As indicated, dissociated [¹⁴C] 14 S was added
as an internal peptide marker to the preparations.

might be due to proteolytic activity, phenylmethyl sulfonyl fluoride, a serine protease inhibitor, was added to all solutions, including the incubation mixture, in an experiment equivalent to that of panel B. This resulted in a two-fold decrease in the level of vitellogenin synthesis as compared to panel B but no change in the polypeptide pattern (panel C).

The two most striking features of these analyses were first that despite the different methods of preparation and different degrees of precaution against proteases, the patterns of all three runs were very similar. The second and more important point is that the peptide pattern was very different from 14 S and showed a preponderance of two larger peptides labeled α and β. In addition, a new peptide, E, appeared between A and D. The molecular weights for β and E, obtained as shown in Fig. 5 and Table 1 by electrophoresis with suitable reference proteins were 179,000 and 108,000, respectively. The value for α was estimated to be 260,000 daltons from Fig. 6 using the known values for β and A for calibration of this region of the gel and assuming a linear relationship between log MW and distance migrated. The actual value could be even greater. The size of α and β in comparison to the largest peptide found in 14 S indicates that they are composed of at least two smaller peptides. Indeed, when the variation in molar ratio of the various peptides shown in this figure as well as in many other gels were compared, it was found that E and C always varied together (the average ratio E/C for the four examples in Fig. 6 is 1.0 ± 0.1, see also Fig. 7), suggesting that both of these peptides were processed from a single precursor of at least 166,000 daltons. A candidate for such a precursor could be β with a calculated MW of 179,000 daltons. The relationship of α to the other peptides is unclear at this time, although its size would allow it to be the precursor to A and B.

In order to see if some processing of vitellogenin occurred during secretion from the fat body cells, a similar analysis was performed on fat body extracts (Fig. 6D). This preparation came from the same organ culture as the sample shown in panel B, and it is clear that essentially the same amount and size of peptides were found. It should be noted, however, that protease inhibitors were not employed in this experiment. Although it is possible that larger peptides were originally formed, they would have had to be rapidly broken down to the same extent during the course of tissue disruption as those which were secreted during the incubation in order to account for the essential similarity between panels B and D. This experiment, however, does not rule out processing of vitellogenin precursors inside the fat body cell.

If, in fact, precursor proteins larger than the mature oocyte subunit are synthesized and secreted from the fat body, they must

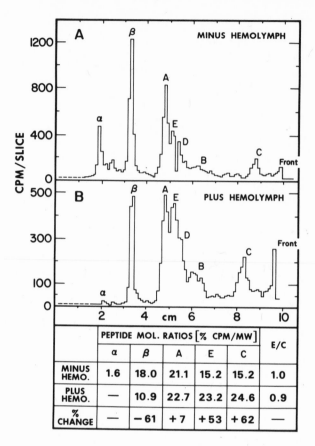

Fig. 7. SDS gel analysis of vitellogenin synthesized in organ
culture in the presence of insect hemolymph. Insect hemolymph was
collected from females with maturing oocytes by injection of Mark's
M-20 medium, homogenized in the medium to break up the coagulum,
and centrifuged at 7,000 x g for 10 min. An equal volume of the
supernatant was added to fresh M-20 medium and the fat body incubated for 5 hr with radioactive leucine. Addition of 10^{-5} M PMSF
did not change the peptide pattern (data not shown). Analysis was
carried out as described in panel B, Fig. 6. Panel A: minus hemolymph in culture medium. Panel B: with hemolymph in the culture
medium. Bottom panel: Summary of the pertinent peptide molar ratios
from all of the gels electrophoresis. Minus hemolymph is an average
of 9 different gels. Plus hemolymph is an average of 3 different
gels.

be subsequently processed into the mature oocyte form outside the
cell. The enzymes involved should be in the hemolymph. To test
this hypothesis, we extracted hemolymph from active females by
injection of M-20 medium and bleeding. After centrifugation, the
supernatant was diluted with an equal volume of medium and used
for incubation of fat body. The vitellogenin which was subsequently
extracted from the medium (Fig. 7, panel B) was partially processed.
Comparison with panel A, in the absence of hemolymph, shows that
band α had practically disappeared, and there were quantitative
changes in the other bands as well. This is most clearly shown by
consideration of the polypeptide relative molar ratios before and
after treatment. While peptide β decreased, peptides E and C
increased to almost the same extent suggesting, as already mentioned, that β may be the precursor to E (and hence to D) and C.
As pointed out above, the molecular weights for these peptides
do not conflict with such a scheme.

Vitellogenin mRNA

The finding of large vitellogenin polypeptide precursors is
of importance for the isolation of specific mRNAs for vitellogenin.
By assuming an average molecular weight of 100 daltons for each
amino acid and 330 daltons for each nucleotide, it is possible to
calculate the approximate molecular weight expected of the two
mRNAs for the large vitellogenin precursors α and β. The largest
polypeptide precursor (α), has a molecular weight of about 2.6 x
10^5 daltons, and therefore, its mRNA should be about 2.6 x 10^6
daltons. Similar calculations suggest the mRNA for β to be about
1.8 x 10^6 daltons. Since at least 50% of the total protein synthesized by the fat body is vitellogenin, sucrose gradient analysis
of poly A containing mRNA should reveal a detectable peak of RNA
sedimenting between 30-45 S.

Preliminary analysis of poly A containing RNA extracted from
active females verify these expectations. The experiment of Fig. 8
shows the presence of a peak of RNA at about 40 S in the fraction
of poly A containing RNA which is most active in translation in
vitro. This was true using both the wheat germ S-30 extract as well
as the ascites cell S-30 extract for translation although the wheat
germ system was generally much more active.

Characterization of the protein products of either the wheat
germ or ascites translation systems by SDS gel analysis or immunoprecipitation have been disappointing. We have not been able to
show the specific synthesis of vitellogenin so far. It appears
that translation is incomplete, and that only such small fragments
of protein are being made that none of the specific antigenic
regions of vitellogenin can be recognized.

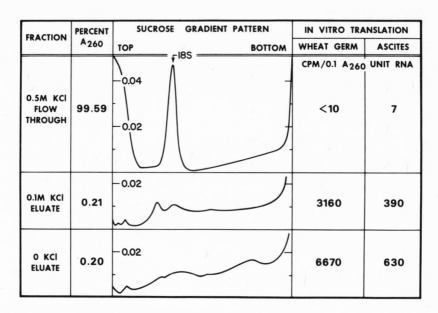

FRACTION	PERCENT A260	SUCROSE GRADIENT PATTERN		IN VITRO TRANSLATION	
		TOP	BOTTOM	WHEAT GERM	ASCITES
0.5M KCl FLOW THROUGH	99.59			<10	7
0.1M KCl ELUATE	0.21			3160	390
0 KCl ELUATE	0.20			6670	630

Fig. 8. Analysis of poly A containing RNA from fat body of females
with maturing oocytes. RNA extracted from fat body of females with
maturing oocytes by the phenol–chloroform procedure was passed over
an oligo–dT cellulose column in 0.01 M Tris pH 7.5, 0.5 M KCl.
Subsequent elutions were performed as indicated with Tris buffer
containing progressively less KCl. The recovery of RNA from the
column was 89%. 0.2 A_{260} units of each fraction were analyzed on
sucrose gradients, and a separate aliquot was tested for the ability
to direct the incorporation of [^{3}H] leucine into a hot–TCA insoluble
form.

DISCUSSION

In this report, we have demonstrated that the JH–induced biosynthesis of vitellogenin in the adult female fat body of Leucophaea
maderae is an excellent model system for the study at the molecular
level of the way JH regulates protein synthesis. Hormone induction
can be initiated at will, protein synthesis can be carried out in
organ culture, and large amounts of a readily characterizable product are formed. JH probably acts directly on the fat body although
until in vitro induction has been achieved, this point will not be
certain. It is also not clear which cell type(s) of the fat body
is activated by JH to produce vitellogenin.

We have, however, clearly shown that Leucophaea fat body is

not induced by a secondary ovarian hormone as is the case in Aedes
aegypti (Fallon et al., 1974; Hagedorn et al., 1975). Decapitated
and ovariectomized adult females were stimulated with JH I and
vitellogenin synthesis was quantitated. Comparative studies with
females not ovariectomized yielded a vitellogenin stimulation
pattern identical to that obtained from the ovariectomized animals.
In both groups, over 80% of the total protein made was vitellogenin
at the peak of stimulation. Moreover, this percentage did not
decline as protein synthesis decreased after 96 hr which might
suggest that the other proteins being made are also concerned with
oocyte maturation and/or egg case development, and are all coordin-
ately regulated by JH. Clearly this aspect needs further study.

As part of our development of this system, we have characterized the subunit nature of both 14 S and 28 S forms of oocyte vitellogenin. 14 S vitellogenin is made up of four polypeptides in the
ration $A_1B_1C_2D_2$. The experimentally determined molecular weights
for these peptides (Table 2) agree with the previously determined
value for 14 S when this stoichiometry is used. 28 S vitellogenin,
on the other hand, consists of only the three peptides A, B, and C
in the ratio $A_1B_3C_2$. Although it was originally suggested (Dejmal
and Brookes, 1972) that the 28 S component is a simple trimer of
the 14 S in view of their relative molecular weights, it is now
clear that this cannot be so since 28 S lacks polypeptide D. This
dilemma can be resolved if D of 14 S is processed by proteolysis
to yield B^D, a polypeptide different from B but of the same size, in
order to allow trimerization to occur. 28 S vitellogenin then
would actually consist of $A_1B_1B_2^DC_2$. This hypothesis would also
explain why 28 S and 14 S vitellogenin do not redistribute to an
equilibrium mixture after isolation, an event which never occurs
since presumably an active process consisting of cleavage of D to
B^D must take place to allow three 14 S components to aggregate.
The splitting of 9,000 daltons of protein from D probably occurs
in the oocyte itself rather than in the follicular space since
in young oocytes a large fraction of the total vitellogenin can
be found in the 14 S form (Brookes and Dejmal, 1968; Koeppe and
Ofengand, 1976).

According to this hypothesis, peptide B of 28 S vitellogenin
should consist of two distinct polypeptides in the ratio of 1:2;
the one present in larger amount being related to D of 14 S and
the other one corresponding to B of 14 S. We are currently
attempting to verify this hypothesis by structural and immunological studies on the isolated polypeptides. Furthermore, the appropriate proteolytic activity should be found in the oocyte. Since
28 S vitellogenin is less soluble than 14 S, this processing activity may make it easier to package vitellogenin in the limited space
of the oocyte.

The earlier stages of biosynthesis of vitellogenin are largely

speculative although it is now clear that at least some of the final
polypeptides A-D are synthesized as larger precursor molecules
(Fig. 6). Although the exact relationships among the precursors
α, β, and E, and the mature polypeptides are unclear, the stoichiometry of 14 S implies that A and B are made by cleavage of a single
precursor, and C and D are made from a separate precursor. Coor-
dinate synthesis of the mature peptides could then be regulated by
controlling the ratio of the two precursors at either the transcriptional or translational level or alternatively, the gene for
the C + D precursor could be duplicated (as illustrated in Fig. 9).
The A + B precursor would then be expected to be at least 205,000
daltons, and the C + D precursor at least 154,000 daltons. The
larger precursor could, therefore, correspond to α, and the smaller
one to β. A highly speculative scheme incorporating these and other
features of the data from Figs. 6 and 7 is illustrated in Fig. 9.
Polypeptide E was placed in the scheme leading from β because of
its constant ratio with C.

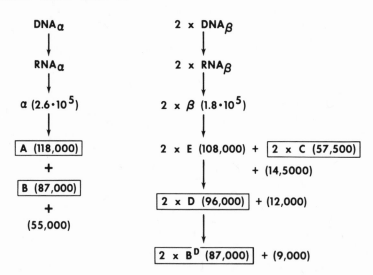

Fig. 9. Hypothetical scheme for synthesis and processing of
vitellogenin.

The failure to find differences in the polypeptide pattern
between vitellogenin isolated from the fat body and that secreted
into the medium suggests that partial processing may already take
place in the fat body and that specific proteolytic cleavage at
the cell membrane does not play any regulatory role in the secretory process. Further processing of vitellogenin can occur in the
hemolymph, as shown by the experiment of Fig. 7, and possibly also
during uptake into the oocyte. Since the secreted vitellogenin
already sediments at 14 S, the basic complex of α β$_2$ must already

have been formed in the fat body, and all subsequent modifications
to the subunits must reflect later proteolytic trimming processes.

The speculation that there exists two vitellogenin precursors
in a 1:2 ratio is of particular significance for the isolation and
quantitation of their specific mRNAs induced by hormone stimulation.
Preliminary experiments designed to isolate and identify specific
size classes of mRNA (Fig. 8) indicated a large peak of poly A
containing RNA in the size range expected for the mRNA for the
vitellogenin precursors (30-40 S). Although we could not show
vitellogenin synthesis in vitro with the RNA fraction containing
this peak, it is apparent that most of the translational activity
is found in this fraction. Additional studies will be needed to
localize the translating activity of the mRNA to the 30-40 S peak,
to characterize the translation products as vitellogenin precursors,
and to compare JH-induced and uninduced females in order to show
that a specific mRNA is being induced by the hormone.

SUMMARY

Juvenile hormone I (JH I) increases the synthesis of vitellogenin in isolated abdomens of virgin female Leucophaea maderae by
more than 50-fold. Synthetic capacity was assayed by pulse-label-
ing the fat body in organ culture at various times after induction.
Companion experiments using ovariectomized isolated abdomens showed
that vitellogenin synthesis was not dependent on secondary stimulation by an ovarian hormone and that in both cases up to 75-85%
of the protein secreted was vitellogenin.

Oocyte vitellogenin exists as a 560,000 dalton monomer complex
(14 S) and a 1,590,000 dalton trimer complex (28 S). The 14 S
complex is found only in young oocytes. By SDS gel electrophoresis,
the 14 S monomer was found to consist of four discrete peptides in
the ratio $A_1B_1C_2D_2$ with molecular weights 118,000, 87,000, 57,500,
and 96,000, respectively. The 28 S trimer contains only the three
peptides A, B, and C in the ratio 3 x $A_1B_3C_2$. These values,
corrected for lipid content, satisfactorily account for the known
molecular weights for each complex. Maturation of 14 S to 28 S in
the oocyte may involve the specific proteolytic conversion of the
14 S polypeptide D to B^D, a polypeptide of the same size as B but
different in sequence. The true composition of 28 S is suggested
to be $A_1B_1B_2^DC_2$.

The 14 S vitellogenin synthesized and secreted by fat body
consists of two large precursor polypeptides as well as several
smaller peptides comparable to the ones extracted from oocytes.
The large polypeptides (α and β) have molecular weights of >260,000
daltons and 179,000 daltons respectively, and either disappear (α)
or their concentrations are greatly reduced (β) upon the addition

of extracted hemolymph enzymes. Concomitantly, two smaller peptides increase proportionately. We speculate that α and β are the true gene products, that they associate to a pro-vitellogenin α β$_2$ complex inside the fat body, and are sequentially processed by proteolytic trimming during maturation. According to this scheme, α is the precursor of A and B, and β the precursor of D and C.

mRNA for the precursors α and β should be 30-40 S. Preliminary results indicate the presence of poly A-rich RNA having molecular weights close to that calculated for precursor protein mRNA and possessing the capacity to stimulate in vitro protein synthesis.

REFERENCES

Both, G.W., Lavi, S., and Shatkin, A.J., 1975, Cell 4:173.

Brookes, V.J., 1969, Develop. Biol. 20:459.

Brookes, V.J., and Dejmal, R.K., 1968, Science 160:999.

Clemens, M.J., 1974, in "Progress in Biophysics and Molecular Biology" (A.J.V. Butler, and D. Noble, eds.), 28, pp. 69-108.

Dejmal, R.K., and Brookes, V.J., 1972, J. Biol. Chem. 247:869.

Doane, W., 1973, in "Developmental Systems: Insects" (S.J. Counce, and C.H. Waddington, eds.), 2, pp. 291-497, Academic Press, New York.

Engelmann, F., 1969, Science 165:407.

Engelmann, F., 1971, Arch. Biochem. Biophys. 145:439.

Engelmann, F., 1974, Amer. Zool. 14:1195.

Engelmann, F., and Friedel, T., 1974, Life Sci. 14:587.

Fallon, A.M., Hagedorn, H.H., Wyatt, G.R., and Laufer, H., 1974, J. Insect Physiol. 20:1815.

Green, M., Zehavi-Willner, T., Graves, P.N., McInnes, J., and Pestka, S., 1976, Arch. Biochem. Biophys. 172:74.

Hagedorn, H.H., O'Connor, J.D., Fuchs, M.S., Sage, B., Schlaeger, D.A., and Bohm, M.K., 1975, Proc. Nat. Acad. Sci. USA 72:3255.

Koeppe, J., and Ofengand, J., 1976, Arch. Biochem. Biophys., 173:100.

Segrest, J.P., and Jackson, R.L., 1972, Methods Enzymol. 28:54.

Weber, K., Pringle, J.R., and Osborn, M., 1972, Methods Enzymol. 26:3.

JUVENILE HORMONE CONTROL OF VITELLOGENIN SYNTHESIS IN LOCUSTA MIGRATORIA

T.T. Chen, P. Couble, F.L. De Lucca, and G.R. Wyatt

Department of Biology, Queen's University, Kingston, Ontario, Canada

INTRODUCTION

The action of juvenile hormone (JH) in controlling insect metamorphosis has been difficult to investigate at the molecular level. This is due to the close interactions with ecdysone as well as the lack of identified JH-controlled proteins or other specific cell products related to the morphogenetic actions of the hormone. Although a theoretical model of JH action at the gene level has been constructed (Williams and Kafatos, 1972), and various other suggestions regarding the mode of action have been put forth (see Wyatt, 1972), these have, as yet, little basis in solid biochemical knowledge.

The second role of JH in regulating reproductive maturation in adult insects (see Doane, 1973) is free of these obstacles to biochemical study. During insect oogenesis, egg precursor proteins (vitellogenins) are synthesized in large amount in the fat body, released into the hemolymph and taken up into the developing oocytes (Telfer, 1965). In many (but not all) species, these processes are under the control of JH. The hormone-induced synthesis of vitellogenin in the fat body of appropriate insect species provides an excellent system for analyzing how this most interesting hormone can control cell function.

The control of vitellogenin synthesis by JH has been investigated in the cockroach Leucophaea maderae (Englemann, 1971; 1974; Koeppe and Ofengand, 1976); however, the fat body of all cockroaches contains abundant intracellular bacterial symbiotes which might complicate experiments at the subcellular level. Studies were also initiated with the Monarch butterfly (Pan and Wyatt, 1971), but

this insect is difficult to maintain in the laboratory. As a
favorable species, we have chosen the African migratory locust,
Locusta migratoria, which is conveniently large, free of symbiotes
and easy to rear. It has previously been shown that removal of
the corpora allata (the source of JH) from adult female locusts
(L. migratoria and Schistocerca gregaria) prevented the maturation
of eggs and lowered the incorporation of amino acids into protein
by the fat body (Hill, 1965; Minks, 1967). However, the identity
of the vitellogenic protein, and the effects of JH on specific
protein synthesis had not been clearly established for these
species (Benz et al., 1970). Recently, with L. migratoria, a
major glycoprotein has been isolated from eggs (Yamasaki, 1974),
and study of yolk proteins has been undertaken in relation to their
uptake into the oocytes, which is also controlled by JH (Bar-Zev
et al., 1975; Gellissen et al., 1976).

     The term vitellogenin has been applied to the yolk precursor
proteins present in insect hemolymph (Pan et al., 1969), and also
to the comparable proteins of vertebrates that are made in the
liver and transported to the ovaries in the plasma (Wallace, 1970).
During the formation of yolk, these proteins may undergo modification. In vertebrates, this includes cleavage into the well-charac-
terized moieties phosvitin and lipovitellin (Bergink et al., 1974;
Wallace and Bergink, 1974; Clemens et al., 1975). In insects,
there may be aggregation and other changes which are not yet well
defined (Dejmal and Brookes, 1972; Koeppe and Ofengand, 1976).
Insect yolk protein has been called lipovitellin (Dejmal and
Brookes, 1972), but in view of its lower lipid content than the
lipovitellins of vertebrates, its content of carbohydrate and its
uncertain homologies, we favor the simple term vitellin. Insect
vitellins (from yolk) share antigenic identity with the corresponding vitellogenins (from fat body or hemolymph) and, because
of their ready availability, are often used by preference as
antigens or for structural study.

                         RESULTS AND DISCUSSION

            Physical and Chemical Properties of Locust Vitellin

     The first requirement was isolation and characterization of
the protein whose synthesis we intended to study. The protein
patterns of hemolymph and egg extracts from adult locusts were
analyzed by acrylamide gel electrophoresis (Fig. 1). In egg
extracts, a major protein component (Rf = 0.36) was identified as
the vitellin, since it reacted with female-specific antiserum.
(Specific antisera, upon which many of the subsequent experiments
depended, were prepared in rabbits with the aid of Freund's adju-
vant; initially egg extracts and later purified vitellin were used

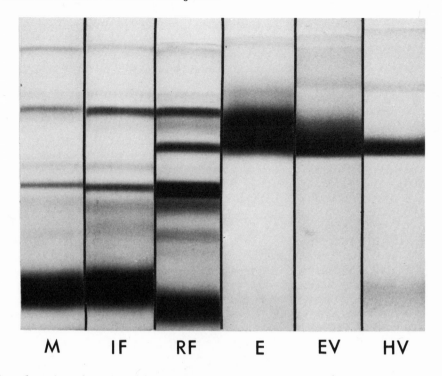

M   IF   RF   E   EV   HV

Fig. 1. Acrylamide gel electrophoresis of locust hemolymph and
egg proteins. Hemolymph samples were allowed to clot at room
temperature for 30 min and the supernatant after centrifugation
was used. Egg proteins were extracted in 0.05 M Tris-HCl (pH 8.1)
containing 0.4 M NaCl and 0.1 mM phenylmethylsulfonyl fluoride
(PMSF). Electrophoresis: 5% acrylamide, Tris-glycine buffer,
pH 8.3. Stain: Coomassie blue. M, male hemolymph; IF, immature
adult female hemolymph; RF, reproductive adult female hemolymph;
E, egg yolk extract; EV, vitellin purified from egg; HV, vitellogenin purified from hemolymph.

as antigens and the resulting sera were exhaustively absorbed with
male hemolymph.) Some slower moving electrophoretic components
include a band identified as an aggregation form of the vitellin,
and several non-specific proteins (present in males also). Hemolymph from reproductively mature adult females shows a band,
absent in the hemolymph of males and young adult females, which
corresponds to vitellin of egg extracts and represents the vitellogenin (Fig. 1). The close relationship of the specific proteins
in egg extracts and hemolymph of reproductively mature females was
further confirmed by acrylamide immunoelectrophoresis. We have
found that it is essential to remove the clotting proteins in
hemolymph samples by brief incubation at room temperature and

centrifugation before electrophoresis since they comigrate with
the vitellogenin. Omitting this step may account for previous
failures to identify an electrophoretic component specific to
adult female hemolymph (Tobe and Loughton, 1967; Benz et al., 1970).

Vitellin has been purified to homogeneity from egg extracts
(Fig. 1) by a procedure involving two steps of chromatography on
DEAE cellulose (Chen and Wyatt, 1976). Some chemical and physical
properties of the purified protein are summarized in Table 1. The
molecular weight was determined by three independent techniques,
with results in satisfactory agreement, giving a mean value of
550,000. The purified protein carries about 10% of lipids, which
consist of phosphatidyl choline, phosphatidyl ethanolamine and
triglycerides. The protein also contains carbohydrate, including
about 12% of mannose. This finding is in good agreement with
Yamasaki's (1974) report of carbohydrate (about 14%) containing
mannose and glucosamine in the ratio 14:2 in a glycoprotein prepared from locust eggs. We have determined the amino acid composition of locust vitellin and found aspartic acid, glutamic acid
and leucine to be the most abundant residues. Vitellogenin has
also been purified from the hemolymph of adult females by the same
procedure with slight modification, and corresponds electrophoret-
ically and immunochemically to the protein purified from eggs
(Fig. 1). L. migratoria vitellin has been independently purified
and characterized in Dr. Emmerich's laboratory (Gellissen et al.,
1976) with results generally in close agreement with ours. Locust
vitellin is very similar to the vitellins and vitellogenins isolated from several other insect species (Dejmal and Brookes, 1972;
Engelmann and Friedel, 1974; Pan and Wallace, 1974; Koeppe and
Ofengand, 1976; Kunkel and Pan, 1976).

To investigate its subunit structure, vitellin was denatured
in sodium dodecyl sulfate (SDS) and β-mercaptoethanol and analyzed
in SDS-acrylamide gels. As shown in the band diagrams in Fig. 2,
eight polypeptides were observed, with molecular weights ranging
from 140,000 - 52,000 daltons. The amounts of these components,
measured by scanning of quantitatively stained gels or by counting
of gel fractions from uniformly labeled protein, did not correspond
to any simple stoichiometry. Furthermore, their proportions were
significantly different in hemolymph vitellogenin and in egg
vitellin, the latter having greater amounts of the smaller components. The relationship of these polypeptides to the native protein
was studied with the aid of mature female fat body maintained in
culture in vitro (Wyatt et al., 1976). After pulse labeling with
a mixture of [^{3}H]-amino acids for 10 min, fat body intracellular
vitellogenin analyzed by SDS gel electrophoresis showed predominant
incorporation into a 260,000 dalton component (Fig. 2a). When the
pulse labeled tissue was chased for 30 and 60 min in medium containing unlabeled amino acids, progressive shifting of the radioactivity to components with molecular weights from 140,000 - 52,000

TABLE 1.   PHYSICAL AND CHEMICAL PROPERTIES OF VITELLIN ISOLATED
FROM LOCUSTA MIGRATORIA EGGS.

Sedimentation coefficient[a] $S^o_{20,w}$ = 17.1S

Molecular weight

      Sucrose gradient centrifugation   5.8 x 10^5

      Gel filtration (Sepharose 4B)   5.6 x 10^5

      Polyacrylamide gel electrophoresis 5.0 x 10^5

      Mean   5.5 ± 0.4 x 10^5

Isoelectric point[b]: 6.9

Lipids: 9.6 ± 0.8%

    Triglyceride

    Phosphatidyl choline

    Phosphatidyl ethanolamine

Carbohydrate content

    Mannose[d]   12.2 ± 0.3%

    Glucosamine[e]   about 2%

---

[a] Measured in the analytical ultracentrifuge with the assistance
of Dr. W.G. Martin, National Research Council Laboratories, Ottawa.

[b] Gellissen et al. (1976).

[c] Extracted with chloroform–methanol (2:1, v/v), determined
gravimetrically and identified by thin layer chromatography on
silica gel.

[d] Identified by thin layer chromatography after hydrolysis in
2 N H_2SO_4 at 100°C for 6 hr; determined by anthrone with mannose
standard.

[e] From Yamasaki (1974) for yolk protein from L. migratoria; not
determined in the present work.

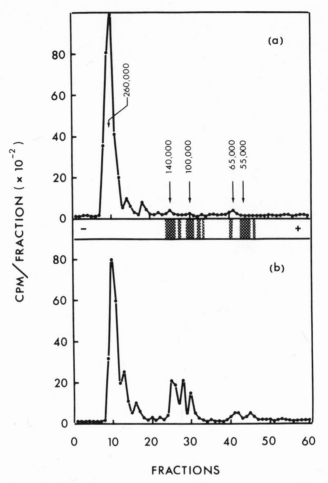

Fig. 2. Processing of vitellogenin polypeptides after synthesis
in fat body in vitro, shown by SDS gel electrophoresis. Fat body
from mature adult females was incubated in 2 ml locust ringer
(Chen and Wyatt, 1976) containing 20 μCi/ml [^{3}H]-amino acids for
10 min, and then chased in medium containing non-labeled amino
acids (Wyatt et al., 1976) for various times. After washing in
locust ringer, tissue was homogenized in 1.5 ml of 0.05 M Tris-
HCl (pH 8.1) containing 0.4 M NaCl, 1 x 10^{-4} M PMSF and 0.2%
Triton X-100. Extracts were centrifuged at 10,000 x g for 10 min
and 120,000 x g for 4 hr. Specific products were isolated by
precipitation with vitellogenin antiserum, the precipitate was
washed twice with 0.05 M phosphate buffer (pH 7.5) containing 0.9%
NaCl, and treated with 0.15 ml of 0.1 M Tris-HCl (pH 8.0) containing
10% glycerol, 3% SDS and 1% β-mercaptoethanol for 5 min at 100°C.
Electrophoresis: 7.5% acrylamide (containing 0.1% SDS), Tris-
glycine buffer (pH 8.8, 0.1% SDS). Gel slices (1 mm) were dissolved
in 0.2 ml 30% H$_2$O$_2$, and counted in 6 ml Triton-X-100 toluene based
scintillator (Patterson and Greene, 1965).

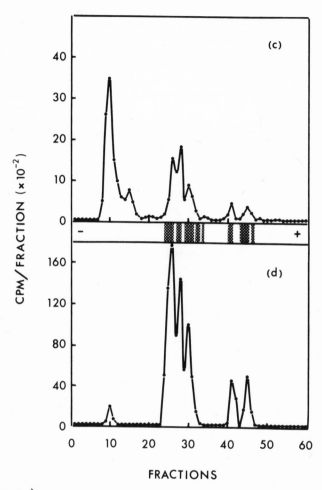

Fig. 2 (cont.)

    (a) Intracellular vitellogenin, 10 min pulse;
    (b) Intracellular vitellogenin, 10 min pulse, 30 min chase;
    (c) Intracellular vitellogenin, 10 min pulse, 60 min chase;
    (d) Vitellogenin recovered from medium after 10 min pulse
        and 3 hr chase.

The band diagrams represent the pattern obtained from vitellin
purified from eggs and denatured and subjected to electrophoresis
in SDS as described above, stained with Coomassie blue.

was observed (Fig. 2b, 2c). Furthermore, the vitellogenin secreted
from the cells and recovered from the medium, after being chased
for 3 hr, consisted of these smaller polypeptides, and contained
only a trace of the 260,000 dalton component, which might originate from injured fat body cells (Fig. 2d). These results indicate
that the complex polypeptide patterns observed after dissociation
by SDS are products of cleavage by proteases within the fat body.
This conclusion may explain the perplexing multiplicity of apparent
subunits in insect vitellins and vitellogenins observed by others
(Koeppe and Ofengand, 1976; Kunkel and Pan, 1976). Similar cleavage into multiple polypeptides is indicated by recent work on vertebrate lipovitellins (Bergink et al., 1974; Clemens, 1974).

The molecular weights of native intracellular and secreted
vitellogenin were determined by sucrose gradient centrifugation
(Fig. 3). The newly synthesized protein had a molecular weight of
260,000 (Fig. 3a), but was converted to a form of about 520,000
daltons, presumably the dimer, before being secreted into the medium (Fig. 3b, 3c). We therefore conclude that the primary translation product of locust vitellogenin is 260,000 daltons, and that
dimerization and proteolytic cleavage (as well as addition of lipid
and carbohydrate) take place within the fat body to yield the
550,000 dalton protein found in hemolymph or eggs.

Control of Vitellogenin Synthesis by Juvenile Hormone

The occurrence and amount of vitellogenin in the hemolymph of
normal adult females were determined by the technique of quantitative immunodiffusion developed by Oudin (1952) (Fig. 4). Vitellogenin is absent from the hemolymph of newly emerged adult females,
and, with locusts under crowded conditions with both sexes present,
first appeared about the 7th day of adult life. The concentration
of vitellogenin then increased rapidly until production was balanced by uptake into the ovaries. At this time, yolk deposition
became evident in the growing oocytes.

To test whether vitellogenin synthesis was controlled by JH,
adult females were allatectomized within one day after emergence.
This permanently prevented maturation of eggs and the production
of vitellogenin, but these processes were restored by repeated
topical administration of pure synthetic JH I. That this involved
induced de novo protein synthesis was shown by the experiment
presented in Fig. 5, in which [³H]-leucine was incorporated, in JH-
treated locusts only, into a product isolated in a sucrose gradient
and precipitated by specific antiserum.

In order to measure precisely the capacity of the fat body of
variously treated locusts to synthesize vitellogenin, we developed
an in vitro assay method (Wyatt et al., 1976). This involves

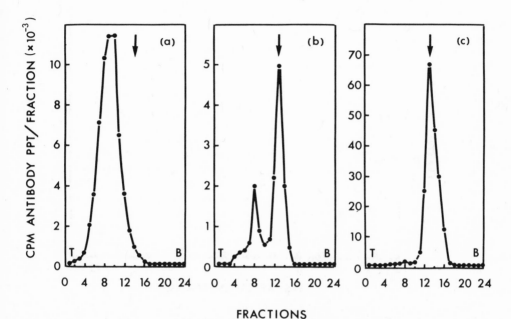

FRACTIONS

Fig. 3. Molecular weight transition of intracellular vitellogenin
in fat body, shown by sucrose gradient centrifugation. Fat body
from reproductive adult females was incubated in locust Ringer
containing 20 μCi/ml [^{3}H]-amino acids for 10 min and then in medium containing unlabeled amino acids (Wyatt et al., 1976).
Extracts were prepared as described in Fig. 2. To 12 ml 5-25%
linear sucrose gradients, 0.5 ml of the tissue supernatant and
0.05 mg of E. coli β-galactosidase (MW 520,000) were added and
the gradients were centrifuged at 120,000 x g for 12 hr at 4°C.
Fractions were collected, β-galactosidase was assayed according to
Craven et al. (1965) (peak centers shown by arrows), and the
vitellogenin was precipitated with antibody and counted as described
previously (Wyatt et al., 1976).

(a) Intracellular vitellogenin, 10 min pulse;
(b) Intracellular vitellogenin, 10 min pulse and 50 min chase;
(c) Vitellogenin recovered from medium, 3 hr label.

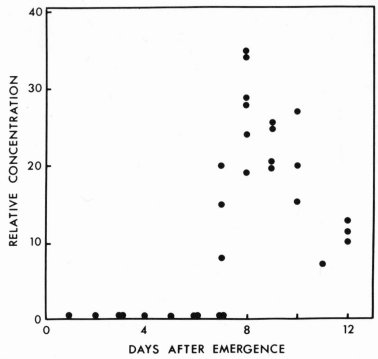

Fig. 4. Level of vitellogenin in hemolymph of normal adult females. The vitellogenin content of hemolymph samples was determined by quantitative immunodiffusion (Oudin, 1952). Relative concentration units represent mg/ml, referred to egg extracts as a standard, assuming that 100% of the protein (determined by biuret reaction) in the latter is vitellin.

incubation of fat body in a synthetic medium containing [^{3}H]-leucine and precipitation of labeled product with specific antiserum. Using this system, we examined the activity of synthetic JH I (C_{18}JH) and the analog ZR-515 (Henrick et al., 1973) applied topically to allatectomized females (Fig. 6a, 6b). With JH I, we found that topical application to the abdominal cuticle in oil or acetone was far more effective than injection. In addition, we adopted a regimen of multiple applications at 8 hr intervals, since the half life of JH I in the locust is extremely short (Erley et al., 1975). Under these conditions, a remarkably sharp dose-response curve was obtained, with a midpoint at a total dose of about 75 µg and maximal effect above 100 µg of JH I (Fig. 6a). At the saturation dose, 40-50% of the total labeled protein output from the fat body was precipitable by specific antiserum, which is close to the proportion (about 60%) obtained with fat body from normal egg-maturing locusts. The synthetic JH mimic, ZR-515, which was expected to be

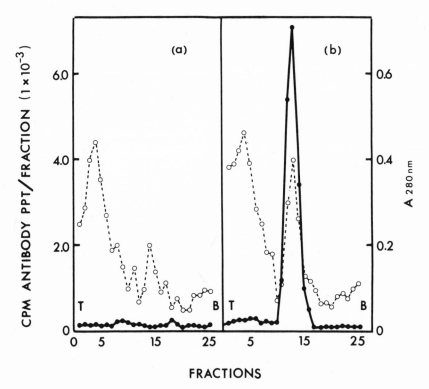

Fig. 5. Demonstration of induced synthesis of vitellogenin by JH.
JH I was applied topically to allactectomized females (10 µg in
5 µl acetone followed by 5 µg every 4 hr for 48 hr). Controls
received applications of 5 µl acetone. Two hr before hemolymph
was taken, 5 µCi of [^{3}H]-leucine (5 Ci/mmol) was injected into
each animal. Hemolymph samples (75 µl) were diluted to 2.0 ml
with 0.05 M Tris-HCl buffer, pH 8.1, containing 0.4 M NaCl and
0.1 mM PMSF, and dialyzed in the same buffer overnight at 4°C.
Samples of 0.5 ml were layered on 12 ml 5-25% linear sucrose
gradients made up in the same buffer, and centrifuged as described
in Fig. 3. Gradients were divided and protein content in each
fraction was measured at 280 nm. [^{3}H]-vitellogenin in each fraction
was precipitated with antiserum and the immune precipitate was
collected and counted as described in Fig. 3. (——•——) radioactivity; (--o--) A$_{280}$ nm. (a) control. (b) JH treated.

more stable (Henrick et al., 1973), was applied topically in a
single dose. When the fat body was tested 72 hr later, a quite
similar curve was obtained (Fig. 6b) and the proportion of specific
antibody-precipitable protein attained about 60% of the total
protein secreted by the fat body. Since the schedules of hormone
treatment in these two experiments are different, the potencies of

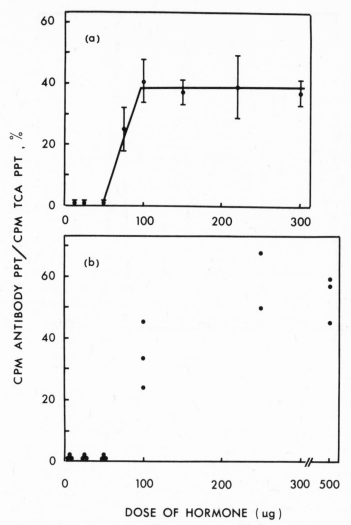

Fig. 6. Dose-response curves for induction of vitellogenin synthesis by JH I and ZR-515 in vivo.

(a) JH I (Eco-Control Inc., Cambridge, Mass.) dissolved in acetone
was applied topically every 8 hr for a period of 32 hr, to give
the total doses shown, and fat body was taken at 36 hr. Each
point shows the mean and S.E.M. for 6 animals.

(b) ZR-515 dissolved in acetone was applied topically in a single
dose, and the vitellogenin synthetic activity was measured 72 hr
later. Each point shows the result from one test animal.

Vitellogenin synthesis was measured in cultured fat body as described previously (Wyatt et al., 1976).

the two hormones cannot be directly compared. The effective doses
are surprisingly large when compared with the doses of the order
of 1 µg found effective for vitellogenin synthesis in cockroaches
(Engelmann, 1974; Kunkel, 1973) or Monarch butterflies (Pan and
Wyatt, 1971).

The remarkably steep slopes of the dose-response curves are
similar to those found for induced vitellogenin synthesis in
Leucophaea (Engelmann, 1974). They contrast with the results from
assays of JH morphogenetic activity in immature insects, in which
the range of doses from minimal to maximal response may extend
over several orders of magnitude (e.g., Riddiford, this volume).
A possible reason for this difference is that, whereas the induction of vitellogenin synthesis is a single cellular process, the
morphogenetic responses to JH may involve many processes differing
in sensitivity to the hormone.

After a single dose of ZR-515 (500 µg) was administered to
allatectomized females, capacity for vitellogenin synthesis in
fat body was evident after 24 hr, maximal within 72 hr and per-
sisted undiminished until at least 120 hr (Fig. 7). This indicates
that this JH analog is rather stable in the locust.

Cytological Development in the Fat Body

The strong stimulation of synthesis of a protein destined for
export in fat body under the influence of JH seemed likely to have
a basis in development of cell structure. The fine structure of
the imaginal female fat body of insects during the reproductive
cycle has been little studied. In the Colorado beetle, Leptinotarsa, De Loof and Lagasse (1970) have shown that allatectomy leads
to proliferation of a peripheral form of fat body rich in glycogen,
lipid and protein granules, and apparent loss of an internal form
specialized for protein synthesis. In the blowfly, Calliphora,
Thomsen and Thomsen (1974) have described changes in the fat body
cells including rough endoplasmic reticulum, Golgi complexes and
protein granules, correlated with the egg maturation cycle.

We have studied the natural sequence of development in fat
body of female locusts during the first 20 days after emergence.
The paragonadal lobes of the fat body were generally used, but
observations of peripheral abdominal fat body showed similar developmental changes. The lobes consist of reticulated strands of
large mononucleate cells. They are surrounded by a collagenous
basement membrane including tracheoblasts; oenocytes are often seen
adhering to it. Analysis of the cytological changes permits us to
recognize three periods which can be correlated with the appearance
of vitellogenin in hemolymph and the development of eggs.

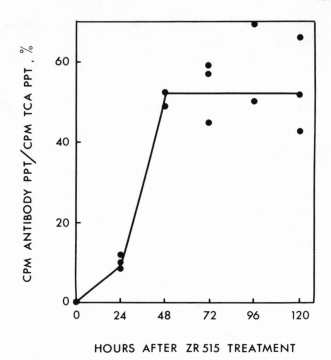

Fig. 7. Time course of induction of vitellogenin synthesis by
ZR-515 in vivo. ZR-515 (500 µg in 50 µl acetone) was applied
topically to allatectomized females and the activity of fat body
in vitellogenin synthesis was determined at time intervals as
indicated. Each point is the result from one animal.

Pre-Vitellogenic Phase. (Approx. days 0-7; Fig. 8a, 8c).
Up to 7 days after the imaginal molt, when vitellogenin is not
detectable and the ovary remains undeveloped, fat body morphology
also remains stable, except for some increase in the fat content.
The cells are mainly occupied by large lipid droplets around which
are areas rich in particulate glycogen. The thin strips of cytoplasm between these inclusions contain some mitochondria, but
ribosomes (free or membrane bound), endoplasmic reticulum and
Golgi are rarely observed. Pinocytotic vesicles are often seen in
folds of the plasma membrane at the outer poles of the cells. The
nuclei are shrunken and deformed between the fat droplets, suggesting little activity.

Vitellogenic Phase. (Approx. days 8-15; Fig. 8b, 8d). About
the 8th day, when vitellogenin first appears in the hemolymph,
initial development of cytomembranes is observed. The cytoplasm
contains numerous ribosomes, both free and bound to newly formed

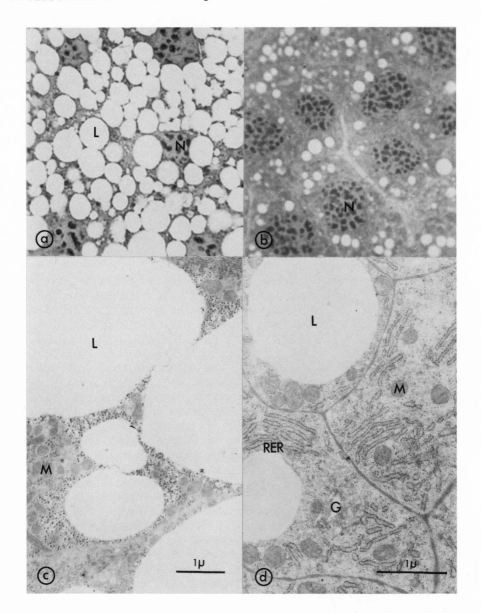

Fig. 8. Fat body structure during normal development of the adult
female locust. Fat body was fixed in 3% buffered glutaraldehyde
and post-fixed in 1% OsO₄. (a), (b) Light micrographs after staining with toluidine blue: (a) One day after emergence; (b) 15 days
after emergence. (c), (d) Electron micrographs of portions of
cytoplasm: (c) day 1; (d) day 15. G, Golgi; M, mitochondria;
L, lipid droplet; Rer, rough endoplasmic reticulum.

membranes, constituting the first saccules of rough endoplasmic
reticulum (RER). These flattened saccules develop into impressive
arrays of RER, and establish morphological relations with newly
developed Golgi complexes. At the same time, the peripheral plasma
membrane forms a complex system of channels by invagination, pen-
etrating deeply into the cell mass. The extracellular space is
filled with a homogeneous material.

The maximal proliferation of cytomembranes in the fat body
cells is reached about the 15th day, when vitellogenesis in the
terminal oocytes is very active. At this stage glycogen particles
are no longer detectable and lipid droplets have decreased dramatically in number and size, indicating a major utilization of
reserves, presumably providing energy and precursors for synthesis
of intracellular structures and export glycolipoprotein. At the
same time, the nuclei have become enlarged and spherical, with
increased basophilic masses, indicating enhanced synthetic activity.

To establish that these changes are related to specific protein synthesis and transport, we have localized vitellogenin intracellularly by means of ferritin conjugated with antivitellin immu-
noglobulin (Fig. 9). Bound ferritin molecules are visible in the
cisternae of the RER (Fig. 9a), Golgi vesicles (Fig. 9b) and the
extracellular channels (Fig. 9c). Such binding is not observed
in male or immature female fat body, nor in mature female tissue
treated previously with antivitellogenin globulin before application of the conjugate (Fig. 9d,e,f). This suggests that vitellogenin, like other exported proteins (Jamieson and Palade, 1967;
Weinstock and Leblond, 1974; Couble et al., 1975), is synthesized
on the ribosomes of the endoplasmic reticulum and then transported
via the Golgi before being secreted from the cell.

End of the First Reproductive Cycle. (Approx. days 16-20).
At this stage, reflecting the conclusion of the first reproductive
cycle, in addition to the elements already described, different
types of lysosomal figures are found. These include dark granules
(presumably cytolysosomes), membrane-bound inclusions showing a
paracrystalline pattern, and vesicle-containing bodies. The
presence of such inclusions must indicate the partial remodelling
of the cells at the end of a phase of synthetic activity.

Allatectomy completely blocks the normal developmental changes.
The cells retain the structure characteristic of the first few days
of adult life, and no proliferation of RER or Golgi takes place,
although the glycogen content becomes lower and lipids accumulate
even above their normal initial amount (Fig. 10a,c). Elevated
lipid after allatectomy has previously been reported from biochemical analyses (Pfeiffer, 1945; Minks, 1967). Application of the
JH analog ZR-515 to allatectomized females leads to a complete
restoration of normal intracellular morphology within 4 to 5 days

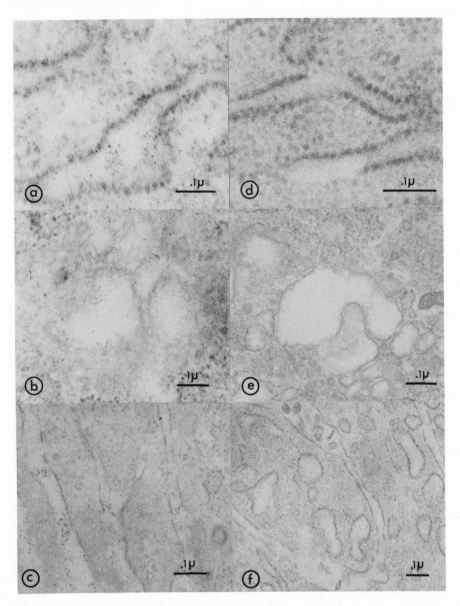

Fig. 9. Localization of vitellogenin in fat body of reproductive adult female. Ferritin molecules are visible (a) in the cisternae of the rough endoplasmic reticulum, (b) inside the Golgi vacuoles, and (c) in channel-like folds of the plasma membrane. Control samples (d,e,f) were treated previously with anti-vitellogenin IgG before the application of ferritin-IgG conjugate. Ferritin-labeled anti-vitellogenin IgG was prepared as described by Kishida et al. (1975). Fat body was fixed with formaldehyde, sectioned and incubated in ferritin-IgG conjugate at 4°C for 48 hr. After washing, the tissue was post-fixed with glutaraldehyde and then with osmium.

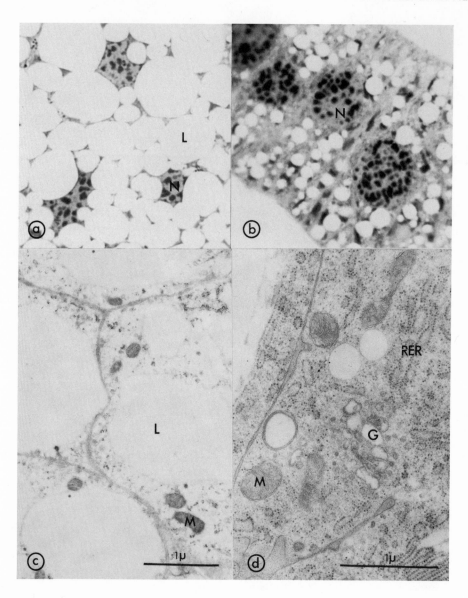

Fig. 10. Fat body structure after allatectomy and treatment with JH analog. Preparations as in Fig. 8. (a), (c) 19 days after allatectomy; (b), (d) 19 days after allatectomy, and 5 days after administration of ZR-515 (500 µg topically in acetone).

(Fig. 10b,d). Besides changes in the nucleus, the RER and Golgi proliferate and material is deposited in the extracellular space.

In summary, the cytological picture suggests control by JH over development of the apparatus for synthesis and processing of protein, based upon activity at the nuclear level.

Demonstration of Vitellogenin Messenger Activity in a Cell-Free Protein Synthesizing System

We have also begun analysis of biochemical changes in adult female locust fat body under the influence of JH. Preliminary experiments indicate a substantial increase in total RNA about day 8, when development of RER is seen and vitellogenin production begins. Changes in polysome profile are also evident, and are currently being analyzed. The hypothesis that the action of JH involves the synthesis of new vitellogenin-specific messenger RNA (Engelmann, 1974) could be tested with the aid of an adequate cell-free protein-synthesizing system which would respond to addition of exogenous messenger. Among several eukaryotic in vitro protein-synthesizing systems recently described, that derived from wheat embryos has advantages of simplicity, low background of endogenous incorporation and reproducibility (David and Kaesberg, 1973; Gozes et al., 1975). We have used this method to assay for the presence of vitellogenin messenger in fat body RNA.

From adult locust fat body, RNA fractions were prepared by sequential phenol extraction at pH 7.4 and at pH 9.0, and the latter, poly A-containing fraction was used (Brawerman, 1974). The requirements of the wheat germ system are shown in Table 2. A 3- to 4-fold stimulation of leucine incorporation into TCA-precipitable material was given by 6 μg of the RNA fraction prepared from male locust fat body. Complete inhibition of incorporation was produced by addition of aurin tricarboxylic acid, a known inhibitor of initiation in protein synthesis, which indicates that the incorporation directed by exogenous locust RNA depends upon initiation of translation (Weeks and Marcus, 1971). The ability of the system to translate RNA from other sources is shown in Table 3 (expt. 1). An 8-fold stimulation of incorporation was given by 5 μg of RNA from Drosophila melanogaster. When purified rabbit globin mRNA was used, a 10-fold stimulation was observed, and SDS gel electrophoresis of the labeled product showed two sharp peaks corresponding to the molecular weights of α and β globin. We therefore believe that this system is capable of faithful translation of exogenous mRNA.

TABLE 2.   REQUIREMENTS FOR WHEAT EMBRYO CELL-FREE PROTEIN SYNTHE-
SIZING SYSTEM DIRECTED BY RNA FROM LOCUSTA MIGRATORIA.

Reaction components	TCA-precipitable radioactivity (cpm/50 µl)
Complete system (+ 6 µg locust ♂ fat body RNA)	42,276
Deletions	
- locust RNA	12,078
- wheat germ tRNA	34,498
- energy regenerating system and 19 amino acids	2,945
Addition	
+ aurintricarboxylic acid (50 µM)	3,232

The complete amino acid incorporation system was prepared from
wheat germ by the method of David and Kaesberg (1973). RNA was
isolated from fat body of mature adult male locusts by phenol
extraction at pH 9.0 after prior extraction at pH 7.4 (Brawerman,
1974). Each tube (50 µl final volume) contained [^{3}H]-leucine
(2 µCi, 50 Ci/mmol). The reaction was carried out at 30°C for
60 min. TCA-precipitable radioactivity was measured by spotting
5 µl of the reaction mixture on Whatman 3MM filter discs and
washing according to Mans and Novelli (1961).

The vitellogenin messenger activity of various fat body RNA
preparations was assayed by precipitation of the polypeptide product with specific vitellogenin antiserum (Table 3). The pH 9.0
RNA fraction from fat body of reproductive mature adult females
directed the synthesis of antibody-precipitable (as well as TCA-
precipitable) products, whereas RNA from fat body of males and
allatectomized females did not. The possibility that specific mRNA
in allatectomized female fat body might remain in the pH 7.4
fraction during extraction was ruled out since this RNA fraction
also failed to direct the synthesis of antibody-precipitable product.
As a further control, to test for possible inhibitory components

TABLE 3.   INCORPORATION INTO SPECIFIC AND NON-SPECIFIC PRODUCTS IN
WHEAT EMBRYO CELL-FREE PROTEIN SYNTHESIZING SYSTEM, DIRECTED BY
RNA FROM LOCUST FAT BODY AND OTHER SOURCES.

| Source of RNA | TCA ppt[a] | Antibody ppt[b] | Antibody ppt[c] (%) |
	(cpm/50 µl)	(cpm/50 µl)	TCA ppt
Experiment 1			
Drosophila melanogaster (adult ♂, 5 µg)[d]	81,776	1,257	1.5
Rabbit globin mRNA (0.7 µg)[e]	150,020	1,140	0.8
Locust mature ♂ (4 µg)	23,033	0	0
Allatectomized ♀ (4 µg)	15,620	0	0
Allatectomized ♀ (12 µg pH 7.4 RNA)	16,380	0	0
Mature ♀ (7.5 µg)	25,185	2,160	8.6
Allatectomized ♀ treated with JH I[f] (8 µg)	43,400	4,013	9.2
Experiment 2			
Mature ♀ (4 µg)	14,770	1,220	8.3
Allatectomized ♀ treated with ZR-515 (3.4 µg)[g]	47,610	2,760	5.8
Allatectomized ♀ treated with ZR-515 (3.4 µg) + mature ♂ (3.2 µg)	36,669	2,360	6.4

    RNA was isolated and the pH 9.0 fraction was assayed in a
wheat embryo cell-free system as described in Table 2.   The amount
of [^{3}H]-leucine (50 Ci/mmol) was 5 µCi per tube in Experiment 1 and
2 µCi in Experiment 2.   Each value is the mean of two determinations.
(See page 526 for footnotes.)

TABLE 3 (Cont.)

[a] Net TCA-precipitable radioactivity counted on Whatman 3MM filters. The average values of controls with no RNA added (16,250 cpm for expt. 1 and 11,490 cpm for expt. 2 per 50 μl reaction mixture) have been subtracted.

[b] Net antibody-precipitable radioactivity, counted on Millipore filters. The average values of controls with no RNA added (4,620 cpm for exp. 1 and 1,520 cpm for expt. 2 per 50 μl reaction mixture) have been subtracted.

[c] Because of differences between counting techniques, these values do not represent true percentages of total protein synthesis.

[d] Prepared with phenol, $CHCl_3$ and SDS at pH 5.0; gift from Dr. H. Lerer.

[e] From Amersham-Searle Corp., Chicago.

[f] JH I, 100 μg in acetone, was applied topically to allatectomized females as described in Fig. 6. RNA was prepared 36 hr after hormone treatment.

[g] ZR-515, 500 μg in acetone, was applied topically to allatectomized females and the RNA was extracted 72 hr after hormone treatment. ZR-515 was a gift from Dr. K. Judy, Zoecon Corporation.

in inactive RNA, RNA from male fat body was added together with active female RNA, and no inhibition of specific synthesis was observed (expt. 2).

Most interestingly, the RNA prepared from fat bodies of allatectomized females treated with JH I (expt. 1) or ZR-515 (expt. 2) contained vitellogenin messenger activity as determined by precipitation of the translation product with vitellin anti-serum. Although the identity of the translation product as vitellogenin polypeptide remains to be confirmed by chemical means, these results strongly indicate the control of synthesis of vitell-ogenin messenger RNA in locust fat body by JH.

SUMMARY AND CONCLUSIONS

From eggs of Locusta migratoria, we have isolated the major female-specific yolk protein, which we call vitellin, and from the hemolymph its precursor, the vitellogenin. These are glyco-lipoproteins of molecular weight 550,000, antigenically identical

but differing in the quantitative proportions among a group of
polypeptides that are released upon denaturation in sodium dodecyl
sulfate. The vitellogenin is produced in the fat body of adult
female locusts, and pulse-chase experiments with fat body in vitro
have shown that it is synthesized as a unit of 260,000 daltons
which then dimerizes and also undergoes cleavage into shorter
polypeptides before release from the cells. There is apparently
further proteolytic cleavage before final deposition in the yolk.
In molecular size and chemical composition, these proteins resemble
those that have been characterized from other insect species.

In the adult female locust, vitellogenin first appears in the
hemolymph on about the seventh day after eclosion. In locusts
allatectomized when newly emerged, its synthesis is permanently
prevented, but can be induced by treatment with JH or analogs.
These are effective when applied topically (JH I in repeated
doses, or ZR-515 in a single dose), and give steep dose-response
curves for activation of fat body to produce vitellogenin during
subsequent incubation in vitro.

Cytologically, the fat body undergoes impressive development
correlated with the acquisition of capacity to synthesize vitellogenin. The nuclei enlarge and increase in basophilia, while lipid
and glycogen deposits diminish, and extensive rough endoplasmic
reticulum as well as Golgi complexes develop in the cytoplasm.
All of these changes are blocked by allatectomy and restored by
treatment with JH I or ZR-515. Associated biochemical changes
include marked increase in total RNA. At the same time, assays
for messenger RNA in the wheat embryo cell-free protein-synthesizing system have shown that specific vitellogenin mRNA is present
in mature female fat body but not in the fat body of males or
allatectomized females. In allatectomized females, however, the
fat body produces vitellogenin mRNA after treatment with JH I or
ZR-515.

The locust vitellogenin system appears to be favorable for
the analysis of the mechanism of JH action in inducing the synthesis of a specific protein. Our initial results suggest action at
the nuclear level, including the synthesis of a new specific mRNA
as well as general build-up of the cellular protein-synthesizing
system. This supports the conclusions from an analogous system
in cockroaches (Engelmann, 1974), and is consistent with mechanisms
of action similar to those proposed for steroid hormones in vertebrates in inducing the synthesis of vitellogenins (Clemens, 1974;
Bergink and Wallace, 1974) and other proteins (O'Malley and Means,
1974). Much remains to be done, however, to establish how JH works
in this and other systems.

ACKNOWLEDGEMENTS

This research was supported by grants from the U.S. National
Institutes of Health (HD-07951), the National Research Council
of Canada, and the Rockefeller Foundation. We thank Francine
Lafleur for willing and skillful technical help.

REFERENCES

Bar-Zev, A., Wajc, E., Cohen, E., Sapir, L., Applebaum, S.W., and
    Emmerich, H., 1975, J. Insect Physiol. 21:1257.
Benz, F., Girardie, A., and Cazal, M., 1970, J. Insect Physiol.
    16:2257.
Bergink, E.W., Wallace, R.A., Van de Berg, J.A., Bos, E.S., Gruber,
    M., and Ab, G., 1974, Am. Zoologist 14:1177.
Brawerman, G., 1974, in "Methods in Enzymology" (K. Moldave and
    L. Grossman, ed.), Vol. 30, pp. 605, Adademic Press, New York.
Chen, T.T., and Wyatt, G.R., 1976, in preparation.
Clemens, M.J., 1974, Progr. Biophys. Mol. Biol. 28:71.
Clemens, M.J., Lofthouse, R., and Tata, J.R., 1975, J. Biol. Chem.
    250:2213.
Couble, P., Prudhomme, J.C., and Daillie, J., 1975, in preparation.
Craven, G.R., Steers, E., Jr., and Anfinsen, C.B., 1965, J. Biol.
    Chem. 240:2468.
David, J.W., and Kaesberg, P., 1973, J. Virol. 12:1434.
Dejmal, R.K., and Brookes, V.J., 1972, J. Biol. Chem. 247:869.
De Loof, A., and Lagasse, A., 1970, Z. Zellforsch. Mikr. Anat.
    106:439.
Doane, W.W., 1973, in "Insects, Developmental Systems" (S.J. Counce
    and C.H. Waddington, ed.) Vol. 2, pp. 291-497, Academic Press,
    New York.
Engelmann, F., 1971, Arch. Biochem. Biophys. 145:439.
Engelmann, F., 1974, Am. Zoologist 14:1195.
Engelmann, F., and Friedel, T., 1974, Life Sci. 14:587.
Erley, D., Southard, S., and Emmerich, H., 1975, J. Insect Physiol.
    21:61.
Gellissen, G., Wajc, E., Cohen, E., Emmerich, H., and Applebaum,
    S.W., 1976, J. Comp. Physiol., in press.
Gozes, I., Schmitt, H., and Littaner, U.Z., 1975, Proc. Nat. Acad.
    Sci. USA 72:701.
Henrick, C.A., Staal, J.B., and Siddall, J.B., 1973, Agric. Food
    Chem. 21:354.
Hill, L., 1965, J. Insect Physiol. 11:1605.
Jamieson, D.J., and Palade, G.E., 1967, J. Cell Biol. 34:577.
Kishida, Y., Olsen, B.R., Borg, R.A., and Prockop, D.J., 1975,
    J. Cell Biol. 64:331.
Koeppe, J., and Ofengand, J., 1976, Arch. Biochem. Biophys., in
    press.

Kunkel, J.G., 1973, J. Insect Physiol. 19:1285.
Kunkel, J.G., and Pan, M.L., 1976, J. Insect Physiol.
    in press.
Mans, R.J., and Novelli, G.D., 1961, Arch. Biochem. Biophys. 94:48.
Minks, A.K., 1967, Arch. neerl. Zool. 17:175.
O'Malley, B.W., and Means, A.R., 1974, Science 183:610.
Oudin, J., 1952, Methods Med. Res. 5:335.
Pan, M.L., and Wyatt, G.R., 1971, Science 174:503.
Pan, M.L., and Wallace, R.A., 1974, Am. Zoologist 14:1239.
Pan, M.L., Bell, W.J., and Telfer, W.H., 1969, Science 165:393.
Patterson, M.S., and Greene, R.C., 1965, Anal. Chem. 37:854.
Pfeiffer, I.M., 1945, J. Exp. Zool. 99:183.
Telfer, W.H., 1965, Ann. Rev. Entomol. 10:161.
Thomsen, E., and Thomsen, M., 1974, Cell Tiss. Res. 152:193.
Tobe, S.S., and Loughton, B.G., 1967, Can. J. Zool. 45:975.
Wallace, R.A., 1970, Biochim. Biophys. Acta 215:176.
Wallace, R.A., and Bergink, E.W., 1974, Am. Zoologist 14:1159.
Weeks, D.O., and Marcus, A., 1971, Biochim. Biophys. Acta 232:671.
Weinstock, M., and Leblond, C.P., 1974, J. Cell Biol. 60:92.
Williams, C.M., and Kafatos, F.C., 1971, in "Insect Juvenile
    Hormones - Chemistry and Action" (J.J. Menn and M. Beroza, eds.)
    pp. 29-41, Academic Press, New York.
Wyatt, G.R., 1972, in "Biochemical Actions of Hormones" (G. Litwack
    ed.) Vol. 2, pp. 384-490, Academic Press, New York.
Wyatt, G.R., Chen, T.T., and Couble, P., 1976, in "Invertebrate
    Tissue Culture - Applications in Biology, Medicine and Agricul-
    ture" (E. Kurstak and K. Maramorosch, ed.), Academic Press,
    New York, in press.
Yamasaki, K., 1974, Insect Biochem. 4:411.

MECHANISM OF STEROID HORMONE ACTION: IN VITRO CONTROL OF GENE

EXPRESSION IN CHICK OVIDUCT CHROMATIN BY PURIFIED STEROID RECEPTOR

COMPLEXES

R.J. Schwartz, W.T. Schrader, and B.W. O'Malley

Baylor College of Medicine, Texas Medical Center
Houston, Texas 77025

INTRODUCTION

Numerous studies have demonstrated that steroid hormones
appear to exert their primary influence at the level of genetic
transcription (O'Malley and Means, 1974). Evidence in endocrine
target tissue systems utilizing estrogen (Gorski and Nicolette,
1963; Hamilton et al., 1963), progesterone (O'Malley and McGuire,
1968a,b), aldosterone (Edelman and Fimognari, 1968), glucocorti-
coids (Kenney and Kull, 1963; Sekeris and Lang, 1964) and vitamin D
(Norman, 1966; Stohs et al., 1967) have demonstrated increased
RNA synthesis after the in vivo localization of administered hormone in the nuclear compartment. Quantitative and qualitative
changes in RNA synthesis usually precede the steroid hormone stimulation of inducible protein biosynthesis in target tissues.
Recent experiments, in fact have conclusively demonstrated that
female sex steroids are capable of inducing a net increase in
specific mRNA molecules in target cells (Chan et al., 1973; Comstock
et al., 1972; Harris et al., 1975).

It thus appears that the pattern of steroid hormone action may
include the following events in a temporal dependent manner: First,
steroids are taken up by the target cells and bind to a specific
cytoplasmic receptor protein; second, the steroid-receptor complex
is translocated into the nuclear compartment; third, receptor-
hormone complex binds to acceptor sites on nuclear chromatin; and
finally, activation of the transcriptional process induces the
appearance of specific new RNA species.

To elucidate further the mechanism of steroid hormone action, it is necessary to examine the process by which biochemical information held by the steroid receptor complex is transferred to the transcriptional apparatus. Of the many possible approaches to this problem, our recent efforts have centered on direct examination of the effects of steroid receptors on chromatin transcription in cell free systems. Initial efforts in our laboratory have been to characterize the interaction of purified RNA polymerase with intact chromatin. The process of initiation of RNA synthesis has received special attention, since this step appears to be regulated by steroids (Schwartz et al., 1975). In this article, we will discuss our recent evidence on the steroid dependent transcription of chromatin in the chick oviduct.

The chick oviduct provides an excellent model system for the hormonal control of cellular differentiation and gene expression. Chronic administration of estrogen over a period of 10 to 14 days results in the growth and differentiation of the chick oviduct. Previous studies from this laboratory have shown that estrogen-mediated growth and differentiation of the chick oviduct involve significant alterations in gene transcription. Competition hybridization experiments have demonstrated qualitative changes in the repetitive sequences of nuclear RNA synthesized during estrogen differentiation (O'Malley and McGuire, 1968b). More recent studies have also demonstrated increased transcription of unique sequence DNA during oviduct growth (Liarakos et al., 1973). Concomitant with steroid-mediated oviduct differentiation, increased levels of nuclear RNA (O'Malley and McGuire, 1968b), RNA polymerase activity (O'Malley et al., 1969; Cox et al., 1973), chromatin template capacity, and changes in chromatin nonhistone proteins (Spelsberg et al., 1973) were also found in the oviduct. Preceding the well documented specific changes in protein synthesis (Rosenfeld et al., 1972; Means et al., 1972; Comstock et al., 1972) previous studies have demonstrated an unequivocal net accumulation of specific, biologically active messenger RNA, coding for ovalbumin (Harris et al., 1972; Chan et al., 1973; Palmiter and Smith, 1973). When estrogen treatment of the chicks is discontinued, a reduction in the overall level of RNA and protein synthesis occurs and the cells ability to synthesize specific proteins such as ovalbumin decreases (Chan et al., 1973; Palmiter, 1973). After 12 days of hormone withdrawal, the rate of ovalbumin biosynthesis is less than 1% of that observed in hormonally stimulated chicks (Harris et al., 1975). If either estrogen or progesterone is readministered (secondary stimulation) there is a rapid increase in the production of ovalbumin mRNA and the induction of the other egg white proteins begins again (Palmiter, 1973). Therefore, the chick oviduct provides a system to study a specific endocrine response in which overall gene expression can be dramatically altered while a specific marker for measuring changes in the transcription rate of a single gene is available. We have therefore undertaken studies of cell free

transcription of chick ovuduct chromatin by exogenous RNA polymerase in order that we might understand the mechanism by which RNA
synthesis is stimulated following hormone uptake by endocrine
responsive tissues.

RESULTS AND DISCUSSION

Formation of Initiation Complexes on Chick DNA and Chromatin

The capacity of oviduct chromatin to serve as a template for
E. coli RNA polymerase has previously been shown to increase following estrogen administration to the immature chick (Cox, 1973;
Spelsberg et al., 1973). While chromatin template activity measurements may generally reflect the amount of DNA sequences made available to RNA polymerase, the components of such a reaction are so
complex that these experiments shed little light on the biochemical
mechanisms of hormone-induced alterations in gene transcription.
The events that may be involved in determining changes in template
activity are: (a) RNA polymerase binding to chromatin, (b) RNA
polymerase search for initiation sites, (c) the formation of stable
RNA polymerase-DNA initiation complexes, (d) the initiation of RNA
chain synthesis, (e) RNA chain propagation rate, (f) the size of
RNA chains, (g) the termination of RNA chain synthesis, and (h)
the reinitiation of the transcriptive process. Potentially, any
of these components can affect the measurement of chromatin template activity. In order to determine the effect of estrogen on
gene transcription in the chick oviduct it appeared necessary to
monitor and control all of the parameters involved in RNA synthesis.

Procedures for measuring the initiation of RNA chains in vitro
were initially adopted from studies in bacterial and bacteriophage
systems (Lill et al., 1970; Bautz and Bautz, 1970). The initiation
of RNA synthesis can be divided into 2 basic processes. First, RNA
polymerase binds randomly and reversibly to DNA to form a series of
nonspecific complexes. However, after sufficient time and at a
proper temperature RNA polymerase proximal to a true initiation
site may form a stable binary complex with DNA (Sipple and Hartman,
1970; Zillig et al., 1970). This complex has undergone a transition
involving the local opening of the DNA duplex structure (Travers
et al., 1973), and is now capable of rapidly initiating an RNA
chain (Hinkle and Chamberlin, 1972). The second process is the
actual initiation step in which RNA polymerase catalyzes the formation of the first phosphodiester bond between 2 nucleoside triphosphates.

The existence of stable RNA polymerase DNA initiation complexes
was elucidated through the use of the drug rifampicin (Umezuwa
et al., 1963; Hartmann et al., 1967). Rifampicin is a competitive

inhibitor of RNA synthesis which acts prior to the formation of the first phosphodiester bond, but which has no effect on RNA chain elongation. RNA chain initiation by RNA polymerase bound to DNA in a stable initiation complex is so rapid that it can occur in the presence of the drug (Hinkel et al., 1972; Mangel and Chamberlin, 1974). The fraction of RNA polymerase in stable complexes can be determined following the simultaneous addition of a mixture of rifampicin and the four ribonucleoside triphosphates (Chamberlin and Ring, 1972). RNA polymerase molecules which are free in solution or randomly bound to DNA will be inhibited by rifampicin. Under these conditions reinitiation of RNA transcription will be completely inhibited and thus each RNA polymerase at a stable initiation site can synthesize one and only one RNA chain. By measuring the number of RNA chains made the rifampicin challenge assay provides a method for quantitating the number of RNA polymerase initiation sites for a given template.

In order to test for the formation of stable RNA polymerase-DNA initiation complexes, increasing amounts of Escherichia coli RNA polymerase were preincubated with a fixed amount of chick DNA at either 0° or 37°C for 15 min. The binary complexes were challenged by the addition of rifampicin (40 µg/ml) together with ribonucleoside triphosphates and allowed to synthesize RNA for an additional 15 min at 37°C. The number of RNA chains synthesized was determined from the amount and length of the RNA produced. The number of chains is equivalent to the number of initiation sites at which RNA polymerase was bound. The results of such analysis for chick DNA is shown in Fig. 1. By adding increasing amounts of RNA polymerase to a fixed amount of template, the total number of available initiation sites can be determined from the saturation level of RNA synthesis. The number of initiation sites available to E. coli RNA polymerase after preincubation at 37°C was calculated to be about 1.3×10^6 per picogram of DNA (based on a RNA number average chain length of 450 nucleotides and a base composition of 40% UMP). This corresponds to an average of one initiation site per 750 base pairs of DNA. If the preincubation is carried out at 0°C, the RNA polymerase is not efficient at formation of stable initiation complexes (Fig. 1), presumably due to the higher temperature necessary for destabilizing the DNA duplex.

The number of initiation sites in estrogen-treated oviduct chromatin was analyzed in an identical procedure as shown in Fig. 2. In the experiment, chromatin (5 µg) was preincubated for 30 min with increasing amounts of RNA polymerase then RNA synthesis was started by the combined addition of nucleotides, rifampicin and heparin (Fig. 2). Heparin was added to inhibit any contaminating RNase activity which if present would decrease the average chain length of the RNA product (Tsai et al., 1975; Cox, 1973). The number of initiation sites found after preincubation with RNA polymerase at 37°C was calculated to be 33,750 per pg DNA (based on a

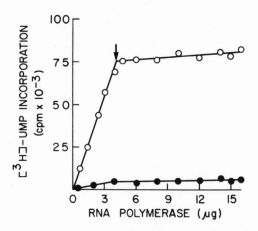

Fig. 1. RNA polymerase saturation curve of chick DNA. E. coli
RNA polymerase (0 to 15 μg) was incubated with 1.5 μg of chick
oviduct DNA in a mixture (0.2 ml) containing 12.5 μmol Tris-HCl
pH 7.9, 0.25 μmol $MnCl_2$, 12.5 μmol $(NH_4)_2SO_4$, 0.5 μmol 2-mercapto-
ethanol, and 0.5 mg/ml bovine serum albumin for 15 min at either
37°C or 0°C. At the end of the preincubation period RNA synthesis
was started with the addition of 37.5 nmol of ATP, GTP, CTP and
[^{3}H]UTP (10 μCi) together with 10 μg of rifampicin and 200 μg of
heparin in 50 μl. The mixture was then incubated at 37°C for
15 min. RNA synthesized was precipitated with 5% trichloracetic
acid, collected on glass fiber filters and counted in scintillation fluid at an efficiency of 19.5%. o---o, preincubated at
37°C; ●---●, preincubated at 0°C.

RNA number average chain length of 750 nucleotides and a base composition of 25% UMP). These data correspond to one initiation site
per 26,000 base pairs of DNA. Thus, chromatin contains only about
2% of the total initiation sites found in deproteinized DNA. Interestingly, formation of the polymerase initiation complexes was found
to be much less dependent on the temperature of preincubation in
chromatin than it was in DNA. One possible explanation for this
phenomenon is that certain chromatin proteins are present at the
initiation sites which help reduce the temperature requirement by
destabilization of the DNA duplex.

Determination of the Number of Initiation Sites
in Estrogen Stimulated Oviduct Chromatin

We utilized the rifampicin challenge assay to determine the
number of RNA polymerase initiation sites, the rate of RNA chain
propagation, and the chain length of RNA synthesized from chromatin

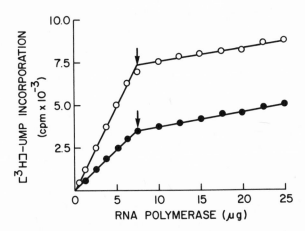

Fig. 2. Temperature independence of the formation of rifampicin
resistant RNA polymerase-chromatin complexes. Experiments were
carried out as described in Fig. 1 with 5 µg of oviduct chromatin
(14 days diethylstilbestrol treated chick oviduct) with 0 to 25 µg
of RNA polymerase, preincubated at either 37°C (o---o) or at 0°C
(•---•).

isolated from various stages during primary estrogen stimulation
of the chromatin oviduct. Chromatin from control and estrogen-
treated chicks were assayed by our rifampicin assay as shown in
Table 1. The chromatin from unstimulated animals had the capacity
to synthesize only 10,000 RNA chains/pg DNA (Table 1). However,
as early as 8 hr following estrogen administration to the chick,
the number of initiated RNA chains increased by 50% over the unstimulated controls. After 1 day of hormone treatment, the number of
initiation sites increased to 33,000 and by day 4 of steroid treatment a maximum of 72,000 initiation sites was detected (Table 1).
Thereafter, the number of initiation sites declined and by 12 days,
when differentiation is complete, a plateau level was reached which
was still 2-fold higher than control values. The increase in transcriptive capacity coincided with the pattern of increased growth
and differentiation previously reported for estrogen stimulation of
the magnum portion of the chick oviduct (Kohler et al., 1969).

During this time of intense transcriptional activity the
parameters of RNA chain propagation rate (∿6 nucleotides/sec) and
the number average chain length of the RNA product (∿700 nucleotides)
did not vary significantly from unstimulated oviduct chromatin
(Table 1). Therefore, the estrogen-induced increase in chromatin
transcription in the oviduct was mainly due to a modulation in the
number of available RNA polymerase initiation sites. However, it
was not clear that an alteration in the number of chromatin

R.J. SCHWARTZ ET AL.

TABLE 1. EFFECT OF PRIMARY ESTROGEN STIMULATION ON INITIATION SITES FOR RNA SYNTHESIS[a].

Days of Hormone Treatment	Size of RNA Product[b]	Rate of Elongation (nucleotides/sec)	pMol of RNA Chains per 5 µg Chromatin	Initiation Site per pg of DNA
0	660	6.0	0.09	10,600
0.33	700	6.0	0.13	16,800
1	810	6.0	0.28	33,500
4	640	6.0	0.60	72,000
8	725	---	0.35	42,200
12	680	---	0.17	21,200
18	700	5.8	0.23	28,600

[a] Oviduct chromatin was isolated from chicks which had received daily injections of 2.5 mg of DES for the indicated periods. Determination of the size of the RNA product, the rate of RNA chain elongation and the number of RNA initiation sites per pg DNA were described elsewhere (Schwartz et al., 1975).

[b] Number average chain length of RNA product in nucleotides.

initiation sites reflected either a direct effect of nuclear bound estrogen on transcription or was simply due to a more indirect route through the differentiation of 3 distinct epithelial cell types from a population of primitive mucosal cells. To overcome this problem, the number of initiation sites for RNA polymerase was measured during estrogen withdrawal and acute (secondary) estrogen stimulation (Tsai, S.Y. et al., 1975), while simultaneous measurements were also made of the level of endogenous nuclear receptor by a [³H]estradiol exchange assay (Anderson et al., 1972). Importantly, estrogen-mediated events in RNA and protein synthesis during secondary estrogen stimulation are not dependent on cell proliferation (Palmiter, 1973; Socher and O'Malley, 1973).

The effect of estrogen withdrawal on both the number of chromatin RNA initiation sites and the level of nuclear estrogen receptors is shown in Fig. 3. The figure shows a distinct similarity in the kinetics of decrease in initiation sites and the level of nuclear-bound receptor molecules. Both parameters decreased to 50% by 2-3 days of withdrawal and declined gradually to a steady-state level by 6 days. This study suggested that as the concentration of nuclear-bound receptors decreased, the number of chromatin DNA sequences available for initiation of RNA synthesis also decreased.

To further investigate the relationship between hormone-receptor complexes and chromatin-dependent RNA synthesis, we compared the effect of acute (secondary) estrogen stimulation on the number of initiation sites for RNA synthesis on chromatin. This parameter was then compared to the level of nuclear-bound estrogen receptors and the intracellular accumulation of ovalbumin mRNA (Tsai, S.Y. et al., 1975). As shown in Fig. 4, the number of initiation sites increased 100% as early as 30 min after readministration of 2.5 mg of DES to withdrawn chicks and reached a plateau at 1-2 hr. Calculations of RNA chain length and propagation rate were also carried out to confirm that the estrogen-dependent effect represented an increase in new RNA chain initiations. The level of nuclear-bound receptors attained its maximum about 20 min after readministration of hormone and reached a steady-state level at 1 hr. Thus, an increase in the concentration of nuclear receptor molecules immediately preceded the increase in initiation sites for RNA synthesis on chromatin prepared from the same nuclei.

For the purpose of comparison, the increase in the intracellular concentration of ovalbumin messenger RNA which occurred during secondary stimulation is also shown in Fig. 4. This measurement was carried out by hybridization analysis using a [³H]DNA probe complementary to ovalbumin mRNA. This [³H]cDNA probe was synthesized using a pure ovalbumin mRNA template and reverse transcriptase prepared from avian myeloblastosis virus (Harris et al., 1973). Secondary estrogen stimulation resulted in a detectable increase in

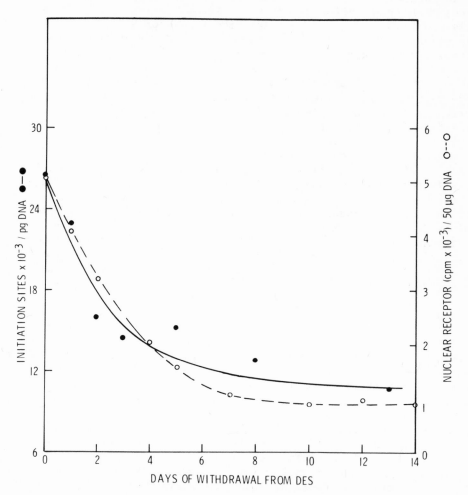

Fig. 3. Effect of estrogen withdrawal on initiation of RNA synthesis and the level of nuclear estrogen receptors. Chicks which had
received daily injection of 2.5 mg of DES for 18 days were withdrawn from hormone treatment for 12 days. The number of initiation
sites in chromatin (●---●) during estrogen withdrawal was determined
as described in Table 1. The level of nuclear receptor (o---o) was
determined by the [³H] estrogen exchange assay as described by
Tsai, S.Y. et al. (1975) and expressed as CPM exchanged per 50 µg
of nuclear DNA.

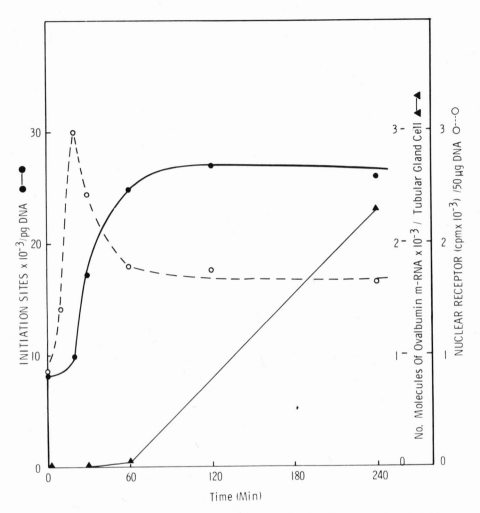

Fig. 4. Effect of secondary estrogen stimulation on the initiation of RNA synthesis and the level of nuclear receptors for estrogen. Chicks which had received daily injections of 2.5 mg of DES for 18 days were withdrawn from hormone treatment for 12 days. Chicks were then given a single 2.5 mg dose of DES and at the indicated times, oviducts were removed and frozen. Chromatin was prepared from a portion of each oviduct sample for the measurement of initiation sites (•---•) as described in Tsai, S.Y. et al. (1975). On the other portion of oviduct, the number of molecules of ovalbumin mRNA per tubular gland cell (▲---▲; Harris et al., 1975a) and the level of nuclear estrogen receptor (o---o) were measured.

ovalbumin mRNA at 30-60 min followed by a linear accumulation
beginning after 1 hr (Tsai, S.Y. et al., 1975). It can be seen
that the number of chromatin initiation sites available to RNA
polymerase increased just prior to the intracellular accumulation
of specific mRNA molecules.

These experiments suggested that estrogen receptor must remain
bound to nuclear chromatin in order to preserve maximal gene transcription, since withdrawal of estrogen leads to a loss of nuclear-
bound receptor and a simultaneous decrease in available sites for
RNA chain initiation. Furthermore, following readministration of
hormone the accumulation of chromatin-bound receptor coincided
with a reappearance of the initiation sites for RNA polymerase.
It is unlikely that new protein synthesis could be required for
"opening" more sites for transcription on chromatin since the
effect occurred within such a short period of time. In addition,
the number of initiation sites for RNA synthesis had almost reached
a maximum by one hr following restimulation, a time when accumulation of even an RNA sequence present in an extremely high cellular
concentration, such as ovalbumin mRNA, was still at a low level
(Fig. 4).

Acute Progesterone Stimulation of Initiation of RNA Synthesis

Although incapable of inducing egg white protein synthesis in
the oviduct during primary stimulation, progesterone not only
reestablishes ovalbumin protein synthesis but appears generally
to mimic estrogen as a secondary stimulant to hormonally withdrawn
chicks (Palmiter, 1973; McKnight et al., 1975). Furthermore, the
similarity in the extent of egg white protein synthesis induced
by progesterone suggests that the time course of induction of RNA
synthesis during secondary stimulation is similar to estrogen and
not influenced by the cytodifferentiation of new tubular gland
cells (Palmiter, 1973). Therefore it was of interest to understand
the manner by which progesterone can substitute for estrogen in
hormone-withdrawn chicks.

Estrogen-treated chicks were withdrawn from hormone treatment
for 14 days and then restimulated with a single injection of progesterone (2.0 mg). Oviduct chromatin was isolated from these
secondary stimulated chicks and assayed by the rifampicin-challenge
technique. The number of initiated RNA chains was calculated as
described above. As shown in Table 2, withdrawn chromatin had the
capacity to support the initiation of 8,700 RNA chains/pg of DNA.
Following a single injection of progesterone, a rapid increase in
the number of initiation sites was found. Within 0.5 hr of hormone
treatment, the number of initiation sites had nearly doubled to a
level of 15,800 sites. After 1 hr of progesterone stimulation a
maximum of 23,000 initiation sites was detected. Thereafter, the
number of initiation sites declined so that by the 24 hr after

TABLE 2. THE EFFECT OF PROGESTERONE ADMINISTRATION *IN VIVO* ON RNA POLYMERASE TRANSCRIPTION OF WITHDRAWN OVIDUCT CHROMATIN *IN VITRO*.

Hours after Progesterone Administration [a]	Size of RNA Product (Nucleotides)	Initial Elongation Rate (Nucleotides/sec) [b]	pmol of RNA Chains Initiated per 5 μg Chromatin [c]	Initiation Sites/ pg DNA
0	820	7.2	0.072	8,600
0.5	730	7.0	0.132	15,900
1.0	750	7.8	0.191	23,000
2.0	750	6.7	0.134	16,100
6.0	800	8.0	0.123	14,800
24.0	750	8.0	0.112	13,500

[a] Chicks were stimulated for 12 days with DES, withdrawn from hormone treatment for 10 days, and then injected with 2 mg of progesterone.

[b] Initial elongation rate was determined for 1 min of chain propagation.

[c] The amount of initiated chains was calculated from the total nucleotides incorporated at the transition point of RNA polymerase titration curves assuming that RNA contained 25% UMP, and divided by an average chain size of 770 nucleotides.

progesterone administration, 13,500 sites were observed. Thus, progesterone mimics estrogen action at the molecular level in the withdrawn chick oviduct.

To investigate further the mechanism by which progesterone augments increased chromatin transcription in withdrawn chicks, we tested the possibility that nuclear progesterone receptors acted at sites on chromatin which were identical to those affected by estrogen. Withdrawn chicks were given single injections containing increasing doses of progesterone. Chromatin was then isolated from oviduct tissue at 1 hr following restimulation and assayed for the number of RNA polymerase initiation sites (Table 3). The number of initiation sites increased in a dose-dependent fashion from a control value of about 9,150 (sites per pg DNA) to a maximum value of ∿22,500 sites at a dose of 1.25 mg of progesterone. We have previously reported a dose-dependent estrogen stimulation of RNA polymerase initiation sites (Tsai, S.Y. et al., 1975). In the present experiments a maximal response occurred 1 hr after the in vivo injection of 1.25 mg of DES and resulted in the formation of ∿23,000 initiation sites (Table 3). Thus the same dose of either steroid was effective in stimulating the same number of initiation sites.

TABLE 3. EFFECT OF THE COMBINED ADMINISTRATION OF PROGESTERONE AND DIETHYLSTILBESTROL ON INITIATION SITES FOR RNA SYNTHESIS IN OVIDUCT CHROMATIN.

Dose of Hormone[a]	RNA Chains Initiated (pmols/5 µg DNA)	Initiation Sites/pg DNA[b]
Control	0.076	9,100
0.25 mg Progesterone	0.135	16,200
1.25 mg Progesterone	0.187	22,500
2.0 mg Progesterone	0.196	23,600
1.25 mg DES	0.191	23,000
1.25 mg Progesterone + 1.25 mg DES	0.180	21,600

[a] Chicks were stimulated for 12 days with DES, withdrawn from hormone treatment for 10 days and then injected with various doses of hormone.
[b] The amount of initiated chain was calculated as described in the legend to Table 1.

We then tested the possibility that these steroids induced
the initiation of transcription at identical sites. Both DES
(1.25 mg) and progesterone (1.25 mg) were administered together to
withdrawn chicks. If estrogen and progesterone acted at different
chromatin RNA initiation sites then the stimulation of RNA synthesis by the two hormones would be additive. The results in Table 3
show that the number of RNA chain initiations was not additive when
both hormones were administered simultaneously. These data suggest
that in the withdrawn chick oviduct, estrogen and progesterone act
at similar chromatin sites to enhance transcription.

A qualitative measure of the intracellular levels of ovalbumin
mRNA was obtained in a cell free translation assay utilizing a
wheat germ lysate and by specific immunoprecipitation of synthesized polypeptide chains with a mono-specific antibody to ovalbumin. RNA extracted from the control withdrawn oviduct contained
no detectable ovalbumin mRNA. However, within 4 hr after progesterone administration to withdrawn chicks, there was a substantial
increase in total translatable messenger RNA and a large increase
in ovalbumin mRNA (Schwartz et al., 1976). Thus, like estrogen,
progesterone induces accumulation of ovalbumin mRNA in oviduct
cells of withdrawn chicks. This increase in the level of ovalbumin
mRNA follows temporarily the steroid-induced increase in chromatin
initiation sites for RNA synthesis. These experiments demonstrate
that during secondary hormonal stimulation of withdrawn chicks,
progesterone is an adequate substitute for estrogen in respect to
its effect on gene expression.

The Purified Progesterone Receptor Stimulates Initiation of Transcription of Oviduct Chromatin In Vitro

During the course of investigating the mechanism of action of
steroid hormone we have identified a progesterone receptor protein
in the chick oviduct (Sherman et al., 1970; O'Malley et al., 1971;
Schrader and O'Malley, 1972). We have recently purified the cytoplasmic progesterone receptor to homogeneity (Kuhn et al., 1975).
The receptor appears to function by complexing with progesterone
in the cytoplasm followed by translocation of the hormone-receptor
complex to the nuclear compartment (Jensen et al., 1968; O'Malley
et al., 1971; Buller et al., 1975a,b). The hormone receptor complex binds then to a limited number of nuclear acceptor sites
which are present in greater numbers in oviduct nuclei than in nontarget cell nuclei (Spelsberg et al., 1971; 1972). Although the
subsequent events are not yet completely defined, stimulation of
nuclear RNA synthesis occurs and leads to mRNA synthesis and ultimately to the induction of synthesis of cell-specific proteins.
Since an increase in RNA chain initiation sites was detected within
30 min of progesterone administration to withdrawn chicks (Table 2)
these results appeared to be consistent with our hypothesis that

steroid receptor complexes enter target cell nuclei and modulate
gene transcription. Nevertheless, there has previously been no
proof that the steroid receptor complex could act directly on
nuclear chromatin to stimulate transcription. Toward this end, we
tested the effect of purified progesterone receptor complexes on
all parameters of transcription using a reconstituted cell free
system which contained the following purified components: progesterone receptor complex, E. coli RNA polymerase, ribonucleotides,
cofactors, and chromatin from withdrawn chicks (Schwartz et al.,
1976).

A fixed concentration of withdrawn oviduct chromatin (5 µg)
was preincubated with increasing quantities of purified progesterone receptor complex (up to 1×10^{-8} M) for 30 min at 22°C. The
cytoplasmic progesterone receptor complex contains mainly 6S dimers
of A (110,000 MW) and B (117,000 MW) subunits and was purified
over 2000 fold by the affinity chromatography procedure of Kuhn
et al., 1975. The chromatin receptor complexes were next incubated
for an additional 30 min with a saturating concentration of RNA
polymerase (15 µg) to allow for the formation of stable initiation
complexes. Finally, rifampicin and nucleotides were added as
before for a 15 min RNA synthesis period. RNA synthesis in the
presence of rifampicin increased in a receptor dose-dependent manner
in which the half maximal stimulation occurred at a progesterone
receptor concentration of 0.5×10^{-8} M (Fig. 5). In this experiment, RNA synthesis was stimulated to a maximum of 62% over control
values at 1.0×10^{-8} M of progesterone receptor complex. At even
higher concentrations of receptors, the effect decreased.

We have previously reported that crude ammonium sulfate precipitates of cytosol fractions of eucaryotic cells can often
spuriously stimulate RNA synthesis by a template independent process which is unrelated to hormone receptor (Buller et al., 1975c).
To rule out such an effect in the present study, we investigated
the sensitivity of rifampicin resistant incorporation of [3H]UMP
into RNA under various conditions. Table 4 shows that RNA synthesized in the presence of receptor and chromatin was completely
dependent upon template and inhibited by actinomycin D. Furthermore, none of the components alone contained significant synthetic
activity, and hence neither the chromatin, polymerase nor receptor
were contaminated with other enzymes which were capable of incorporating [3H]UTP into acid insoluble material. In these experiments,
the progesterone receptor preincubated with chromatin in the presence
of E. coli RNA polymerase increased RNA synthesis 50% over control
values. With the addition of α-amanitin, a potent inhibitor of
eucaryotic RNA polymerase II, there was little effect on the progesterone-receptor directed stimulation of RNA transcription. These
data showed that the stimulation of RNA synthesis was not due to
the activation of endogenous RNA polymerase which could have been
a contaminant in chromatin preparations but rather was due to

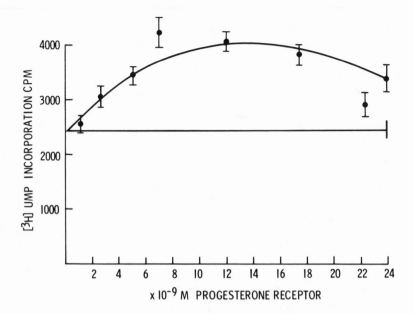

Fig. 5. Effect of purified progesterone receptor complexes on oviduct chromatin RNA synthesis in vitro. Chromatin (5 µg) from withdrawn oviducts was incubated with increasing concentrations of purified progesterone receptors for 30 min at 22°C. RNA polymerase (15 µg) was then added for an additional 30 min. RNA synthesis was started by the addition of nucleotides, rifampicin and heparin as described in the legend to Fig. 1. The amount of RNA synthesized was determined by [³H]UMP incorporated into trichloracetic acid precipitable material (•---•). The straight line represents the level of constitutive RNA synthesis in the absence of added receptors.

transcription by the added E. coli enzyme. Nevertheless it should be noted that receptor mediated stimulation of RNA synthesis also occurred in the presence of exogenously added oviduct RNA polymerase II (Schwartz et al., 1976b). Moreover, Table 4 also shows that the increased RNA synthesis was dependent upon the native structure of the hormone-receptor complex since neither free progesterone nor boiled receptor was effective in stimulating RNA synthesis. Importantly, the purified progesterone receptor does not contain any detectable proteolytic or nuclease activity which would spuriously affect the chromatin template (Schwartz et al., 1976). Thus, the stimulation of [³H]UMP incorporation is derived through a chromatin template-dependent process which is directly mediated by an intact progesterone-receptor complex.

TABLE 4. EFFECT OF PROGESTERONE RECEPTOR COMPLEXES ON RNA SYNTHESIS FROM OVIDUCT CHROMATIN IN VITRO.

Components added to Assay	[^{3}H]UMP Incorporated into RNA/5 μg Chromatin DNA[a] CPM	Percentage Activity
Background:		
RNA polymerase only	230	8
Progesterone-receptor complex plus RNA polymerase	400	14
Chromatin alone	200	7
Control:		
Chromatin plus RNA polymerase	2730	100
Chromatin plus boiled progesterone-receptor complex	2860	105
Chromatin plus 10^{-8} M progesterone alone	2430	89
Experimental:		
Chromatin plus RNA polymerase plus 10^{-8} M progesterone-receptor complex	4100	150
plus α-amanitin (10 μg)	3740	137
plus Actinomycin D (10 μg)	320	12

[a] RNA synthesis was assayed in the presence of nucleotides and rifampicin as described in Figs. 1 and 5.

Quantitation of RNA initiation sites induced by receptor steroid complexes was determined by measuring the total incorporation
of nucleotides in the presence of rifampicin and divided by the
number average chain length of the RNA product. From the data in
Table 5, it was calculated that withdrawn chromatin, in this experiment, had the capacity to code for 10,000 RNA chains per pg of
DNA. After an in vitro incubation with progesterone-receptor
complexes, the capacity to initiate RNA synthesis was increased to
15,200 sites at 1 x 10^{-8} M receptor. There were no significant
differences in either the size of the RNA product or in the initial
rate of elongation. Thus, the purified receptor preparation had
stimulated the RNA initiation events, rather than elongation steps.

The stimulatory effect of the progesterone receptor on RNA
chain initiation on chromatin was next verified by a different
assay. Quantitation of the number of 5'-termini present in the
population of nascent RNA chains was measured by the incorporation
of $\gamma[^{32}P]$GTP into RNA (Maitra and Hurwitz, 1965). We have previously shown that 57% of the initial 5'-tetraphosphate nucleosides (A + G) from the RNA synthesized from oviduct chromatin, is
guanosine (Schwartz et al., 1975). As shown in Table 6 the number
of 5'-tetraphosphate nucleosides incorporated into nascent RNA
chains also increased 50% in the presence of progesterone receptor.
The incorporation of $\gamma[^{32}P]$GTP into the 5'-termini of RNA substan-
tiates our observations that the progesterone receptor stimulates
initiation of transcription in the cell-free assay by increasing
the number of chromatin initiation sites for RNA synthesis available to RNA polymerase.

The kinetics for the stimulation of chromatin transcription
in vitro revealed at T1/2 of 15 min (Schwartz et al., 1976).
This value is close to the optimal time for receptor binding to
chromatin at 22°C and similar to the kinetics of receptor appearance in nuclei in vivo and in vitro following progesterone administration (O'Malley et al., 1971; Buller et al., 1975b). A maximal increase in transcription was found to occur within 30 min
after the addition of receptor. Previous studies have demonstrated
a quantitative tissue specificity for binding of oviduct progesterone receptor complexes to nuclei and to chromatin (Buller et al.,
1975; Jaffe et al., 1975). It was, therefore, of interest to
investigate the ability of the receptor to modify the transcriptional response of various chromatins and DNA (Fig. 6).

At receptor concentrations which produced maximal stimulation,
there were 3,700 initiation sites per pg of DNA in oviduct chromatin but only 750 and 300 additional receptor-induced sites in liver
and erythrocyte chromatin respectively (Fig. 5, Table 7). Progesterone receptor did not significantly increase the number of newly
initiated RNA chains on purified chick DNA (Table 7). It thus
appears that the progesterone receptor-mediated effect is at least

TABLE 5. EFFECT OF PROGESTERONE RECEPTOR ON THE SIZE AND RATE OF RNA CHAIN SYNTHESIS ON OVIDUCT CHROMATIN.

Concentration of Progesterone Receptor	Total Incorp. of Nucleotide[a] (pmol)	Initial Elongation Rate[b] (Nucleotides/sec)	Number Average Chain Size[c] (Nucleotides)	RNA Chains Initiated (pmol/5 μg Chromatin DNA)	Initiation Sites/ pg DNA
No Receptor	62	7.9	739	0.083	10,000
1×10^{-8} M	91	7.8	720	0.126	15,200

[a] RNA was synthesized in the presence of rifampicin and nucleotides as described in Figs. 1 and 5 and corrected for base composition of RNA assuming 25% UMP.

[b] Measured in first minute of synthesis.

[c] Measured by sucrose gradient analysis of RNA synthesized in 15 min.

TABLE 6. EFFECT OF PROGESTERONE RECEPTOR ON THE INCORPORATION OF $\gamma[^{32}P]$GTP AT THE 5' END OF RIFAMPICIN RESISTANT RNA CHAINS.

Treatment	$[^{32}P]$GTP (pmol)[a]	5' Termini (pmol)[b]	RNA Chains/ pg of DNA
Control	0.048 ± 0.002	0.086	10,300
1 x 10⁻⁸ M Progesterone Receptor	0.073 ± 0.003	0.131	15,700

[a] Withdrawn chick oviduct chromatin (5 µg) and 1 x 10^{-8} M progesterone receptor complex were preincubated for 30 min at 22°C. RNA polymerase (15 µg) and the assay components described previously in Fig. 1 were added for an additional 30 min. RNA synthesis was then started by the addition of 85.2 µCi of $\gamma[^{32}P]$GTP (5000 CPM/pmol), together with nucleotides, rifampicin and heparin. After 10 sec at 37°C the transcription reaction was stopped by pipetting a 125 µl sample on DEAE filter paper and the filter were washed according to the procedure of Roeder (1974). The incorporated $\gamma[^{32}P]$GTP was identified as guanosine tetraphosphate after degradation of the RNA product with alkali (Schwartz et al., 1976).

[b] Total number of initiated chains was estimated by the value of guanosine 5'tetraphosphate nucleosides multiplied by 1.8, a factor for total chain starts.

partly dependent on the presence of tissue specific proteins in the oviduct chromatin. These proteins may be related to the non-histone chromosomal proteins which convey quantitative tissue specificity for receptor binding and comprise a vital part of the chromatin "acceptor sites" for steroid hormone-receptor binding (Spelsberg et al., 1971; 1972).

As shown previously in Table 3 both estrogen and progesterone caused a rapid increase in the level of oviduct chromatin initiation sites when administered in vivo. These two steroid hormones were shown to act at similar RNA initiation sites since there was no additive response when they were administered together. If the hormone-receptor complex acts at sites which are similar to the in vitro sites affected by the in vivo administration of hormone, we should observe no in vitro effect of progesterone receptor on chromatins which are isolated from chicks pre-treated with in vivo injections of hormone. As Table 7 reveals; no discernible increase

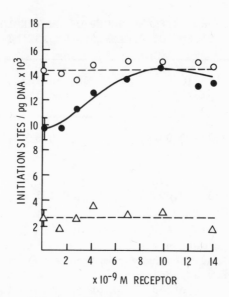

Fig. 6. Tissue specificity for the progesterone receptor's effect
on initiation sites. Chromatin from oviduct (●---●) erythrocyte
(Δ---Δ) or liver (o---o) were incubated with increasing doses of
the purified progesterone receptor. Conditions for RNA synthesis
were identical to those described in Fig. 5. Data are expressed
as initiation sites per pg chromatin DNA.

in transcription was observed following in vitro incubation of
progesterone-receptor with oviduct chromatin prepared from chicks
pre-treated with progesterone and estrogen. In fact, an inhibition
was detected which cannot yet be explained adequately. These data
also suggest that in withdrawn chromatin, the progesterone receptor acts in vitro at sites which are similar to those which are
affected by in vivo hormone treatment.

The Progesterone Receptor Stimulates Synthesis of
Ovalbumin mRNA Sequences from Chromatin In Vitro

The data presented above strongly support our proposal that
direct transcriptional control is the primary locus of steroid
hormone action. Since progesterone can induce ovalbumin in withdrawn chicks, it was conceivable that the purified progesterone
receptors could induce the ovalbumin gene in isolated oviduct
chromatin. Such a demonstration would strongly reaffirm the notion
that the in vitro receptor-chromatin interaction mimics the events
in vivo. To test this, we transcribed oviduct chromatin in vitro
in the presence of bacterial RNA polymerase. The RNA synthesized

TABLE 7. TISSUE SPECIFICITY FOR THE EFFECT OF PROGESTERONE
RECEPTOR ON SITES FOR RNA CHAIN INITIATION.

Source of Chromatin	Addition of Receptor	Number of RNA Chains Synthesized/pg Chromatin DNA[a]	Number of New Chains/ pg DNA
Oviduct (withdrawn)[b]	−	9,800	3,700
	+	13,500	
Liver[b]	−	14,300	750
	+	15,000	
Erythrocyte	−	2,500	300
	+	2,800	
Oviduct (stimulated)[c]	−	24,800	
	+	18,700	---
DNA[d]	−	190,000	620
	+	191,000	

[a] Chromatin (5 µg) was incubated with progesterone receptor
(1 x 10^{-8} M) for 30 min at 22°C. RNA synthesis was performed as
described in Fig. 5. Initiation sites were determined by using
the appropriate average RNA chain length for chromatin-directed
synthesis (750 nucleotides) and 450 nucleotides for DNA.

[b] Chicks were stimulated for 14 days with diethylstilbestrol and
then withdrawn from hormone treatment for 12 days.

[c] Chicks withdrawn from hormone were injected with DES (1.25 mg)
and progesterone (1.25 mg) for 1 hr.

[d] Chick DNA concentration was 1 µg.

was isolated and reacted with [^{3}H]DNA complementary to ovalbumin
mRNA (cDNA$_{ov}$; Harris et al., 1975b). The tritiated cDNA$_{ov}$ hybridization to mRNA$_{ov}$ sequences in the RNA samples constituted a
specific probe to estimate the concentration of these sequences.
This method is essentially the same as described by Young et al.
(1974) and Gilmour et al. (1973). Briefly, a fixed concentration
of [^{3}H]cDNA$_{ov}$ is allowed to hybridize with an excess of either pure
mRNA$_{ov}$ standards or with RNA synthesized in $vitro$. From the

kinetics of the $cDNA_{OV}$-$mRNA_{OV}$ reassociation reaction at various RNA concentrations, the $mRNA_{OV}$ concentration in the extract can be determined. When RNA synthesized from oviduct chromatin was tested using this assay, $mRNA_{OV}$ was indeed synthesized in vitro (Harris et al., 1975b). Ovalbumin mRNA synthesis required addition of RNA polymerase and was only detected in the chromatin prepared from estrogen-stimulated oviducts. These results implied that the ovalbumin gene in chromatin is accessible to RNA polymerase during hormone stimulation, but is inaccessible or "repressed" prior to hormone administration and following hormone withdrawal (Harris et al., 1975b).

To extend these studies further, we then asked whether the purified progesterone receptor complex could directly stimulate the transcription of the ovalbumin gene in a cell free system as shown in Table 8. Bulk amounts of RNA were synthesized from both control withdrawn chromatin and from chromatin incubated in the presence of progesterone receptor (1×10^{-8} M). Both RNA preparations were assayed for complimentary sequences to $[^3H]cDNA_{OV}$. The RNA synthesized in the presence of receptor-hormone complex contained a 10-fold enrichment of $mRNA_{OV}$ sequence as compared to that found in untreated withdrawn chromatin controls (Chang et al., 1976). It thus appears that a steroid-receptor complex may act directly on chromatin to enhance the number of initiation sites for RNA synthesis; leading to the synthesis of specific mRNA for induced proteins.

CONCLUSIONS

The mechanism of steroid hormone action may be related to the distinctive properties of the 6S progesterone receptor dimer (O'Malley et al., 1972; Schrader et al., 1975). The 6S receptor dimer is composed of A and B subunits which have different and unique properties. The B subunit binds to the nonhistone protein-DNA complexes of oviduct chromatin but not to pure DNA, while the A subunit binds to pure DNA, but poorly to chromatin (O'Malley et al., 1972; Schrader et al., 1972). Accordingly, these observations have led to the suggestion that the A subunit could be the actual gene regulatory protein and the B subunit could specify where the A protein is to localize (O'Malley et al., 1972; Schrader et al., 1975). In the absence of the B component of the dimers, the A subunit alone should encounter difficulty in localizing the specific initiation sites (genes) it is to regulate, while the B subunit alone should be totally inactive as a transcriptional stimulant. Consistent with these predictions, we have found that purified A subunit protein (Coty et al., 1976) was capable of stimulating transcription on withdrawn chromatin, but only at much higher concentrations ($\sim$10-50-fold) than that required for the intact dimer (Buller et al., 1976b). The isolated B subunit

TABLE 8. IN VITRO SYNTHESIS OF OVALBUMIN mRNA FROM CHICK OVIDUCT CHROMATIN.

Source of Chromatin	Progesterone Receptor (1×10^8 M)	Chromatin in Reaction (μg DNA)	RNA Synthesized (μg)	Percent $mRNA_{ov}$ in RNA	Pg of mRNA Synthesized ($\times 10^{-3}$)	pg $mRNA_{ov}$/ per μg DNA
Withdrawn Oviduct	-	400	125	0.0015	1.9	0-4.8
Withdrawn Oviduct	+	400	135	0.015	20.0	50.0

Chromatin was preincubated with progesterone receptor for 30 min at 22°C. Bulk RNA synthesis was performed at 22°C as described by Chang et al., 1976. The purified RNA was hybridized with $cDNA_{ov}$ (1.5 μg) as described previously (Harris et al., 1975).

(Schrader et al., 1976) was ineffective in stimulating transcription
from oviduct chromatin at any concentration tested (Buller et al.,
1976b). These observations are consistent with a model in which
one part of the progesterone receptor dimer (B-subunit) acts as a
binding site specifier to localize the dimer in certain regions
of chromatin, while the other subunit (A-subunit) is effective
in destabilizing a portion of the chromatin DNA so that new sites
are available to RNA polymerase for initiation of RNA synthesis.

In summary, the results presented in this manuscript are
inconsistent with models of steroid hormone action which postulate
a requirement for RNA or protein intermediates to be induced which
then exert secondary effects on transcription. Rather, the data
presented here support our proposal that direct, positive regulation of nuclear gene transcription is the mechanism of steroid
hormone action in eucaryotic cells.

ACKNOWLEDGEMENTS

This work was supported by National Institutes of Health
Grants HD-8188, HD-7857, HD-7495-02, and a Ford Foundation Grant
for the Cell Biology Department, Baylor College of Medicine,
Houston, Texas.

REFERENCES

Anderson, J., Clark, J.H., and Peck, Jr., E.J., 1972, Biochem. J.
    126:561.
Bautz, E.K.F., and Bautz, F.A., 1970, Nature 226:1219.
Buller, R.E., Toft, D.O., Schrader, W.T., and O'Malley, B.W.,
    1975a, J. Biol. Chem. 250:801.
Buller, R.E., Schrader, W.T., and O'Malley, B.W., 1975b, J. Biol.
    Chem. 250:809.
Buller, R.E., Schwartz, R.J., and O'Malley, B.W., 1975c, Biochem.
    Biophys. Res. Commun., in press.
Buller, R.E., Schwartz, R.J., Schrader, W.T., and O'Malley, B.W.,
    1976, J. Biol. Chem., submitted for publication.
Chan, L., Means, A.R., and O'Malley, B.W., 1973, Proc. Nat. Acad.
    Sci. USA 70:1870.
Chang, C., Schwartz, R., and O'Malley, B.W., in preparation.
Chamberlin, M.J., and Ring, J., 1972, J. Mol. Biol. 70:221.
Comstock, J.P., Rosenfeld, G.C., O'Malley, B.W., and Means, A.R.,
    1972, Proc. Nat. Acad. Sci. USA 69:2377.
Coty, W., Schrader, W.T., and O'Malley, B.W., 1976, J. Biol. Chem.,
    submitted for publication.
Cox, R.F., 1973, J. Biochem. 39:49.
Cox, R., Haines, M., and Carey, N., 1973, Eur. J. Biochem. 32:513.

Edelman, I.S., and Fimognari, G.M., 1968, Res. Progr. Horm. Res.
    24:1.
Fujii, T., and Villee, C., 1968, Endocrinology 82:436.
Gilmour, R.S., and Paul, J., 1973, Proc. Nat. Acad. Sci. USA 70:
    3440.
Gorski, J., and Nicollette, J.A., 1963, Arch. Biochem. Biophys.
    103:418.
Hamilton, T.H., Widnell, C.C., and Tata, J.R., 1965, Biochim.
    Biophys. Acta. 108:168.
Harris, S.E., Rosen, J.M., Means, A.R., and O'Malley, B.W., 1975a,
    Biochemistry 14:2072.
Harris, S.E., Schwartz, R.J., Tsai, M.J., Roy, A., and O'Malley,
    B.W., 1975b, J. Biol. Chem., in press.
Hartmann, H., Nonikel, K.O., Knusel, F., Nuesh, J., 1967, Biochim.
    Biophys. Acta 145:843.
Hinkle, D.C., and Chamberlin, M.J., 1972, J. Mol. Biol. 70:157.
Jaffe, R.C., Socher, S.H., and O'Malley, B.W., 1975, Biochim.
    Biophys. Acta 399:403.
Jensen, E.V., Suzuki, T., Kawashima, W., Stumpf, E., Jungblut, E.R.,
    and DeSombre, E., 1968, Proc. Nat. Acad. Sci. USA 59:632.
Kenny, F.T., and Kull, F.J., 1963, Proc. Nat. Acad. Sci. USA 50:493.
Kohler, P.O., Grimely, P.M., and O'Malley, B.W., 1969, J. Cell
    Biol. 40:8.
Kuhn, R.W., Schrader, W.T., Smith, R.G., and O'Malley, B.W., 1975,
    J. Biol. Chem. 250:4220.
Lill, H., Lill, U., Sippel, A., and Hartmann, G., 1970, in Lepetit
    Colloquium on RNA Polymerase and Transcription (Silvestri,
    L.G., ed.) p. 55, North Holland Publishing Co., Amsterdam.
Liarakos, C.D., Rosen, J.M., and O'Malley, B.W., 1973, Biochemistry
    12:2809.
Mainwaring, W.I.P., Wilce, P.A., and Smith, A.E., 1974, Biochem. J.
    139:513.
Maitra, U., and Hurwitz, J., 1965, Proc. Nat. Acad. Sci. USA 54:
    815.
McKnight, S.G., Pennequin, P., and Schimke, R.T., 1975, J. Biol.
    Chem. 250:8105.
Means, A.R., Comstock, J.P., Rosenfeld, G.C., and O'Malley, B.W.,
    1972, Proc. Nat. Acad. Sci. USA 69:1146.
Norman, A.W., 1966, Biochem. Biophys. Res. Comm. 23:335.
O'Malley, B.W., and McGuire, W.L., 1968a, J. Clin. Invest. 47:654.
O'Malley, B.W., and McGuire, W.L., 1968b, Proc. Nat. Acad. Sci. USA
    60:1527.
O'Malley, B.W., and Means, A.R., 1974, Science 183:610.
O'Malley, B.W., McGuire, W.L., Kohler, P.O., and Korenman, S.G.,
    1969, Rec. Progr. Horm. Res. 25:105.
O'Malley, B.W., Toft, D.O., and Sherman, M.R., 1971, J. Biol. Chem.
    246:1117.
O'Malley, B.W., Spelsberg, T.C., Schrader, W.T., Chytil, F., and
    Steggles, A.W., 1972, Nature 235:141.

Palmiter, R.D., 1973, J. Biol. Chem. 248:8260.

Palmiter, R.D., and Smith, L., 1973, Mol. Biol. Rep. 1:129.

Roeder, R.G., 1974, J. Biol. Chem. 249:241.

Schrader, W.T., and O'Malley, B.W., 1972, J. Biol. Chem. 247:51.

Schrader, W.T., Toft, D.O., and O'Malley, B.W., 1972, J. Biol. Chem. 247:2401.

Schrader, W.T., Buller, R.E., Kuhn, R.W., and O'Malley, B.W., 1974, J. Steroid Biochem. 5:989.

Schrader, W.T., Kuhn, R.W., and O'Malley, B.W., 1976, J. Biol. Chem., submitted for publication.

Schrader, W.T., Heurer, S.S., and O'Malley, B.W., 1975, Biol. of Reprod. 12:134.

Schwartz, R.J., Tsai, M.J., Tsai, S.Y., and O'Malley, B.W., 1975, J. Biol. Chem. 250:5175.

Schwartz, R.J., Kuhn, R., Buller, R.E., Schrader, W.T., and O'Malley, B.W., 1976, J. Biol. Chem., submitted for publication.

Schwartz, R.J., Tsai, M.J., Schrader, W.T., and O'Malley, B.W., 1976, in preparation.

Sekeris, C.E., and Lang, N., 1964, Life. Sci. 3:625.

Sherman, M.R., Corvol, P.L., and O'Malley, B.W., 1970, J. Biol. Chem. 245:6085.

Sipple, A.E., and Hartman, G.R., 1970, Eur. J. Biochem. 16:152.

Socher, S., and O'Malley, B.W., 1973, Dev. Biol. 30:401.

Spelsberg, T.C., Steggles, A.W., and O'Malley, B.W., 1971, J. Biol. Chem. 246:4188.

Spelsberg, T.C., Steggles, A.W., Chytil, F., and O'Malley, B.W., 1972, J. Biol. Chem. 247:1368.

Stohs, S.J., Zall, J.E., and DeLuca, H.F., 1967, Biochemistry 6:1304.

Travers, A., Baillie, D.L., and Pedersen, S., 1973, Nat. New Biol. 243:161.

Tsai, M.J., Schwartz, R.J., Tsai, S.Y., and O'Malley, B.W., 1975, J. Biol. Chem. 250:5165.

Tsai, S.Y., Tsai, M.J., Schwartz, R.J., Kalimi, M., Clark, J.H., and O'Malley, B.W., 1975, Proc. Nat. Acad. Sci. USA 72:4228.

Umezawa, H., Mizuno, S., Uamasaki, H., and Hitta, K., 1968, J. Antibiot. 21:234.

Young, B.D., Harrison, P.R., Gilmour, R.S., Birnie, G.D., Hell, A., Humphries, S., and Paul, J., 1974, J. Mol. Biol. 84:555.

Zillig, W., Zechel, D., Rabussay, D., Schachner, M., Setha, U.S., Palm, P., Heil, A., and Seifert, W., 1970, Cold Spring Harbor Sympos. Quant. Biol. 35:47.

LIST OF PARTICIPANTS

BASSI, S.
    Benedictine College, Atchison, Kansas, U.S.A.
BIDLINGMEYER, B.
    Waters Associates, Chicago, Illinois, U.S.A.
BIESSELS, H.
    Rijksuniversiteit te Utrecht, The Netherlands
BOLLENBACHER, W.E.
    Northwestern University, Evanston, Illinois, U.S.A.
BOWERS, W.S.
    Cornell University, Geneva, New York, U.S.A.
BRECHBUHLER, H.
    Ciba-Geigy, Basle, Switzerland
BRIDGES, T.
    Hoffman LaRoche, Vero Beach, Florida, U.S.A.
CAMPBELL, W.
    Ciba-Geigy, Vero Beach, Florida, U.S.A.
CAVENEY, S.
    University of Western Ontario, London, Canada
CHEN, T.
    Queen's University, Kingston, Ontario, Canada
CRUICKSHANK, P.
    FMC Corporation, Princeton, New Jersey, U.S.A.
DAHM, K.
    Texas A & M University, College Station, Texas, U.S.A.
DAVEY, K.
    York University, Downsview, Canada
de KORT, C.
    Agricultural University, Wageningen, The Netherlands
DORN, S.
    Maag A.G., Dielsdorf, Switzerland
DUNN, P.
    University of Chicago, Chicago, Illinois, U.S.A.
EMMERICH, H.
    Technische Hochschule, Darmstadt, Germany
ENGELMANN, F.
    University of California, Los Angeles, California, U.S.A.

FERKOVICH, S.
　　U.S.D.A., Gainesville, Florida, U.S.A.
FRISTROM, J.
　　University of California, Berkeley, California, U.S.A.
GASSER, R.
　　Ciba-Geigy, Basle, Switzerland
GILBERT, L.I.
　　Northwestern University, Evanston, Illinois, U.S.A.
GOODMAN, W.
　　Northwestern University, Evanston, Illinois, U.S.A.
GRANGER, N.
　　University of California, Irvine, California, U.S.A.
HAMMOCK, B.D.
　　University of California, Riverside, California, U.S.A.
HAYS, D.
　　Ciba-Geigy, Greensboro, North Carolina, U.S.A.
JOHNSTON, P.
　　Northwestern University, Evanston, Illinois, U.S.A.
JUDY, K.
　　Zoecon Corporation, Palo Alto, California, U.S.A.
KATULA, K.
　　Northwestern University, Evanston, Illinois, U.S.A.
KATZENELLENBOGEN, B.
　　University of Illinois, Urbana, Illinois, U.S.A.
KATZENELLENBOGEN, J.
　　University of Illinois, Urbana, Illinois, U.S.A.
KOEPPE, J.
　　University of North Carolina, Chapel Hill, North Carolina,
　　U.S.A.
KRAMER, K.
　　U.S.D.A., Manhattan, Kansas, U.S.A.
KRISHNA KUMARAN, A.
　　Marquette University, Milwaukee, Wisconsin, U.S.A.
LANZREIN, B.
　　Universitat Bern, Bern, Switzerland
LAUFER, H.
　　University of Connecticut, Storrs, Connecticut, U.S.A.
LAW, JOHN
　　University of Chicago, Chicago, Illinois, U.S.A.
LEZZI, M.
　　Eidgen. Tech. Hoch., Zurich, Switzerland
MENN, J.
　　Stauffer Chemical Co., Mountain View, California, U.S.A.
NOWOCK, J.
　　Northwestern University, Evanston, Illinois, U.S.A.
OBERLANDER, H.
　　U.S.D.A., Gainesville, Florida, U.S.A.
OFENGAND, J.
　　Roche Institute of Molecular Biology, Nutley, New Jersey,
　　U.S.A.

O'MALLEY, B.W.
    Baylor College of Medicine, Houston, Texas, U.S.A.
PETER, M.
    Texas A & M University, College Station, Texas, U.S.A.
POSTLETHWAIT, J.
    University of Oregon, Eugene, Oregon, U.S.A.
PRATT, G.
    A.R.C., Sussex, England
RIDDIFORD, L.
    University of Washington, Seattle, Washington, U.S.A.
ROBERTS, B.
    Monash University, Clayton, Victoria, Australia
SANBURG, L.
    U.S.D.A., Manhattan, Kansas, U.S.A.
SCHNEIDERMAN, H.
    University of California, Irvine, California, U.S.A.
SCHOOLEY, D.
    Zoecon Corporation, Palo Alto, California, U.S.A.
SCHWARZ, M.
    U.S.D.A., Beltsville, Maryland, U.S.A.
SEDLAK, B.J.
    Smith College, Northampton, Massachusetts, U.S.A.
SHUMAKER, W.
    Waters Associates, Chicago, Illinois, U.S.A.
SIDDALL, J.
    Zoecon Corporation, Palo Alto, California, U.S.A.
SMITH, S.
    Northwestern University, Evanston, Illinois, U.S.A.
SRIDHARA, S.
    Northwestern University, Evanston, Illinois, U.S.A.
STAAL, G.
    Zoecon Corporation, Palo Alto, California, U.S.A.
TOBE, S.
    University of Toronto, Toronto, Canada
TOLMAN, R.
    National Institutes of Health, Bethesda, Maryland, U.S.A.
TROST, B.
    University of Wisconsin, Madison, Wisconsin, U.S.A.
VINCE, R.
    Northwestern University, Evanston, Illinois, U.S.A.
WEIRICH, G.
    Texas A & M University, College Station, Texas, U.S.A.
WHITE, B.
    Queen's University, Kingston, Ontario, Canada
WHITMORE, D.
    University of Texas, Arlington, Texas, U.S.A.
WHITMORE, E.
    Arbrook, Inc., Arlington, Texas, U.S.A.
WIELGUS, J.
    Northwestern University, Evanston, Illinois, U.S.A.

WILLIAMS, C.M.
    Harvard University, Cambridge, Massachusetts, U.S.A.
WILLIS, J.
    University of Illinois, Urbana, Illinois, U.S.A.
WILSON, T.
    Northwestern University, Evanston, Illinois, U.S.A.
WRIGHT, J.
    U.S.D.A., College Station, Texas, U.S.A.
WYATT, G.
    Queen's University, Kingston, Ontario, Canada
WYSS, C.
    Eidgen. Tech. Hoch., Zurich, Switzerland
YAGI, S.
    Tokyo University of Education, Tokyo, Japan
ZURFLUEH, R.
    Socar A.G., Dubendorf, Switzerland
ZVENKO, H.
    Northwestern University, Evanston, Illinois, U.S.A.

(Compiled by Stan Smith and Jack Wielgus,
Northwestern University.   JH = juvenile hormone.)